Lecture Notes in Bioengineering

Lecture Notes in Bioengineering (LNBE) publishes the latest developments in bioengineering. It covers a wide range of topics, including (but not limited to):

- Bio-inspired Technology & Biomimetics
- Biosensors
- Bionanomaterials
- Biomedical Instrumentation
- Biological Signal Processing
- Medical Robotics and Assistive Technology
- Computational Medicine, Computational Pharmacology and Computational Biology
- Personalized Medicine
- Data Analysis in Bioengineering
- Neuroengineering
- Bioengineering Ethics

Original research reported in proceedings and edited books are at the core of LNBE. Monographs presenting cutting-edge findings, new perspectives on classical fields or reviewing the state-of-the art in a certain subfield of bioengineering may exceptionally be considered for publication. Alternatively, they may be redirected to more specific book series. The series' target audience includes advanced level students, researchers, and industry professionals working at the forefront of their fields.

Indexed by SCOPUS and Springerlink. The books of the series are submitted for indexing to Web of Science.

More information about this series at http://www.springer.com/series/11564

Albert A. Rizvanov · Bikesh Kumar Singh ·
Padma Ganasala
Editors

Advances in Biomedical Engineering and Technology

Select Proceedings of ICBEST 2018

Springer

Editors
Albert A. Rizvanov
Kazan Federal University
Kazan, Tatarstan Republic, Russia

Bikesh Kumar Singh
Department of Biomedical Engineering
National Institute of Technology Raipur
Raipur, India

Padma Ganasala
Gayatri Vidya Parishad College
of Engineering
Visakhapatnam, Andhra Pradesh, India

ISSN 2195-271X ISSN 2195-2728 (electronic)
Lecture Notes in Bioengineering
ISBN 978-981-15-6328-7 ISBN 978-981-15-6329-4 (eBook)
https://doi.org/10.1007/978-981-15-6329-4

This Springer imprint is published by the registered company Springer Nature Singapore Pte Ltd.
The registered company address is: 152 Beach Road, #21-01/04 Gateway East, Singapore 189721, Singapore

Organising Committee of ICBEST-2018

Chief Patron

Prof. A. M. Rawani, Director, NIT Raipur

Patron

Prof. S. Sanyal, TEQIP III Coordinator, NIT Raipur

Internal Adivsory Committee

Dr. Shrish Verma, Dean Academics, NIT Raipur
Dr. A. P. Rajimwale, Dean Faculty Welfare, NIT Raipur
Dr. P. Diwan, Dean, Student Welfare, NIT Raipur
Dr. (Mrs.) Shubhrata Gupta, Dean, R & C, NIT Raipur
Dr. G. D. Ramtekkar, Dean, P & D, NIT Raipur
Dr. P. Y. Dhekne, Registrar, NIT Raipur

Chairman

Dr. Bikesh Kumar Singh, Department of Biomedical Engineering, NIT Raipur

Conference Secretaries

Dr. Neelamshobha Nirala, Department of Biomedical Engineering, NIT Raipur
Dr. Arindam Bit, Department of Biomedical Engineering, NIT Raipur
Dr. Saurabh Gupta, Department of Biomedical Engineering, NIT Raipur

Contents

About the Editors

Albert A. Rizvanov (Ph.D., Dr. Sci.) graduated from Kazan State University, Russia (biology, microbiology) in 1996. After completing his Ph.D. (2003) in cellular and molecular biology at the University of Nevada, Reno, USA, he undertook his Dr. Sci. (2011) in biochemistry (Habilitation) at Kazan Federal University (KFU), Russia. Currently, Albert Rizvanov is a Professor and Director of the Center for Precision and Regenerative Medicine, Institute of Fundamental Medicine and Biology, KFU. He is the head of the Open Lab Gene and Cell Technologies Laboratory, Director of the Department of Exploratory Researches of Pharmaceutical Research and Education Center and head of the Center of Excellence "Regenerative Medicine". Additionally, he is the Vice-Director of the Strategic Academic Unit "Translational 7P Medicine" as part of the government program of competitive growth ("5–100 Program") and the corresponding member of the Tatarstan Academy of Sciences, Russian Federation. Albert is an author on more than 300 peer-reviewed journal articles, 3 book chapters, and 22 patents, he has successfully supervised 15 Ph.D. and 2 Dr. Sci. dissertations, and is the head of the biochemistry, microbiology, and genetics dissertation committee at KFU. He is the principal investigator of more than 50 grants supported by NATO, British Council, Russian Science Foundation, Russian Foundation for Basic Research and other Russian government federal programs and industry contracts. His fields of expertise include regenerative medicine, precision medicine, gene and cell therapy, molecular neurobiology, molecular virology, cancer diagnostics and therapy. In 2019 Albert Rizvanov became an Honorary Professor of Fundamental Medicine at the Faculty of Medicine and Health Sciences, University of Nottingham, UK.

Bikesh Kumar Singh (Ph.D.) is Assistant Professor in Department of Biomedical Engineering at National Institute of Technology Raipur, Raipur (Chhattisgarh) India. He obtained his B.E. (Electronics and Telecommunication Engineering) Gold Medalist and M.Tech. (Electronics and Telecommunication Engineering) Honors from Pt. Ravishankar Shukla University, Raipur. He received his Ph.D. in Biomedical Engineering from National Institute of Technology Raipur, Raipur (Chhattisgarh) India. He has published more than 70 research papers in various international and national journals and conferences. He is active reviewer and has reviewed several research articles of reputed International Journals. He has teaching and research experience of 12 years. He has been Head of the Department of Department of Biomedical Engineering for 5 years. He is member of International Professional Societies such as IEEE (Senior member) & IACSIT and also of many National Professional bodies like CSI India, IETE India, ISCA India and IEI India. He has received several awards like Chhattisgarh Young Scientist Award, IETE Gowri Memorial Award, IEI Young Engineer Award. He has delivered several expert talks in the area of Machine Learning Applications. He has organized several workshops and international conference in area of Biomedical Engineering, Machine Learning and Softcomputing. His research interest includes applications of machine learning and artificial intelligence in medical image analysis, biomedical signal analysis, computer aided diagnosis, computer vision and cognitive science.

Padma Ganasala (Ph.D.) is currently working as Associate Professor in the Department of Electronics and Communications Engineering, Gayatri Vidya Parishad College of Engineering, Visakhapatnam, India. She had received her Ph.D. in Medical Image Fusion from Indian Institute of Technology Roorkee (IIT-ROORKEE). She is a recipient of MHRD fellowship during studies. To her credit, she possesses several publications in reputed international journals and conferences. She has reviewed many journal papers published by prestigious journals and conferences. She is a Life Member of ISTE. Her research interests include medical image processing and analysis, biomedical signal processing, machine learning and deep learning.

Extraction and Phytochemical Analysis of *Coccinia indica* Fruit Using UV-VIS and FTIR Spectroscopy

Alok Sharma, Bidyut Mazumdar, and Amit Keshav

Abstract The phytochemical analysis of *Coccinia indica* fruit extract was performed employing UV-VIS and FTIR Spectroscopy. Extractions of phytochemicals were carried out using different solvents, selected on the basis of polarity viz. ethanol, methanol, and chloroform. Antioxidant activity of extracts was measured by the DPPH method where ascorbic acid was used as standard. The UV-VIS spectroscopy revealed the characteristic peaks for different phytochemicals present in the extract. The FTIR analysis helped to identify the presence of different functional groups which ultimately leads to the confirmation of existence of phytochemicals in the extract. The phytochemicals thus extracted and identified have major applications in biotechnology, food processing, and pharmaceutical industries.

Keywords *Coccinia indica* · Phytochemicals · FTIR spectroscopy · UV-VIS spectroscopy · DPPH

1 Introduction

Coccinia indica is a perennial and creeping plant belonging to Cucurbitaceae family. It is also known as Ivy Gourd, Kundru (Chhattisgarhi), Kovakai (Tamil), Tindora. It is found extensively throughout the Indian Subcontinent. The fruit is berry shaped, green when unripe, and becomes orange-red when ripens. The fruits of this plant are the main constituent of the regular meal in Indian culinary. It has been looked forward for its herbal and medicinal properties as mentioned in Ayurveda. *C. indica* is also known for its antidiabetic, anti-obesity, antimicrobial, antifungal, antileishmanic, antioxidant, antihypertensive, antitussive, antiulcer, analgesic, antipyretic, antianaphylactic, and anticancer properties (Sakharkar and Chauhan 2017; Singha et al. 2007). The fruits are believed to have numerous phytochemicals which include antioxidants, phenolic compounds, flavonoids, alkaloids, terpenoid, tannins,

A. Sharma (✉) · B. Mazumdar · A. Keshav
Department of Chemical Engineering, NIT Raipur, Raipur, India
e-mail: alok.sharma00512@gmail.com

A. A. Rizvanov et al. (eds.), *Advances in Biomedical Engineering and Technology*, Lecture Notes in Bioengineering,
https://doi.org/10.1007/978-981-15-6329-4_1

saponins, glycosides, etc. (Kumar et al. 2014). This current research was focused on exploring the phytochemicals present in *C. indica* using different analysis methods.

2 Material and Methods

Fresh fruits of *C. indica* were purchased from the local market of Raipur Chhattisgarh (India). It was properly rinsed with water to remove the physical impurities from the surface. Afterwards, the fruits were chopped and sliced by a kitchen slicer to get the uniform slices prior to sun drying. The sliced fruits were subjected to sun drying for 2 days to remove the indigenous water from it. The dried fruit was ground in a mixer-grinder. Solvent extraction was studied using ethanol, methanol, chloroform which are of analytical grade and obtained from Merck, India. Ltd. Antioxidant activity was measured using 1, 1-diphenyl-2-picryl-hydrazyl (DPPH) obtained from Merck, India. Ltd.

Solvent extraction of phytochemicals was done using the 250 ml Soxhlet apparatus for 12 h using ethanol, methanol, and chloroform as solvents, respectively. About 25 g of dried fruit powder was taken into the thimble to perform the extraction. The extract thus obtained was filtered (Whatman Filter Paper No. 1) and concentrated to get the crude extract, which was then stored at refrigerated temperature until further use.

The extracts were analyzed for the identification and characterization of the phytochemicals by UV-VIS Spectrophotometer (Shimadzu UV-1800) and FTIR (Bruker). The extracts were diluted to the extent of 1:10 for the same solvent prior to UV-VIS Spectrometry. FTIR Spectroscopy was done for identification of the functional groups present in the extract. The KBr thin disc was formed for this analysis, which was made by mixing small amount of *C. indica* extract with dry potassium bromide. Further, the disc was placed over the sample cup of diffuse reflectance accessory. The extracts were analyzed using FTIR Spectrophotometer where IR spectrum was within 4000–400 cm^{-1}. The results thus obtained from UV-VIS and FTIR were recorded.

The antioxidant activity was evaluated by DPPH assay as mentioned by Baba and Malik (Baba and Malik 2015). About 3.8 ml of freshly prepared DPPH solution was taken and 200μL of extract was added for each solvent viz. ethanol, methanol, and chloroform. Further the reaction mixture was incubated in dark for 1 h at room temperature. The measurement of absorbance was done at 517 nm wavelength by UV-VIS spectrometer. Ascorbic acid was used as positive control. The DPPH activity was calculated by the formula mentioned below:

$$\text{DPPH Activity} = \frac{Control\,absorbance - Sample\,absorbance}{Control\,absorbance} \times 100$$

3 Result and Discussion

The extraction of phytochemicals from *C. indica* fruits was found to be more effective for chloroform than ethanol and methanol. The UV-VIS Spectroscopy shows the different absorbance peaks obtained for the respective phytochemical which is also mentioned in literatures. The UV-Vis spectroscopy dictates that the phytochemicals are present in the extract based on the λ_{max} values for each phytochemical. Table 1 depicts the absorbance values (λ_{max}) of phytochemicals with respect to the wavelength, for ethanol, methanol, and chloroform, respectively. Chlorophyll a, Chlorophyll b, Taraxerol, β amyrin, Lupeol, and Sitosterol were the phytochemicals detected in the ethanolic and methanolic extracts of *C. indica* fruit (Wang et al. 2007; Laphookhieo 2004; El-Alfy et al. 2011; Quilitzsch et al. 2005; Jain and Bari 2010; Chung and Hahn 2005). Similar phytochemicals were detected in the chloroform extract. β carotene was only detected in the extracts of chloroform. The λ_{max} identified phytochemicals was mentioned in the literatures (Okoye and Daniel 2014; Khanra et al. 2014; Miller et al. 1936; Mallick 2014).

The FTIR spectroscopy was used to identify the functional groups of phytochemicals present in the extracts of *C. indica* fruit. The peak values obtained by the FTIR spectroscopy validate the presence of particular functional groups that a particular phytochemical contains. The results of FTIR peak values and functional groups have been illustrated in Table 2 and Figs. 1, 2 and 3.

The FTIR spectrum profile confirms the presence of functional groups in ethanol, methanol, and chloroform extracts of *C. indica.* Amides were detected at 3325, 3326, and 2839 cm^{-1} wavenumber in ethanol and methanol extracts. Alkanes were detected in ethanol and chloroform extracts at 2923, 2924, and 2852 cm^{-1} wavenumber.

Alkene was observed in methanol and chloroform extracts at 1646 and 1658 cm^{-1}. 1° amine was detected at 1653 and 1631 cm^{-1} in ethanol and chloroform extracts. Aromatics were identified in all the solvents at 1043, 868, 1447, 1406, 1462, 772, and 686 cm^{-1}. Nitro groups were detected in extracts of ethanol and methanol at 1378 and 1554 cm^{-1}.

Table 1 Phytochemicals identified in different extracts by UV-Vis spectroscopy

S. No	Phytochemical	Ethanol extract		Methanol extract		Chloroform extract	
		λ_{max}	Abs	λ_{max}	Abs	λ_{max}	Abs
1	Chlorophyll b	665	0.485	665	0.495	665	0.617
2	Chlorophyll a	606	0.150	606	0.170	607	0.203
3	Taraxerol	535	0.234	535	0.230	536	0.261
4	β amyrin	409	2.188	409	2.158	436	1.317
5	β carotene					412	1.978
6	Lupeol	314	4.0	314	3.96	322	1.364
7	Sitosterol	269	4.0	269	3.984	268	1.495

Table 2 FTIR peak values of identified functional groups in different extracts

Ethanol		Methanol		Chloroform	
Wave number cm^{-1}	Functional group	wave number cm^{-1}	Functional group	wave number cm^{-1}	Functional group
3325	Amides	3326	Amides	3425	Phenols
2969	Alkanes	2839	Amides	2924	Alkanes
2923	Alkanes	1646	Alkene	2852	Alkanes
1653	1°amines	1447	Aromatic	1742	Esters
1378	Nitro compound	1406	Aromatic	1658	Alkene
1085	Aliphatic Amines	1017	Aliphatic Amine	1631	1°amine
1043	Aliphatic Amines	541	Alkyl halide	1554	Nitro compound
878	Aromatic	518	Alkyl halide	1536	Unknown
525	Alkyl Halides	510	Alkyl halide	1462	Aromatic
514	Alkyl Halides			1382	Alkane
				1219	Aliphatic Amine
				1032	Aliphatic Amine
				772	Aromatic
				686	Aromatic

The wavenumber 1085, 1017, 1219, and 1032 cm^{-1} denoted the presence of aliphatic amine in all the extracts. Alkyl halides were present in ethanol and methanol extracts at the wavenumber of 525, 514, 518, and 510 cm^{-1}.

Moreover, phenols and esters were identified only at 3425 and 1742 cm^{-1} respectively in chloroform extracts only. The functional groups thus detected in the different extracts leads to the assumption that a variety of phytochemicals are present in the *C. indica* fruits.

DPPH test is useful to determine the radical scavenging activity of extraction. The method relies on the decrease in absorption of DPPH solution after addition of antioxidant. The standard for this test is done using ascorbic acid.

DPPH has red color and degree of discoloration indicates the scavenging potential of the antioxidant. DPPH radical scavenging activity was calculated using the absorbance values obtained by spectrophotometer. The ethanol extracts showed 60.15% of activity, while it was 60.46% and 61.07% for methanol and chloroform, respectively.

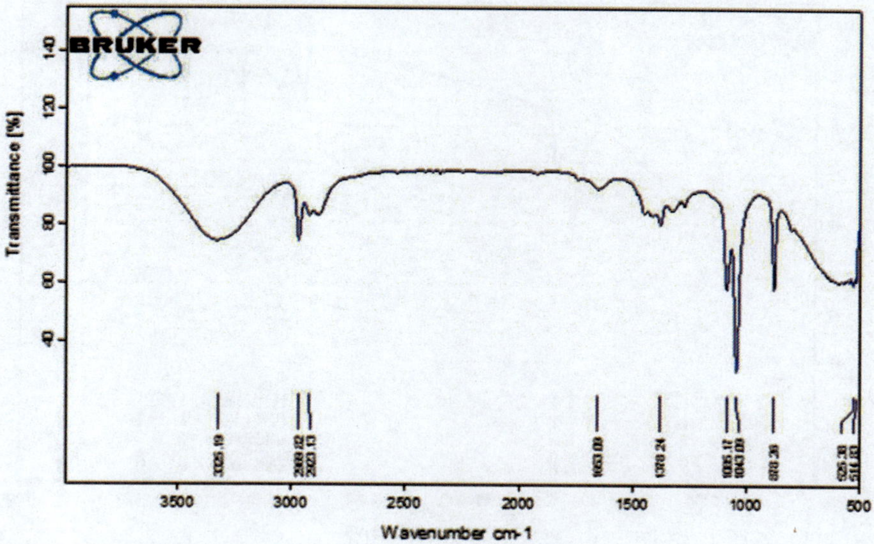

Fig. 1 FTIR spectra of ethanolic extract of *C. indica* fruit

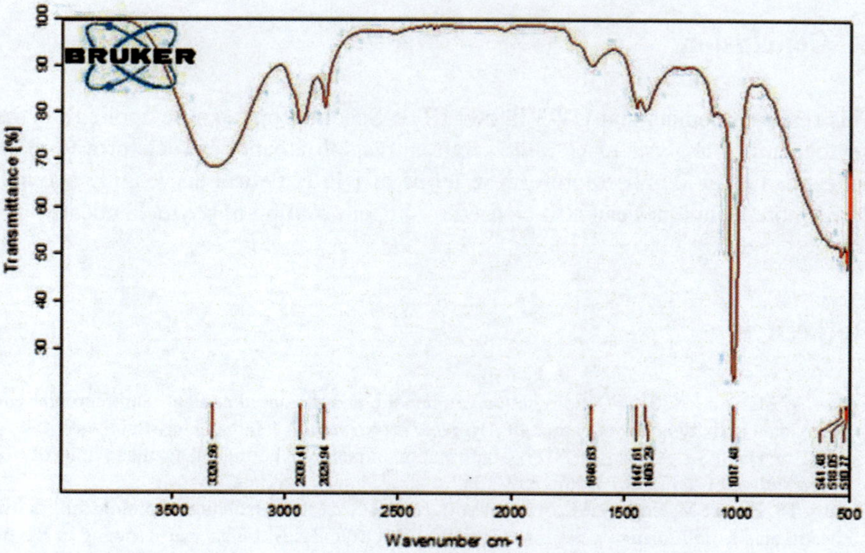

Fig. 2 FTIR spectra of methanolic extract of *C. indica* fruit

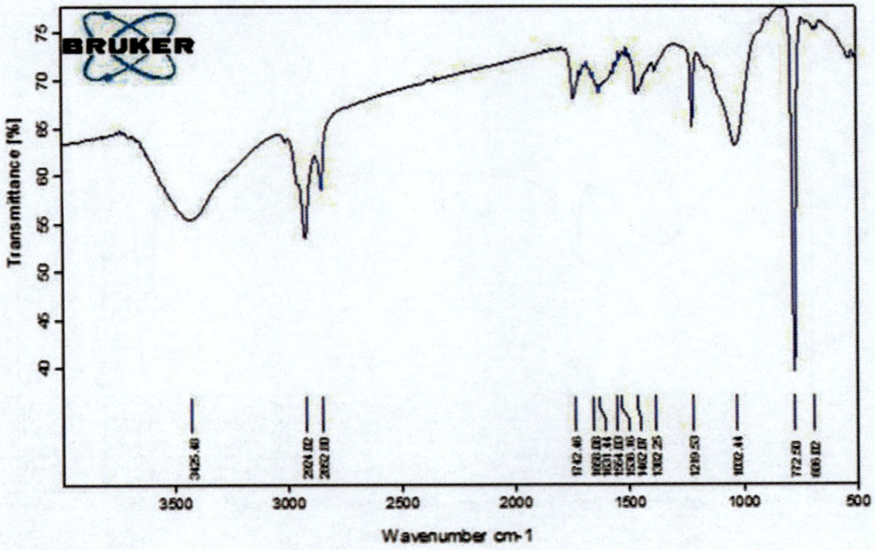

Fig. 3 FTIR spectroscopy of chloroform extract of *C. indica* fruit

4 Conclusion

This research depicts that UV-VIS and FTIR Spectroscopy can be applied for the phytochemical analysis of *C. indica* fruit extracts. Methanol extracts proved to be better than ethanol and chloroform in terms of DPPH radical scavenging activity. Some more techniques can also be used for better profiling of phytochemicals.

References

Baba SA, Malik SA (2015) Determination of phenolic and flavonoid content, antimicrobial and antioxidant activity of a root extract of *Arisaema jacquemontii*. J Taibah Univ Sci 9:449–454

Chung I-m, Hahn S-j, Ahmad A (2005) Confirmation of potential herbicidal agents in hulls of rice *Oryza sativa*. J Chem Ecol 31(6):1339–1352

El-Alfy TS, Ezzat SM, Hegazy AK, Amer AMM, Kamel GM (2011) Isolation of biologically active constituents from *Moringa peregrina* (Forssk.) Fiori. (family: Moringaceae) growing in Egypt. Pharma Magazine 7(26):109–115

Jain PS, Bari SD (2010) Isolation of Lupeol, Stigmasterol and Campesterol from petroleum ether extract of woody stem of *Wrightia tinctoria*. Asian J Plant Sci 9(3):163–167

Khanra R, Dewanjee S, Dua TK, Sahu R, Gangopadhyay M, De Feo V, Zia-Ul-Haq M (2014) *Abroma augusta* L. (Malvaceae) leaf extract attenuates diabetes induced nephropathy and cardiomyopathy via inhibition of oxidative stress and inflammatory response. J Trans Med 13:1–14

Kumar M, Alok S, Jain SK, Dixit VK (2014) Macroscopial, anatomical and physico-chemical studies on fruits of *Coccinia indica* wight and Arn. (Cucurbitaceae). Asian Pac J Trop Dis 4(1):S121–S128

Laphookhieo S, Karalai C, Ponglimanont C (2004) New Sesquiterpenoid and Triterpenoids from the Fruits of *Rhizophora mucronate*. Chem Pharm Bull 52(7):883–885

Mallick SS, Dighe VV (2014) Detection and estimation of alpha-Amyrin, beta-Sitosterol, Lupeol, and n-Triacontane in two medicinal plants by high performance thin layer chromatography. Adv Chem 1–7

Miller ES, Inney GM, Zscheile FP Jr. (1936) Absorption spectra of alpha and beta carotenes and lycopene. Plant Physiol. 9(1):375–381

Okoye NN, Ajaghaku DL, Okeke HN, Ilodigwe EE, Nworu CS, Okoye FBC (2014) Beta-Amyrin and alpha-amyrin acetate isolated from the stem bark of *Alstonia boonei* display profound anti-inflammatory activity. Informa Healthcare 52(11):1478–1486

Quilitzsch R, Baranska M, Schulz H, Hoberg E (2005) Fast determination of carrot quality by spectroscopy methods in the UV-VIS, NIR and IR range. J Appl Bot Food Qual 79:163–167

Sakharkar P, Chauhan BS (2017) Antibacterial, antioxidant and cell proliferative properties of *Coccinia grandis* fruits. Avicenna J Phytomed 7(4):295–307

Singha G, Gupta P, Rawat P, Purib A, Bhatiab G, Maurya R (2007) Antidyslipidemic activity of polyprenol from *Coccinia grandis* in high-fat diet-fed hamster model. Phytomedicine 14:792–798

Wang L, Zhang C, Wu F, Deng N (2007) Photodegradation of aniline in aqueous suspensions of microalgae. J Photochem Photobiol B Biol 87:49–57

Numerical Simulation Method to Predict Air Flow and Contaminant Control in a Multiple Bed Intensive Care Unit of Hospital

Arvind Kumar Sahu, Shobha Lata Sinha, and Tikendra Nath Verma

Abstract A hospital's physical pattern is a vital part of its contamination management measures to reduce the danger of spread of any communicable disease. Recent and rising communicable diseases challenges as higher public prospects and awareness of care connected problems, a lot of thought has got to incline to the layout of the hospital. In present study, airflow simulation of a multi-patients ICU room has been carried out using FLUENT version 15 CFD software. For simulation of airflow standard k-epsilon turbulence model used with high-quality unstructured mesh. Five different cases of multiple staff orientation have been studied to look at the infection between every patients and additionally for medical staff. In whole cases, inlet fresh airflow temperature (273 K) and airflow rate (0.2 m/s) are held constant. An average of 9.78 min time was taken by mobile contaminants to leave ICU room. The location of air inlet and outlet holds good air ventilation as particles coming out from the mouth of patient moves out of ICU in most of the instances and hospital staff orientations.

Keywords Room airflow · Buoyancy · Recirculation zone · k-epsilon model · Room airflow · Breathing

1 Introduction

Intensive care units of hospital require excellent medical supervision staff i.e., physician and nurses for well caring of critically ill patients. ICU of hospitals is commonly proposed for multi patients, these ICUs or wards are economically well-organized in which critically ill patients can be placed along one hall or in two rows, which facilitates less time for supervision. Airflow analysis of multiple patient wards or ICU room's ventilation plays a vital role on health of patients and medical staffs as well. In the present investigation, numerical simulation of ICU room for multiple patients has been carried out. The objective of current study is to envision the effect

A. K. Sahu (✉) · S. L. Sinha · T. N. Verma
Department of Mechanical Engineering, NIT Raipur, Raipur, India
e-mail: arvindkumarsahu25@gmail.com

A. A. Rizvanov et al. (eds.), *Advances in Biomedical Engineering and Technology*, Lecture Notes in Bioengineering,
https://doi.org/10.1007/978-981-15-6329-4_2

of contaminant which is getting out from the mouth of patients on different orientations of hospital staff to recover out the safest orientation to prevent hospital staffs and patients from infection. A total of 5 different cases of two staff orientations have been studied to examine the infection between each patient and also for the staff. For all cases, airflow rate and temperature of inlet air are kept constant. Various case studies using CFD for fluid flow in ventilations of multiple/single bed hospitals were performed. Contaminant distribution considered with variable air volume (VAV) and heating, ventilation and air conditioning (HVAC) and systems in hospitals. The effect of the location of the infected patients in transmitting air-borne diseases with hospitals using CFD is also presented (Sinha et al. 2000; Prakash and Ravi Kumar 2015; Verma and Sinha 2015a,b; Senthilkumar and Raju 2016). It is found that CFD applications are beneficial to assist and recognize the suitability of the ventilation system in renewal of hospital design to fulfill the latest engineering standards (Chow and Yang 2003).

An additional study also revealed that the respiratory events such as breathing, sneezing, talking, and coughing were the main source of transfer of contaminants. The study uses a model of equations using various parameters like rate of flow, area of mouth orifice opening, etc., and proposed that the model be used for describing source of contaminant transfer due to talking and breathing (Gupta et al. 2010). The airflow patterns of an operating room (OR) during opening and foot traffic are studied. Even though OR has slightly higher pressure than other adjacent rooms, a small volume of air still enters during a cycle of door opening and closing even without any person entering the room. The study has revealed a higher volume of air enters the OR if the person enters the OR (Villafruela et al. 2013). The evaporation and condensation of expiration droplets and their size (coughing and speaking) have been found to have negligible impact on usual droplet size from human beings (13.5 μm from coughing and 16.0 μm from speaking) for average expiration velocity of 11.7 m/s and 3.9 m/s for coughing and speaking, respectively (Chao et al. 2009).

Contaminant distribution in an office environment of 6.6 m (L) × 3.7 m (W) × 2.6 m (H) dimension with air conditioning and mechanical ventilation was also studied. Tracer gas (SF_6) is used for simulation of contaminants on a model room and CFD was used for validating the results. The study revealed that the pattern of contaminant dispersion depends greatly on the velocity flow field. The layout of various objects like furniture also influences the pattern of airflow and contaminant. Another study suggests that CFD can be used effectively in predicting the spatial distribution of bio-aerosol in indoor environments like hospitals. The study was conducted at three different layouts—empty, single bedroom, and two bedrooms. Deposition of the particles has no correlation with relative surface concentration and source distance but partition among the patients proved to be effective in reducing cross contamination among patients. (Cheong et al. 2003; King et al. 2013). Authors have performed various experiments in different regions using the respective environment conditions. The authors in this paper have performed real time study by means of experimental and numerical solution using environment conditions of the hospital in Raipur (21.2514° N, 81.6296° E, 298 m altitude above sea level), Chhattisgarh (Sahu 2018; Verma 2018,2014; Verma and Sinha 2013).

2 Governing Equations

To study air spread pattern inside ICU model room FLUENT version 15 CFD tool has been used. GAMBIT preprocessing software is used to create high-quality unstructured grids. Grid independence test carried out before simulation and after GIT computation domain has 5.2×10^6 grids. Turbulence in airflow is considered at the entry of ICU model room and K-epsilon turbulence model is used for processing airflow in computational domain. Here airflow in an ICU model room is described by mass, momentum, and energy conversion equations which are basically quantity of fluid per second, newton's second law of motion, and thermodynamics first law, respectively. Mathematically it is expressed as (Patankar 1980).

$$\frac{\partial}{\partial t}(\rho\phi) + \mathrm{div}(\rho u\phi) = \mathrm{div}\left(\mathcal{T}_\phi \mathrm{grad}\phi\right) + S_\phi \quad (1)$$

Motion of mobile contaminant particles are tracked using newton's second law of motion (Patankar 1980).

$$\frac{\partial u_p}{\partial t} = F_i, F_D = \frac{18\mu C_d Re}{24\rho_p d_p^2} \quad (2)$$

$$\frac{du_p}{dt} = F_D\left(u_i - u_p\right) + \frac{g(\rho_p - \rho)}{\rho_p} + F_x \quad (3)$$

where F_i represents external forces exerted on the particle and F_D represents drag force (N), C_d is coefficient of drag, ρ is Mass density in kg/m^3, μ is dynamic viscosity in Ns/m^2 and F_x is additional forces exerted on particles. P denotes for particle and i is Particle identifier. The above equation is integrated with particle tracking module of FLUENT software and used for tracking mobile contaminant.

The following assumptions have been used during computation:

- The cross section of beds are considered rectangular;
- Walls, floors, and roof of ICU model room are considered well insulated;
- Lying position of patients is the east–west direction;
- At a time only one patient is considered to be producing mobile contaminant;
- Shape of mobile contaminant particle is considered solid spherical;
- No contaminants recoil on solid walls;
- Motion of one mobile contaminant is considered for clarity of particle tracking in figures;
- Heat and mass transmission between mobile contaminant and air inside ICU room is neglected;
- Diameter of mobile contaminant is assumed to be circular and uniform.

3 Problem Statement

Figure 1 shows the outline of five patient ICU model room with 5 different orientation of hospital staff. The outline of ICU room has two rectangular inlet vents and two outlet vents for entrance of fresh air and elimination of sick air from the model ICU room. The ICU model room is inspired by actual ICU room of a hospital which is situated at BALCO region of Korba C.G. position of inlet, outlet vents, and lying arrangement of ICU model room is considered same as actual. Height, width, and length of ICU model room are 3.0 m, 5.8 m, and 6.3 m, respectively and sizes of all inlet and outlet vents are 0.6 m $\times$ 0.4 m.

The location of both inlet vents is kept constant i.e., 2.3 m overhead the ground, inlet 1 is 1.4 m and inlet 2 is 3.8 m ahead the east wall. The positioning of staff is based on literature surveys and surveys of various hospitals of Chhattisgarh. Inlet airflow rate in ICU room is considered 0.2 m/s. The temperature of fresh air is considered 293 K for all five cases and air properties are taken as per this temperature i.e., 293 K (Table 1).

Temperature of different walls has been selected from the ISHRAE handbook-2007 for Raipur (C.G.) region (Table 2).

4 Result and Discussion

Figure 2a, b shows the variation in velocity vector for two staff orientation case 1, where the temperature of fresh air stream is considered as 293 K, inlet air velocity is considered 0.2 m/s. The fresh and unpolluted air stream enters from inlet vents and throws well into ICU room. After throwing it mixes and spreads well with the existing air of ICU room. After appropriate distribution in ICU room, the air mixture drops near to opposite walls and moves out from the outlet vents.

Figure 2a shows airflow pattern on plane 1.7 m ahead of east wall, it is clear from Fig. 2a that one recirculation zone is formed near south wall. Figure 2b shows variation in velocity vector on plane 4.1 m ahead of east wall, it is clear from the Fig. 2b one recirculation zone is found between patient 2 and patient 3 and smaller air circulation zone found at right corner near to north wall.

Figure 3a, b shows the temperature contour on the plane 1.7 and 4.1 m ahead of east wall for two hospital staff orientation case 1.Colors in Fig. 4a–e show time for contaminant to leave ICU in minutes. Here red color stands for maximum and blue stands for minimum contaminant leaving time. The temperature of the roof and different walls are considered different due to the difference in incident solar radiation, which is clear from Fig. 3a, b. Mixing of fresh air with existing air is found appropriate and uniform at occupied zone. Temperature boundary layers are formed near to walls of ICU room due to temperature variation in walls (Table 1). Also small temperature variation due to airflow rate is found at the entry of airflow and right corner of the ICU room.

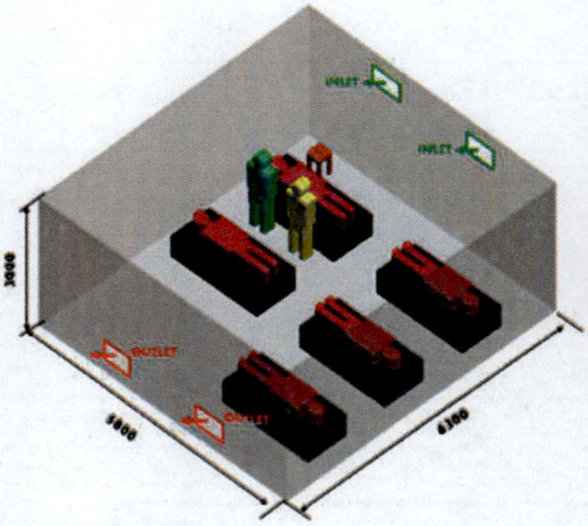

(a) Hospital staff orientation 1

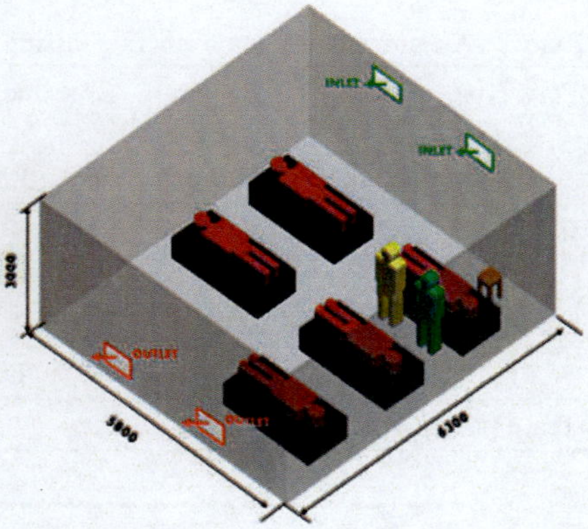

(b) Hospital staff orientation 2

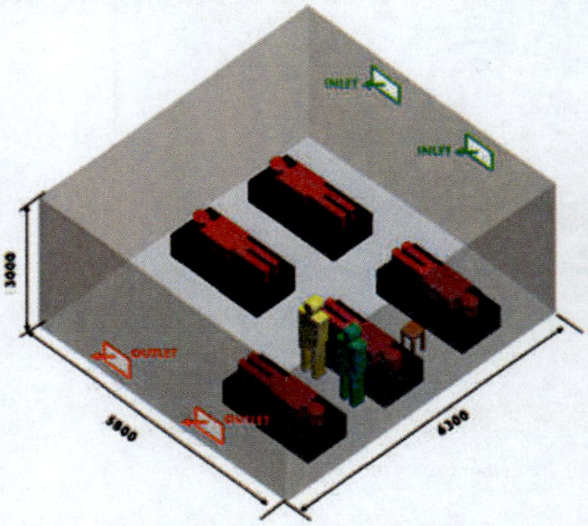

(c) Hospital staff orientation 3

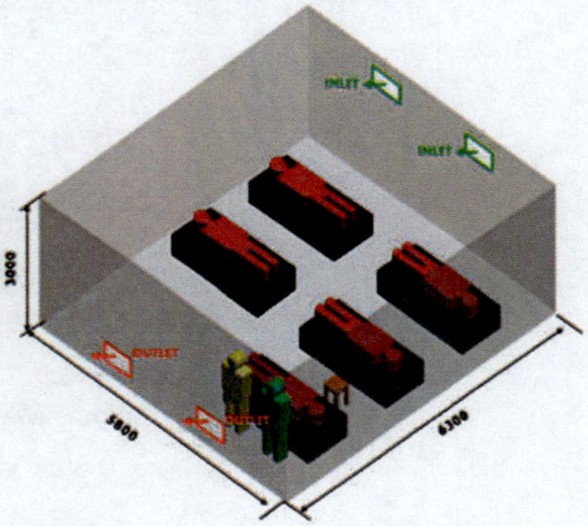

(d) Hospital staff orientation 4

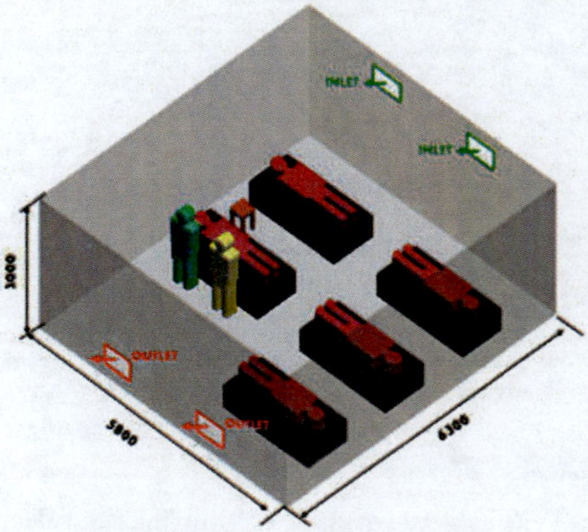

(e) Hospital staff orientation 5

Fig. 1 Five-bed ICU room with different hospital staff orientations

Table 1 Air properties at atmospheric pressure

Flow type	Temp	Density	Specific heat	Thermal conductivity	Dynamic viscosity
	(K)	(kg/m^3)	(J/kg-K)	(W/m-K)	(kg/m-s)
Cold	293	1.204	1007	0.02514	1.83×10^{-05}

Table 2 Temperature of walls

Walls	North	South	East	West	Celling	Floor
Temp (K)	290	301	304	298	321	296

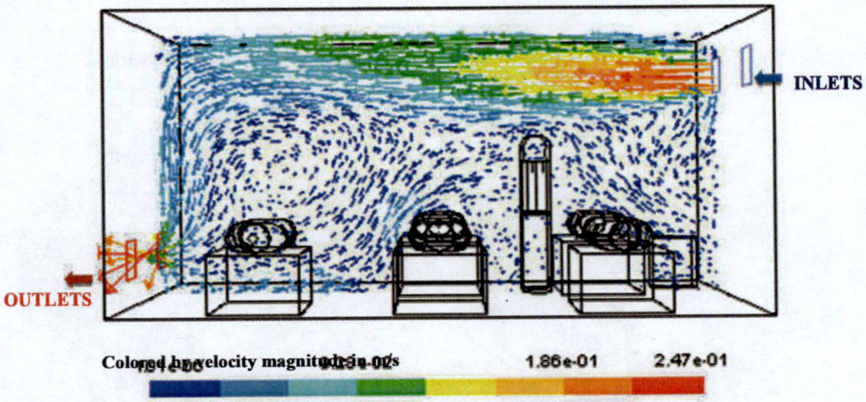

(a) Velocity vector at plane x=1.7 m (inlet air velocity 0.2 m/s)

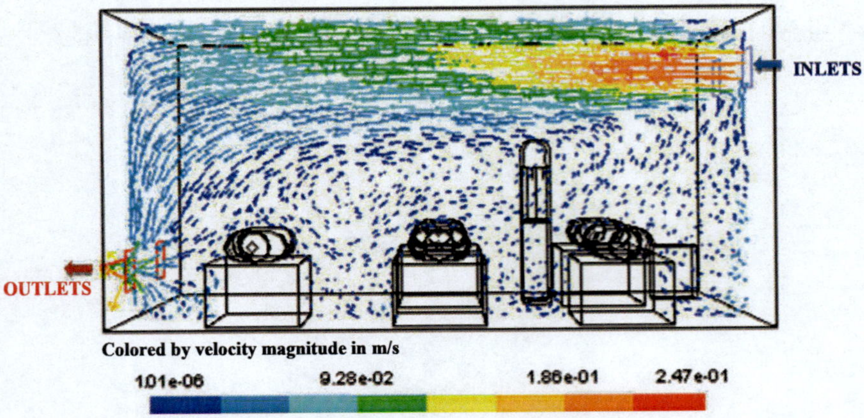

(b) Velocity vector at plane x=4.1 m (inlet air velocity 0.2 m/s)

Fig. 2 **a, b** Velocity vector for two hospital staff orientation 1 (ACH = 2.05, Re = 12590)

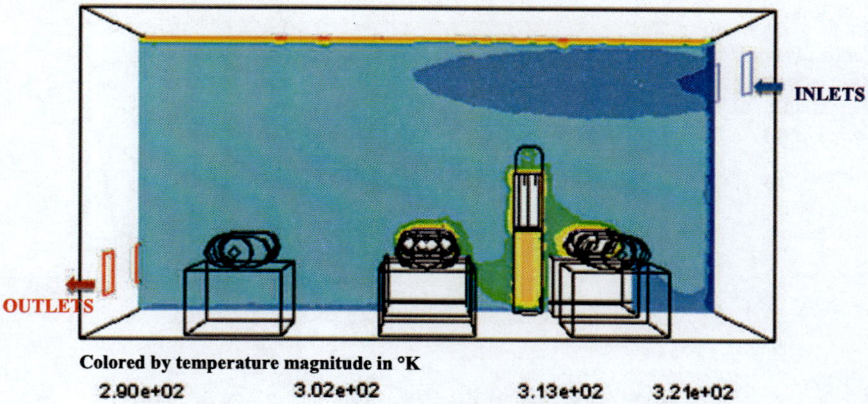

(a) Temperature contour at plane x=1.7 m (inlet air velocity 0.2 m/s)

(b) Temperature contour at plane x=4.1 m (inlet air velocity 0.2 m/s)

Fig. 3 **a, b** Temperature contour for two hospital staff orientation 1 (ACH = 2.05, Re = 12590)

Figure 4a–e shows the movement of mobile contaminant for two staff orientation case 1, for different particle which emergences from the mouth of ill patients in the ICU room. Here, the profile of motion of the contaminated particles which is coming out from patients are not affecting other patients and occupants near the patients. Table 3 shows the time required for elimination of mobile contaminant from ICU room. It is clear from Table 3 that elimination time for the mobile contaminant from patient 1 is comparatively less than other patients.

Fig. 4 **a–e** Movement of mobile contaminant through patients for hospital staff orientation 1 (ACH = 2.05, Re = 12590)

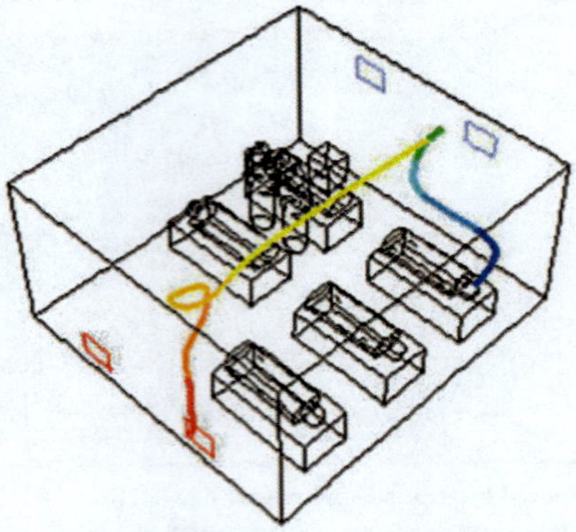

(a) Movement of mobile contaminant through patient 1

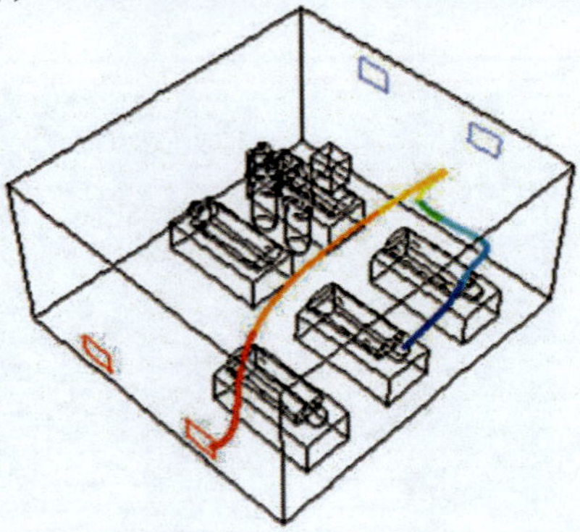

(b) Movement of mobile contaminant through patient 2

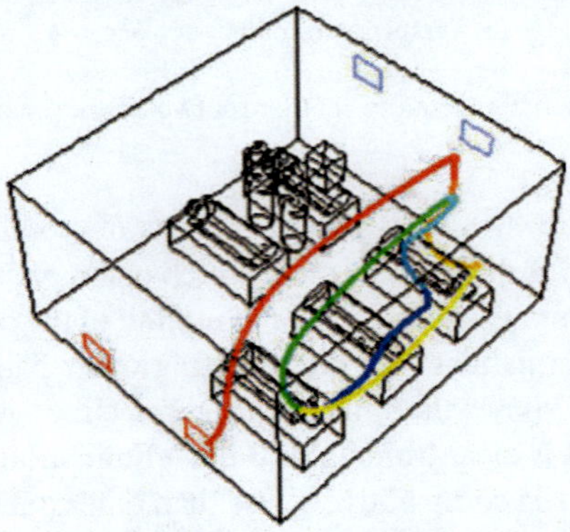

(c) Movement of mobile contaminant through patient 3

Fig. 4 (continued)

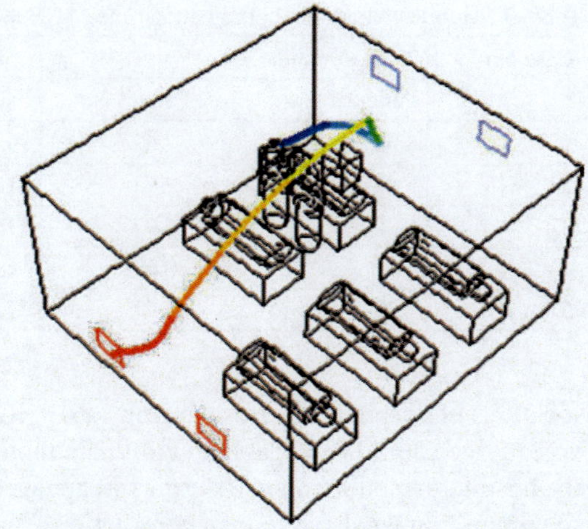

(d) Movement of mobile contaminant through patient 4

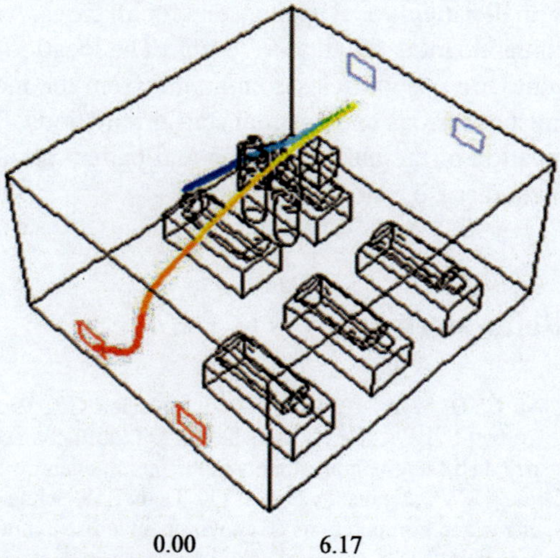

(e) Movement of mobile contaminant through patient 5

5 Conclusion

The numerical simulations of airflow have been carried out for ICU room. Control volume method is used to solve energy and Navier Stokes equations by utilizing FLUENT software. It is a realistic flow problem which brings into account the outcome of multiple patients and multiple staff as obstacles in a room. It is seen that ventilation execution is fully influenced upon airflow rate and its entry and exit

Table 3 Mobile contaminant elimination time (ACH = 2.05, Re = 12590)

Case No	Time in minute				
	Patient 1	Patient 2	Patient 3	Patient 4	Patient 5
1	8.58	7.62	19.50	5.13	5.13
2	5.08	7.78	11.87	6.88	4.75
3	6.30	6.97	11.05	16.67	28.50
4	4.80	20.67	6.42	9.60	7.08
5	6.17	7.77	17.50	5.70	6.90

locations of hospital. The fresh airflow rate in ICU should be a sufficient amount and spread rate should be capable to eliminate mobile contaminants from all portion of the hospital/isolation rooms. The air spread rate ought to be inside with worthy limits of commotion level and human consolation. Standard k-epsilon turbulence model is rigidly valid for turbulent flows. In this investigation, basic dimensions like length, height, and width of ICU of hospital are kept constant. To compare effect of various orientation of hospital staff, the location of inlet vents, outlet vents, and beds along with ill patients are kept constant for all cases. Average time for elimination of mobile contaminant is found as 9.78 min. The location of air inlet and outlet holds great air ventilation as particles coming out from the mouth of patients move out of ICU in most occasions and hospital staff orientations. By perception, it is prompted to keep location of the nurture-station and patient caring staff closed to inlet vents to keep them disease free.

References

Chao CYH, Wan MP, Morawska L, Johnson GR, Ristovski ZD, Hargreaves M, Mengersen K, Corbett S, Li Y, Xie X, Katoshevski D (2009) characterization of expiration air jets and droplet size distributions immediately at the mouth opening. J Aerosol Sci 40:122–133

Cheong KWD, Djunaedy E, Poh TK, Tham KW, Sekhar SC, Wong NH, Ulah MB (2003) Measurements and computations of contaminant's distribution in an office environment. Build Environ 38:135–145

Chow TT, Yang XY (2003) Performance of ventilation system in a nonstandard operating room. Build Environ 38(12):1401–1411

Gupta JK, Lin CH, Chen Q (2010) Characterizing exhaled airflow from breathing and talking. Indoor Air 20:31–39

King MF, Noakes CJ, Sleigh PA, Camargo-Valero MA (2013) Bio-aerosol deposition in single and two-bed hospital rooms, a numerical and experimental study. Build Environ 59:436–447

Patankar SV (1980) Numerical heat transfer and fluid flow. McGraw Hill, Washington

Prakash D, Ravi Kumar P (2015) Analysis of thermal comfort and indoor air flow characteristics for a residential building room under generalized window opening position at the adjacent walls. Int J Sustain Built Environ 4:42–57

Sahu AK, Sinha SL, Verma TN (2018) Numerical simulation of air flow to ventilate intensive care unit of hospital, computer application in education & research for science and technology. Int Res Public House Delhi 131–138

Senthilkumar M, Raju NMS (2016) transient effect of window position on naturally ventilated room with various vent cross section area. Int J Adv Eng Technol

Sinha SL, Arora RC, Roy S (2000) Numerical simulation of two dimensional room air flow with and without buoyancy. Energy Build 32(1):121–129

Verma TN, Sinha SL (2013) Contaminant control in intensive care unit (ICU) using CFD modeling. Int J Mechan Industr Eng 3:121–125

Verma TN, Sinha ST (2014) Contaminant control in intensive care unit of hospital. Appl Mechan Mater 2486–2490

Verma TN, Sinha SL (2015) Trajectory of contaminated particle in intensive care unit of hospitals using numerical modelling. Int J Des Manuf Technol 9(1)

Verma TN, Sinha SL (2015) Numerical simulation of contaminant control in multi-patient intensive care unit of hospital using computational fluid dynamics. J Med Imaging Health Inform 5:1–5

Verma TN, Sahu AK, Sinha ST (2018) Numerical simulation of air pollution control in hospital. Energy Environ Sustain 185–206

Villafruela JM, Olmedo I, Ruiz M, de Adana, Méndez C, Nielsen PV (2013) CFD analysis of the human exhalation flow using different boundary conditions and ventilation strategies. Build Environ 62:191–200

A Simple Robust Image Processing Algorithm for Analysis of Static Foot Pressure Intensity Image to Detect Foot Risk Areas in Diabetic Patients

Hari S. Nair, Navya Thomas, and R. Periyasamy

Abstract According to WHO, about 422 million people are affected with diabetics and among them 15% of diabetic patients are associated with foot problems. Due to peripheral neuropathy, the patients are unable to sense pressure pain which eventually creates calluses. Thus the aim of this study is to effectively find the risk areas which may form ulcers in the foot. The physical components required for this study are a camera which captures high-quality images of foot and a laptop capable of processing image by using MATLAB (The Mathworks Inc., USA). The algorithm which we developed extracts the high-intensity areas in the foot which are actually the areas where the patient exert high pressure compared to other areas in the foot. This method was evaluated by assessing the foot for 5 normal patients and 5 diabetic patients. It is found that this method is capable of giving the riskier areas of the foot and thereby further prevention of foot ulcer can be done. Our method is time-efficient, cost-efficient, and space-efficient and thus can be easily installed in any hospital for diabetic foot analysis. In most of the cases, the metatarsal head region and plantar medial tubercle region are mainly affected by calluses. By proper diagnosis and daily care of the foot can reduce the risk of ulcer in diabetic patients.

Keywords Callus · Ulcers · Metatarsal head region · Plantar tubercle region

Abbreviations

WHO World Health Organization
USA United States of America
CPU Central Processing Unit

H. S. Nair (✉) · N. Thomas
Department of Biomedical Engineering, National Institute of Technology Raipur, Raipur, India
e-mail: harisn2391@gmail.com

R. Periyasamy
Department of Instrumentation and Control Engineering, National Institute of Technology Trichy, Trichy, India

A. A. Rizvanov et al. (eds.), *Advances in Biomedical Engineering and Technology*, Lecture Notes in Bioengineering,
https://doi.org/10.1007/978-981-15-6329-4_3

1 Introduction

Diabetes being a serious chronic disease needs global attention. The global prevalence of diabetics among adults over 18 years of age has risen from 4.7% in 1980 to 8.5% in 2014 that is about 422 million people (Mathers and Loncar 2002). About 15% of people affected with diabetes are associated with foot problems (Dayananda and Kiran 2014). Diabetic foot problems such as ulcerations, infections, and gangrene are the major cause of hospitalization in diabetic patients (Ingrid and Steven 2006). Diabetic foot ulcer is a common complication which is a major source of morbidity and it is showing an increasing trend over previous decades (Leila et al. 2015).

Since the sensory neurons are affected in diabetes, patients are unable to sense pressure pain and microtrauma in the foot resulting in tissue breakdown and formation of cavity at the level of epithelial layer (Prabhu et al. 2001). Peripheral neuropathy and arterial occlusion are mainly the causes of diabetic foot ulcers. Diabetic neuropathy will develop in 50% of type1 and type 2 patients with diabetes (Hajieh et al. 2013). The foot ulcer once developed leads to an increased risk ulcer progression that may lead to amputation. Studies show that the rate of lower limb amputation in patients with diabetes mellitus is 15 times higher than those without diabetes. The multiple risk factors which are associated with the development of foot ulcer as per recent studies are gender (male), duration of diabetes longer than 10 years, advanced age of patients, high body mass index, and other comorbidities such as retinopathy, diabetic peripheral neuropathy, peripheral vascular disease, glycated hemoglobin level, foot deformity, high plantar pressure, infections, and inappropriate foot self-care habits (Leila et al. 2015). Poorly fitting shoes, poor foot care often cause foot deformation that can lead to the formation of callus. The formation of calluses is clearly been associated with foot ulceration (Pavicic and Korting 2006). More than 70% of patients who have developed diabetic foot ulcer experience an exacerbation of the disease in the next 5 years. The ulcer usually appears in the same extremity or the extremity of the opposite side; at least a quarter of these ulcers do not heal (Bijan et al. 2013). In this study, the image of diabetic foot is captured and by proper processing of the image, the possibility to predict the risk areas on the foot that may eventually develop to a foot ulcer is studied. By proper caring of the risk areas, infection can be prevented in diabetic patients. Thus by proposing a method which can effectively show the risk areas on the foot, foot ulcer formation can be prevented.

Previously, plantar pressure is measured in laboratories or in hospitals which require heavy equipment such as pressure platforms and mats. Currently in-shoe pressure measurements are being under research which is less bulky and easy to use as compared to the previous one. But both these devices lack the ability to measure the plantar pressure in real-life situations. These methods will be able to pressure hotspots in case of laboratory situations only. It is also important to note that these techniques need bulky devices which tend to increase the cost of setting up in hospitals and also require more efficient and qualified technicians.

Lei Wang et al. (Lei et al. 2015) propose a method for the assessment of diabetic foot ulcers by capturing the RGB image of diabetic foot using a smartphone and the wound assessment algorithms include: (1) mean-shift-based image segmentation, which groups all image pixels into a number of homogeneous regions; (2) a fast method for detecting the largest connected component, to recognize the foot outline; and (3) final wound boundary determination achieved by analyzing the internal and external boundaries of the foot outline. After image processing, a wound healing score ranging from 1 to 10 is given to each subject. This technique can be applied only after the wound has occurred and early wound analysis is not proposed in this method.

Dayananda and Kiran (2014) developed an image processing algorithm for monitoring the insole wear patterns. In this study, advanced Gabor filters were evaluated to detect the skin irregularities by using a photograph of the insole captured by a smartphone. This method used fuzzy set clustering for clustering the image. As this method used an ordinary smartphone for image capture which effectively reduces the cost but eventually reduces the image quality and thereby loss of image data may occur.

Dmitry et al. (2011) proposed a method to predict the tissues which are at the risk of ulceration by using hyperspectral tissue oximetry. Type1 and type2 diabetes mellitus subjects that are under risk of ulceration undergoes hyperspectral image along tissue oximetry. The data are retroactively analyzed and an ulceration prediction index is developed. They then developed an image processing algorithm based on the ulceration index. This algorithm is then capable to predict the tissue under the risk of ulceration. This method requires complex hardware and also the patients are asked for several visits for the process.

2 Methods

2.1 Foot Image Data Acquisition

Foot images are acquired by using a suitable camera that can capture high-quality images. Two images are captured, one by placing the foot and the other without the foot which can be used as a reference image. It is done so that the noises that may creep in can be erased using the reference image. These images can be easily transferred to a computer in which the next step processing takes place.

The captured image is then saved using the name of the patient along with the date in which the image is captured for future reference. Clinicians can afterwards use these images to compare the current situation of the patient and can determine the stage of healing of the ulcer. This research protocol is a small part of large study which was carried out in accordance with the guidance of the Institutional Ethical Committee NIT Raipur and all the participants have signed the inform consent.

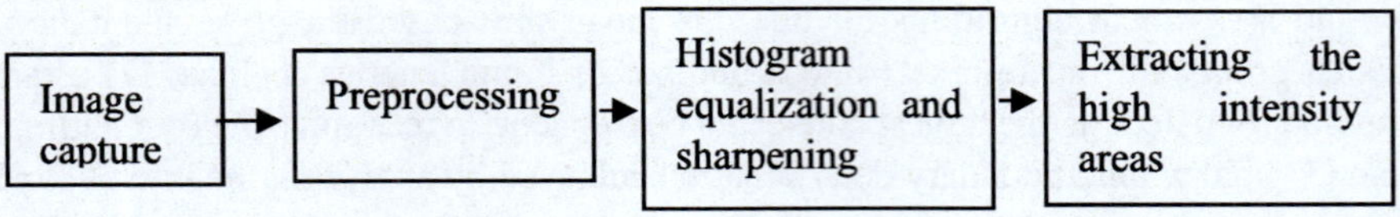

Fig. 1 Flow diagram for the process

2.2 *Image Processing Algorithms*

For processing the images, we are using MATLAB (The Mathworks Inc., USA). During wound image processing, the risk areas are found out by 4 modules: (1) background correction; (2) histogram equalization; (3) sharpening the images; (4) finding the risk areas by extracting the high-intensity areas from the processed image Fig. 1.

The captured images first need to be converted to grayscale for further processing. The image is first made to be free from background noises. It can be removed by subtracting the foot images from their respective reference images and filling the holes using "imfill" function available in MATLAB. These pre-processed images need histogram equalization which can be done by using "adapthisteq" function followed by sharpening of the edges by using "imsharpen" function. It is found that the risk areas are of higher intensities than the other parts of foot due to the formation of calluses. Thus the higher intensity regions are extracted from the images. The processed image can be used to conclude the extent of risk for getting ulcer. From the subjects whom we used for data acquisition, it is found out that in case of diabetic foot (having risk of ulcer), the processed image stresses on some part of the image while this phenomenon is absent in the case of normal foot. The processed images of both normal foot and diabetic foot along with the captured images are shown in Figs. 2 and 3.

3 Results

The results after processing are shown in Figs. 1 and 2 both in the cases of normal patients and diabetic patients. From Figs. 1 and 2, diabetic patients are under greater risk of foot ulcer. In the case of diabetic patients the processed image highlights some regions in the foot which are confirmed as risk areas. Thus for those patients who are under risk, proper diagnosis and daily care of the foot can prevent ulcer. It is also found that the risk areas are mainly in the first, second, and third metatarsal head and in the plantar medial tubercle region of the foot. In the algorithm, the "limit" can take any value between 0 and 255. But the preferred value must lie between 225 and 245. The output images for different values are shown in Figs. 1 and 4.

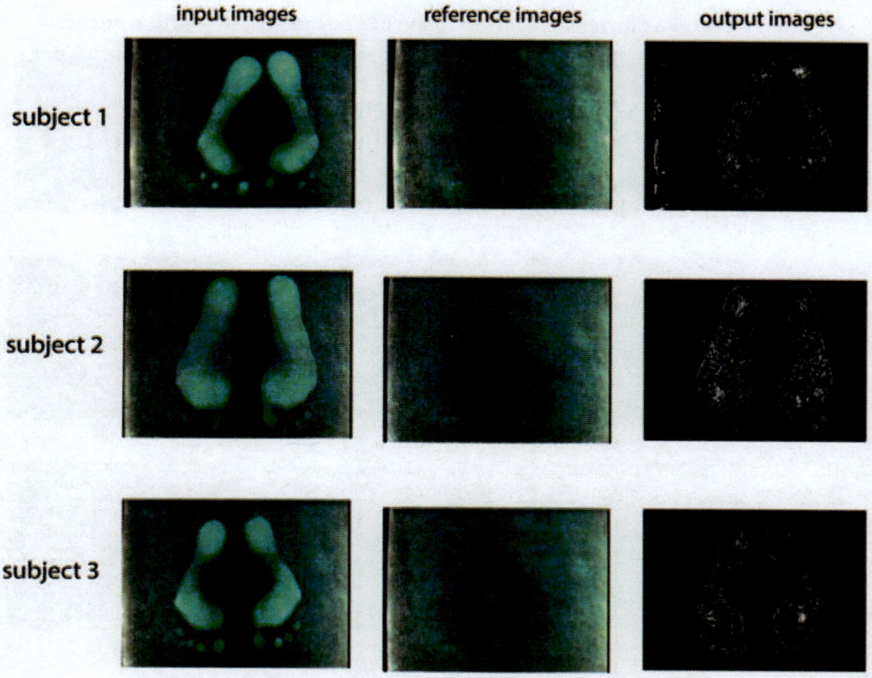

Fig. 2 Input image, reference image and processed images of normal patients

From Fig. 3, it shows that as the limit value increases, the highlighted area decreases. For limit = 225, the image shows even the low-risk areas while for limit = 235, image shows the areas which needs attention and for limit = 245, the areas present (if any) is of greater risk that is at the verge of ulcer.

4 Conclusion

The image processing algorithm which we developed can efficiently give the risk areas in the foot of a diabetic patient. The patient should be educated regarding the foot ulcer and the chance he/she may get affected by it. With proper technologies it is possible to trace the walking pattern of the patient, then he/she can be trained to change his walking pattern so that the pressure is evenly distributed throughout his/her foot. This can prevent foot ulcer to an extent. For those patients who are under risk can be asked for frequent visits so that condition of the foot can be observed regularly by comparing the images with the images of the previous visit.

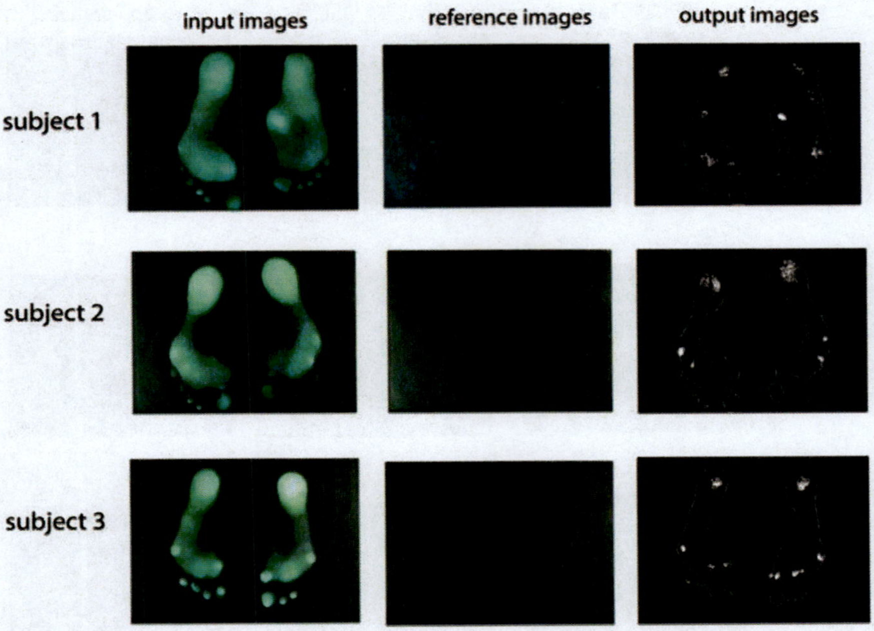

Fig. 3 Input images, reference images and processed images of diabetic patients

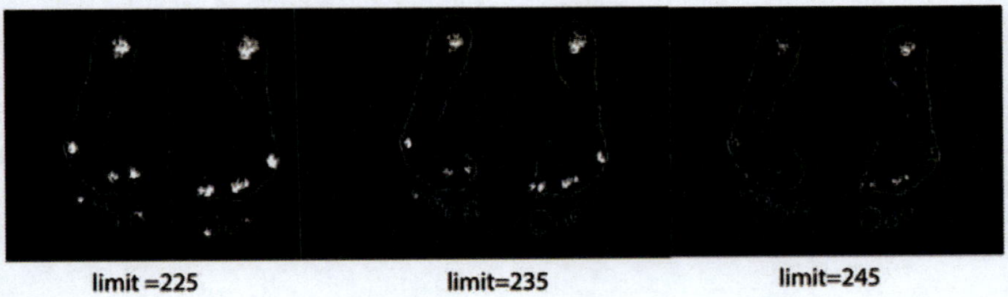

Fig. 4 Output images for different values of limit

The total computing time for the algorithm implemented on a laptop with CPU (Intel i7 3.5 GHz) is nearly 1.1 to 1.2 s for images in dimension 640 * 480 pixels. Thus this method is very time-efficient for calculating the risk areas. As the required equipment comprises of an infrared camera and a laptop, it is cost-efficient as well as space-efficient compared to other clinical instruments. Thus it is easy to install it in any hospital for treating foot ulcers.

Appendix

The image processing algorithm works as follows:

```
clc,clear;

%reading the original image and reference image

he=imread('input foot image.bmp');

he1=imread('reference image.bmp');

%converting to grayscale

he=rgb2gray(he);

he1=rgb2gray(he1);

%background correction

he3=he-he1;

he4=abs(he3);

he5=he4>mean2(he4);

b=imfill(he5,'holes');

b=double(b);

he=double(he);

c=b.*he;

c=uint8(c);

%histogram equalisation

d=adapthisteq(c,'cliplimit',0.05);

e=imsharpen(d,'amount',5);

%extracting hish intensity regions

limit=235;

f=e>limit;

f=f*255;

%edge capturing technique

BW1 = edge(he3,'prewitt');

BW1=BW1*255;

g=BW1+double(f);

figure,imshow(uint8(g));
```

References

Bijan I, Fariborz K, Alireza E, Gholamreza A (2013) Prevention of diabetic foot ulcer. Int J Prev Med 4(3):373–376

Dayananda KJ, Kiran KP (2014) Analysis of foot sole image using image processing algorithms. In: 2014 IEEE global humanitarian technology conference—South Asia satellite (GHTC-SAS), Trivandrum, pp 57–63

Dmitry Y, Aksone N, Kevin S, Laurent P (2011) Assessing diabetic foot ulcer development risk with hyperspectral tissue oximetry. J Biomed Opt 16(2):026009

Hajieh S, Leila Y, Seyed ML (2013) Risk assessment of patients with diabetes for foot ulcers according to risk classification consensus of international working group on diabetic foot (IWGDF). Pak J Med Sci 29(3):730–734

Ingrid K, Steven E (2006) Evaluation and treatment of diabetic foot ulcer. Clin Diabetes 24(2):91–93

Leila Y, Morteza N, Sara A (2015) Literature review on the management of diabetic foot ulcer. World J Diabetes 6(1):37–53

Lei W, Peder CP, Diane MS, Bengisu T, Emmanuel A, Ron I, Qian H (2015) An automated assessment system of diabetic foot ulcers based on wound area determination, colour segmentation, and healing score evaluation. J Diabetes Sci Technol 10(2):421–428

Mathers CD, Loncar D (2006) Projections of global mortality and burden of disease from 2002 to 2030. PLoS Med 3(11):e442

Pavicic T, Korting HC (2006) Xerosis and callus formation as a key to the diabetic foot syndrome: dermatologic view of the problem and its management. J Dtsch Dermatol Ges 4(11):935–941

Prabhu KG, Patil KM, Srinivasan S (2001) Diabetic feet at risk: a new method of analysis of walking foot pressure images at different levels of neuropathy for early detection of plantar ulcers. Med Biol Eng Comput 39(3):288–93

Heavy Metal Ions Detection by Carbon Paste Electrode as an Electrochemical Sensor

Arti Mourya, Bidyut Mazumdar, and Sudip K. Sinha

Abstract Voltammetry techniques such as cyclic voltammetry and electrochemical impedance spectroscopy are used to quantify metal ions in aqueous solutions. In this paper, we have reported electrochemical detection of lead (II), cadmium (II), and Copper (II) ions based on carbon paste electrode using graphite powder. Potassium ferricyanideK3 [Fe(CN)6] solutions as benchmark media to check redox reactions. The Detection limit was 6.56×10^{-9} M investigated by cyclic voltammetry. Effects of pH and supporting electrolytes were also measured.

Keywords Carbon paste electrode · Cyclic voltammetry · Electrochemical impedance spectroscopy · Heavy metal

1 Introduction

Water contamination is one of the major environmental issues and between the wide diversity of contaminants heavy metal ions are one of them Momodu (2010) Heavy metal contamination damages the ecosystem due to higher toxicity even at minute quantity. Cadmium (II), Lead (II), and Copper (II) are classified as heavy metal ions which enter in water bodies through various industrial operations. Elevated levels of heavy metal ions can cause behavioral changes, liver damage, and impair intelligence (Migliorini 2017; Rajawat 2014). Hence, the recognition and monitoring of toxic metals from polluted sites and aquatic ecosystems is an important analytical task that is needed by society.

Various analytical methods and techniques are used to determine the heavy metal ions at low concentrations like atomic absorption/emission spectrometry

A. Mourya · B. Mazumdar (✉)
Department of Chemical Engineering, NIT Raipur, Raipur, India
e-mail: bmazumdar.che@nitrr.ac.in

S. K. Sinha
Department of Metallurgy Engineering, NIT Raipur, Raipur, India

A. A. Rizvanov et al. (eds.), *Advances in Biomedical Engineering and Technology*, Lecture Notes in Bioengineering,
https://doi.org/10.1007/978-981-15-6329-4_4

(AAS/AES), and atomic fluorescence spectrometry (AFS). Among them electroanalytical method is superior, because of its high sensitivity, eco-friendly, selectivity, easy data readout, quick response time, cost-effective, and easy handling (Vasileva 2017; Rajesh 2017). Selecting suitable material for electrochemical senor decides good and reliable detection for heavy metal ions (Bagheri 2013). Graphite is a unique and versatile material in electrochemical systems, has been utilized for decades in sensors, batteries, fuel cells, and electrochemical capacitor Panice (2014). In this reporting, we study on simulated aqueous phase (with lead, cadmium, and copper) by carbon paste electrode (CPE) incorporated with voltammetric methods.

2 Material and Methods

2.1 Reagents

All the experiments are carried out with the help of analytical grade chemicals without any further purification. Graphite powder, chemicals (size < 40 μm), paraffin liquid, and aluminum powder 98% were purchased from Loba chemicals. $K_3Fe(CN)_6$ was purchased from Sigma-Aldrich. Cadmium nitrate, lead nitrate, cupric nitrate, potassium nitrate 98% (KNO_3), and NaOH chemicals are obtained from Merck.

2.2 Preparation of Electrode as a Sensor

For the preparation of CPE, 70% graphite and 30% paraffin oil binder was mixed in mortar and pestle. The paste was left overnight for self homogenization. After that, the resulting paste was filled into an insulin syringe (electrode holder). A thin copper wire was inserted through the opposite end of the electrode body for the electrical contact. And the surface of the electrode gets smoothed on A4 paper until it had a mirror finish. After each experiment, a few millimeters (mm) of carbon paste was extruded from the electrode holder. Further it was polished by 0.05 micron alumina powder slurry, and washed well by distilled water and ethanol (Ranđelović 2017).

3 Result and Discussion

3.1 Cyclic Voltammetry

To check the electrochemical properties of CPE, cyclic voltammetry (CV) technique was used for 5×10^{-3} mol L^{-1} $K_3Fe(CN)_6$ and 1 mol L^{-1} KNO_3 as an electrolyte. Figure 1a shows the cyclic voltammogram at different scan rates with

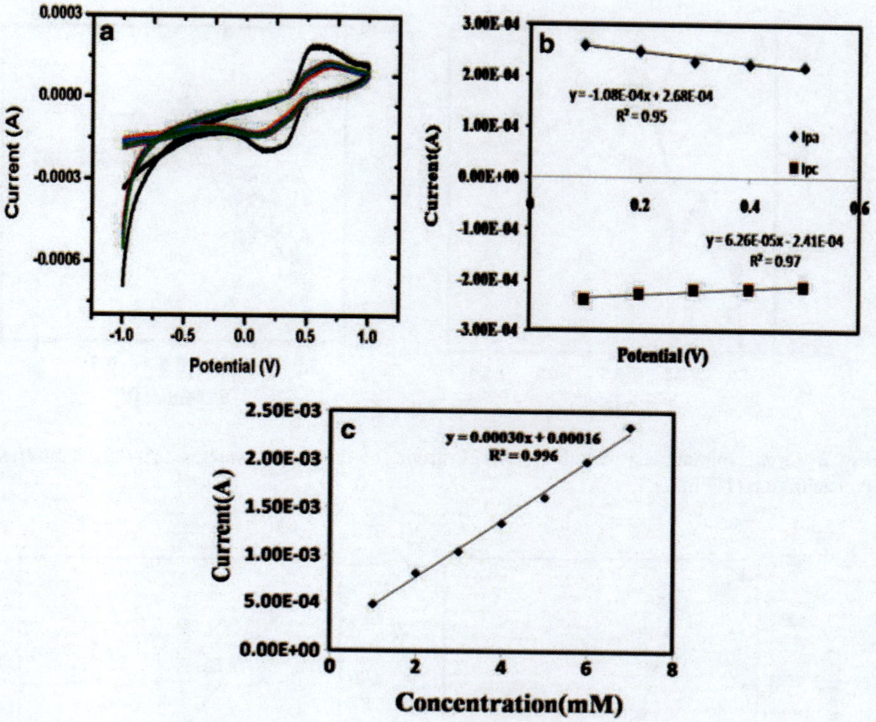

Fig. 1 **a** Cyclic voltammogram of CPE for 5 × 10^{-3} mol L^{-1} $K_3Fe(CN)_6$ and 1 mol L^{-1} KNO_3 at different scan rates. **b** potential versus current at different scan rates (0.1–0.5 mV/s), **c** calibration curve at different concentration of $K_3Fe(CN)_6$

single anodic and cathodic peak currents. A plot between potential versus current Fig. 1b for different scan rates shows a linear equation with 0.95 and 0.97 regression values. Randles–Sevcik equation was used to calculate the electrode surface area: Ip = (2.69 × 105)n^3/2AD1/2V1/2, where, A surface area of the working electrode (cm^2), Ip peak current (A), D diffusion coefficient, (cm^2/s), C initial concentration (mol/L), n number of electrons involved in the reaction, and V scan rate (V/s). Further, Fig. 2a, b shows cyclic voltammograms for the simultaneous determination of the lead, cadmium, and copper ions at a different scan rates (10–50 mV/s) and concentrations (1–7 mM) in pH of 5.23. To investigate the reaction kinetics, the effect of scan rate on the electrochemical redox behavior, CV shows three different anodic and cathodic peaks for lead (II), cadmium (II), and copper (II) ions. It can be clearly seen that the anodic and cathodic peak current increased with increasing the concentrations of heavy metal ions.

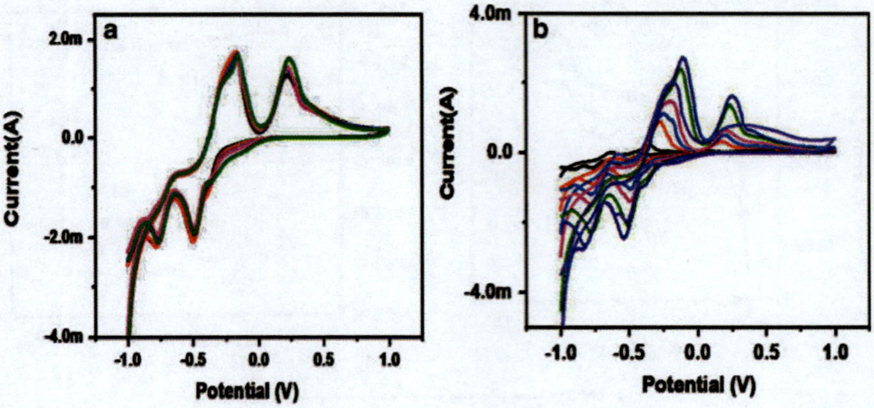

Fig. 2 Cyclic voltammetry of carbon paste electrode at **a** different scan rates (0.1–0.5), **b** different concentration (1-7 mM).

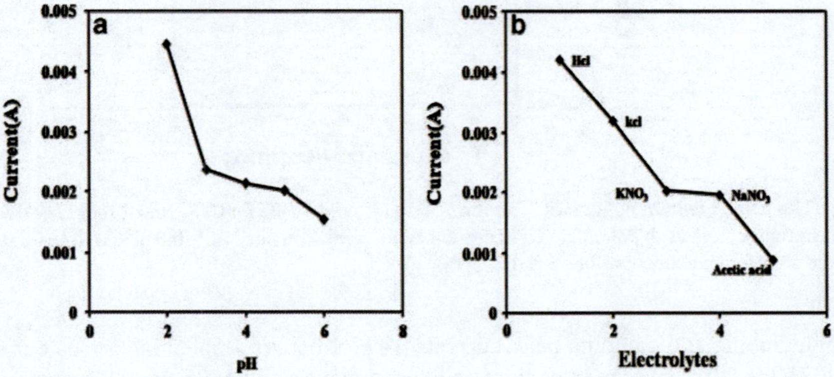

Fig. 3 **a** Effect of pH and **b** electrolytes on cyclic voltammogram

3.2 *Effect of Operating Parameters*

Effect of pH &Electrolyte. The voltammetric output was investigated between the pH range of 2.0 -6.0 in 1 M KNO_3 electrolytic solution with 5 mM of Pb, Cd, and Cu. As expected (Fig. 3a), higher current signals were observed at lower pH. However, the peak current was decreased because of the hydrolysis of heavy metal ions as the pH reaches higher value from 2.0 to 6.0 (Xiong 2016). The electrochemical behaviors of the heavy metal ions are different in different electrolytes such as HCl, KCL, KNO_3, $NaNO_3$, and acetic acid. Among these electrolytes (Fig. 3b), HCL shows the higher current signals due to the occurrence of well-defined peaks with the largest cathodic peak current of 4 mA.

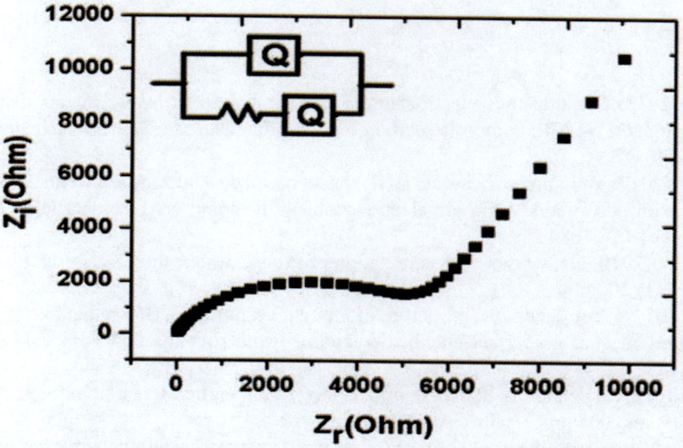

Fig. 4 Electrochemical impedance spectroscopy of carbon paste electrode containing 5.0 mM Pb, Cd, and Cu solution with 1 M KNO_3

Table 1 Impedance parameter by CPE for Pb(II), Cd (II), and Cu (II)

Parameters	Q-Yo	Q-n	R	Q-Yo	Chi squared	Electric circuit
CPE	3.252E-8	0.8	5590	0.0001203	2.370e-02	(Q(RQ)

3.3 Electrochemical Impedance Spectroscopy (EIS)

The EIS was performed after a (70 s) open circuit potential, and 0.1–100000 Hz frequency ranges. The Nyquist plot (Fig. 4) represents the electrical equivalent circuit (inset) used in the EIS fitting and impedance parameters of the CPE are shown in Table1. From the results obtained, CPE showed a semicircle in lower frequency and a straight line with higher range. According to open literature such behavior shows a charge transfer limited process and its value is equal to the diameter of the semicircle.

4 Conclusion

CPE was effectively developed for electrochemical detection of lead (II), cadmium (II), and copper (II) ions.The CV results show successive enhancement in the peak currents (Ip) with a constant potential. As expected, higher current signals were observed at lower pH value of 2. (Q(RQ) electrical circuit was a best fit model for electrochemical impedance spectroscopy, performed to check the interface properties of the prepared electrode. Finally the developed sensor was successfully applied for heavy metal ion detection in aqueous solution by cyclic voltammetry technique.

References

Bagheri H (2013) Simultaneous electrochemical determination of heavy metals using a triphenylphosphine/MWCNTs composite carbon ionic liquid electrode. Sensors Actuators B Chem 186:451–546

Migliorini F (2017) Voltammetric cadmium (II) sensor based on a fluorine doped tin oxide electrode modified with polyamide 6/chitosan electrospunnanofibers and gold nanoparticles. Microchim Acta 184(4):1077–1084

Momodu MA (2010) Heavy metal contamination of ground water: the case study. Res J Environ Earth Sci 2(1):39–43

Panice LB (2014) Electrochemical properties of the hexacyanoferrate (II)–ruthenium (III) complex immobilized on silica gel surface chemically modified with zirconium (IV) oxide. Mater Sci Eng B 188:78–83

Penka V (2017) Application of starch-stabilized silver nanoparticles as a colorimetric sensor for mercury(II) in 0.005 mol/L Nitric Acid. J Chem 1–10

Rajawat DS (2014) Trace determination of cadmium in water using anodic stripping voltammetry at a carbon paste electrode modified with coconut shell powder. J Anal Sci Technol 5(1):19

Rajesh K (2017) Design and development of graphene intercalated V_2O_5 nanosheets based electrochemical sensors for effective determination of potentially hazardous 3, 5–Dichlorophenol. Mater Chem Phys 199:497–507

Ranđelović MS (2017) Electrocatalitic behaviour of serpentinite modified carbon paste electrode. J Electroanal Chem 801:338–344

Shrivastava A (2011) Methods for the determination of limit of detection and limit of quantitation of the analytical methods. Chronicles Young Sci 2(1):21

Xiong W (2016) Development of gold-doped carbon foams as a sensitive electrochemical sensor for simultaneous determination of Pb (II) and Cu (II). Chem Eng J 284:650–656

Median Filtering Detection Using Markov Process in Digital Images

Saurabh Agarwal and Satish Chand

Abstract Digital images are prevalent medium for information representation. The wide acceptability of images also leads to some problems. Few people spread forged images for their own interests. This type of practice questions the credibility of images. To ensure the credibility, forensic investigation is carried out to ensure the genuineness of images. Detection of median filtering is one of the important forensic analyses. Median filtering has several uses in antiforensics. It is applied to hide the impression of image forgery due to its nonlinear nature. Some techniques are evolved to countering these antiforensics i.e., median filtering. These techniques can classify median filter operated and non-median filter operated images. In this paper, we compare three techniques that are based on Markov process. We find that performance of these techniques varies according to the size of database.

Keywords Image forensics · Median filtering detection · Antiforensic

1 Introduction

From earlier time, pictures are the preferable medium of information representation. In this digital era, most of the modern gadgets are equipped with camera. In spite of this, with the help of high-speed internet, images can be communicated anywhere. Sometimes peoples float forge images for their own malicious interests. We can see the latest example of image forgery popular on twitter November 1, 2017 in Fig. 1. Figure 1a is a pristine image from the scene of movie "Kya Yeh Sach Hai" and Fig. 1b is a forged image. In this forged image, it is shown that a senior police officer kneeling and touching the feet of Home Minister Rajnath Singh. However, it is found

S. Agarwal (✉)
ASET, Amity University, Noida, India
e-mail: saurabhnsit2510@gmail.com

S. Chand
Jawaharlal Nehru University, Delhi, India

A. A. Rizvanov et al. (eds.), *Advances in Biomedical Engineering and Technology*, Lecture Notes in Bioengineering,
https://doi.org/10.1007/978-981-15-6329-4_5

Fig. 1 An example of pristine and fake image

(a) Pristine Image

(b) Fake Image

that some opponent morphed Rajnath Singh's face onto the face of the actor playing the minister role in the scene of movie "Kya Yeh Sach Hai."

To ensure the credibility of images, forensic analysis of images is performed. For creating the realistic forge image, various post-processing operations are performed on forged image like resampling, contrast enhancement, histogram equalization, etc. Some of these post-processing operations have also been used to hide the forgery artifacts and this process is called antiforensics. The forensic analysis of various antiforensics operations is necessary to identify image forgery. Out of these operations, detection of median filtering is crucial and difficult due to its nonlinear nature.

The importance of median filtering motivates researchers to develop several methods for detecting it. The first well known method (Kirchner and Fridrich 2010) detects the median filtering by considering streaking artifacts in histogram bins of first-order difference images. This method fails to provide good accuracy on compressed images. The method (Pevny 2010) is known as Subtractive Pixel Adjacency Matrix (SPAM) applicable for many applications like Steganalysis is also applied for median filtering detection. It gives good accuracy on compressed images also. In this method Markov process is applied on first-order difference images. Markov process provides Subtractive Pixel Adjacency Matrix (SPAM) features on adjacent difference pairs. The method (Chen 2013) known as Global and Local Feature set (GLF) extracts local correlation of neighboring pairs on first- and second-order difference image pixels. The authors have shown some results on noisy images and claim better results from previously available techniques. The technique (Shen 2014) utilizes Rotation-Invariant Local Binary Pattern (RI-LBP) texture descriptor for median filtering detection. Similarly, the technique (Zhang 2014) uses higher order Local Ternary Pattern (LTP) texture descriptor to identify median filtering. These techniques (Shen 2014; Zhang 2014) provide average results on low resolution and compressed images. In (Ravi et al. (2015)) transition probability features are extracted from image by utilizing the correlation of spatial domain quantization noise. This method provides good results on non-compressed and low compressed images. Further method SPHO (Agarwal 2016) based on SPAM considered higher order pixel differences for improving the detection accuracy. The method (Agarwal 2016) provides better results on low resolution and highly compressed images.

In this paper, we assess the performance of methods (Agarwal 2016); Chen 2013; Pevny 2010) on different sizes of image databases. This paper primarily emphasizes on the effect of image database size on median filtering detection accuracy. We have chosen these methods (Agarwal 2016; Chen 2013; Pevny 2010) because their performance is somehow comparable.

2 Feature Construction

To differentiate between median filtered and non-filtered images, relevant features need to be extracted from images. In SPAM Pevny (2010), features are extracted using Markov process on first-order difference arrays. In SPHO Agarwal (2016) Markov process is applied on first, second, and third order difference arrays. The technique GLF TECH (Chen 2013) extracts two types of feature sets i.e., global probability feature set and local correlation feature set. In which global probability feature set extraction process. It can be seen that somehow these three methods are based on Markov process. Therefore, here we discuss briefly about Markov process when applied on difference arrays.

First-order backward difference array in horizontal direction is defined for a 2–D image X of size m × n as:

$$D^h_{u,v} = X_{u,v} - X_{u+1,v} \tag{1}$$

Where, $u \in \{1, 2, \ldots, m-1\},\ v \in \{1, 2, \ldots, n\}$.

Like first-order difference array in horizontal direction, difference array in forward direction can be found. Similarly, forward and backward differences for major diagonal, minor diagonal, and vertical directions can be identified.

Now, Markov process of second order is applied on these difference arrays. The Markov process of second order is used in further discussions and experiment works. Markov process of second order provides relevant size and high-quality discriminant features.

The second-order Markov process is defined as.

$$M_{p,q,r} = pr(D_{u+2,v} = P|Du + 1, v = q.D_{u,v} = r) \tag{2}$$

where Pr represents probability $p, q, r \in \{-T, \ldots, T\}$ and for $\Pr(D_{u+1,v} = q, D_{u,v} = r) = 0, M_{p,q,r} = 0$.

Here T is the threshold parameter. In thresholding process, all values greater than T is replaced by T and all values smaller than −T is replaced by −T in difference array D. Therefore, we have numbers in the range of {−T, −T + 1, ..., 0,..., T−1, T}. This helps in obtaining small size feature array. The feature array is found by concatenation of two feature arrays. The first feature array contains the Markov features of horizontal and vertical difference 2–D arrays and the second feature array contains Markov features of major diagonal & minor diagonal difference 2–D arrays, in forward and backward directions. Finally, the second-order Markov process provides the feature array of size 2 * (2T + 1)3.

3 Experimental Setup and Results

In this section, we will discuss image databases, classifier, and experimental results.

3.1 Datasets

We evaluate the performance of SPAM Pevny (2010), SPHO (Agarwal 2016), and GLF TECH (Chen 2013) on combined databases of BOWS2 Bas and Furon (2008), UCID (Schaefer and Stich 2003), and NCI (Liu and Chen 2014). The BOWS2 grayscale image database contains 10,000 images of size 512 × 512 pixels. The UCID color image database consist 1,338 images of size 384 × 512 pixels. The NCI color image database has 5,150 images of size 256 × 256 pixels. These databases have images of several types like indoor, outdoor, natural, objects, fruits, foods, textures, etc. We crop the images by taking central block of size 64 × 64 pixels from all three used databases (Bas and Furon 2008; Liu and Chen 2014; Schaefer and Stich 2003). From these images, two sets of images are produced. The first set is created by applying JPEG compression of quality 50. The second set is created by applying median filter of size 3 × 3 and JPEG compression of quality 50. These sets are named as set 1 and set 2, respectively. From these two sets, we select equal number of images for performing experiments.

3.2 Classifier

We have applied Linear Discriminant Analysis (LDA) classifier to distinguish between median and non-filtered images. LDA classifier and Markov process both consider the conditional joint distribution. Therefore, LDA classifier is more appropriate choice than Support Vector Machine (SVM) classifier. In experiments, 50% images are considered for training and rest 50% for testing from the set of images of both types i.e., median and non-filtered images set. We have shown the experimental results in terms of detection accuracy (AC) and ROC curve with classification error (Pe). Detection accuracy (AC) in percentage (%) is described as:

$$AC = \frac{(TP + TN) * 100}{(TP + TN + FP + FN)} \tag{3}$$

where, TP-True Positive, TN-True Negative, FP-False Positive, and FN-False Negative. Classification error (Pe) = 1 − (AC/100).

3.3 Experimental Results

First, we have shown the results by taking 500 non-filtered images from set 1 and 500 median filtered images from set 2, randomly. As can be seen from Fig. 2a SPAM method for T = 2 provides the least classification error i.e., 9.44. Only GLF TECH method is comparable with SPAM with 11.21 classification error. Method SPHO for T = 3 performs worst with 31.52 classification error (Pe). In the second experiment, we have taken 1000 images from each set. It is clearly seen from Fig. 2b, method SPHO for T = 2 and SPAM for T = 2 perform decently with Pe = 5.45 & 6.32, respectively. Similarly, we evaluate the performance by considering 2000, 3000, 4000, & 5000 from both sets i.e., set 1 & set 2. As evident from Figs. 2c–f method SPHO T = 2 performs best.

As can be seen from Fig. 3 SPHO for T = 2 gives best results for more than 2000 images consistently. The performance of SPAM for T = 2, 3 is also fair. GLF TECH performance is consistent but detection accuracy is least in most of the cases. The performance of SPHO for T = 3 is worst for small number of images.

4 Conclusion

High dependency and usage of digital images make it more vulnerable. Strong antiforensics techniques are required to compete with forgers. One of the popular antiforensics operations is median filtering. In many cases, detection of median filtering is crucial for forensic investigation of images. A number of methods have

Fig. 2 Median filtering detection techniques ROC curves with classification error

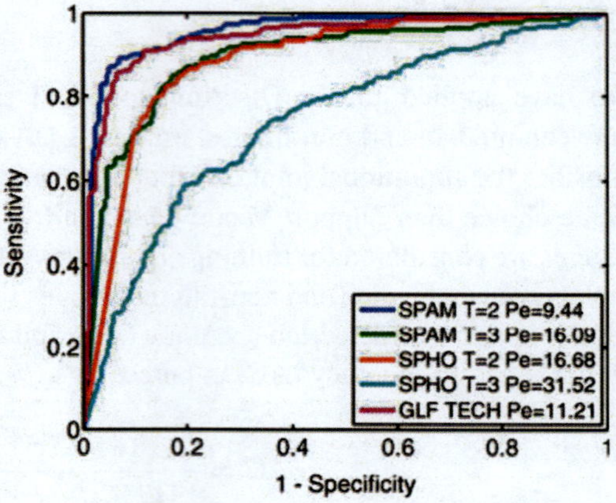

(a) Performance comparison for 500 images of set 1 and set 2

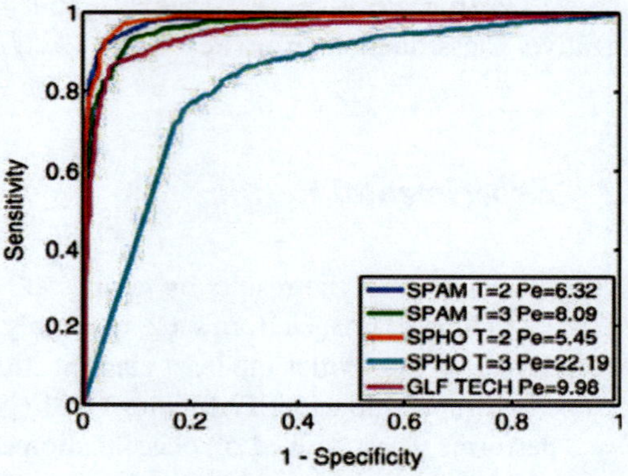

(b) Performance comparison for 1000 images of set 1 and set 2

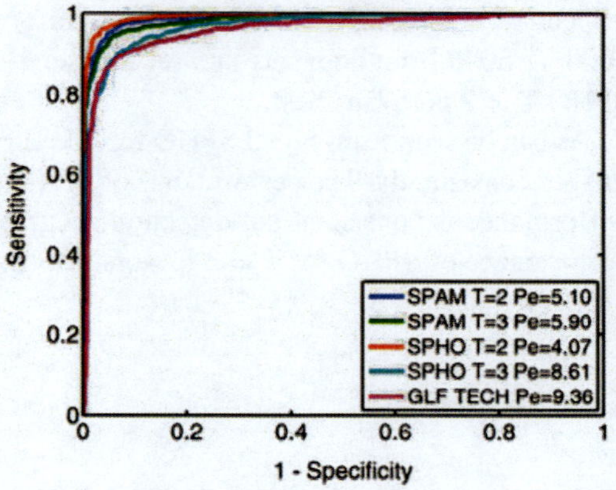

(c) Performance comparison for 2000 images of set 1 and set 2

Fig. 2 (continued)

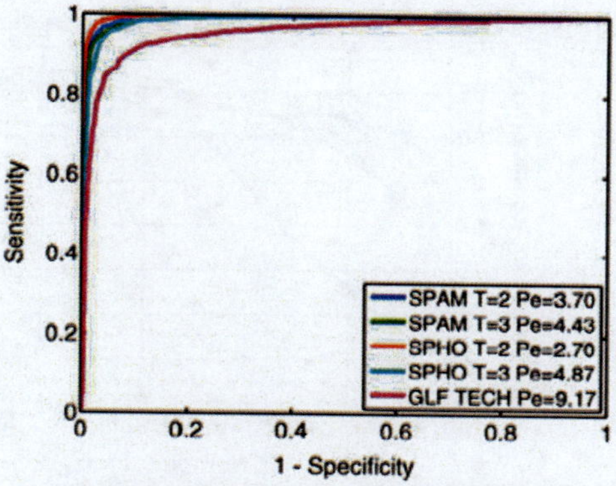

(d) Performance comparison for 300 images of set 1 and set 2

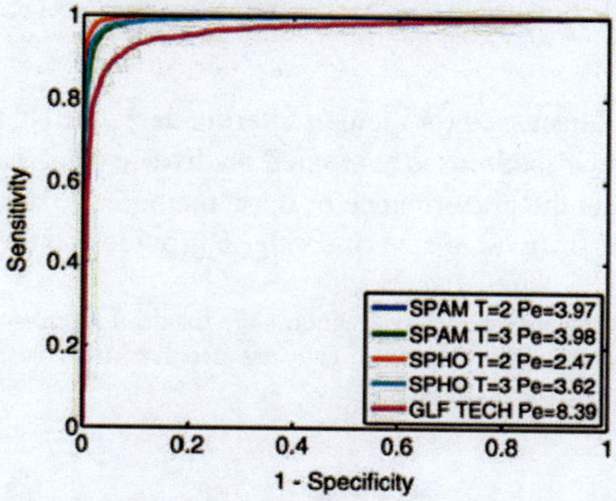

(e) Performance comparison for 4000 images of set 1 and set 2

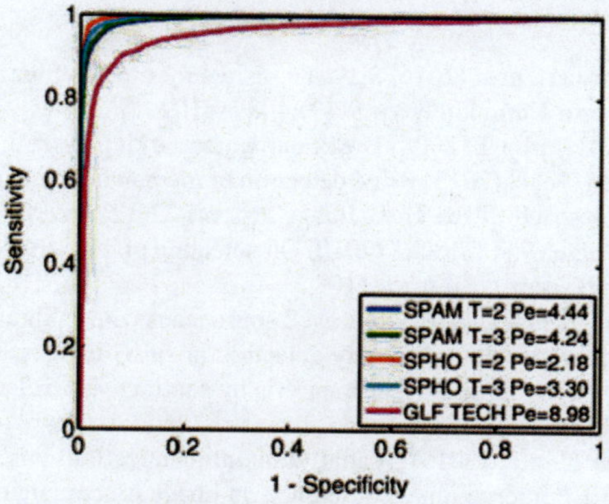

(f) Performance comparison for 5000 images of set 1 and set 2

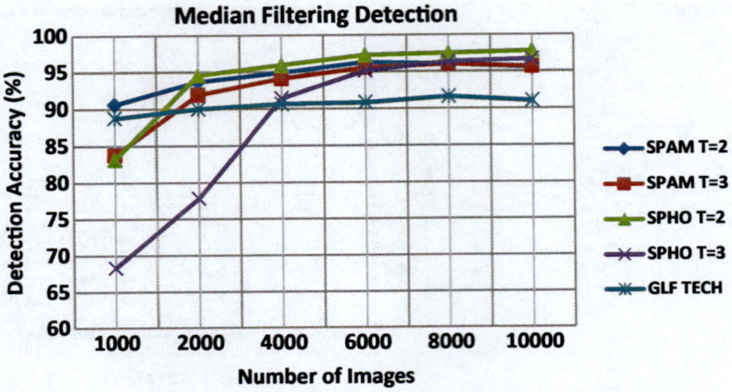

Fig. 3 Comparative analysis of methods with respect to number of images

been discussed for median filtering detection. In this paper, the performance of three popular methods is compared on diverse databases. We find that the size of database affects the performance of these methods. In most of the cases, classification model build from large size of database provides better detection accuracy.

Acknowledgements The authors are thankful to creator of BOWS2 [2], UCID [8] and NCI [5] for sharing their databases. They are also thankful to authors of SPAM [6], and GLF TECH [3] for providing their code.

References

Agarwal S et al (2016) SPAM revisited for median filtering detection using higher-order difference. Secur Commun Netw 9(17):4089–4102

Bas P, Furon T (2008) Break our watermarking system. Available https://bows2.ec-lille.fr/2nd

Chen C et al (2013) Blind detection of median filtering in digital images: a difference domain based approach. IEEE Trans Image Process 22(12):4699–4710

Kirchner M, Fridrich J (2010) On detection of median filtering in digital images. In: Media forensics and security II, p 754110

Liu Q, Chen Z (2014) Improved approaches with calibrated neighboring joint density to steganalysis and seam-carved forgery detection in JPEG images. TIST 5:4

Pevny T et al (2010) Steganalysis by subtractive pixel adjacency matrix. IEEE Trans Inf Forensics Secur 5(2):215–224

Ravi H et al (2015) Spatial domain quantization noise based image filtering detection. In: 2015 IEEE international conference on image processing (ICIP), IEEE, pp 1180–1184

Schaefer G, Stich M (2003) UCID—an uncompressed colour image database. SPIE, Storage Retr Methods Appl Multimed 5307:472–480

Shen Z et al (2014) Blind detection of median filtering using linear and nonlinear descriptors. Multimed Tools Appl 75(4):2327–2346

Zhang Y et al (2014) Revealing the traces of median filtering using high-order local ternary patterns. IEEE Signal Process Lett 21(3):275–280

Differential of EMG Activity of Selected Calf Muscle During DLHR Exercise in Relation to Performance Level

Monika, L. M. Saini, and Saravjeet Singh

Abstract Muscular fatigue is described as a condition when the ability of muscles to contract and produce force is reduced under sustained contraction. The quantification of muscular fatigue by surface electromyography (EMG) provides a noninvasive method for easily accessing and measuring the physiological processes occurring during sustained muscular work. Double leg heel raise (DLHR) exercise is performed for the strengthening of calf muscles which generally gets weak following immobilization after injury or surgery. It is used as evaluation process/test by physiotherapists/clinicians/others associated with rehabilitation of athletes/nonathletes to strengthen their lower-body muscles and connective tissues after joint-related injury. The study purpose is to compare the effect of sustained DLHR exercise in relation to performance level among males. The EMG activity of Gastrocnemius Lateral (GSL) and Gastrocnemius Medial (GSM) muscles of both legs are considered for this study because they are the dominating calf muscle. Here frequency domain and time-domain features were chosen for extracting necessary information from the EMG signal using the algorithm developed in MATLAB. Understanding from the findings is significant in order to achieve optimum muscular strength and strength endurance development whereas reducing the probability of training-related injuries to sportspersons.

Keywords Gastrocnemius lateral (GSL) · Double leg heel raise · Right leg (RL) · Fatigue · Mean frequency

Monika (✉) · L. M. Saini
School of Biomedical Engineering, NIT Kurukshetra, Kurukshetra, Haryana, India
e-mail: monikanitadhiman@gmail.com

S. Singh
Department of Biomedical Engineering, DCRUST Murthal, Murthal, Haryana, India

A. A. Rizvanov et al. (eds.), *Advances in Biomedical Engineering and Technology*, Lecture Notes in Bioengineering,
https://doi.org/10.1007/978-981-15-6329-4_6

1 Introduction

In the past decade, fatigue is explained as an exercise-induced, decrease in the muscle ability to produce power/force, if or not the task can be continued (Barry and Enoka 2007). When we exercise intensely/carry on similar kind of motion activity, then muscle of our body fatigue and as a result, we become to lose coordination, cannot train efficiently, work efficiency potential decreases for serious injury (Sakurai et al. 2010). So, muscle fatigue is essential information during muscular strength building or working, and is necessary to evaluate quantitatively. The sEMG i.e. surface electromyography (electrical activities measured over the skin surface during muscle contractions) signal analysis helps in prosthetic myoelectric control, clinical diagnosis, ergonomics, sports biomechanics, and evaluating the muscle fatigue (Karthick et al. 2018).

The Gastrocnemius muscle (overlying the soleus) is a broader calf muscle which makes half of the calf muscle. It runs down back of leg, from behind knee to the Achilles tendon in the heel. The gastrocnemius muscle consists of 2 heads: lateral and medial, join soleus at tendo-Achilles and inserts distally on the calcaneus. Gastrocnemius crosses 2 joints, working for plantar flexion of the ankle and flexion of knee. Muscles cross two joints, like gastrocnemius muscle, are especially susceptible to injury. Gastrocnemius injuries may happen at distal muscle–tendon junction at the head of medial muscle. The medial head of gastrocnemius may fracture during jumping or sprinting activities. Affected athletes suddenly have stabbing pain and may feel like their back of the calf has been struck directly with a ball. Shield Against Calf Sprains—According to a survey conducted by Runner's World (in 2011), calf sprains were the 2nd most common injury between injured runners. When calf muscles are not trained regularly, it loses strength and conditioning which is necessary for it in order to support athletic activities and as a result chance of injury increases. During running, walking, or jumping, our calf muscles, mainly the gastrocnemius muscle generates power (California 2005). In sports like volleyball and basketball, stronger calf muscles may be advantageous for the player. Strengthening of calf muscle is necessary in order to prevent any injury from accidentally turning of foot outwards or inwards.

Double leg heel raise (DLHR) exercise is performed for the strengthening of calf muscles (California 2005) which generally gets weak following immobilization after injury or surgery. Injuries, like Ankle fractures and Achilles tendon rupture, usually require a concerted calf stretching effort during recovery period. This exercise also strengthens the tibialis posterior muscle, which gets weak/dysfunctional in patients having acquired adult flatfoot deformity.

The study purpose is to compare the effect of sustained DLHR exercise in relation to performance level among males. The EMG activity of Gastrocnemius Lateral (GSL) and Gastrocnemius Medial (GSM) muscles of both legs are considered for this study because they are the dominating calf muscle. Here frequency domain and time-domain features were chosen for extracting necessary information from the EMG signal using the algorithm developed in MATLAB. We used EMG signal because it reflects the degree of muscular activation (Hopkins et al. 1999). Also, EMG signal is widely used in the human movement analysis where myoelectric signals produced by muscles are analyzed for muscular function investigation (Rainoldi et al. 2004). Understanding from the findings is significant in order to achieve optimum muscular strength and strength endurance development whereas reducing the probability of training-related injuries to sportspersons (Kaur et al. 2016).

2 Material and Methodology

Five healthy sports people engaged in different sports and six normal people were recruited as volunteers (age $= 21.5 \pm 1.87$, weight $= 59 \pm 10.21$ kg, height $=$ 172.96 ± 8.55 cm). Criteria of Inclusion for the study were subjects above 18 years of age, nonalcoholic, willing to engage in the study, and not having any other medical diseases which might hamper their nerves and muscles functioning, in order to make them unfit for participation in the study. Before their participation in the study, each member was described about the motive and protocol to be followed. In order to make them familiar with the study, the participants were asked to experience few pre-data collection trails. As per ethical guidance and needs, all participants provided voluntary written informed consent before their participation in the study.

2.1 Data Acquisition

EMG data were obtained using a 4-channel EMG BIOPAC Inc. MP 100 system (Gain: 5–50,000, CMRR: 110 dB at 50/60 Hz and Input Impedance: 2 MΩ) in a quiet laboratory room for all the subjects. Before data acquisition, skin of subjects was first shaved to remove hairs and then cleaned with cotton carrying alcohol for minimizing skin impedance and enhancing signal acquisition. Then, Disposable electrodes (7.5 mm diameter) were placed over the Gastrocnemius Lateral (GSL) and Gastrocnemius Medial (GSM) muscles of both legs of the participants based on Seniam (European Recommendations for Surface Electromyography) (Disselhorst-klug 1999). The EMG data were obtained from both the legs simultaneously. Reference electrode was put on the ankle of the subject to act as ground and for safety demands of the equipment (Konrad 2006). Inter-electrode separation (center to center) was kept 20 mm. The schematic of electrode placement is shown in Fig. 1. After the subject preparation, they were instructed to perform Double Leg Heel Raise (DLHR) exercise for 120 s in order to become familiar with the protocol. The sampling frequency of data acquisition software (Acqknowledge3.9.9, BIOPAC Systems Inc.) was set to 2000 Hz in order to avoid the aliasing effects as per Nyquist criteria.

Data Segmentation. The study group of male athletes and nonathlete subjects was divided into high and low performance groups, respectively. These groups were given

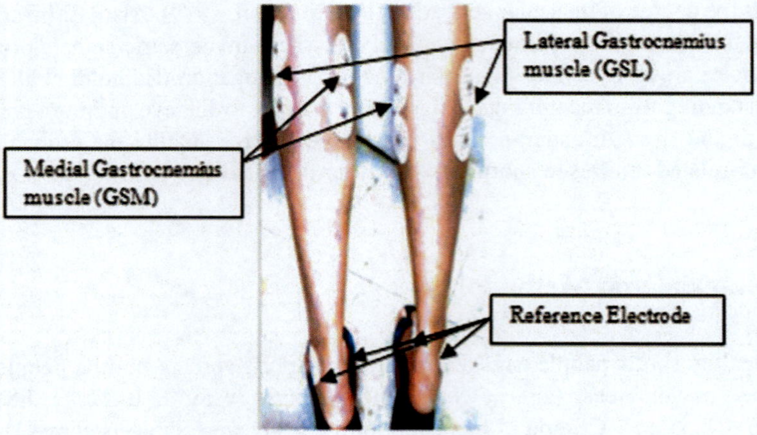

Fig. 1 Electrode placement and muscles

name as Male high performance (MHP) and Male low performance (MLP). All these groups were able to perform a minimum of 120 s DLHR exercise, hence for better understanding the influence of performance on EMG signal, data is further divided into 40, 80, and 120 s.

Data processing. Acquired raw EMG signals from different subjects were quantified into MATLAB. The signal processing methodologies applied to the raw EMG signals include two parts namely preprocessing and feature extraction as depicted in Fig. 2. The Digital filters were used in MATLAB for EMG signal preprocessing (Kaur et al. 2016). For the removal of 50 Hz noise interference from EMG signal, Notch filter was applied. To remove other noise sources from EMG signal, cascaded high pass filter (20 Hz) and low pass filter (450 Hz) were applied (Luca et al. 2010). Time-domain and Frequency-domain features were extracted from the filtered EMG signal namely Root Mean Square (RMS), Integrated EMG (IEMG), Median Frequency (MDF), and Mean Frequency (MNF), respectively for each subject. The reason for choosing these features in this study is that these are the most often selected features for the determination of changes due to fatigue in muscles.

Integrated EMG (IEMG). It represents the sum of the absolute value of the EMG signal amplitude (Phinyomark et al. 2012). IEMG relates to the EMG signal firing point and mathematically represented as:

$$X_i(t) = \int X_R(t)dt \tag{1}$$

where $X_i(t)$ is integration of signal and $X_R(t)$ is the rectified signal.

Increase of IEMG value during contraction of muscle corresponds to higher muscle fiber recruitment for a fixed external force. Therefore, its analysis is necessary in order to determine the fatigue development in muscle fiber(s) during contraction.

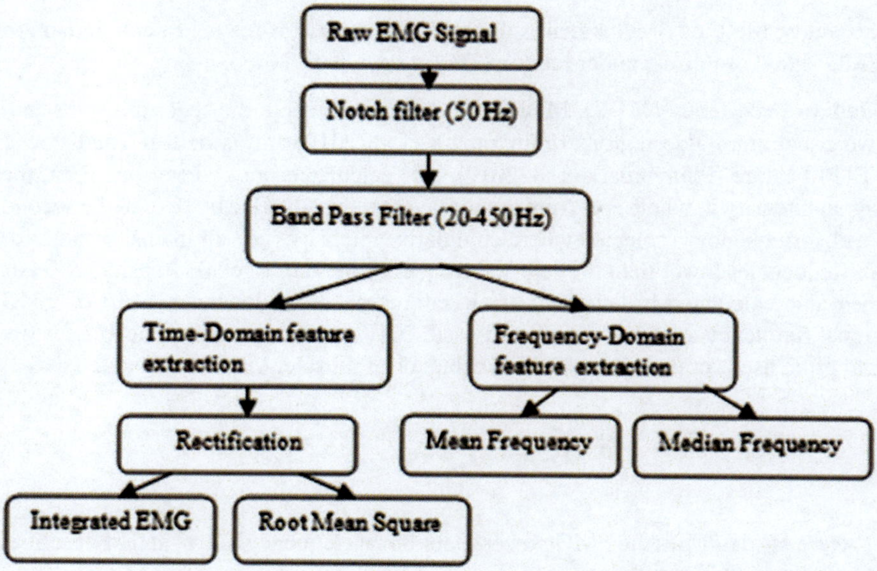

Fig. 2 EMG signal processing steps

Root Mean Square (RMS). This feature is most advanced and is used to analyze EMG signal. RMS is framework as amplitude modulated Gaussian random process which corresponds to non-fatiguing contractions and constant force (Phinyomark et al. 2016). RMS is the square root of mean power of EMG signal for a given time period and reflects the signal mean power. This feature is considered to be analyzed here because it shows the extent of physiological activity in motor unit during sustained contraction of muscle (Konrad 2006). Mathematically it is computed as:

$$X_r(t) = \sqrt{\frac{1}{T2 - T1}\int_{T1}^{T2} X_R(t)^2} \quad (2)$$

Mean Frequency (MNF). Mean Frequency is defined as the summation of product of EMG power spectrum and the frequency divided by the total sum of spectrum intensity (Phinyomark et al. 2010) and can be written as:

$$\sum_{m=1}^{M}(f_m P_m) / \sum_{m=1}^{M}(P_m) \quad (3)$$

Here f_m is the spectrum frequency at frequency bin m, P_m describes the EMG power spectrum at frequency bin m, and M is the frequency bin length. This feature is used in the estimation of muscle fatigue in EMG signal and hence is analyzed in this study. MNF of EMG signal is double when muscle is resting in comparison to EMG signal of muscle under fatigue (Thongpanja et al. 2013).

Median Frequency (MDF). MDF is the frequency where the spectrum splits into two equal amplitude regions (Phinyomark et al. 2010) and is half of Total Power (TTP) Feature (Phinyomark et al. 2012). It is calculated in two levels as: First, the signal intensity in whole spectrum is summed and then divided by two. In the second level, a frequency is selected where cumulative intensity (i.e., all intensity values of all frequencies lower than the selected frequency including focal intensity) exceeds the value calculated in level 1. Muscle fatigue causes a downward shift of EMG signal frequency spectrum (Pincivero et al. 2018). It is the most applicable feature and often used for describing fatigue behavior of muscle. MDF is calculated as:

$$\sum_{m=1}^{MDF} P_m = \sum_{m=MDF}^{M} P_m = \frac{1}{2}\sum_{m=1}^{M} P_m \tag{4}$$

Here P_m describes the EMG power spectrum at frequency bin m, and M describes the frequency bin length.

3 Result and Discussion

IEMG and RMS values of GSL and GSM muscles of Left Leg (LL) and Right Leg (RL) are given below. It demonstrates performance difference for male athletes and nonathletes with respect to time-domain features of both the muscles for both legs. Higher performance groups were having higher mean value for IEMG and RMS in terms of performance difference along with time progression in both muscles of subjects during DLHR exercise (Fig. 3).

Mean Frequency and Median Frequency values of GSL and GSM muscles of Left Leg (LL) and Right Leg (RL) are shown below. It presents performance difference for male athletes and nonathletes in the experimental protocol with respect to the GSL and GSM muscles activity for both legs. Lower performance groups were having higher mean value for MNF and MDF (except MDF of GSMof Right Leg (RL) where Higher performance groups were having higher mean value) in terms of performance difference along with time progression in both muscles of subjects during DLHR exercise (Fig. 4).

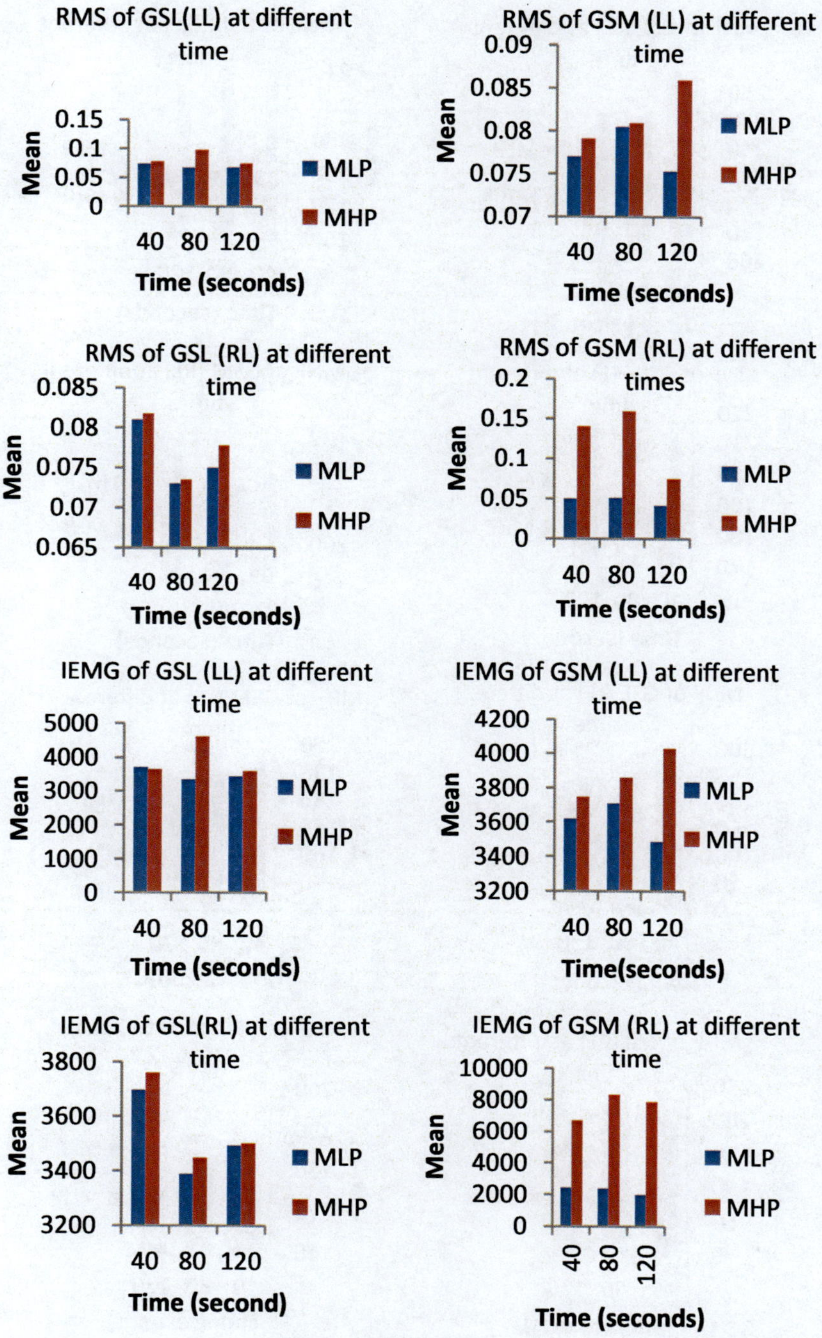

Fig. 3 Comparison of Time-domain features (RMS and IEMG) in terms of performance level

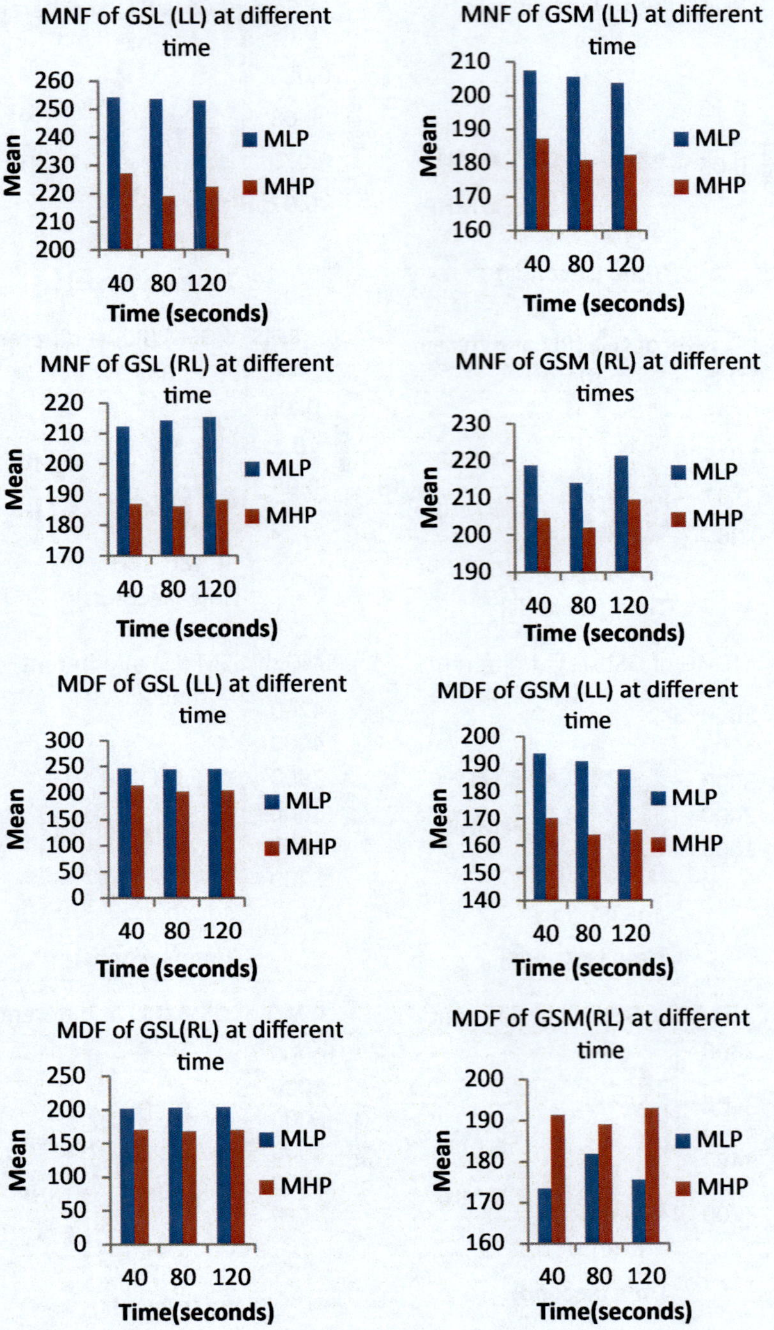

Fig. 4 Comparison of Frequency-domain features (MNF and MDF) in terms of performance level

4 Conclusion

Difference in EMG activity of left and right leg, muscle to muscle as well as Time-domain and Frequency-domain features extracted were observed. The Performance level of male athletes and nonathletes exhibits insignificant EMG activity differences of selected muscles. At last, we conclude that understanding from the findings is significant in order to achieve optimum muscular strength and strength endurance development whereas reducing the probability of training-related injuries to sportspersons.

References

Barry BK, Enoka RM (2007) The neurobiology of muscle fatigue: 15 years later. Integr Comp Biol 47(4):465–473

California S (2005) Biomechanics of the heel-raise exercise biomechanics, Apr 2014

De Luca CJ, Gilmore LD, Kuznetsov M, Roy SH (2010) Filtering the surface EMG signal: movement artifact and baseline noise contamination. J Biomech 43(8):1573–1579

Disselhorst-klug C (1999) European recommendations for surface ElectroMyoGraphy, pp 8–11

Hopkins JT, Ingersoll CD, Sandrey MA, Bleggi SD (1999) An electromyographic comparison of 4 closed chain exercises 34(4):353–357

Karthick PA, Ghosh DM, Ramakrishnan S (2018) Surface electromyography based muscle fatigue detection using high-resolution time-frequency methods and machine learning algorithms. Comput Methods Programs Biomed 154:45–56

Kaur M, Nara S, Shaw D, Bhatia D (2016) EMG asymetricity of selected knee extensor muscles in sustained Squat posture (a Yogic posture) of athletes in relation to their gender and performance 7:1–6

Konrad P (2006) The ABC of EMG, Mar 2006

Phinyomark A, Hirunviriya S, Limsakul C, Phukpattaranont P (2010) Evaluation of EMG feature extraction for hand movement recognition based on Euclidean distance and standard deviation. In: 2010 international conference on electrical engineering, computer, telecommunications and information technology (ECTI-CON), pp 856–860

Phinyomark A, Phukpattaranont P, Limsakul C (2012) Feature reduction and selection for EMG signal classification. Expert Syst Appl 39(8):7420–7431

Phinyomark A, Limsakul C, Phukpattaranont P (2009) A novel feature extraction for robust EMG pattern recognition 1:71–80

Pincivero DM et al (2018) Influence of contraction intensity, muscle, and gender on median frequency of the quadriceps femoris 99004:804–810

Rainoldi A, Melchiorri G, Caruso I (2004) A method for positioning electrodes during surface EMG recordings in lower limb muscles 134:37–43

Sakurai T, Toda M, Sakurazawa S, Akita J, Kondo K, Nakamura Y (2010) Detection of muscle fatigue by the surface electromyogram and its application 43–47

Thongpanja S, Phinyomark A, Phukpattaranont P, Limsakul C (2013) Mean and median frequency of EMG signal to determine muscle force based on time-dependent power spectrum 51–56

Advanced Encryption Standard Algorithm in Multimodal Biometric Image

Sharmila S. More, Bhawna Narain, and B. T. Jadhav

Abstract Cryptography is becoming slowly but surely a more important feature of computer security. In every field of technology, there are major issues of security but these issues overcome by using cryptographic algorithms. Biometric images are used as input and then the features like shape, size, edge, and texture are extracted. This is done by applying feature extraction algorithms and advanced encryption standard algorithm on Fingerprints, Iris, faces, and palm print simultaneously. In this paper, we have explained the information about multimodal biometrics, limitations of the DES algorithm on the encryption process of multimodal biometric images, Solution to these problems, and the role of AES in multimodal. We have discussed the implementation of the AES algorithm by using MATLAB on parameters of captured images according to age and gender. Parameters such as key size, input size, time taken, simulation, memory requirement, CPU usage. Major issues in multimodal biometrics such as matching algorithm, time delay, FAR are also discussed.

Keywords Cryptographic algorithm · Advanced encryption standard · Biometric traits · FAR, FRR · Data encryption standard · Cipher text

S. S. More (✉)
MATS School of Information Technology, MATS University, Raipur, India
e-mail: sharmilamore22@gmail.com

B. Narain
MATS University, Raipur, India
e-mail: dr.bhavana@matsuniversity.ac.in

B. T. Jadhav
RIRD, Satara, Maharashtra, India
e-mail: btj21875@gmail.com

A. A. Rizvanov et al. (eds.), *Advances in Biomedical Engineering and Technology*,
Lecture Notes in Bioengineering,
https://doi.org/10.1007/978-981-15-6329-4_7

1 Introduction

Biometrics is used in computer science as a form of identification and access control. It is also used to identify individuals in groups (Shinde and Shinde 2014). The biometric process always goes into two modes i.e., Enrollment, Verification (Authentication)/Identification (Recognition) modes. The biometric implementation process consists of a sensor, feature extraction, matcher, and system database. Biometrics is becoming an important international standard as an authentication technology providing security controls in every field.

Biometrics involves the use of physiological and behavioral characteristics to provide the identification of individuals as applied to physical and network security within a business. All the types of a biometric system are passes to these modes. If the first time an individual uses a biometric system then it is called the enrollment process and then biometric information is captured and stored. And Second time it is verified or recognized. In the verification or authentication mode, the system performs a one-to-one comparison of a captured biometric with a specific template stored in a biometric database to verify the individual is the person they claim to be. In recognition or identification mode the system performs a one-to-many comparison against a biometric database in an attempt to establish the identity of an unknown individual. There are lots of technical differences between every biometric type e.g., in the iris recognition system we analyze features of a colored ring of the eye and shape of the eyes, eyebrows, nose, lips, etc., measured in Facial recognition system, etc. (More et al. 2017). Instead of all the types of biometric systems, we captured images of fingerprint, face, and iris and palm print recognition. We apply a feature extraction algorithm on that images and applying the advanced cryptographic algorithm on that captured images by using MATLAB tool for analyzing the parameters such as key size, input size, time taken, simulation, memory requirement, CPU usage.

(1) Fingerprint Recognition System: In the fingerprint recognition system we check the uniqueness of fingerprints is due to the series of ridges and furrows on the fingers. We also consider the arch, loop, and whorl of the captured fingerprint. For the fingerprint recognition system, we can use a previously defined feature extraction algorithm i.e., Canny Edge Detection/Gaussian Filter/Gaussian mixture/Gabor Filter Algorithm.

(2) Face Recognition System: Facial recognition techniques measure facial characteristics. It measures the position, shape, and size of facial features such as eyebrows, eyes, nose, lips, and chin (More et al. 2017). In Face Recognition System we can use a previously defined feature extraction algorithm i.e., Gabor Filter Algorithm.

(3) Iris Recognition System: The iris recognition technique analyzing feature of a colored ring of the eye (More and Jadahv 2016). Gabor filter algorithm is used for feature extraction of Iris recognition.

(4) Palmprint Recognition System: To determine the geometry features, principal line features, and wrinkle features we use processing techniques and applying these on the captured image (Sree Rama Murthy kora et al. 2012). For feature extraction of palmprint recognition, we can use Gabor Filter for texture analysis. Figure 1 shows sample images of different biometric traits.

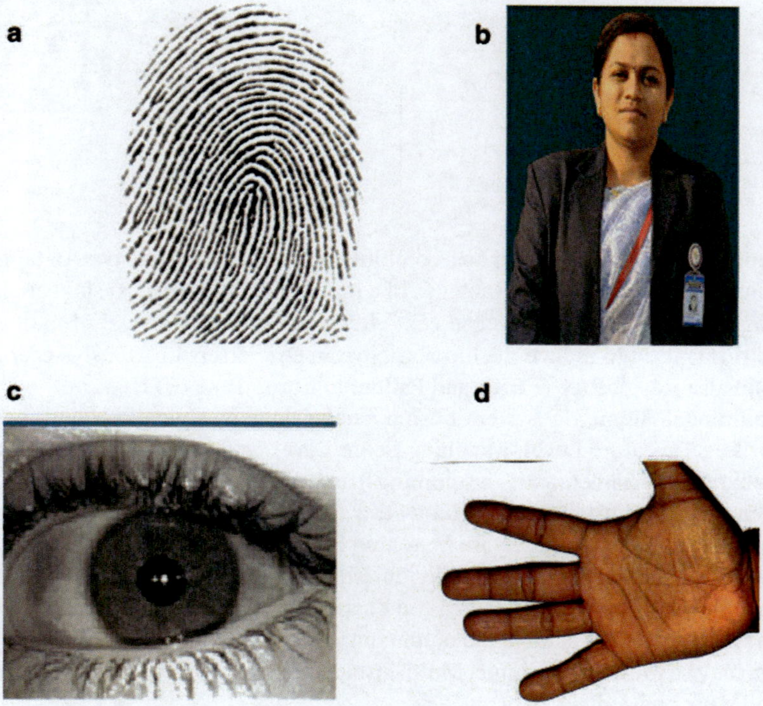

Fig. 1 **a** Fingerprint, **b** face, **c** iris, **d** palmprint

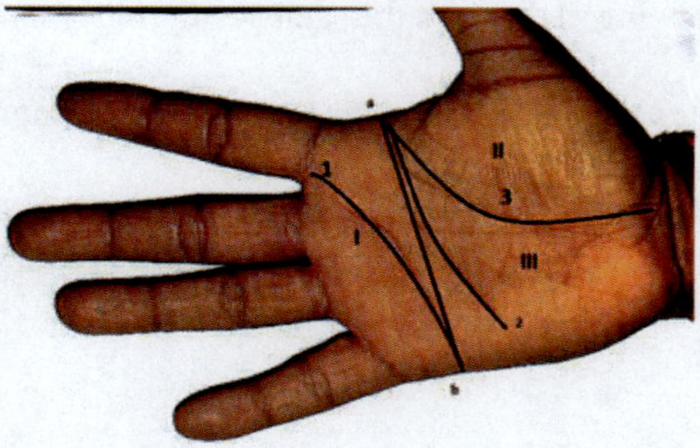

Fig. 2 Definitions of palm print: principal lines (I-heart line, 2-head line, 3-life line, regions (I-finger-root region, II-inside region, and III- outside region) and datum points. (a, b-endpoints, o-their midpoints)

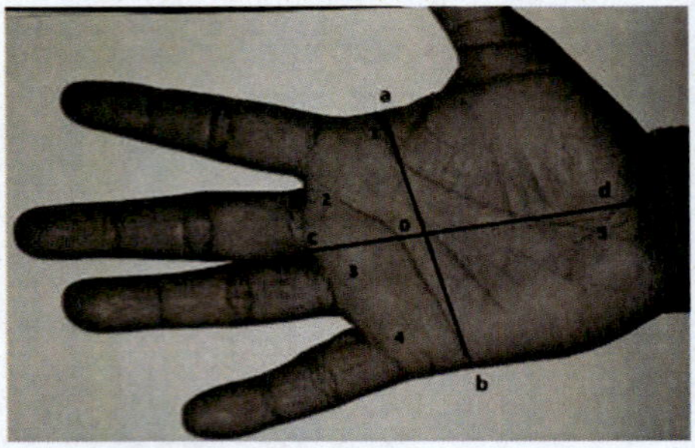

Fig. 3 Geometry features and delta points features of a palm print, where a–d is the perpendicular bisector of segment a–b, and points 1–5 are delta points

2 Multimodal Biometrics

Multimodal biometrics is based on a combination of more than one type of biometric modalities or qualities. The advantages of a multimodal biometric system are higher accuracy, security, universality, and cost-effectiveness, etc. The goal of multimodal biometrics is used to reduce the biometric parametric errors i.e., False Accept Rate (FAR), False Reject Rate (FRR), and Failure to Enroll Rate (FTE).

Multimodal Biometric System Fusion can be done by four levels and these are Sensor Level, Feature Level, Matching Score Level, and Decision Level. In sensor Level, biometric characteristics are coming from sensor level. In Feature Level fusion, the signal coming from different biometric channels is first proposed and feature vectors are extracted separately. In Matching Score Level combining the feature, we process them separately and the individual matching score is found. And lastly at decision Level, each modality is first pre-classified independently. The final classification is based on the fusion of the output of the different modalities. Multibiometric system may be Multi-algorithmic, Multi-instance, Multi-sensorial. We use this image by era (More and Jadhav 2017).

Data encryption standard limitations are:

1. The key selected in the rounds is weak because during the splitting of keys to two half and swapping, it displays the same result if they have continuous 1s and 0s.
2. The S Box creates the same output from the different inputs on permutations. We can call these as Semi Weak keys.
3. If the message is encrypted with a particular key and it takes 1s complement of that encryption will be the same as that of the encryption of the compliment message and compliment key.

2.1 Role of Multimodal Biometrics in AES

Cryptography is becoming an increasingly important feature of computer security (More et al. 2017). The proposed method gives the security to the whole system by

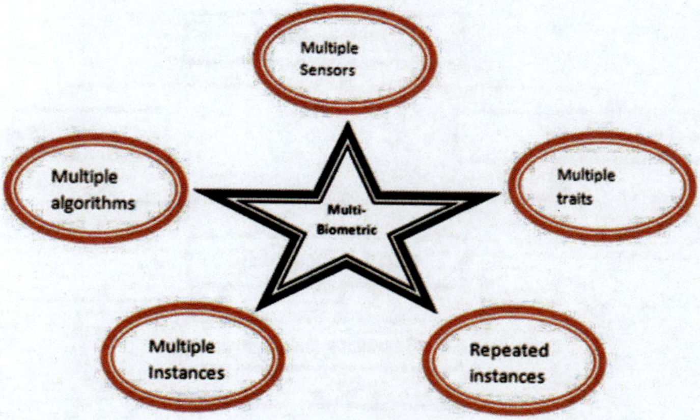

Fig. 4 Classification of a multimodal biometric system

using fingerprint, face, Palm, and facial features as a key in a cryptosystem. In the proposed model we use multimodal biometric features. Biometric template protection is one of the important issues in deploying a practical biometric system. To tackle this problem, many algorithms have been reported in recent years, most of them applying to use fingerprint, face, Palm, and face biometric. Since the contents and representation of every template are different than others. The template protection algorithm of one biometric trait cannot be directly applied to others. Moreover, we believe that no single template protection method can satisfy the diversity, revocability, security, and performance requirements (Bala and Joanna 2014). Data fusion is the process of integrating multiple data sources to produce more consistent, accurate, and useful information than that provided by any individual data source (Fig. 5).

(a) **AES Algorithm for Image Processing**. The algorithm consists of combinations, permutations, and substitution between the images to be encrypted and the key is applied on both the encryption & decryption process. AES Algorithm is already developed we apply this on multimodal biometric images.
(b) **Tools Used For Digital Image Processing**. MATLAB, Scilab, and Octave are widely used by engineers and scientists in both industry and academia for performing numerical computations and for developing and testing mathematical algorithms and image processing with related applications (Narain et al. 2013). In our work, we use the MATLAB tool for processing images, which is a high-performance language for technical computing. It is used in computation, visualization, and programming environment.
(c) **Implementations of AES Using Multimodal Biometric**: Encrypted data set of sample images of fingerprint, face, Iris, and Palmprint are shown in Fig. 6.

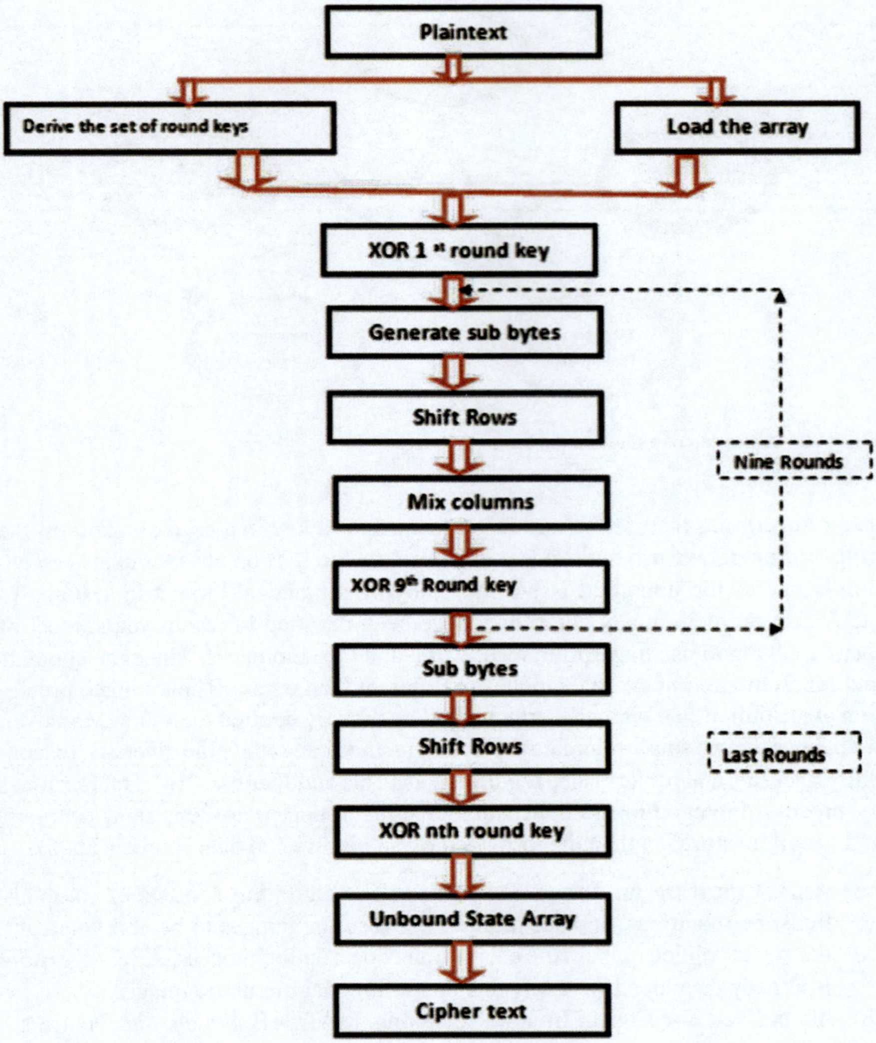

Fig. 5 Working on the AES algorithm

3 Methodology

3.1 Proposed Architecture

The multimodal biometric system designed consists of six modules as in Fig. 1.

- Fingerprint analysis module.
- Iris analysis module.

	1	2	3	4	5	6	7	8	9
1	255	255	255	254	254	254	254	253	253
2	255	255	255	254	254	254	254	253	253
3	255	255	255	254	254	254	254	253	253
4	255	255	255	254	254	254	254	253	253
5	255	255	255	254	254	254	254	253	253
6	255	255	255	254	254	254	254	253	253
7	255	255	255	254	254	254	254	253	253
8	255	255	255	254	254	254	254	253	253
9	255	255	255	254	254	254	254	253	253
10	255	255	255	254	254	254	254	253	253
11	255	255	255	254	254	254	254	253	253
12	255	255	255	254	254	254	254	253	253
13	255	255	255	254	254	254	254	253	253
14	255	255	255	254	254	254	254	253	253
15	255	255	255	254	254	254	254	253	253
16	255	255	255	254	254	254	254	253	253
17	255	255	255	254	254	254	254	253	253
18	255	255	255	254	254	254	254	253	253
19	255	255	255	254	254	254	254	253	253
20	255	255	255	254	254	254	254	253	253
21	255	255	255	254	254	254	254	253	253

a) Fingerprint-I1

	1	2	3	4	5	6	7	8	9
1	0	114	116	116	115	115	115	115	115
2	114	116	116	116	115	116	116	116	116
3	115	116	116	116	115	116	116	116	117
4	115	116	116	116	115	115	116	116	116
5	115	117	117	117	117	117	116	116	116
6	114	115	117	117	116	115	116	116	116
7	114	116	116	117	117	117	116	116	116
8	114	116	116	117	117	116	116	116	116
9	116	116	117	117	117	117	116	116	116
10	116	116	116	116	116	116	116	116	116
11	115	116	116	116	116	117	117	117	117
12	115	115	116	116	116	116	117	117	117
13	115	116	116	116	116	116	117	117	118
14	115	116	116	116	116	116	116	117	117
15	115	116	116	116	116	115	116	116	117
16	115	116	116	116	115	115	115	115	116
17	115	116	116	116	116	116	115	115	115
18	115	116	116	116	116	116	115	115	115
19	115	116	114	113	113	113	113	114	115
20	113	114	113	113	112	113	113	113	115
21	111	113	113	112	113	113	113	113	113

b) Face-I2

Fig. 6 Encrypted data set of fingerprint, face, iris, palm print sample images

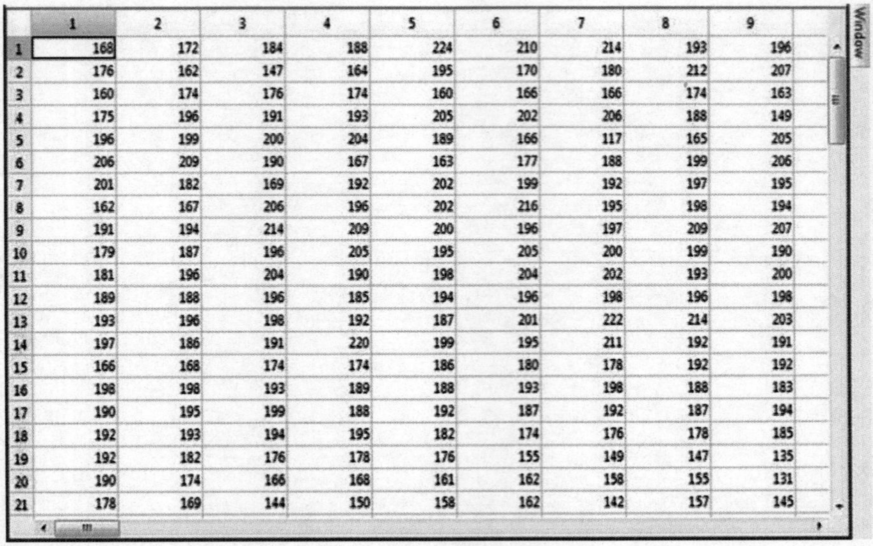

	1	2	3	4	5	6	7	8	9
1	168	172	184	188	224	210	214	193	196
2	176	162	147	164	195	170	180	212	207
3	160	174	176	174	160	166	166	174	163
4	175	196	191	193	205	202	206	188	149
5	196	199	200	204	189	166	117	165	205
6	206	209	190	167	163	177	188	199	206
7	201	182	169	192	202	199	192	197	195
8	162	167	206	196	202	216	195	198	194
9	191	194	214	209	200	196	197	209	207
10	179	187	196	205	195	205	200	199	190
11	181	196	204	190	198	204	202	193	200
12	189	188	196	185	194	196	198	196	198
13	193	196	198	192	187	201	222	214	203
14	197	186	191	220	199	195	211	192	191
15	166	168	174	174	186	180	178	192	192
16	198	198	193	189	188	193	198	188	183
17	190	195	199	188	192	187	192	187	194
18	192	193	194	195	182	174	176	178	185
19	192	182	176	178	176	155	149	147	135
20	190	174	166	168	161	162	158	155	131
21	178	169	144	150	158	162	142	157	145

c) Iris-I3

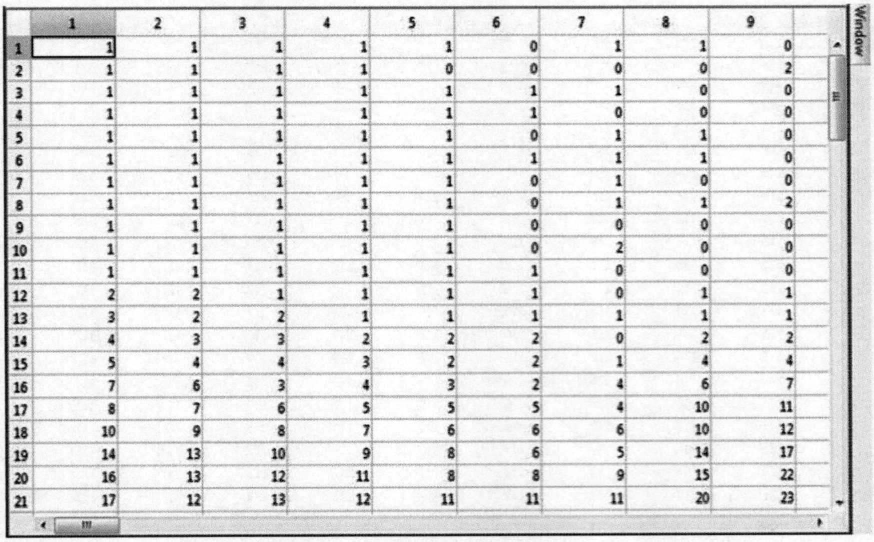

	1	2	3	4	5	6	7	8	9
1	1	1	1	1	1	0	1	1	0
2	1	1	1	1	0	0	0	0	2
3	1	1	1	1	1	1	1	0	0
4	1	1	1	1	1	1	0	0	0
5	1	1	1	1	1	0	1	1	0
6	1	1	1	1	1	1	1	1	0
7	1	1	1	1	1	0	1	0	0
8	1	1	1	1	1	0	1	1	2
9	1	1	1	1	1	0	0	0	0
10	1	1	1	1	1	0	2	0	0
11	1	1	1	1	1	1	0	0	0
12	2	2	1	1	1	1	0	1	1
13	3	2	2	1	1	1	1	1	1
14	4	3	3	2	2	2	0	2	2
15	5	4	4	3	2	2	1	4	4
16	7	6	3	4	3	2	4	6	7
17	8	7	6	5	5	5	4	10	11
18	10	9	8	7	6	6	6	10	12
19	14	13	10	9	8	6	5	14	17
20	16	13	12	11	8	8	9	15	22
21	17	12	13	12	11	11	11	20	23

d) Palmprint-I4

Fig. 6 (continued)

32x6 double

	1	2	3	4	5	6
2	1.5093	0.4457	1.3184	0.4664	0.4222	0.3500
3	-0.4552	-0.4592	-0.4611	-0.4640	-0.4603	-0.4625
4	-0.4551	-0.4575	-0.4609	-0.4626	-0.4586	-0.4604
5	3.1938	3.0596	3.1588	2.9594	2.9531	2.8734
6	0.6278	0.5117	0.6574	0.4454	0.3533	0.3500
7	-0.4555	-0.4591	-0.4613	-0.4640	-0.4604	-0.4625
8	-0.4547	-0.4578	-0.4606	-0.4625	-0.4582	-0.4605
9	-0.4545	-0.4578	-0.4603	-0.4626	-0.4588	-0.4609
10	-0.4552	-0.4588	-0.4611	-0.4637	-0.4599	-0.4621
11	-0.4554	-0.4589	-0.4612	-0.4638	-0.4600	-0.4621
12	-0.4550	-0.4584	-0.4608	-0.4634	-0.4596	-0.4617
13	-0.4550	-0.4584	-0.4608	-0.4633	-0.4595	-0.4616
14	-0.4552	-0.4587	-0.4611	-0.4636	-0.4599	-0.4620
15	-0.4554	-0.4590	-0.4612	-0.4638	-0.4599	-0.4620
16	-0.4546	-0.4580	-0.4609	-0.4634	-0.4595	-0.4614
17	1.2159	1.8355	1.3738	1.9827	2.0170	2.1277
18	-0.0026	0.1595	0.0113	0.1614	0.1398	0.1724
19	-0.4554	-0.4591	-0.4614	-0.4640	-0.4604	-0.4625
20	-0.4545	-0.4578	-0.4602	-0.4624	-0.4583	-0.4605
21	1.2159	1.8355	1.3738	1.9827	2.0170	2.1277
22	-0.0330	0.0962	0.0097	0.1634	0.1742	0.2056

Fig. 7 Decrypted dataset

- Palm print analysis module.
- Face analysis module
- Conversion and Fusion.
- AES Encryption/Decryption module.

The generation of secure biometric keys with the help of multimodal biometrics such as iris, fingerprint, face, and palm print is done as shown in Fig. 8.

3.2 Module Implementation

The minutiae points are extracted from the fingerprint image, texture features from iris image, ROI score from palm print and shape of the eyes, eyebrows, nose, lips, etc.,

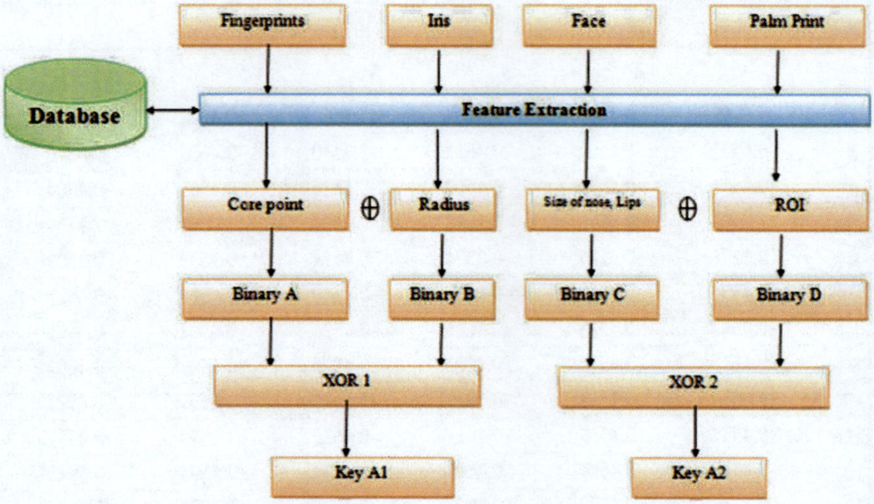

Fig. 8 Generation of Secure biometric keys

measured in facial recognition system which is generated by the biometric system. The features are now converted into respective decimals. The decimals are converted into binaries and all the four binary follows XOR operation to generate the combined cryptographic key. The key is later compressed to Hexadecimal value which can act as the encryption key. The encryption key is now used for the DES encryption and decryption process. Again encryption and decryption is followed based on the two ciphers generated.

Module 1: Biometrics Sensing.

This module helps to recognize the biometric information of the users via sensors, cameras.

- Images are generated which are further passed to module two for evaluation.
- The Module helps to collect the information of human biometrics
- The information to be collected are Fingerprint, Palm print, Iris, and Face.

Module 2: Preprocessing Feature Extraction (Joseph and Parthiban 2016).

- It helps to extract the features from human biometrics to generate biometric key
- The features are extracted in the form of decimals which are then used to convert the binary value
- Different techniques are followed for each biometric
- At first, the image is enhanced, followed by thinning, segmentation.

Module 3: Normalization and Fusion Description (Joseph and Parthiban 2016)

- This module helps to normalize the data or information gathered in the form of features to the type which can be used to create or generate the key.
- The features are then fused by following the XOR operation of the biometric values obtained.

Module 4: Generation of Keys Description

- The above module and the current one are integrated to generate binary biometric ciphers key
- The two binary keys generated are then considered as input to the next module of encryption.

Module 5: AES Encryption and Decryption Description

- The module follows the process of encryption and decryption by using the AES encryption process.
- The binary keys generated from the above module are passed as inputs to generate encrypted ciphers.

4 Results and Discussion

Analysis of about 5 different samples is followed and detailed evaluation can be seen from the table All the features are extracted and normalized in the form of binary values which later follows the proposed algorithm. All the values are passing as input to the proposed technique which provides better security due to 4 levels of multimodal biometrics (Isaac et al. 2018). This paper combines the scores based on the fusion of Iris, Fingerprint, face, and Palmprint data to generate biometric cryptographic keys.

The analysis is done and it provides information about the performance and calculates approximately measures of the combined biometric techniques. The Iris, Fingerprint, face, and Palmprint data are collected from about 5 individuals and used for evaluation. Scores for each biometric trait are generated respectively. The calculation of analysis parameters such as key size, input size, time taken, simulation, memory requirement, CPU usage are estimated, and also FAR, FRR.

The following figure shows the expected outcomes. In Fig. 9 we enter the ID of any person and then the preprocessing and enhancement process is done in Figs. 10 and 11. After that in Fig. 11 we apply Gabor and PCA filter for feature extraction to

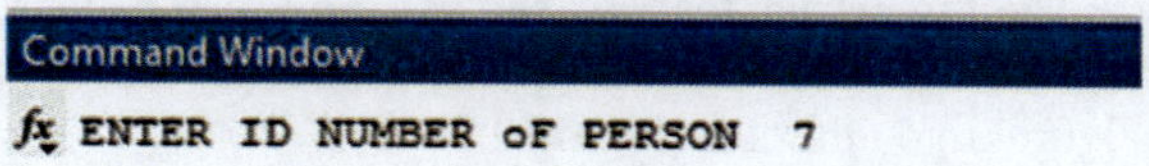

Fig. 9 First step of entering input

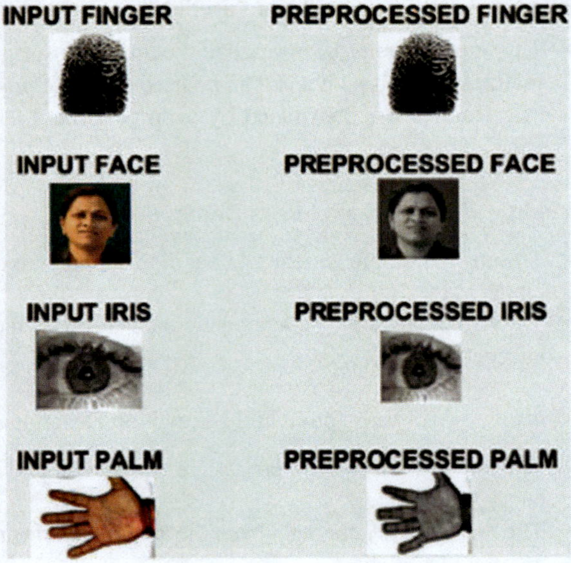

Fig. 10 Output of the program—preprocessing on the multimodal images

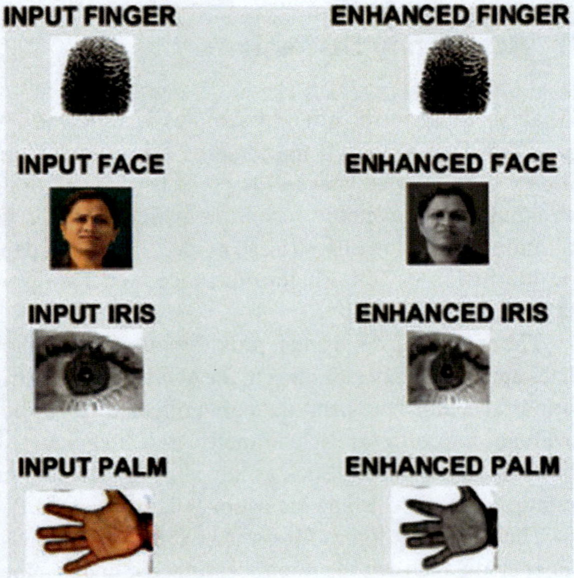

Fig. 11 Output of the program—enhancement process of fingerprint, face, iris, and palmprint

display a graphical representation of it in Fig. 12 and identifying or authenticate a person.

The biometrics features for Iris, Fingerprint, face, and Palmprint are collected separately according to age and gender in the future. Then we apply the AES algorithm on these multimodal images.

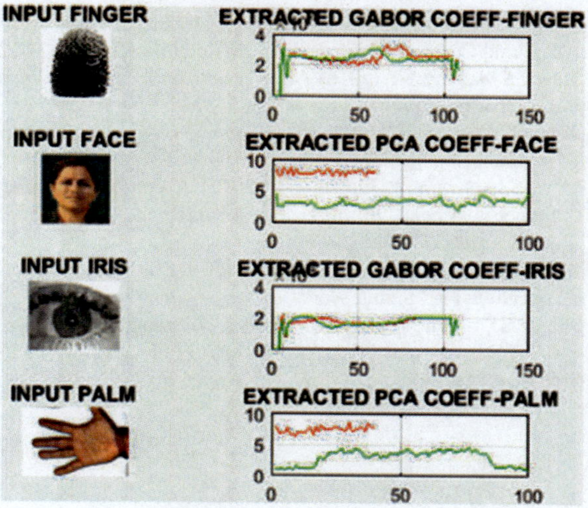

Fig. 12 The output of the program—graphical representation of multimodal images

5 Conclusion

In this paper, we have discussed the security issues of multimodal biometrics of the existing system which used the DES algorithm, but it has some limitations and this is overcome in the proposed system and gives more security in biometrics. The AES algorithm is already developed algorithm, here we apply it on the multimodal biometrics and checking the medium performance and then compare to the proposed system methodology as well as the algorithm by the experimental results and tabulations to justify the results. In the future, we will be trying to implement on a large dataset and for different cryptographic algorithms RSA, Blowfish, M-RSA, and proposed algorithm. This algorithm implementation time and matching algorithm's time taken for completion should be minimized and try to avoid the false acceptance rate and false rejection rate in the system. And we will also try to compare the analysis of parameters of these algorithms with the proposed algorithm in the future.

Ethical Clearance We are putting sample images since all the images shown in the study are images of one of the authors and no other participants were involved in the study. Hence, ethical approval was not required. Results shown are obtained using five samples (different orientations) of one of the authors of this paper. In the future, the study will be extended on a new dataset after obtaining approval of the institutional ethical committee with informed consent.

References

Bala BK, Joanna JL (2014) Multi modal biometrics using cryptographic algorithm. Eur J Acad Essays 1(1):6–10

Isaac RA, Kathera A, Venkatachalam KH, Raj MT, Gokulnath G (2018) Spoofing Detection for fingerprint, palm-vein and facial recognition using deep representation 6(4):169–174

Joseph T, Parthiban L (2016) Multimodal biometric based authentication for ensuring data security in Cloud Computing. J Chem Pharm Sci 9(4). ISSN: 0974-2115

More SS, Jadahv BT (2016) An overview on technologies used in biometric system. Int J Innov Res Comput Commun Eng 4(2):365–373

More SS, Jadhav BT (2017) Comparative study of biometric devices. Int J Innov Res Comput Commun Eng 5(2):1302–1309

More SS, Narain B, Jadahv BT (2017) A comparative analysis of unimodal and multimodal biometric systems. In: International conference on innovative trends in engineering science and management (ITESM-2017)

Narain B, Zadgaonkar AS, Kumar S (2013) Impact of digital image processing on research and education. Natl Semin Work

Shinde SJ, Shinde J (2014) Biometrics: overview and potential use for E-governance services. Int J Adv Res Comput Sci Softw Eng 4(6):1145–1151

Sree Rama Murthy kora, Verma P, Kashyap Y (2012) Palmprint recognition: palm print

In-Silico Construction of Hybrid ORF Protein to Enhance Algal Oil Content for Biofuel

Mohit Nigam, Ruchi Yadav, and Garima Awasthi

Abstract As a renewable resource of biodiesel, algae has received global attention and hence our primary concern is to explore the new potential technologies for increasing yield of algal biofuel. The aim of our study was to develop a hybrid ORF, which will increase the oil producing capacity of algae on its expression as a functional protein. Utilizing BLAST and intense literature survey, 6 oil producing algal genes were selected which belonged to different superfamilies (NADB_Rossmann superfamily, RfaB superfamily, Aldo_ket_red superfamily, PP-binding Superfamily, LPLAT superfamily, and Acyl-ACP TE superfamily). Further, using the conserved regions of these superfamilies, the hybrid ORF was constructed by vector NTI tool. Since protein structure is an important aspect to verify protein sequence for its stability and existence in nature, multi-template homology modeling was done using Schrödinger software suite version 10.4.018. The five templates (1i24_A, 1pz1_A, 2jjm_A, 2m5r_A, 5 × 04_A) showing highest similarity with the hybrid ORF were selected for multi-template homology modeling. The comparative three-dimensional structure prediction of the constructed hybrid ORF was done using Phyre^2 server which is based on the principles of ab initio method. Structure verification of both the resulting models i.e., model generated by Schrödinger software and model predicted by Phyre^2 server is done using Ramachandran plot and identified that functional region has been modeled accurately having 87.3% and 91% residues in allowed region, respectively. The metabolic pathway analysis of all the selected genes also verified, that they all are involved in lipid biosynthesis. This shows that the hybrid ORF designed can be potential tool to increase oil content in

M. Nigam · R. Yadav · G. Awasthi (✉)
Amity Institute of Biotechnology, Amity University Uttar Pradesh, Lucknow, UP, India
e-mail: gawasthi@lko.amity.edu

M. Nigam
e-mail: mnigs302@gmail.com

R. Yadav
e-mail: ryadav@lko.amity.edu

A. A. Rizvanov et al. (eds.), *Advances in Biomedical Engineering and Technology*,
Lecture Notes in Bioengineering,
https://doi.org/10.1007/978-981-15-6329-4_8

algae for biofuel production. Further in vitro analyses are required to study physiochemical properties of the protein and its effectiveness after constructing genetically engineered microorganism.

Keywords Homology modeling · Schrödinger software · Phyre^2 server · Microalgae · Biofuel

1 Introduction

Biofuel research is currently an area of immense interest due to the increase in global energy demand. Biofuel is center of attention due to the emerging economies, increase in global oil prices, reduces environmental impact, foreign exchanges, and benefits to rural sector. Research on multiple approaches has been carried out currently for the use of microorganisms in the production of various types of feedstock for the biofuel production (Elshahed 2010; Demirbas 2009).

Microalgae represents an exceptionally diverse but highly specialized group of microorganisms adapted to various ecological habitats. Algae is a good source for many products like carbohydrates, essential fatty acids, pigments, food supplements, fertilizer, pharmaceutical, and biofuel (Hemaiswarya et al. 2011).

Many microalgae have the ability to produce high amounts of triacylglycerols (TAG) under photo-oxidative stress or different adverse environmental conditions that can be utilized as feedstock for producing biodiesel (Hu et al. 2008; Chisti 2007, 2008).

Algae production is facing several challenges like water, nutrient, and contamination, which can overcome by using heterotropic and mixotropic algae. These algae are well known to grow faster and produce high oil content as feedstock for production biofuel compared to photoautotrophic cells (Miao and Wu 2004; Cerón García et al. 2000). The concept of strain improvement, leads to explore new ways based on genomics and transcriptomic information so as to improve the growth conditions of algae and increase lipid content that can help in reducing the cost of biofuel production. Many approaches are taken to enhance triacylglycerol (TAG) content in microalgae include over-expression of genes for biosynthesis of TAG, lipid catabolism inhibition, and interference with many other pathways (Radakovits et al. 2010).

Genome engineering is gradually increasing its steps in algae research (Banerjee et al. 2018). In the present study in silico algal genes consortium is developed for expression of hybrid ORF protein to enhance the oil content in algae and will be fuel precursor molecules in TAG biosynthesis pathway.

2 Material and Methods

Literature survey was done by journals and finally by NCBI genomic and proteomic databases for the identification of oil producing algal genes. 6 genes from different strains of algae were selected and its functional protein information was collected.

Superfamily (Conserved regions) of these proteins was identified by BLASTp and then using these conserved regions of functional protein sequences, Hybrid ORF was constructed. Hybrid sequence verification and clone designing for the hybrid ORF was done using vector NTI tool (Guoqing and Moriyama 2004).

Schrödinger software suite version 10.4.018 was used for multi-template homology modeling of the hybrid sequence. The 3D structure of hybrid protein was modeled and structure verification was done using Ramachandran plot (Schrödinger 2011, 2014; Ivanov et al. 2009; Ferrara and Edgar 2007).

Ab initio method for protein structure prediction was also done by using Phyre^2 server. The modeled structure was compared with the three-dimensional structure predicted using homology modeling method (Kelley et al. 2015). The best structure was identified for further analysis.

The metabolic pathways of the selected algal genes were studied using KEGG database. Pathway study was done to identify the function and molecular mechanism of selected genes.

3 Result and Discussion

3.1 Identification of Algal Gene and Its Functional Protein

Oil producing algal gene was searched through intensive literature survey and NCBI genomic and proteomic databases. Total of 6 genes and their functional proteins were identified and screened by literature survey as mentioned in Table 1. The 5 gene SQD1, SQD2, ACP1, CGLD24, and FAT1 were reported in *Chlamydomonas reinhardtii* only F751_2396 was identified in *Auxenochlorella protothecoides*. The protein sequences were saved in FASTA format and functional protein information was retrieved.

3.2 Hybrid ORF Construction Using Selected Conserved Regions of Superfamilies

Protein screening of selected gene was done on the basis of different superfamilies, using BLASTp tool. The 6 target genes were further classified according to their superfamilies to which they belong, that were NADB_Rossmann superfamily, RfaB

Table 1 Algal oil producing genes and its functional protein information

S. no.	Gene symbol	Protein	Organisms	Accession no.
1.	SQD1	UDP-sulfoquinovose synthase	*Chlamydomonas reinhardtii*	XP_001697898.1
2.	SQD2	Sulfolipid synthase	*Chlamydomonas reinhardtii*	XP_001689662.1
3.	F751_2396	D-arabinose 1-dehydrogenase	*Auxenochlorella protothecoides*	XP_011397567.1
4.	ACP1	Acyl-carrier protein	*Chlamydomonas reinhardtii*	XP_001699275.1
5.	CGLD24	Diacylglycerol acyl transferase	*Chlamydomonas reinhardtii*	XP_0016931C9.1
6.	FAT1	Acyl carrier protein thioesterase	*Chlamydomonas reinhardtii*	XP_001696619.1

superfamily, Aldo_ket_red superfamily, PP-binding Superfamily, LPLAT superfamily, and Acyl-ACP TE superfamily. Conserved sequences of these superfamilies were used for the designing of hybrid ORF using vector NTI tool. The selected superfamilies were studied to identify ORF start-stop positions and length of ORF, which was used for designing hybrid ORF as mentioned in Table 2. In silico techniques were used to study properties of hybrid sequence and its stability for cloning and expression in vector.

3.3 *Hybrid ORF Clone Designed Using Vector NTI Tool Kit*

Vector NTI tool was used for the clone of designed hybrid ORF, also analysis and verification of cloned sequence was done. The length of hybrid ORF was 5175 bp and the gene sequence of hybrid ORF is mentioned below.

Complete Sequence of Constructed Hybrid ORF Designed for the Biosynthesis of Lipid in Algal Strain.

>hy_ORF _ length_5175 bp

Table 2 Selected conserved regions of superfamilies for designed hybrid ORF

S. no.	Organisms	Super family	Protein	Conserved region sequence	ORF start position	ORF stop position	Length of ORF	Accession no.
1.	*Chlamydomonas reinhardtii*	NADB_Rossmann superfamily	UDP-sulfoquinovose synthase	STVRQATSSVRAASR ATSVKVQATPATVE KATAPAGSLSSNGA GTRVMIIGGDGYCG WATALHLSARGYEV CIVDNLCRRQFDLQL GLDTLTPIATIHDRV RRWGEVSGKHISLQI GDICDWEFLSQAFTS FKPNHVVHFGEQRS APYSMIDRQKAVFT QHNNVIGTINVLFAI KELQPDCHMVKLGT MGEYGTPNIDIEEGY ITINHNGRTDTLPYP KQGNSFYHLSKIHDS TNMLFTCKAWKIAA TDLNQGVVYGVRTD ETMADPLLLNRYDY DGIFGTALNRFVVQ AAVGHPLTVYGKGG QTRGFLDIRDTVRCI QLAIDNPAPKGEMR VYNQFTEQFSVNQL AEIVEREGKKLGLNV EVTKVPNPRVELEEH YYNAKCTKLRDLGL QPHLLADSMIDSLLE FAVTYKDRVRHELIK PAVDWRKTGVKVN TMGAAV	40	477	439	XP_001697898.1

(continued)

Table 2 (continued)

S. no.	Organisms	Super family	Protein	Conserved region sequence	ORF start position	ORF stop position	Length of ORF	Accession no.
2.	*Aureococcus anophagefferens*	RfaB superfamily	UDP-sulfoquinovose:DAG sulfoquinovosyltransferase	RPSKLPTTACRAAAT SDTDTPQRKVALLV EPTPFTHVSGYANRF QEMLKHLERRGDVV AVATPDDVPEAPASF GKFAVTTLGGFRFRP WYPEICLSLDLDGAA LQMIRDLDPDVVHA SSPGFLAVAALRRA GQGAERKPLLLSYH THIPVYVRKYASWV PFIEKTTWALLRAVH NRADLTIATSPQIRD ELLANGVTAIERVGV WNKGIDTDRFHPKF RSDAARARMTSGHP GDKLAVYVGRLGVE KRIDELRGVLEAIPEL RLALVGAGPAEPGL RETFADVADRVVFT GLLRGDELSAAFAS ADVFLMPSDSETLGF VVLESMASGVPVVG CRAGGIPNLIDDDQE GATGRLHAVGDVAE IAELTRGLLDDAPKR DAMGAAARAEAER WDWASSGETLRADS YGAAIRNFAAR	22	430	415	XP_009034698.1

(continued)

Table 2 (continued)

S. no.	Organisms	Super family	Protein	Conserved region sequence	ORF start position	ORF stop position	Length of ORF	Accession no.
3.	*Auxenochlorella protothecoides*	Aldo_ket_red superfamily	L-galactose dehydrogenase	RPLGSTGLEISIIGFG ASPLGNVFGDVHQD TATEAVRTAFDLGIT LFDTSPFYGLTKSED VLGQALRDAGLPRD QFVLATKVGRYGQD TFDFSGPRVTRSVEE SLERLHTSYIDLIQVH DMEFGSLDQIIAETL PALQRLKEKGLVRHI GITGLPLACFQYVLD RVPCGTVDVVLSYC HYTLCDQSLGRILPY LESKAVGVINASVLG MGLLTPHGPPAWHP APAELQAAARAAAR AADAHAVDLPKLAT MFSVANPGIATHLIG FSTPDQVRNAVHAV LQAQGLEENAQAEQ EERAMVEIREILAGT AQVTWPSGLPEN	8	327	320	XP_011397567.1

(continued)

Table 2 (continued)

S. no.	Organisms	Super family	Protein	Conserved region sequence	ORF start position	ORF stop position	Length of ORF	Accession no.
4.	*Chlamydomonas reinhardtii*	PP-binding superfamily	acyl-carrier protein	KDSVTERVLHVTKH FEKIDASKVSPAASF EKDLGLDSLDVVEL VMALEEEFGLEIPDA EADKIASVGDAINYI CSNP	49	125	77	XP_001699275.1
5.	*Chlamydomonas reinhardtii*	LPLAT superfamily	diacylglycerol acyl transferase	AAAYFPTRVVVTDP EAFRTDRGYLFGFCP HSALPIALPIAFATTA SPLLPKELRGRTHGL ASSVCFSAPIVRQLY WWLGVRPATRQSIS GLLDKCRARKVAVL VPGGVQEVLNMEHG KEVAYLSSRTGFVRL AVQHGAPLVPVWAF GQTRAYSWFRPGPP LVPTWLVERISRAAG AVPIIFHGRGMFGQY GTPLMPHREPLTIVV GRPIPVPELAPGQLE PEPEVLAALLKRFTD DLQALYDKHKAQFG VPKGEELVI	81	326	258	XP_001693189.1

(continued)

Table 2 (continued)

S. no.	Organisms	Super family	Protein	Conserved region sequence	ORF start position	ORF stop position	Length of ORF	Accession no.
6.	*Chlamydomonas reinhardtii*	Acyl-ACP TE superfamily	acyl carrier protein thioesterase	SFREEHRIRGYEVSP DQRATIVTVANLLQE VAGNHAVGMWGRT DEGFASLPSMKDYN LLFVMTRLQVRMYE YPKWGDVVAVETYF TEEGRLAFRREWKL MDVATGKLLGAGTS TWVTINTATRRLSKL PEDVRKRFLRFAPPS SVHILPPEETKKKLQ DMPKYELPGQVQSA QQVARRADMDMNG HINNVTYLAWTLESL PERVMSGGYKMQEI ELDFKAECTAGNAIE AHCNP	90	318	234	XP_001696619.1

AGCACCGTGCGCCAGGCCACCAGCAGCGTGCGCGCCGCCAGCCGCGCCACCAGCGTGA
AGGTGCAGGCCACCCCCGCCACCGTGGAGAAGGCCACCGCCCCCGCCGGCAGCCTGAG
CAGCAACGGCGCCGGCACCCGCGTGATGATCATCGGCGGCGACGGCTACTGCGGCTGG
GCCACCGCCCTGCACCTGAGCGCCCGCGGCTACGAGGTGTGCATCGTGGACAACCTGT
GCCGCCGCCAGTTCGACCTGCAGCTGGGCCTGGACACCCTGACCCCCATCGCCACCATC
CACGACCGCGTGCGCCGCTGGGGCGAGGTGAGCGGCAAGCACATCAGCCTGCAGATCG
GCGACATCTGCGACTGGGAGTTCCTGAGCCAGGCCTTCACCAGCTTCAAGCCCAACCAC
GTGGTGCACTTCGGCGAGCAGCGCAGCGCCCCCTACAGCATGATCGACCGCCAGAAGG
CCGTGTTCACCCAGCACAACAACGTGATCGGCACCATCAACGTGCTGTTCGCCATCAAG
GAGCTGCAGCCCGACTGCCACATGGTGAAGCTGGGCACCATGGGCGAGTACGGCACCC
CCAACATCGACATCGAGGAGGGCTACATCACCATCAACCACAACGGCCGCACCGACAC
CCTGCCCTACCCCAAGCAGGGCAACAGCTTCTACCACCTGAGCAAGATCCACGACAGC
ACCAACATGCTGTTCACCTGCAAGGCCTGGAAGATCGCCGCCACCGACCTGAACCAGG
GCGTGGTGTACGGCGTGCGCACGACGAGACCATGGCCGACCCCCTGCTGCTGAACCGC
TACGACTACGACGGCATCTTCGGCACCGCCCTGAACCGCTTCGTGGTGCAGGCCGCCGT
GGGCCACCCCCTGACCGTGTACGGCAAGGGCGGCCAGACCCGCGGCTTCCTGGACATC
CGCGACACCGTGCGCTGCATCCAGCTGGCCATCGACAACCCCGCCCCCAAGGGCGAGA
TGCGCGTGTACAACCAGTTCACCGAGCAGTTCAGCGTGAACCAGCTGGCCGAGATCGT
GGAGCGCGAGGGCAAGAAGCTGGGCCTGAACGTGGAGGTGACCAAGGTGCCCAACCC
CCGCGTGGAGCTGGAGGAGCACTACTACAACGCCAAGTGCACCAAGCTGCGCGACCTG
GGCCTGCAGCCCCACCTGCTGGCCGACAGCATGATCGACAGCCTGCTGGAGTTCGCCGT
GACCTACAAGGACCGCGTGCGCCACGAGCTGATCAAGCCCGCCGTGGACTGGCGCAAG
ACCGGCGTGAAGGTGAACACCATGGGCGCCGCCGTGATGATCCGCTACCTGGTGGAGG
CCGGCTGCCAGGTGCTGGTGGTGACCACCGGCGCCGGCTACACCCTGCCCGGCGTGGA
CGCCAGCAGCTTCCGCGAGCAGCCCGAGACCTTCGCCGGCGCCCGCGTGGTGAGCGCC
CTGAGCTTCGGCTGCCCCTGGTACCTGCAGGTGCCCCTGACCTTCGCCCTGAGCCCCCG
CATCTGGCGCGAGGTGCGCGACTTCCAGCCCGACCTGATCCACTGCAGCAGCCCCGGC
GTGATGGTGTTCGCCGCCAAGCTGTACGCCTGGCTGCTGAAGAAGCCCATCGTGCTGAG

CTACCACACCCACGTGCCCAGCTACCTGCCCAAGTACGGCATCAGCTACCTGGTGCCCG
CCATGTGGGGCTTCCTGCGCATGCTGCACGCCACCGCCCACCTGACCCTGACCGTGAGC
CCCGCCATGGTGGACGAGCTGGTGACCAACCGCGCCGTGAACAGCGCCAAGCAGGTGC
AGGTGTGGAAGAAGGGCGTGGACAGCGAGGTGTTCCACCCCCGCTTCCGCAGCGCCGC
CATGCGCGAGCGCCTGACCGGCGGCCAGCCCGACCGCCCCACCCTGGTGTACGTGGGC
CGCCTGGGCTTCGAGAAGAACCTGTTCTTCCTGCGCGAGGTGCTGCAGCGCAACCCCGG
CGTGGGCCTGGCCTTCGTGGGCGACGGCCCCGCCCGCCAGGAGCTGCAGGCCGCCTTC
AAGGGCACCCCCACCCAGTTCCTGGGCATGCTGCACGGCGAGGACCTGAGCGCCGCCT
ACGCCAGCAGCGACATCTTCGTGATGCCCAGCGAGAGCGAGACCCTGGGCTTCGTGGT
GCTGGAGGCCATGGCCAGCGAGCTGCCCGTGGTGGCCGTGCGCGCCGGCGGCATCCCC
GACATCATCACCCCCGGCGACAGCGGCGTGACCGGCTTCCTGTACGAGCCCGGCGACG
TGGACAAGGCCGCCGAGCTGATCCAGCAGCTGGCCGCCGACGCCCAGCTGCGCAGCCG
CGTGGGCATCCGCGCCCGCCAGGAGGTGGCCAAGTGGGACTGGCGCGCCGCCACCATG
CACCTGCTGAACGTGCAGTACCCCATCGCCATGGCCGCCGCCGCCGCCCAGTACGGCG
AGGCCCTGGGCCGCGTGCAGTGGCTGCCCGCCCAGGACGCCCTGGCCGCCCAGCCCCC
CCAGCGCCCCCTGGGCAGCACCGGCCTGGAGATCAGCATCATCGGCTTCGGCGCCAGC
CCCCTGGGCAACGTGTTCGGCGACGTGCACCAGGACACCGCCACCGAGGCCGTGCGCA
CCGCCTTCGACCTGGGCATCACCCTGTTCGACACCAGCCCCTTCTACGGCCTGACCAAG
AGCGAGGACGTGCTGGGCCAGGCCCTGCGCGACGCCGGCCTGCCCCGCGACCAGTTCG
TGCTGGCCACCAAGGTGGGCCGCTACGGCCAGGACACCTTCGACTTCAGCGGCCCCCG
CGTGACCCGCAGCGTGGAGGAGAGCCTGGAGCGCCTGCACACCAGCTACATCGACCTG
ATCCAGGTGCACGACATGGAGTTCGGCAGCCTGGACCAGATCATCGCCGAGACCCTGC
CCGCCCTGCAGCGCCTGAAGGAGAAGGGCCTGGTGCGCCACATCGGCATCACCGGCCT
GCCCCTGGCCTGCTTCCAGTACGTGCTGGACCGCGTGCCCTGCGGCACCGTGGACGTGG
TGCTGAGCTACTGCCACTACACCCTGTGCGACCAGAGCCTGGGCCGCATCCTGCCCTAC
CTGGAGAGCAAGGCCGTGGGCGTGATCAACGCCAGCGTGCTGGGCATGGGCCTGCTGA
CCCCCCACGGCCCCCCCGCCTGGCACCCCGCCCCCGCCGAGCTGCAGGCCGCCGCCCGC
GCCGCCGCCCGCGCCGCCGACGCCCACGCCGTGGACCTGCCCAAGCTGGCCACCATGT
TCAGCGTGGCCAACCCCGGCATCGCCACCCACCTGATCGGCTTCAGCACCCCCGACCAG
GTGCGCAACGCCGTGCACGCCGTGCTGCAGGCCCAGGGCCTGGAGGAGAACGCCCAGG
CCGAGCAGGAGGAGCGCGCCATGGTGGAGATCCGCGAGATCCTGGCCGGCACCGCCCA
GGTGACCTGGCCCAGCGGCCTGCCCGAGAACAAGGACAGCGTGACCGAGCGCGTGCTG
CACGTGACCAAGCACTTCGAGAAGATCGACGCCAGCAAGGTGAGCCCCGCCGCCAGCT
TCGAGAAGGACCTGGGCCTGGACAGCCTGGACGTGGTGGAGCTGGTGATGGCCCTGGA
GGAGGAGTTCGGCCTGGAGATCCCCGACGCCGAGGCCGACAAGATCGCCAGCGTGGGC
GACGCCATCAACTACATCTGCAGCAACCCCGCCGCCGCCTACTTCCCCACCCGCGTGGT
GGTGACCGACCCCGAGGCCTTCCGCACCGACCGCGGCTACCTGTTCGGCTTCTGCCCCC
ACAGCGCCCTGCCCATCGCCCTGCCCATCGCCTTCGCCACCACCGCCAGCCCCCTGCTG
CCCAAGGAGCTGCGCGGCCGCACCCACGGCCTGGCCAGCAGCGTGTGCTTCAGCGCCC
CCATCGTGCGCCAGCTGTACTGGTGGCTGGGCGTGCGCCCCGCCACCCGCCAGAGCATC
AGCGGCCTGCTGGACAAGTGCCGCGCCCGCAAGGTGGCCGTGCTGGTGCCCGGCGGCG

TGCAGGAGGTGCTGAACATGGAGCACGGCAAGGAGGTGGCCTACCTGAGCAGCCGCAC
CGGCTTCGTGCGCCTGGCCGTGCAGCACGGCGCCCCCTGGTGCCCGTGTGGGCCTTCG
GCCAGACCCGCGCCTACAGCTGGTTCCGCCCCGGCCCCCCCTGGTGCCCACCTGGCTG
GTGGAGCGCATCAGCCGCGCCGCCGGCGCCGTGCCCATCATCTTCCACGGCCGCGGCA
TGTTCGGCCAGTACGGCACCCCCCTGATGCCCCACCGCGAGCCCCTGACCATCGTGGTG
GGCCGCCCCATCCCCGTGCCCGAGCTGGCCCCCGGCCAGCTGGAGCCCGAGCCCGAGG
TGCTGGCCGCCCTGCTGAAGCGCTTCACCGACGACCTGCAGGCCCTGTACGACAAGCA
CAAGGCCCAGTTCGGCGTGCCCAAGGGCGAGGAGCTGGTGATCAGCTTCCGCGAGGAG
CACCGCATCCGCGGCTACGAGGTGAGCCCCGACCAGCGCGCCACCATCGTGACCGTGG
CCAACCTGCTGCAGGAGGTGGCCGGCAACCACGCCGTGGGCATGTGGGGCCGCACCGA
CGAGGGCTTCGCCAGCCTGCCCAGCATGAAGGACTACAACCTGCTGTTCGTGATGACCC
GCCTGCAGGTGCGCATGTACGAGTACCCCAAGTGGGGCGACGTGGTGGCCGTGGAGAC
CTACTTCACCGAGGAGGGCCGCCTGGCCTTCCGCCGCGAGTGGAAGCTGATGGACGTG
GCCACCGGCAAGCTGCTGGGCGCCGGCACCAGCACCTGGGTGACCATCAACACCGCCA
CCCGCCGCCTGAGCAAGCTGCCCGAGGACGTGCGCAAGCGCTTCCTGCGCTTCGCCCCC
CCCAGCAGCGTGCACATCCTGCCCCCCGAGGAGACCAAGAAGAAGCTGCAGGACATGC
CCAAGTACGAGCTGCCCGGCCAGGTGCAGAGCGCCCAGCAGGTGGCCCGCCGCGCCGA
CATGGACATGAACGGCCACATCAACAACGTGACCTACCTGGCCTGGACCCTGGAGAGC
CTGCCCGAGCGCGTGATGAGCGGCGGCTACAAGATGCAGGAGATCGAGCTGGACTTCA
AGGCCGAGTGCACCGCCGGCAACGCCATCGAGGCCCACTGCAACCCC

Construction of Hybrid ORF and its Restriction Enzyme Analysis. Hybrid ORF was constructed showing 6 conserved regions of selected superfamilies. These were further analyzed for restriction enzyme analysis. The restriction sites were identified in hybrid DNA sequence for many restriction enzymes as shown in Fig. 1. These restriction enzymes and their sites help in construction of hybrid ORF for further in silico or in vitro cloning and analysis.

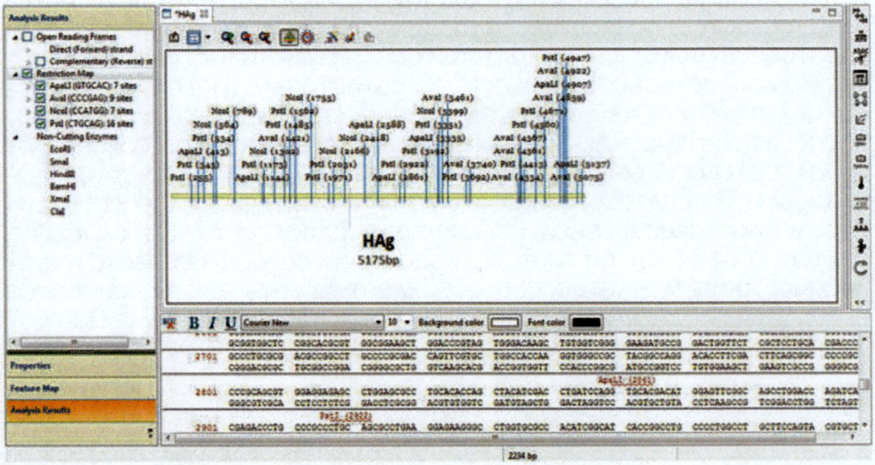

Fig. 1 Restriction enzyme mapping of constructed hybrid ORF

Identification and Analysis of Hybrid ORF using ORF Finder. The hybrid ORF identification was done using ORF finder. The hybrid ORFs are marked with directional arrow in the sequence panel. According to the observation the arrows were solid, which display complete ORFs. As the analysis was done for complete ORF, which was run on default parameters and the minimum size base pair was specified as 150. The start codons were ATG GTG and stop codons were TAA TGA TAG for the ORF finder analysis. The positions of ORFs were marked in the sequence, which is shown in Fig. 2. The ORF finder also translates ORF in amino acid sequence, which is visible in Fig. 3.

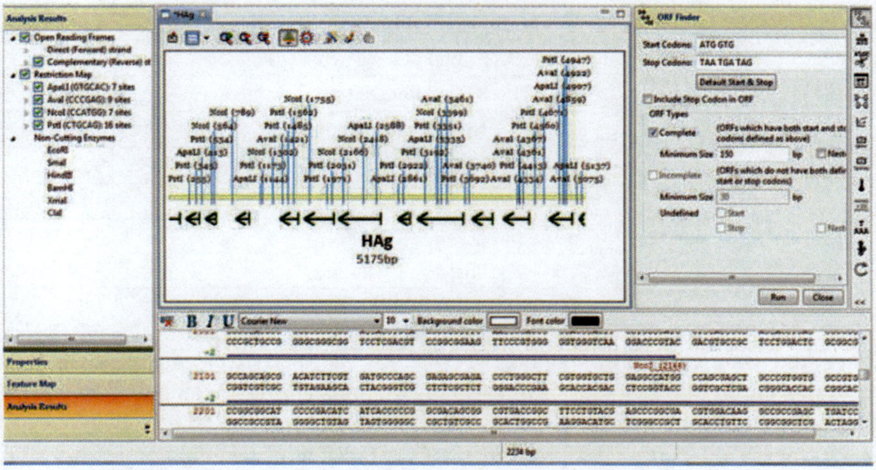

Fig. 2 Identification and analysis of ORF in hybrid sequence using ORF Finder

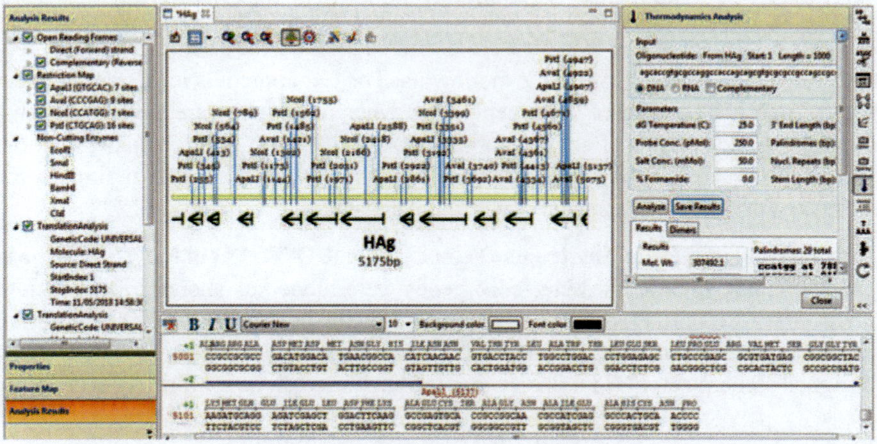

Fig. 3 Thermodynamics analysis of hybrid ORF as predicted from vector NTI tool

Table 3 Thermodynamics analysis for the Hybrid ORF on various parameters

S. no.	Parameters of thermodynamics analysis	Results
1.	dG Temperature(C)	25.0
2.	Probe Conc. (pMol)	250.0
3.	Salt Conc. (mMol)	50.0
4.	% Formamide	0.0
5.	Therm. Tm.	100.0
6.	GC Content	67.6%
7.	%GC Tm	87.0
8.	Stem Length (bp)	3
9.	Palindromes (bp)	6
10.	3' End dG	−18.2
11.	3' End Length (bp)	7
12.	Nucl. Repeats (bp)	4
13.	Mol. Wt.	307492.1
14.	dH	−8640.2
15.	dG	−2322.1
16.	dS	−21184.8

Thermodynamics Analysis of Hybrid ORF. The thermodynamic analysis was done to verify the hybrid clone as well as to check the stability by studying important thermodynamics parameters as shown in Fig. 3 (Table 3).

Vector NTI® *Express* designer calculates two different melting temperatures for DNA/RNA oligonucleotides, which is thermodynamic Tm (Therm. Tm) and %GC Tm. The analysis was carried out at default parameters like probe concentration, 250 pM and salt concentration, 50 mM for thermodynamic calculation. The GC content of designed hybrid ORF was 67.6% and % GC Tm was 87.0, which was higher. GC content higher than 60% is considered for the gene design, protein expression, and primer design for PCR experiments. The GC content is important because the GC pair is having three hydrogen bonds, whereas AT pairs are having only two hydrogen bonds, which affect the stability of DNA. The GC content also affect the secondary structure of mRNA and annealing temperature in PCR experiments for template DNA (BIC, Homepage, GC Calculator 2018).

Cloned Hybrid ORF in Invitrogen Vector pENTR/D-TOPO of length 2580 base pair (circular form). The selected 6 genes were cloned as shown in Fig. 4, into invitrogen vector pENTR/D-TOPO of length 2580 bp as the small vector is preferred for the contraction of multigene sequence cloning and it also provides stability to the designed clone. Vector NTI tool constructed the desired clone for further verification.

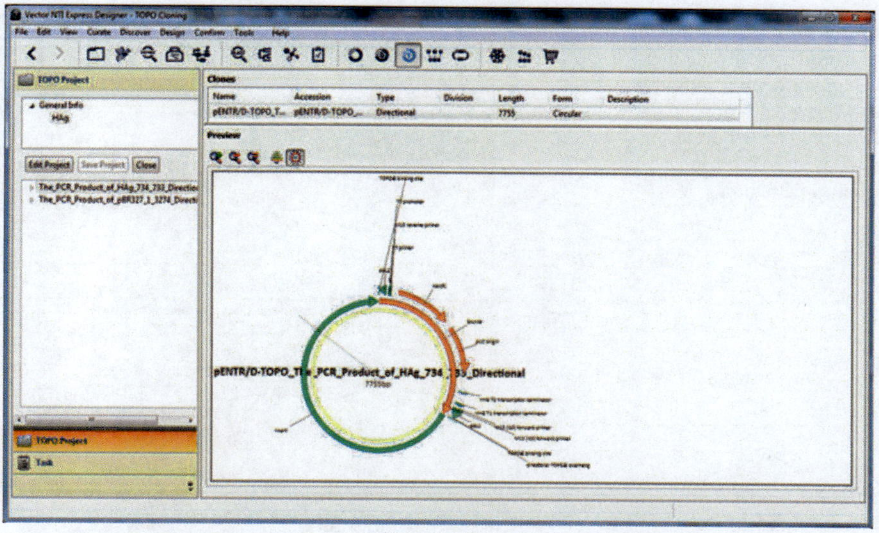

Fig. 4 Cloned Hybrid ORF in vector pENTR/D-TOPO (circular form) using TOPO cloning method

3.4 *Multi-template Homology Modeling of Hybrid ORF Protein Using Schrödinger Software*

Hybrid ORF construction was done and the hybrid protein structure was modeled using Schrödinger software based on multi-template homology modeling method. 1i24_A, 1pz1_A, 2jjm_A, 2m5r_A, 5 × 04_A, are the 5 selected templates which were showing similarity with the hybrid ORF sequence and were used to model the hybrid protein as shown in Fig. 5.

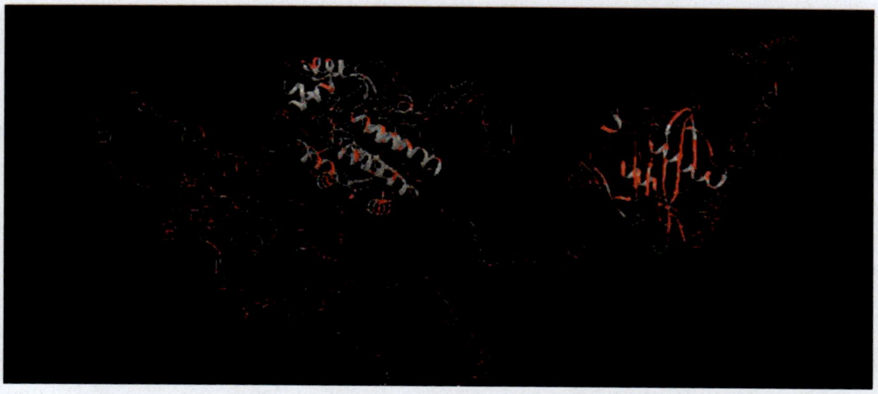

Fig. 5 Protein structure of hybrid ORF. Protein structure prediction was done using multi-template homology modeling method as included in Schrödinger software

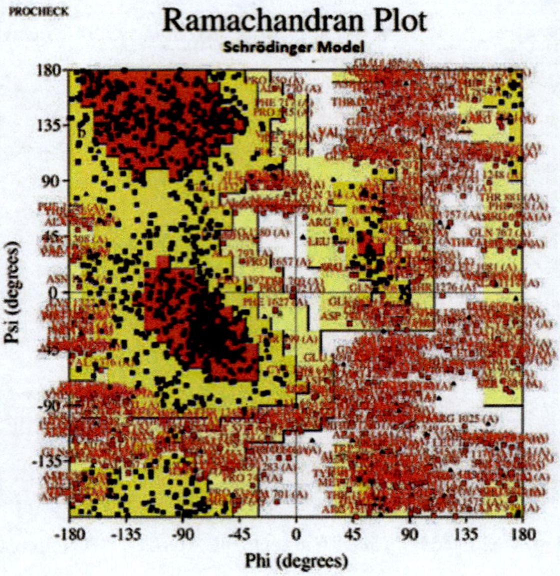

Fig. 6 Ramachandran plot of hybrid protein structure using Schrödinger software showing 87.3% of residues in favored and allowed regions and 12.7% in disallowed region

Structure verification was done using Ramachandran plot (Ramachandran et al. 1963; Kleywegt and Jones 1996; Lovell et al. 2003; Ho and Brasseur 2005; Wang et al. 2009) as shown in Fig. 6, showing 87.3% of residues in favored and allowed regions and 12.7% in disallowed region.

The protein structure obtained from multi-template homology modeling using Schrödinger software was not showing appropriate results, so the ab initio method was used for protein structure prediction using Phyre^2 server.

3.5 Protein Structure Prediction of the Constructed Hybrid ORF by Phyre^2 Server

Phyre^2 server was used for the comparative 3-D structure prediction of the constructed hybrid ORF sequence, which is based on ab initio method. A comparatively better model was predicted using Phyre^2 server and protein structure was shown in Fig. 7. The three dimensional structures of proteins are predicted, on the basis of amino acid sequence by using ab initio method.

Structure verification of predicted protein was done using Ramachandran plot as shown in Fig. 8. Ramachandran plot shows that 91% of residues lie in favored region (red area) and 9% in allowed region (yellow area). This Ramachandran plot shows secondary structure like α helix and β sheet in the fully allowed part (favored region) and outer limit (allowed region). The generously allowed regions are located as pale greenish yellowish area. The disallowed regions (white area) generally involve

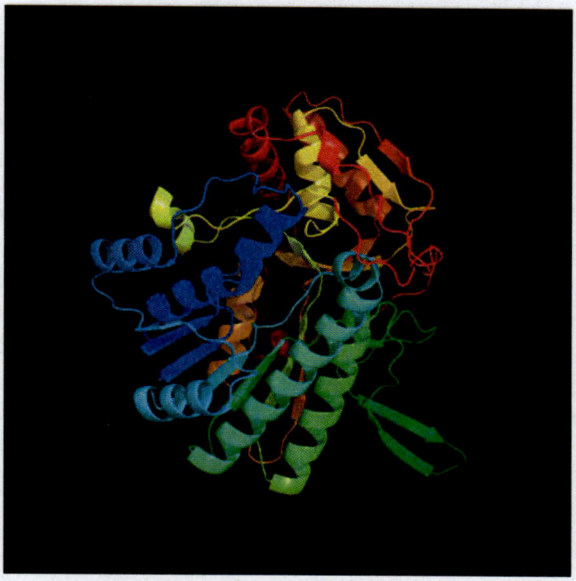

Fig. 7 Protein structure of hybrid ORF build using ab initio method by Phyre[2] server

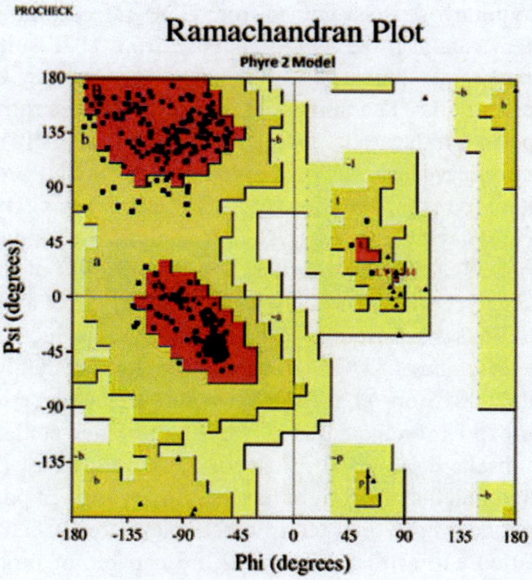

Fig. 8 Ramachandran plot of predicted 3-D protein model using Phyre[2] server showing 91% of residues in favored region and 9% in allowed regions

steric hindrance between the side chains of one amino acid with the backbone of the succeeding amino acid. The glycine is an exception since it lacks the side chain responsible for the clash and can adopt phi and psi angles in disallowed region of the Ramachandran plot, which is acceptable (Cragg and Newman 2013).

Metabolic Pathway Analysis of Algal Genes Selected for Designing of Hybrid ORF. These selected genes are involved in glycerolipid metabolism and metabolic pathways, which are supporting fatty acid biosynthesis. These genes are involved in lipid biosynthesis (SQD1 and SQD2), glucose catabolism (F751_2396), enhance the lipid content even in stressed condition (CGLD24), support growth of fatty acid chain (ACP1), and increases lipid synthesis through protein–protein interaction (FAT1). These selected genes favor lipid biosynthesis during its metabolic pathway and are not involved in any other biosynthesis pathway except ACP1 and CGLD24. Acyl-carrier protein of ACP1 gene is also involved in polyketide secondary metabolite biosynthesis. Polyketides are medically important compounds, used as antibiotics, anticancer agents, immune suppressants, and cholesterol-lowering agents (SWISS-MODEL | Course | Secondary Structure and Backbone Conformation 2018). DGAT protein of CGLD24 is involved in triacylglycerol metabolism in higher eukaryotes for physiologic processes like intestinal fat absorption, lipoprotein assembly, etc. (Wang et al. 2007). Almost the other function of ACP1 and CGLD24 gene deals with the fatty components. The details of all the selected gene regarding the metabolic pathway is mentioned in Table 4.

Mechanism of Selected Gene during Metabolic Pathway. Lipid biosynthesis is having two steps; the first step reaction is catalyzed by SQD1 protein in which UDP-sulfoquinovose is assembled from UDP-glucose and sulfite. The second step SQD2 protein catalysis the sulfoquinovose from UDP-sulfoquinovose to diacylglycerol (Sanda et al. 2001; Yu et al. 2002). D-arabinose 1-dehydrogenase catalyzes D-arabinose into D-arabinono-1,4-lactone, in heterotrophic conditions this protein helps in synthesis of fatty acids (Gao et al. 2014). Acyl carrier protein (ACP) is a conserved carrier of acyl intermediates and provides substrates to enzymes, which are responsible for fatty acid biosynthesis (Byers and Gong 2007). Diacylglycerol acyltransferase (DGATs) catalyze the formation of diacylglycerol from 2-monoacylglycerol and fatty acyl-CoA. DGATs also catalyze the terminal step in triacylglycerol synthesis by using diacylglycerol and fatty acyl-CoA as substrates. DGAT has two isoforms DGAT1 and DGAT2, which is reported in plant and animals. The deficiency of DGATs accumulates less triacylglycerol (Zou et al. 1999; Smith et al. 2000; Stone et al. 2004), whereas triacylglycerol increases, as DGAT enzyme over-expression increases in plants (Andrianov et al. 2010; Xu et al. 2008), animals (Kamisaka et al. 2010; Liu et al. 2009) and yeast (Kamisaka et al. 2007). It has been found that most algae have multiple copies of putative DGAT2s genes, whereas single genes are reported in other eukaryotes (Chen and Smith 2012). It has been reported earlier that microalgae are capable of producing triacylglycerols (TAG) in stress condition. Phospholipid: Diacylglycerol acyltransferase (PDAT) catalyzes TAG synthesis in green microalga *Chlamydomonas reinhardtii* through two pathways: transacylation of diacylglycerol (DAG) with acyl groups from phospholipids and galactolipids and DAG:DAG transacylation. PDAT also possesses acyl hydrolase activities using TAG, phospholipids, galactolipids, and cholesteryl esters as substrates (Kamisaka et al. 2007). The fatty acid acyl carrier protein (ACP) along with thioesterase (TE) undergoes protein–protein interactions and results in fatty acid

Table 4 Metabolic pathways of genes responsible for algal oil production

S. no.	Gene symbol	Micro organism	Protein	Protein function in lipid biosynthesis	Protein function in any other biosynthesis	KEGG pathway id	Name of pathway	References
1.	SQD1	*Chlamydomonas reinhardtii*	UDP-sulfoquinovose synthase	Converts UDP-glucose and sulfite to the sulfolipid head group precursor UDP-sulfoquinovose (Sulfolipid biosynthesis)	–	Cre 00561, cre 00520	Glycerolipid metabolism, Amino sugar and nucleotide sugar metabolism	Sanda et al. 2001)
2.	SQD2	*Chlamydomonas reinhardtii*	Sulfolipid synthase	Catalyzes the transfer of the sulfoquinovose moiety from UDP-sulfoquinovose to diacylglycerol during sulfolipid biosynthesis	–	Cre 00561, cre 01100	Glycerolipid metabolism, Metabolic pathways	Yu et al. 2002)
3.	F751_2396	*Auxenochlorella protothecoides*	D-arabinose 1-dehydrogenase	Catalyzes the oxidation of D-arabinose, L-xylose, L-fucose and L-galactose in the presence of $NADP^+$	–	Apro 00053, apro 01100, apro 01110	Ascorbate and aldarate metabolism, Metabolic pathways, Biosynthesis of secondary metabolites	Gao et al. 2014)

(continued)

Table 4 (continued)

S. no.	Gene symbol	Micro organism	Protein	Protein function in lipid biosynthesis	Protein function in any other biosynthesis	KEGG pathway id	Name of pathway	References
4.	ACP1	*Chlamydomonas reinhardtii*	Acyl-carrier protein	Carrier of the growing fatty acid chain in fatty acid biosynthesis	Polyketide secondary metabolite biosynthesis	Cre 00190, cre 01100	Oxidative phosphorylation, Metabolic pathways	Byers and Gong 2007)
5.	CGLD24	*Chlamydomonas reinhardtii*	Diacylglycerol acyl transferase	Transferring acyl groups other than amino-acyl groups (transferase activity)	DGAT is important in higher eukaryotes for physiologic processes involving triacylglycerol metabolism such as intestinal fat absorption, lipoprotein assembly, adipose tissue formation, and lactation	Cre 00561, cre 01100	Glycerolipid metabolism, Metabolic pathways	Patent and https:, , patecm, patent, US9771605, last accessed 2018, 12, 14, 2018; Cases et al. 1998)
6.	FAT1	*Chlamydomonas reinhardtii*	Acyl carrier protein thioesterase	Plays an essential role in chain termination during de novo fatty acid synthesis	-	Cre 00061	Fatty acid biosynthesis	Blatti et al. 2012)

hydrolysis within the algal chloroplast (Blatti et al. 2012). The metabolic pathway analysis verifies that selected genes sequences for the construction of hybrid ORF protein are involved in lipid biosynthesis.

4 Conclusion

The in silico construction of genetic engineered microorganism was done by constructing hybrid ORF which was capable of expressing functional protein to enhance oil content in algae. The hybrid ORF protein was modeled by Schrödinger software and Phyre^2 server. Structural verification was done by Ramachandran plot, which shows that both the constructed hybrid protein models are stable, with 87.3% residues in favored region by Schrödinger software and 91% residues in favored region by Phyre^2 server. These comparative results suggest that Phyre^2 web-based homology modeling method is preferred over multi-template homology modeling done by Schrödinger software. The stability of constructed hybrid ORF protein was also verified by thermodynamic analysis, through which GC content estimated was 67.6%. The estimated GC content favors the stability of hybrid ORF protein. The metabolic pathway analysis also verified that all the selected genes used for designing hybrid ORF are majorly involved in lipid synthesis pathways. These results justify the construction of hybrid ORF protein, which will have the potential to serve as an effective tool for designing of genetically engineered microorganisms to enhance high cellular oil content in algae. This complete in silico analysis will be a great support for construction of genetically engineered microorganisms during in vitro studies. The wet laboratory studies will be conducted in near future for the in vitro verification of hybrid ORF, constructed to enhance oil content in algal strains. This study will provide an immense support in the area of biofuel research.

Acknowledgements We like to acknowledge tools and databases for bioinformatics analysis present at Amity Institute of Biotechnology, Amity University Uttar Pradesh, Lucknow campus for conducting this study and also our immense gratitude and deep regard to all those who directly or indirectly helped us to successfully complete this work. This research project is not funded by any specific grant from funding agencies in the public, commercial, or nonprofit sectors.

References

Andrianov V, Borisjuk N, Pogrebnyak N, Brinker A, Dixon J, Spitsin S, Flynn J, Matyszczuk P, Andryszak K, Laurelli M, Golovkin M, Koprowski H (2010) Tobacco as a production platform for biofuel: overexpression of Arabidopsis DGAT and LEC2 genes increases accumulation and shifts the composition of lipids in green biomass. Plant Biotechnol J 8:277–287

Banerjee A, Banerjee C, Negi S, Chang JS, Shukla P (2018) Improvements in algal lipid production: a systems biology and gene editing approach. Crit Rev Biotechnol 38(3):369–385

BIC, Homepage, GC Calculator (2018) https://www.biologicscorp.com/tools/GCContent/#.XBOU4NszbIU. Accessed 14 Dec 2018

Blatti JL, Beld J, Behnke CA, Mendez M, Mayfield SP, Burkart MD (2012) Manipulating fatty acid biosynthesis in microalgae for biofuel through protein-protein interactions. PLoS ONE 7(9):e42949. https://doi.org/10.1371/journal.pone.0042949

Byers DM, Gong H (2007) Acyl carrier protein: structure-function relationships in a conserved multifunctional protein family. Biochem Cell Biol. 85(6):62–649

Cases S, Smith SJ, Zheng YW, Myers HM, Lear SR, Sande E, Novak S, Collins C, Welch CB, Lusis AJ, Erickson SK, Farese RV Jr (1998) Identification of a gene encoding an acyl CoA:diacylglycerol acyltransferase, a key enzyme in triacylglycerol synthesis. Proc Natl Acad Sci 95(22):13018–13023

Cerón García MC, Fernández Sevilla JM, Acién Fernández FG, Grima EM, Camacho FG (2000) Mixotrophic growth of Phaeodactylum tricornutum on glycerol: growth rate and fatty acid profile. J Appl Phycol 12:239–248

Chen J, Smith AG (2012) A look at diacylglycerol acyltransferases (DGATs) in algae. J Biotechnol 162(1):28–39

Chisti Y (2007) Biodiesel from microalgae. Biotechnol Adv 25(3):294–306

Chisti Y (2008) Biodiesel from microalgae beats bioethanol. Trends Biotechnol 26(3):126–131

Cragg GM, Newman DJ (2013) Natural products: a continuing source of novel drug leads. Biochim Biophys Acta 1830(6):3670–3695

Demirbas A (2009) Political, economic and environmental impacts of biofuels: a review. Appl Energy 86:108–117

Elshahed MS (2010) Microbiological aspects of biofuel production: current status and future directions. J Adv Res 1(2):103–111

Ferrara P, Edgar J (2007) Evaluation of the utility of homology models in high throughput docking. J Mol Model 13:897–905

Gao C, Wang Y, Shen Y, Yan D, He X, Dai J, Wu Q (2014) Oil accumulation mechanisms of the oleaginous microalga Chlorella protothecoides revealed through its genome, transcriptomes, and proteomes. BMC Genomics 10:15–582

Guoqing L, Moriyama EN (2004) Vector NTI: a balanced all-in-one sequence analysis suite. Brief Bioinform 5:378–388

Hemaiswarya S, Raja R, Ravi Kumar R, Ganesan V, Anbazhagan C (2011) Microalgae: a sustainable feed source for aquaculture. World J Microbiol Biotechnol 27:1737–1746

Ho BK, Brasseur R (2005) The Ramachandran plots of glycine and pre-proline. BMC Struct Biol 5(14). doi.org/10.1186/1472–6807–5–14

Hu Q, Sommerfeld M, Jarvis E, Ghirardi M, Posewitz M, Seibert M, Darzins A (2008) Microalgal triacylglycerols as feedstocks for biofuel production: perspectives and advances. Plant J 54:621–639

Ivanov AA, Barak D, Jacobson AK (2009) Evaluation of homology modeling of G protein-coupled receptors in light of the A_{2A} adenosine receptor crystallographic structure. J Med Chem 52(10):3284–3292

Kamisaka Y, Tomita N, Kimura K, Kainou K, Uemura H (2007) DGA1 (diacylglycerol acyltransferase gene) overexpression and leucine biosynthesis significantly increase lipid accumulation in the Deltasnf2 disruptant of Saccharomyces cerevisiae. Biochem J 408:61–68

Kamisaka Y, Kimura K, Uemura H, Shibakami M (2010) Activation of diacylglycerol acyltransferase expressed in Saccharomyces cerevisiae: overexpression of Dga1p lacking the N-terminal region in the Deltasnf2 disruptant produces a significant increase in its enzyme activity. Appl Microbiol Biotechnol 88:105–115

Kelley LA, Mezulis S, Yates CM, Wass MN, Sternberg MJE (2015) The Phyre2 web portal for protein modelling, prediction and analysis. Nat Protoc 10(6):845–858

Kleywegt GJ, Jones TA (1996) Phi/psi-chology: Ramachandran revisited. Structure 4:1395–1400

Liu L, Shi X, Bharadwaj KG, Ikeda S, Yamashita H, Yagyu H, Schaffer JE, Yu YH, Goldberg IJ (2009) DGAT1 expression increases heart triglyceride content but ameliorates lipotoxicity. J Biol Chem 284:36312–36323

Lovell SC, Davis IW, Arendall WB III, de Bakker PIW, Word JM, Prisant MG, Richardson JS, Richardson DC (2003) Structure validation by Cα geometry: ϕ, ψ and Cβ deviation. Proteins 50:437–450

Miao X, Wu Q (2004) High yield bio-oil production from fast pyrolysis by metabolic controlling of Chlorella protothecoides. J Biotechnol 110:85–93

Radakovits R, Jinkerson RE, Darzins A, Posewitz MC (2010) Genetic engineering of algae for enhanced biofuel production. Eukaryot Cell 9:486–501

Ramachandran GN, Ramakrishnan C, Sasisekharan V (1963) Stereochemistry of polypeptide chain configurations. J Mol Biol 7:95–99

Sanda S, Leustek T, Theisen MJ, Garavito RM, Benning C (2001) Recombinant Arabidopsis SQD1 converts UDP-glucose and sulfite to the sulfolipid head group precursor UDP-sulfoquinovose in vitro. J Biol Chem 276:3941–3946

Schrödinger L (2011) Schrodinger software suite. Schrödinger, LLC, New York

Schrödinger P (2014) Version 3.5. LLC, New York, NY

Smith SJ, Cases S, Jensen DR, Chen HC, Sande E, Tow B, Sanan DA, Raber J, Eckel RH, Farese RV Jr (2000) Obesity resistance and multiple mechanisms of triglyceride synthesis in mice lacking DGAT. Nat Genet 25:87–90

Stone SJ, Myers HM, Watkins SM, Brown BE, Feingold KR, Elias PM, Farese RV Jr (2004) Lipopenia and skin barrier abnormalities in DGAT2-deficient mice. J Biol Chem 279:11767–11776

SWISS-MODEL | Course | Secondary Structure and Backbone Conformation (2018) https://swissmodel.expasy.org/course/text/chapter1.htm. Accessed 14 Dec 2018

US Patent (2018) https://patents.google.com/patent/US9771605. Accessed 14 Dec 2018

Wang Y, Xu HY, Zhu Q (2007) Progress in the study on mammalian diacylgycerol acyltransgerase (DGAT) gene and its biological function. Yi Chuan 29(10):72–1167

Wang Y, Xiao J, Suzek TO, Zhang J, Wang J, Bryant SH (2009) PubChem: a public information system for analyzing bioactivities of small molecules. Nucleic Acids Res 37(2):623–633

Xu J, Francis T, Mietkiewska E, Giblin EM, Barton DL, Zhang Y, Zhang M, Taylor DC (2008) Cloning and characterization of an acyl-CoA-dependent diacylglycerol acyltransferase 1 (DGAT1) gene from Tropaeolum majus, and a study of the functional motifs of the DGAT protein using site-directed mutagenesis to modify enzyme activity and oil content. Plant Biotechnol J 6:799–818

Yu B, Xu C, Benning C (2002) Arabidopsis disrupted in SQD2 encoding sulfolipid synthase is impaired in phosphate-limited growth. Proc Natl Acad Sci 99(8):7–5732

Zou J, Wei Y, Jako C, Kumar A, Selvaraj G, Taylor DC (1999) The Arabidopsis thaliana TAG1 mutant has a mutation in a diacylglycerol acyltransferase gene. Plant J 19:645–653

ANFIS Detects the Changes in Stressful Patterns of Sleep EEG

Prabhat Kumar Upadhyay and Chetna Nagpal

Abstract An automated analysis and detection of sleep electroencephalogram (EEG) and stress levels have been performed in this work on pre-recorded EEG data. Based on physiological indicators and powers of sub-band frequencies computed through wavelet transform, features were extracted. With the help of these features, physiological changes in the subjects have been investigated to frame the fuzzy rules to differentiate chronic and acute stress from their respective control groups. Identification of sleep stage is followed by stress level detection using fuzzy logic. The proposed system uses Mamdani fuzzy model and adaptive neuro-fuzzy inference system (ANFIS), that achieves an accuracy of 89.7% for acute stress and 88.0% for chronic stress as compared to their control groups.

Keywords Sleep EEG · Stress · Wavelet transform · Fuzzy logic · ANFIS · Mamdani fuzzy model

1 Introduction

Study on the effect of heat stress on human nervous system has become an essential component to understand the cause of many psychiatric problems which arise due to the hot environment as one of the natural stress markers. It has been observed that animals respond by activating several processes pertaining to neurophysiology when exposed to hot environments. The influencing factors such as intensity, duration, adaptations to hot environment, play the key role in metabolic change, and obtain the level of thermoregulatory activity, which impacts the functioning of all animals (Menon and Dandiya 1969). Sleep plays an important role in ensuring good physical health. An average adult utilizes 7–8 h per day sleeping during which the sleep cycle,

P. K. Upadhyay
Department of EEE, Birla Institute of Technology, Mesra, Ranchi, India
e-mail: uprabhatbit@gmail.com

C. Nagpal (✉)
Department of EEE, Birla Institute of Technology, Offshore Campus, Ras Al Khaimah, UAE
e-mail: chetnakochhar0@gmail.com

A. A. Rizvanov et al. (eds.), *Advances in Biomedical Engineering and Technology*, Lecture Notes in Bioengineering,
https://doi.org/10.1007/978-981-15-6329-4_9

which consists of various stages of sleep, repeats four to six times. Due to heat stress, alterations in sleep cycles take place. Some of these changes have been noticed to be permanent and some are transitory. Sleep-related disorders can affect the emotional and mental state. Reflection of these sleep disorders appears in the form of deviations in normal sleep cycles (Menon and Dandiya 1969).

A normal sleep stage consists of five stages: NREM-1 (Non rapid eye movement-1), NREM-2, NREM-3, NREM-4, and REM sleep. NREM-1 or Awake is the stage between wakefulness and drowsiness, NREM-2, 3, and 4 or Slow wave sleep (SWS) is characterized by decrease in responsiveness to environment, REM (Rapid eye movement is characterized by rapid eye movement and also referred to as dreaming phase. The American Association of Sleep Medicine (AASM) characterizes the different stages of sleep based on different physiological changes during these stages (Grigg-Damberger 2012). Due to the computational ability and learning mechanism of artificial neural networks and fuzzy logic in detecting the sleep patterns, many researchers have been using these soft-computing tools in automated analysis of EEG. Neural network with supervised and unsupervised learning mechanisms have been applied while analysing sleep EEG of animal model and detecting heat stress (Subasi et al. 2005). On the same animal model, EEG power spectrum was also classified which differentiates the depressed subjects from control groups (Upadhyay et al. 2010a). Literature reveals that very few works have been carried out that differentiates normal candidates from heat-stressed conditions with the help of soft-computing techniques.

Electroencephalogram (EEG) responses are non-stationary signals which are generated randomly in a response to stimulus (Morstyn et al. 1983). Electromyogram (EMG) and Electrooculogram (EOG) activities are also recorded as polysomnographic parameters to study the sleep patterns. Previous research works suggest that frequency spectrum of EEG covers a wide range of frequencies (Patil and Patil 1998; Tagluk et al. 2009). These frequency bands have been named as delta (0.5–4 Hz), theta (4–8 Hz), alpha (8–12 Hz), and beta (12–40 Hz). In order to train the proposed adaptive system in the classification of sleep stages, a set of feature vectors from sub-band frequencies are extracted from the recorded polysomnograms with the help of which Awake, Slow wave sleep (SWS) and REM were classified (Aboalayon et al. 2016). Using such classification approach, diagnosis of sleep-based disorders becomes quite easy and reliable for clinicians and medical professionals. Many researchers have carried their studies in the field of sleep EEG where artificial neural networks along with other statistical methods have been extensively applied (Aboalayon et al. 2016). However, the advent of fuzzy system which is a rule-based technology has made the analysis of such problems even simpler.

Several reports have been published in past few years which investigate how the brain electrical activities get affected when heat is induced externally. The study correlates the changes in EEG pattern as a function of ambient temperature. Due to an increase in the ambient environmental heat, body temperature increases either artificially or spontaneously and as a result of which, an increase in EEG frequency has been observed. Acute heat stress is developed when the subject is kept under stress for a smaller period whereas if stress is applied for a long time, chronic stress is developed in the subject (Upadhyay et al. 2010b).

The proposed work involves soft-computing techniques such as fuzzy logic and neural networks to classify the sleep patterns and types of stress induced due to exposure of high heat. The aim of this study is to develop an automated medical investigative system to detect the anomalies in neuronal functioning under the effect of high environmental heat conditions. Alterations in powers of different frequency sub-bands were evaluated with respect to the normal subjects. Further, fuzzy rules which have been framed in terms of variation in power were used to predict the stressed spectra as acute or chronic.

2 Materials and Methods

2.1 *Features*

Percentage change in band power as well as relative band power has been computed for each frequency band, which acts as important features. Relative band power indicates what amount of any frequency sub-band of sleep EEG constitutes in the entire frequency band. It is obtained when absolute power of a particular band is divided by the integrated power computed for the whole spectrum under consideration. Quantitative presence of frequency components in sleep EEG helps in creating fuzzy sets for adaptive fuzzy system (Yang et al. 2002). This formulates the base of determining if a specific frequency constitutes a small or large share of the frequency band. It has been observed that frequency sub-bands—alpha and beta cover a major part of awake state, delta is largely found in slow wave sleep, and presence of theta is found to be very low in awake state. Apart from the features extracted from EEG data set, some statistical features were also extracted from the other two channels i.e., EOG (Electrooculogram) and EMG (Electromyogram) because in order to predict sleep stages, recordings of EOG and EMG activities along with EEG play a vital role to identify the sleep events (Tagluk et al. 2009; Jansen 1985). Mean of the absolute values of EOG and EMG have been calculated for any given sample and considered as one of the features.

2.2 *Neuro-fuzzy System*

The Neuro-Fuzzy implementations on biomedical applications are successful as it combines two approaches i.e., ANN and fuzzy set theory. ANFIS utilizes the computational properties of neural network with fuzzy rule base engine which gives the significant results to model the nonlinear systems. This approach tunes the entire model and minimizes the error between the calculated output and the desired output by learning the input feature vectors through adjusting the weight vectors. Hybrid learning algorithm is used to identify the parameters of Sugeno model (ANFIS) which is an adaptive system. In order to train the function parameters of FIS and emulate the training data set, the algorithm follows a mixture of back propagation steepest descent method and the least-squares method.

2.3 Rules for Manual Scoring

EEG signal consists of a range of frequency bands which are used for the purpose of clinical diagnosis. Manual scoring rules (Tagluk et al. 2009; Estévez et al. 2002; Hobson 1969) were followed to obtain the fuzzy rules to deal with the fuzziness inherently present in the system. In order to classify sleep EEG, following rules has been laid down. Sleep stage will be recognized as:

a. **Awake**: If the powers of alpha and beta sub-bands are high, power of theta sub-band is low and EMG is high.
b. **SWS**: If the power of delta sub-band is high, EMG activity is low.
c. **REM**: If the power of alpha and beta sub-band is high, power of theta sub-band is low and EOG is high.

2.4 Observations on Fuzziness in Stress

As a result of externally induced heat stress, subjects undergo physiological changes. These changes may be temporary or permanent, based on the types of heat exposure given to the subject. Several research works have been done in this field where authors have studied the consequence of heat stress in terms of changes in sleep pattern (Upadhyay et al. 2010b; Sinha and Ray 2004). With the help of these changes observed during the experiment, membership functions for input and output fuzzy variables have been designed and the fuzzy rules have been framed to classify stress. When the subjects are under exposure to heat stress for longer period (chronic stress), the changes in sleep patterns were observed to be different than when they were exposed to high environmental heat for shorter duration. These changes in sleep EEG were recorded for chronic and acute situations. In both the cases, variation in powers of delta, theta, alpha, and beta sub-bands were studied to analyze the changes in power with respect to their respective control groups in all frequency bands to study the influence of environmental heat on stress level. It was observed that there has been a noticeable change in powers of these frequency bands relative to their control groups where subjects were not exposed to any heat stress. For awake, SWS and REM stages, variation in powers of these four frequency sub-bands has been tabulated below (Tables 1, 2 and 3).

Table 1 Stress classification (fuzzy rules for awake)

	Chronic	Acute
∂ power	↓	No change
θ power	No change	No change
α power	No change	No change
β power	No change	↑

Table 2 Stress classification (fuzzy rules for SWS)

	Chronic	Acute
∂ power	↓	↑
θ power	No change	↑
α power	↓	↑
β power	↑	↓

Table 3 Stress classification (fuzzy rules for REM)

	Chronic	Acute
∂ power	No change	No change
θ power	↑	↑
α power	↓	↑
β power	↓	↑

3 Methodology

The present study consists of the following four steps which are explained in the respective sections.

a. Data acquisition and organization
b. Features extraction
c. Sleep stage classification
d. Stress level classification
e. Result interpretation

3.1 Data Recording

This research work uses the same animal model as used by Sinha (2008). Male Charles Foster adult healthy rats of weight between 180–200 g have been selected as subjects in the experiment for data recording under stress and normal conditions. Since neurological development in rats is identical to human being, many researchers have used rats to study their behavioral pattern (Sinha 2003; Kirov and Slavianka 2002). Exposure of heat stress was given to the subjects under the following clinical environment:

a. Control group: The temperature of the BOD incubator in which rats were kept, was maintained at 23 ± 1 °C.
b. Acute stress: The temperature of the BOD incubator for a single day was maintained at 38 ± 1 °C for continuous four hours. Data was recorded on the next day.
c. Chronic stress: The temperature of the BOD incubator was maintained at 38 ± 1 °C for one hour daily for the period of 21 days. Recording was done on the 22nd day.

3.2 Feature Selection

From the continuous recording of EEG signal under stress and normal conditions, the processed data is saved in an interval of 30 s. Features are extracted from these datasets and stored along with their labels. After adjusting baseline and bandpass filtering of the signal, power is calculated for each epoch and power spectrum is analyzed. For each sub-band frequencies, relative band power and percentage change in mean band power were obtained. Therefore, total ten features from each sample dataset were extracted: mean band power delta (pdelta), mean band power of theta (ptheta), mean band power of alpha (palpha), mean band power of beta (pbeta), relative band power of delta (rdelta), relative band power of theta (rtheta), relative band power of alpha (ralpha), relative band power of beta (rbeta), mean of EMG (mEMG), and mean of EOG (mEOG).

3.3 Sleep Stage Classification

Classification of sleep stages have been performed efficiently with the help of Sugeno type fuzzy model which uses six inputs as shown in Fig. 1. Feature vectors contain relative powers of delta, theta, alpha, and beta as well as mean of EEG and mean of EMG. A fuzzy object is created and these extracted features have been added to it. Each feature vector is associated with a Gaussian bell membership function. Universe of discourse for these two fuzzy sets representing high and low is obtained by observing the range of the respective input vectors.

Since the output which is obtained from the fuzzy system ranges from 0 to 1, range of such outputs as 0–0.33, 0.33–0.66, and 0.66–1 were labeled as Awake, SWS, and REM, respectively. From the feature space, 70% of data is utilized for training the fuzzy system whereas the rest 30% is used for testing. While training, it was observed that the percentage error reduces drastically and after some epochs, the algorithm saturates. It implies that any further training would not make much difference in the weights adaptation of neuro-fuzzy system. In order to extract band power and relative power using wavelet transform, samples length of 30 s has been separately processed. All visually examined samples were used to train and test the performance

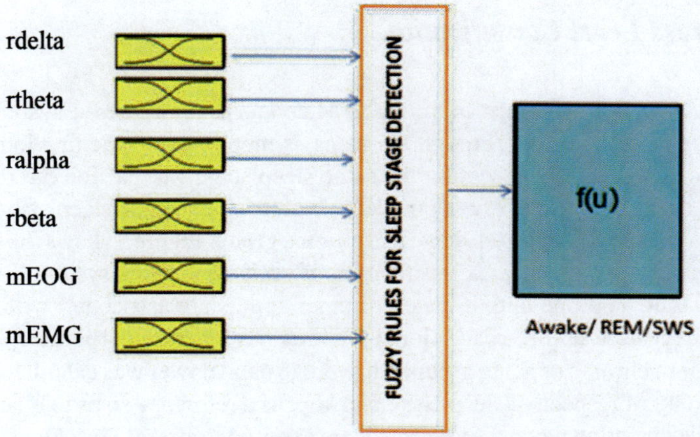

Fig. 1 Fuzzy system for sleep stage classification

of the system. The output for a given set of input vectors is shown in Fig. 2 for the sleep stage classification. The change in any one of the inputs in rule base engine reflects the change in output. Each input is associated with the three membership functions as low, medium, and high. An example of this has been illustrated in Fig. 2 where it produces REM as output because the calculated output which is 0.947, falls in the range of REM as per the criteria mentioned above.

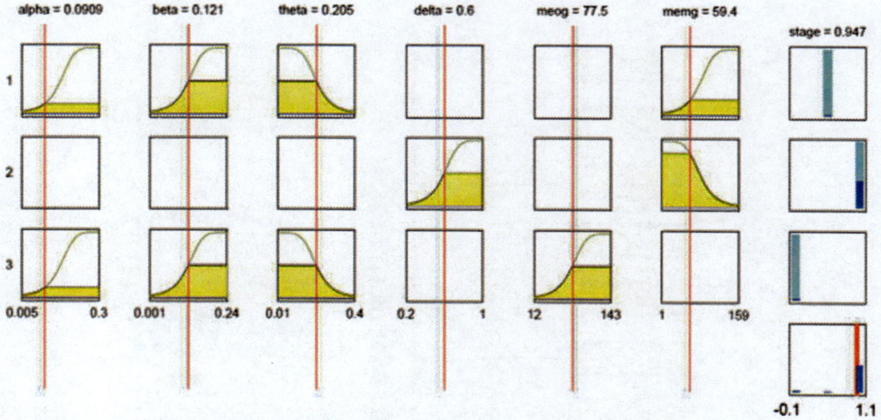

Fig. 2 Sleep stage classification using Sugeno model

3.4 Stress Level Classification

Stress classification is performed through a Mamdani fuzzy inference system. Having labeled the feature vectors through the steps as mentioned in the previous section, separation of features representing the same sleep stage was carried out for each of the stress spectra. Further, a fuzzy model is built to classify the sleep stages such as Awake, SWS, and REM into an acute and chronic group. Figure 3 shows the fuzzy rule implementation for the stress classification of each stage. This system is presented with ten inputs and one output, where first six inputs are associated with the sleep stage classification and four additional features have been incorporated to classify the stress into chronic or acute group. An average band power was calculated for each control group of respective sleep stage and kept as a reference so as to quantitatively analyze the deviation when the subjects were exposed to stress. The change (increase or decrease) in-band power is reflected by the deviation of band power in acute and chronic cases for respective sleep stage.

The percentage change in band powers is further fed to the input of the fuzzy system. In order to represent the percentage change, two fuzzy membership functions with linguistic variables—high and low have been used where the universe of discourse was properly taken from the available input fuzzy sets. Based on the nature of input data set and in view of the inherent advantages of a triangular membership function, function *'trimf'* was used to express the fuzzy sets low and high.

The fuzzy rules mentioned in Tables 1, 2 and 3 were added in the fuzzy system for stress detection and classification of Awake, SWS and REM stages. In

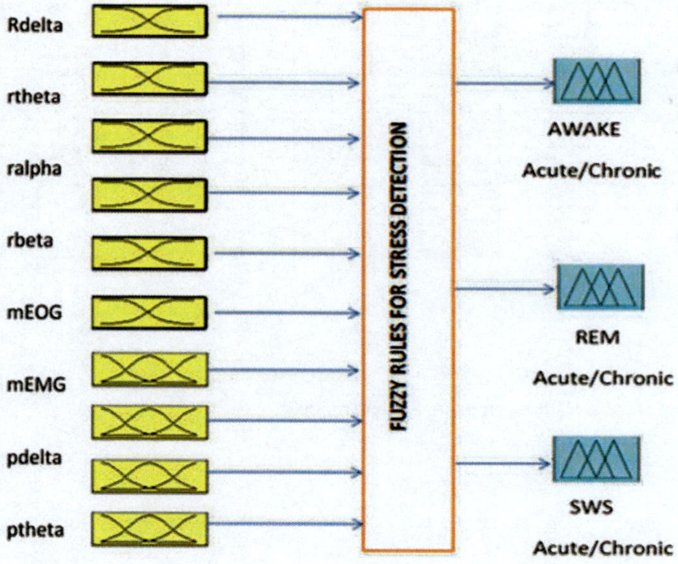

Fig. 3 Fuzzy system used for stress detection

Table 4 Stress classification

Percentage change in band power	AWAKE		SWS		REM	
	Chronic	Acute	Chronic	Acute	Chronic	Acute
pdelta	L	X	L	H	X	X
ptheta	X	X	X	H	H	H
palpha	X	X	L	H	L	H
pbeta	X	H	H	L	L	H

Fig. 4 Accuracy breakdown

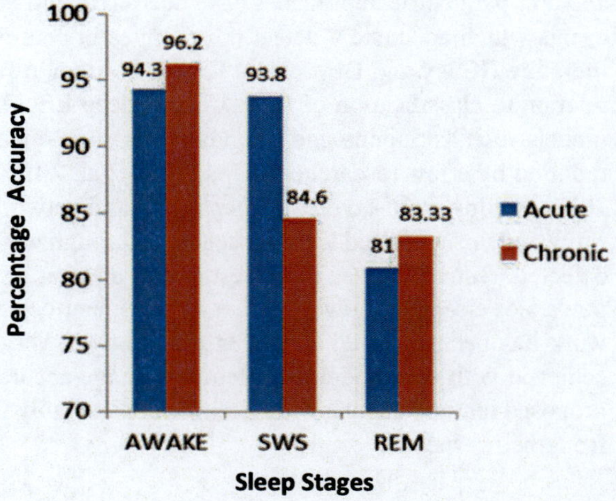

Table 4, percentage change in delta, theta, alpha, and beta band power in comparison to their respective control group have been presented, where 'L' shows the decrease in percentage power and 'H' shows the increase in percentage power.

4 Results

A plot as shown in Fig. 4 shows the performance of stress classification in which percentage accuracy of two stress classes belonging to three sleep stages have been clearly depicted.

5 Discussion

Many approaches such as fuzzy logic, neural network (Sukanesh and Harikumar 2007; Yang et al. 2002), SVM (support vector machine) (Koley and Dey 2013), RBF (Radial basis function), etc., have been implemented by researchers for the detection of sleep stages. The rules for manual scoring which was established by R&K were followed in ANFIS and the results achieved an accuracy of 88.5% [30]. The modified rule which was established by AASM has been followed in this study and has resulted to perform the sleep stage classification with an accuracy of 90.65%. After sleep stages have been classified, stress level classification has been done. Use of wavelet transform as a preprocessor and as an analyzer too has been very common in the analysis of problems such as sleep EEG analysis, spike detection, sleep spindles, epileptic seizures detection, event-related potentials, etc., (Heiss et al. 2002) (Sukhorukova et al. 2010). Several linear and nonlinear classifiers with parametric and non-parametric approaches have been frequently used to classify different EEG events which adequately detect the changes in neuronal functioning as reported in literature (Koley and Dey 2011). However, literature does not witness any work in relation to classification of heat-stressed sleep EEG spectra with the help of fuzzy models such as Sugeno and Mamdani models. Previous works in the same area as reported by a few researchers (Upadhyay et al. 2010a; Sinha and Ray 2004; Sinha 2008) employ multi-layered perceptron neural network with back propagation algorithm and unsupervised learning schemes through which heat-stressed patterns were detected. But so long as the investigation on alteration in sub-band frequency and powers of sleep EEG owing to hot environment using fuzzy logic is concerned, no work has been reported so far. In this study, a very encouraging result has been achieved with this model in which percentage accuracy is obtained as 89.2%. The proposed method enables the system to considerably discriminate the stress patterns from their respective control.

6 Conclusion

In this paper, we have successfully classified stressed pattern from normal using neuro-fuzzy concepts. Classification results are found to be in agreement with visual scoring. This concept may further be extended to study the effects of other stress stimuli such as noise, immobility, pollutants, etc., on brain electrophysiology. Application of other sophisticated DSP tools such as s-transform may be applied to investigate the phases of EEG signal, if required for brain signal analysis. With this approach, an online automated diagnostic system can be developed to classify or identify the abnormalities in brain-functions.

References

Aboalayon KAI et al (2016) Sleep stage classification using EEG signal analysis: a comprehensive survey and new investigation, entropy

Estévez P, Held C, Holzmann C, Perez C, Pérez J, Heiss J, Garrido M, Peirano P (2002) Polysomnographic pattern recognition for automated classification of sleep-waking states in infants. Med Biol Eng Comput 40(1):105–113

Grigg-Damberger M (2012) The AASM scoring manual four years later. J Clin Sleep Med

Heiss J, Held C, Estevez P, Perez C, Holzmann C, Perez J (2002) Classification of sleep stages in infants: a neuro fuzzy approach. IEEE Eng Med Biol Mag 21(5):147–151

Hobson A (1969) A manual of standardized terminology, techniques and scoring system for sleep stages of human subjects. Electroencephalogr Clin Neurophysiol 26(6):644

Jansen B (1985) Feature extraction methods for EEG analysis. Electroencephalogr Clin Neurophysiol 61(3):S222

Kirov R, Moyanova S (2002) Distinct sleep-wake stages in rats depend differentially on age. Neurosci Lett 322(2):134–136 (Web)

Koley B, Dey D (2011) An ensemble system for automatic sleep stage classification using single channel EEG signal. Comput Biol Med 42(12):1186–1195

Koley B, Dey D (2013) Automatic detection of sleep apnea and hypopnea events from single channel measurement of respiration signal employing ensemble binary SVM classifiers. Measurement 46(7):2082–2092

Menon MK, Dandiya PC (1969) Behavioural and brain neurohormonal changes produced by acute heat stress in rats: influence of psychopharmacological agents. Eur J Pharmacol 8:284–291

Morstyn R, Duffy FH, McCarley RW (1983) Altered topography of EEG spectral content in schizophrenia. Electrocephalogr Clin Neurophysiol 56:263–271

Patil SA, Patil SP (1998) Computerised EEG analyser. IETE Tech Rev 15(6):503–507

Sinha RK (2003) Artificial Neural Network detects changes in electro-encephalogram power spectrum of different sleep-wakes in an animal model of heat stress. Med Biol Eng Comput 41:595–600

Sinha RK (2008) EEG power spectrum and neural network based sleep-hypnogram analysis for a model of heat stress. J Clin Monit Comput 22:261–268

Sinha RK, Ray AK (2004) An assessment of changes in open-field and elevated plus-maze behavior following heat stress in rats. Irani Biomed J 8:127–133

Subasi A et al (2005) Automatic recognition of vigilance state by using a wavelet-based artificial neural network. Springer, 25 Jan 2005

Sukanesh R, Harikumar R (2007) Analysis of fuzzy techniques and neural networks (RBF&MLP) in classification of epilepsy risk levels from EEG signals. IETE J Res 53(5):465–474

Sukhorukova S, Ofoghi V, Saleem U, Muecke A, Philippe BM, Huda B, Lévy G (2010) Automatic sleep stage identification: difficulties and possible solution. In: Proceedings of the 4th Australasian workshop on health informatics and knowledge management (HIKM 2010), Brisbane, Australia

Tagluk M, Sezgin N, Akin M (2009) Estimation of sleep stages by an artificial neural network employing EEG, EMG and EOG. J Med Syst 34(4):717–725

Upadhyay PK, Sinha RK, Karan BM (2010) Predicting heat-stressed EEG spectra by self-organising feature map and learning vector quantizers SOFM And LVQ based stress prediction. J Biomed Sci Eng 03(05):529–537

Upadhyay PK, Sinha RK, Karan BM (2010) Detection and analysis of the effects of heat stress on EEG using wavelet transform—EEG analysis under heat stress. J Biomed Sci Eng 405–414

Yang E-J, Shin D-S, Kim E-S (2002) The characteristic analysis of EEG artifacts. J Fuzzy Log Intell Syst 12(4):366–372 (Web)

Recent Advances in Deep Learning Techniques and Its Applications: An Overview

Abhishek Hazra, Prakash Choudhary, and M. Sheetal Singh

Abstract Learning with images and their classification, segmentation, localization, annotation, and abnormally detection is one of the current challenging and exciting task for the researchers. Recently deep learning techniques give excellent performance in Object Detection, Speech Recognition, Abnormality Detection, Business Analysis, and almost all other domains. But one important implication of deep learning techniques can found in Medical Image Analysis. Deep learning techniques beat the human-level performance and come with a better solution in the medical domain. Among different deep learning techniques Convolutional Neural Network, Recurrent Neural Network, Long Short-Term Memory, Deep Belief Network models are topmost priority for the researchers. In this paper, we briefly examine different application area of deep learning techniques and some current state-of-the-art performances of it. Moreover, we also discuss some of the limitations of Deep Learning techniques. As expected this paper creates a clear understanding of Deep Learning techniques and its applications.

Keywords Deep learning · CNN · RNN · Health care

1 Introduction

Neuroscience researchers examine that visual representation of brain can be done in two pathways, Dorsal pathway and Ventral pathway. The information of location movement observation follows the dorsal pathway and detection, color, texture shape,

A. Hazra
Indian Institute of Technology (ISM), Dhanbad 826004, Jharkhand, India
e-mail: abhishekhazra.18dr0018@cse.iitism.ac.in

P. Choudhary (✉) · M. Sheetal Singh
National Institute of Technology Manipur, Imphal 795004, India
e-mail: choudharyprakash@nitmanipur.ac.in

M. Sheetal Singh
e-mail: msheetalsingh@live.com

A. A. Rizvanov et al. (eds.), *Advances in Biomedical Engineering and Technology*, Lecture Notes in Bioengineering,
https://doi.org/10.1007/978-981-15-6329-4_10

such information follow the ventral pathway. In 1959 Hubel and Wiesel (Hubel and Wiesel 1968) first observe that visual cortex's in brain mainly responsible for detecting lights. Inspired by this innovation LeCun et al. (Le Cun et al. 1990) in 1990 first make a handwritten character recognition system. This was the first neural network architecture which tried to learn and recognize characters. Since 90s many researchers come with their deep advance architecture which solves many real-time applications. AlexNet (Krizhevsky et al. 2012), VGGNet (Simonyan and Zisserman 2015), ResNet (He et al.2015c), models break all the previous performance in various domains.

Recently Deep Learning Techniques come with a better solution for analyzing different kinds of data. The idea of deep learning has a very old history. Because of its high computation power and colossal amount of data, deep learning techniques were not so popularly used back then. But in late 20s deep learning techniques accelerate its performance with the help of Graphical Processing Unit and massive amount of data. Deep Learning techniques gives state-of-the-art performance in almost all the domains like Object Detection, Speech Recognition, Fraud Detection, Face Recognition, Sentiment Analysis. Currently, deep learning techniques give excellent performance in medical image analysis. From 2015, research in deep learning for medical domain has increases exponentially. The number of research papers, journals, and articles in this domain are increasing day by day. It means this field is gaining interests gradually at a very rapid rate. There are special issues in almost all the journals with deep learning as a keyword. These are the primary motivation for us to work in this particular field. In this paper, we briefly examine different application area of deep learning techniques. Our search list contains "Deep Learning" either in the title or the keyword in various articles, journals good conference proceedings which are mainly focused on this field only. In this review, we try to scrutinize all the popular related papers published till 30 March 2018. We expect that this paper makes a brief overview of deep learning techniques in all the domains.

Our main aim for this review:

- To show deep learning techniques and its performance.
- To display current research Scenarios.
- To highlight some of the current research challenges.

The rest of the paper structured as follows. In Sect. 2 we introduce some essential deep learning techniques. Section 3 describes some popular application area of deep learning techniques. Section 4 narrates the deep learning implementation tools. Section 5 shows an essential overview discussion in medical imaging. Finally, conclusion and feature work were speculated in Sect. 6.

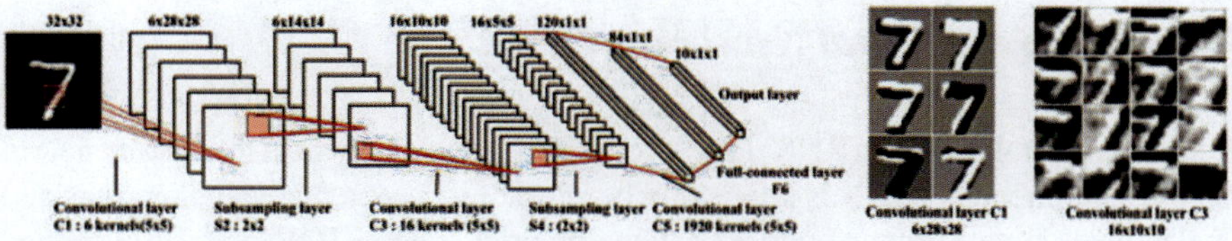

Fig. 1 Famous LeNet-5 network architecture with intermediate visual representation

2 Overview of Deep Learning Techniques

In recent times deep learning techniques are widely used in all the domains like Object Detection (Yoo et al. 2015), Sentiment Analysis (Wang et al. 2016c), Medical Image Analysis (Lo et al. 1995), Speech Recognition (Waibel et al. 1990), Self-Driving car, Automatic Machine Translation, Automatic Text Generation (He et al. 2015b), advertising, and many more. With the advancement of GPU based systems, several deep learning techniques are also introduced to address different kinds of problems. Convolutional Neural Network, Recurrent Neural Network, Restricted Boltzmann Machine, LSTM, Deep Autoencoder networks are current state of the learning algorithm. In this section, our main aim is to address some of the most popular deep learning techniques which create a huge impact on current research.

2.1 Convolutional Neural Network

Convolutional Neural network (CNN) (Vincent et al. 2008) is one of the most popular learning algorithms in computer vision field. Currently, many researchers come with their individual CNN architecture though there is a similarity between all the networks. Basically convolutional neural network consists of four types of layers. Convolutional layer, Activation layer, Pooling layer, and Fully Connected layer. LeNet-5 as presented in Fig. 1, was the first generalizes neural network architecture (Le Cun et al. 1990) which is still popularly used in current times. Convolutional layer takes information from the input data and produces a feature map with the help of kernels. The number of convolutional layers is vary from architecture to architecture. Generally first level of convolutional layers learns low-level features like dark and bright pixels, second layer of convolutional layers may learn horizontal edges, vertical edges, next level of convolutional layer learns some more complex functions like ears, nose mouth. As the number of layers increased neural network learns even more complex functions like face, object, and characters. These feature maps are passed through a nonlinear activation function which gives acceleration to the CNN to understand complex functions. Finally one or more number of fully connected layers which summarize the learnable information and put into a softmax classifier. The softmax classifier gives the output probability of each and every class for the given input.

2.2 Recurrent Neural Network

Recurrent neural network (RNN) (Poultney 2006) is one of the most interesting neural network architecture. RNN is interesting because of its use in many applications and is also notable performance under challenging applications. RNN is mostly trained by a sequence of data like sentence and make subsequent similar sentences which are most likely used in chatbots. RNN is widely used in several applications like image captioning, generating review, generating feedback, generating Music. Feedforward neural networks are not designed for sequence/time-series data, hence results with time series/sequence data are inadequate and moreover, they cannot design for storing memory. To address this problem, recurrent neural network was designed. Recurrent neural networks are the type of networks designed for capturing information from sequence/time series data. In RNN sequences is feed as current input, calculate a hidden state, and compute the output. For the next time step, the network takes new information as well as information from the previously hidden state to compute the current hidden state to predict the next output. Finally, a loss functions to improve the accuracy of RNN. This types of networks use in time series prediction like weather forecasting, Stock prediction, and sequence of data generating application like music, video.

2.3 Long Short-Term Memory

Long short-term memory (LSTM) is a type of recurrent neural network. LSTM is the next logic step in the progression of neural network Learning. It is technique of learning sequence of data or video frame and capable of learning long-term dependencies. One interesting idea of weighted self-loop to introduce path where Gradient flow for a long time in Long short-term memory. By addition of self-loops current hidden layers are controlled by previously computed hidden layers. Even for fixed parameters, the time scale of integration can be changed according to the input time is the output of this model. Different researchers found that LSTM networks are incredibly successful in many applications such as speech recognition, music generation, machine translation, image captioning, handwritten recognition.

There are several deep learning techniques like Deep Autoencoder (Salakhutdinov and Hinton 2009; Rifai et al. 2011; Masci et al. 2011; Chen et al. 2013), Boltzmann Machine (Younes 1999; Center Berkeley 2016), and Deep belief networks are also popularly used in various domains.

3 Applications of Deep Learning Techniques

Currently, deep learning techniques are giving an excellent performance in Action Recognition, Significant Data Analysis, Sentiment Analysis, Medical Image Analysis, Character Recognition, Image Classification, Object Detection, Object Tracking, Pose Estimation, Visual Salient Detection Sense leveling, Speech Recognition, Natural Language Processing, Remote Sensing. In this segment, we try to introduce some of the current research application of deep learning techniques as illustrated in Fig. 2.

3.1 Action and Gesture Recognition

One of the interesting application of Deep learning techniques is in action recognition. Most of the Companies are using some action recognition for their internal security purpose. Because of its high demand and current challenges attract deep learning researchers in this field. This field has been examining from last decades and reported huge progress within the computer vision field. RNN (Le Cun et al. 1990), LSTM (Rowley et al. 1998), 3D convolutional neural network (Yang et al. 2017a), pertained features are the topmost priorities of deep learning researchers. Mainly three types of network model were used for action and texture recognition. 3D convolutional layer (de Brebisson and Montana 2015; Gao et al. 2015; Lo et al. 1995; Chen et al. 2017; Gao 2016; Tarando et al. 2016; Zhu et al. 2017; Xu et al. 2016b; Dittrich et al. 2011), motion-based input feature (Alexe et al. 2012; Zhao et al. 2016; Xu et al. 2016b; Chen et al. 2017; Hinton and Salakhutdinov 2006) and temporal methods which is the combination of 2d or 3d CNN networks. Though RNN is one of the important deep learning architecture particularly used for this task, this kind of network suffers from short-term memory loss. To address this problem LSTM (Anavi et al. 2015) was introduced. LSTM works in the inner layer of RNN. B-RNN (Goodfellow et al. 2014), H-RNN (He et al. 2015a), D-RNN (Janowczyk et al. 2017) are some extended, modified version of LSTM. Moreover fusion-based deep learning techniques (Lo et al. 1995; Pluim et al. 2003; Ngo et al. 2017; Yan et al. 2014; Chen et al. 2017; Center Berkeley 2016) are also popularly used for action recognition.

3.2 Deep Learning for Big Data

Big Data refers considerable amount of datasets (Philip Chen and Zhang 2014) which can synthesize specific patterns. Deep learning techniques are widely used for analyzing big data and succeeded to find certain hidden pattern that was impossible so far. Proper knowledge plays a critical role for success in many companies as well. This need can be satisfied by combining this two domain: Deep Learning and Big Data.

Big companies like Facebook, Google, and Yahoo used deep learning techniques and getting benefited from it. The analysis of big data can be subdivided into three phase: Big Data processing, Big Data storage, Big Data management. For better decision making we need large and good quality of data which requires data preprocessing (Hinton and Salakhutdinov 2006; Witten et al. 2016; Riabov and Liu 2006; Han et al. 2014; Siddiqa et al. 2016; Michael and Miller 2013; McAfee et al. 2012). Data cleansing, transmission sequencing are some of the intermediate steps of data processing. Storing big data in PT scale is not a feasible solution for researchers and interesting communities. Though recent advances of cloud computing anyhow reduce some problem. The main interesting thing in it is to create a storage management system which provides enough data and utilizes information retrieval (Dittrich et al. 2011), replication, indexing are the intermediate steps of storing big data (Li et al. 2008; Deng et al. 2014; Chen 2010) processing is one of the challenging tasks. There are several processing issues in managing big data. Recently AI companies invested a huge amount of money in big data processing (Buza et al. 2014; Porkar 2012; Jafari et al. 2016; Waibel et al. 1990). For addressing such problem, many machine learning (ML) researchers come with their handcraft feature learning techniques but fails to give a good result in practical aspects. But deep learning techniques give a better solution for handling both labeled and unlabeled datasets.

3.3 Deep Learning for Sentiment Analysis

Deep Learning Techniques (Morin and Bengio 2005; Mikolov et al. 2013a, b; Mnih and Kavukcuoglu 2013; Moraes et al. 2013; Johnson and Zhang 2015) are also useful for analyzing emotions. Though understanding of human emotion and explain it in terms of words is a challenging task for the computer vision researchers. Sometimes words are not enough to correctly explain our emotions as some emotions has no language translation. But deep learning techniques assistance to understand human emotional data which helps to take optimal decisions. There are mainly two basic approaches of sentiment analysis. Lexicon-based approach and AI-based Approach. In lexicon-based approach for given sentence words are split into small tokens also knows as tokenization. Bag of words is the count the number of frequencies of each word. Based on this it decides positive and negative sentences. AI-based deep learning techniques are the current trend research in sentiment analysis. For a large dataset deep learning techniques were trained and also be applied for real-time applications. CNN (Kalchbrenner et al. 2014; Kim 2014; dos Santos and Gatti 2014; Wang et al. 2016b, c; Guggilla et al. 2016; Mishra et al. 2017; Bengio et al. 2013; Qian et al. 2015), RNN (Tang et al. 2015; Guan et al. 2016; Yu and Jiang 2016) LSTM (Tang et al. 2016; Salakhutdinov and Hinton 2009; Qian et al. 2017; Li et al. 2017; Wang et al. 2015c; Huang et al. 2017; Le and Mikolov 2014; Glorot et al. 2011) models are popularly used in this task. Though deep learning-based sentiment analysis is a hard process of computation but these techniques give better result than traditional techniques. Document-level Sentiment classification (Wang et al. 2015c; Williams and Zipser 1989; Liu and Zhang 2017; Masci et al. 2011), Sentence level sentiment classification (Loshchilov and Hutter 2016; Wang et al. 2016c), aspect level sentiment classification (Liu and Zhang 2017; Yang et al. 2017b) are some of the intermediate steps of sentiment analysis. Large social media companies like Facebook twitter google has deep learning-based approaches for analyzing customers perspective.

3.4 *Deep Learning for Medical Image Analysis*

Analysis of medical images and their classification localization segmentation annotation abnormally detection are one of the current research interest. Since 2014 after the development of GPU based systems deep learning techniques give excellent performance in medical domain. Many researchers collect their data and make available for research purpose. Different research shows that CNN (Suk and Shen 2016; de Brebisson and Montana 2015; Choi and Jin 2016; Zhang et al. 2015a; Birenbaum and Greenspan 2016; Brosch et al. 2016) based deep learning models are most widely used in medical engineering. Apart from CNN, RBM, RNN (Stollenga et al. 2015; Andermatt et al. 2016) Autoencoder based models are also popularly different health care applications like brain image analysis(Sarraf and Tofighi 2016; Chen et al. 2016; Ghafoorian et al. 2016a, b), retinal image analysis (Gulshan et al. 2016; Zilly et al. 2017; Chen et al. 2015; Abràmoff et al. 2016; Lu et al. 2016a; van Grinsven et al. 2016; Gulshan et al. 2016; Gao et al. 2015), chest x-ray image analysis (Lo et al. 1995; Anavi et al. 2015; Anavi et al. 2016; Lin et al. 2014; Vaillant et al. 1994; Hwang et al. 2016; Kim and Hwang 2016; Rajkumar et al. 2017; Yang et al. 2017a), CT chest x-ray image analysis (Wang et al. 2017; Charbonnier et al. 2017; Shen et al. 2015a; Chen et al. 2017; Dou et al. 2017; Setio et al. 2016; Sun et al. 2016; Anthimopoulos et al. 2016; Christodoulidis et al. 2017; Gao 2016; Tarando et al. 2016; van Tulder and de Bruijne 2016; Avendi et al. 2016), pathology image analysis (Xie et al. 2016; Wang et al. 2016e; Xu et al. 2016a, b; Chang et al. 2017; Çiçek et al. 2016; Chen et al. 2017; Janowczyk et al. 2017; Hubel and Wiesel 1968), cardiac image analysis (Emad et al. 2015; Ngo et al. 2017; Poudel et al. 2016; Tran 2016; Prasoon et al. 2013), abdominal image analysis (Li et al. 2015; Vivanti et al. 2015; Wang and Gupta 2015; Yu et al. 2017; Zhu et al. 2017; Zhao et al. 2016), musculoskeletal image analysis (Shen et al. 2015b; Suzani et al. 2015; Antony et al. 2016). Figure 2 shows the pictorial application area of deep learning techniques in medical science.

3.5 *Deep Learning for Text Detection and Recognition*

Character and text recognition is one of the current time research and had been studied in the computer vision field from long time. Optical character recognition also popularly known as OCR recognition is one of the fundamental academic research. CNN is the main building block architecture for recognition of characters. We also divide this

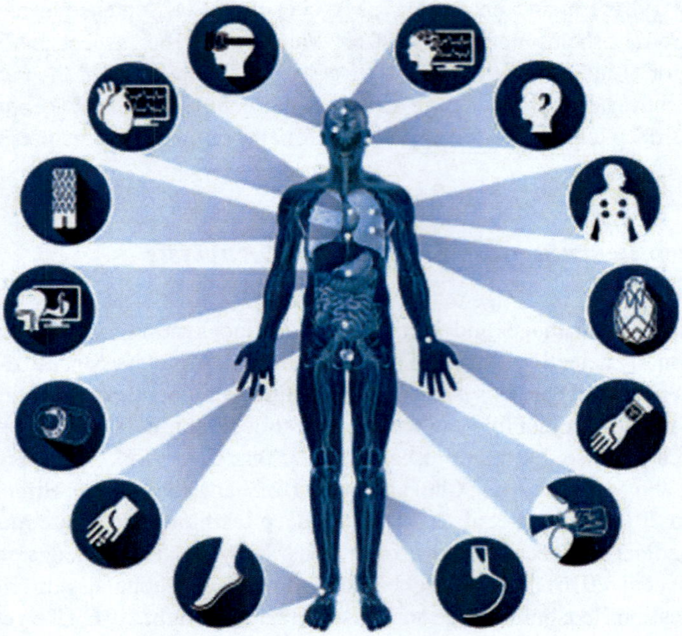

Fig. 2 Application area of deep learning techniques in health care sector (Masci et al. 2011)

task into three subcategories: Text detection, text recognition from small region, and combination of text detection and recognition. CNN models are widely used for text detection (Zhang et al. 2015b). There are several standard handwritten and optical character available in almost all the languages in worldwide. One improvement of this work is the combination of CNN and Maximally Stable External Regions (MSER) (Goodfellow et al. 2014; Zhang et al. 2015b), Bag-of-Words (BoG) and CNN based sliding Non-Maximal suppression (NMS) (He et al. 2015b) based CNN structures are also popularly used. Similar to text detection, Text recognition is also popular research area Good fellow et al. (Shi et al. 2015) proposed a multilevel CNN classifier for character recognition from multidigit input string. Conditional Random Fields (CRF) based CNN, feature extraction based CNN (Gers et al. 2000), Sliding window-based LSTM (Jaderberg et al. 2014), and feature extraction based text recognition (Jaderberg et al. 2015) are also some popular techniques of text recognition. End-to-End text spotting with bounding box (Lawrence et al. 1997; Simard et al. 2003) is also the popular research interest in computer vision field. According to Ethnologies catalog of world languages, there is 6909 number of registered script language exist and most of the counties have their own official languages. Thus character recognition field has its own separate interest. Currently, automatic character recognition is used in machine translation, postal systems, identification recognition, image translations.

3.6 Deep Learning for Image Classification

Deep Learning Techniques gives a tremendous performance for classification of the object from a large dataset. Though CNN was used for image classification long back (He et al. 2015b), but it creates a remarkable performance in recent times with the advances of Graphical processing units and a large amount of data (He et al. 2015b; Lawrence et al. 1997; Everingham et al. 2014; Deng et al. 2009). In 2012 AlexNet creates a huge impact for mage classification of large-scale images which also wins the ILSVRC 2012 challenge. Taking this motivation of the work many researchers take their interest in increasing the classification accuracy by tuning the hyper parameters in the neural network. Several researchers come up with their new classification technique which sometimes works well. Hierarchy based image classification is a common technique for classifying a large class of images (Wang et al. 2015a). Hierarchy of CNN in discriminate feature learning for sharing their hierarchy of information to share among the classes (Xiao et al. 2014). Fine-gained feature learning (Yan et al. 2014), trained hierarchal network (Nilsback and Zisserman 2008), embedding CNN into a subcategory of the hierarchy methods are also popularly used to reducing the classification error. Subcategory image classification datasets (Yu and Grauman 2014; Yang et al. 2015) also takes lots of interest. CNN (Uijlings et al. 2013), R-CNN (Pluim et al. 2003; Lin et al. 2015b) Deep LAC (Lin et al. 2015a) based object part classification is also popularly used. Create a subnetwork, localize a region, and estimate the predictive class (Krause et al. 2015) also helps to improve classification accuracy. Both supervised and unsupervised learning techniques on annotated data are popular to fine-tuned (Dalal and Triggs 2005) the class. Ensemble the localization (Rowley et al. 1998), co-segmentation (Rowley et al. 1998), leveling by simplicity, visual attention based CNN models are also popularly used for image classification.

3.7 Deep Learning for Object Detection

One of the current time computer vision problems is object detection. There are many research issues for detecting objects from video or images. Though CNN based object detection techniques started in early 90s. However, due to lack of computational power and a small amount of data breaks the progress of CNN-based system. Recently in 2012 after the huge success in ImageNet challenge (Deng et al. 2009) this field gets back interested from the research community. In earlier times CNN based object detection (Lin et al. 2014; Vaillant et al. 1994) using sliding window were so popular, but these techniques require high computational power which makes them unreliable for massive datasets. Like VOC (He et al. 2015b), IMAGENET (Lawrence et al. 1997), MSCOCO (Alexe et al. 2012). To address this issue, Object proposal based technique introduced. Different literature (Carreira and Sminchisescu 2012; Pluim et al. 2003) shows that object proposal based techniques are the most generic measure of the test, a generic window is used to propose whether an object present or not., then passes it to next level of generic detection to understand objects are belonging to the same class or not. Region-based CNN (R-CNN) (Sermanet et al. 2013) is one of the popular objection technique. A pertained CNN is used on a selective search to extract feature and SVM used to classify objects. Several improvements were done to improve the performance. Feature extraction (He et al. 2015a; Carreira and Sminchisescu 2012; Sermanet et al. 2013), SPP net (Yoo et al. 2015), pyramidal R-CNN (Felzenszwalb et al. 2010), bounding box (Redmon et al. 2016), bootstrapping (Liu et al. 2015), Yolo, SDD, top-down search (Gidaris and Komodakis 2015) methods are introduced for better performance in dynamically challenging environments (Liu et al. 2015; Loshchilov and Hutter 2016; Lu et al. 2016b).

3.8 Deep Learning for Object Tracking

Another success of deep learning techniques can be found in object tracking. CNN and RNN based models are popular in this particular task. CNN based object tracking, target specific (Li et al. 2014) object tracking, temporal adaptation mechanism (Li et al. 2014), tracking based (Plis et al. 2014), similarity-based visual tracking (Hong et al. 2015), are most popular. In almost all the small/big companies and institutes use some kinds of tracking system for detecting persons, counting vehicles, finding missing elements, video surveillance.

4 Software and Implementation Tools

Table 1 shows some of the deep learning implementation tools. Keras, Tensorflow, Theano, PyTorch tools are widely used for implementation of AI techniques. Most of the tools use python as their underlying framework. The number of libraries for supporting the python is increased with the acceleration of GPU based systems. One of the main reasons for development of deep learning techniques is the Nvidia Corporation. Almost all the researchers use GPU based systems for accelerating their training time. In Table 1 we try to introduce the interdependencies of different learning techniques.

5 Discussion Overview

Our research query also exploits one common problem "what are the best possible ways of training a neural network." To find this answer to the question, we examine some breakthrough performances and general intuitions for understanding a neural network. We found out that there are mainly two ways we can train our neural network: first is to create own neural network architecture and the second is by using

Table 1 Some of the popular deep learning implementation tools

Tools	Platform	Support	Interface
Caffe (Williams and Zipser 1989)	Windows, Linux, Mac OSX	CNN, RNN	Python, C++, Matlab, Cuda
Tensorflow (Salakhutdinov and Hinton 2009)	Windows, Linux, Mac OSX, Android	Almost support all deep learning techniques	Python
Theano (Younes 1999)	Windows, Linux, Mac OSX	Almost support all deep learning techniques	Python, Cuda
Torch (Microsoft 2016)	Windows, Linux, Mac OSX	Almost support all deep learning techniques	Lua
Keras (Delakis and Garcia 2008)	Windows, Linux, Mac OSX	Almost support all deep learning techniques	Cross-platform, Cuda
PyTorch (Xu and Su 2015)	Linux, Mac OSX	Almost support all deep learning techniques	Python, C, Cuda

transfer learning. In the previous section, we already examine some unique deep learning architecture. Now in this segment, we will try to understand the transfer learning techniques. There are mainly three ways of using pre-trained model and train neural network: Fixed feature extractor, Fine-tuning the model, and pertained the model(Choudhary and Hazra 2019).

Fixed feature extraction: It is one of the early ML algorithms. First, use a technique which summarize the features and then apply on a classifier for predicting output levels. Also, the same way we can train neural network at first choose a convolutional neural network trained on big dataset like ImageNet and by removing last fully connected layer the network can be treated as a fixed feature extractor. Once the feature was extracted then the neural network trains on a classifier for new dataset.

Fine-Tuning the ConvNet: Another important strategy in deep learning is not only retrained as the classifier over new dataset but also to replace and fine-tune the learning experiences of the neural network. It may also be possible to train all the layers or keep some of the earlier layer fixed and fine-tune the upper layers. One notable thing to mention, the earlier layer of convolutional layer contain more generic low-level information's which can be advantageous for new dataset. Different experiments show that layer-wise fine-tuning of a ConvNet for a big data performs better than making a neural network from the sketch. There are certain intuitions when and how to fine-tune a network, deciding choosing a perfect transfer learning technique on a new dataset is a bit challenging task. There are several strategies one should take care, but the important two are the size of the new dataset and similarity between old and new datasets. In the lower level of ConvNet contain a lower level of generic information and upper level of the network contain more specific information related to the dataset. Some thumb rule for fine-tuning the new dataset are:

- If the dataset is small in size, then fine-tuning a ConvNet over a small dataset will not be a good idea as the deep neural network may suffer from overfitting problem. Hence, using a linear classifier on a small dataset might be a good idea.
- If the dataset is large and there is a similarity between two datasets then using a pre-trained model will give more confidence not to be overfitting the network, hence chances of increasing the performance of the network.
- If the new data is small and differ from original data, then using a linear classifier may not always work, instead use of support vector machine classifier may be beneficial for new dataset as the network contain data specific information.
- If the data is large and differs from original data, then fine-tuning a residual neural network sometimes helpful because it is found out that exploring vanishing gradient can lead some problem for weight updation. Even though making a neural network from scratch also works depending on the dataset.

Pretrain Models: Training a neural network on a large image dataset like ImageNet may take ~2–3 weeks for training on a Search Results Web results Graphics processing unit (GPU) based systems. Researchers sometimes release their final work for helping others. Using a pretrain model a fuse of different deep neural network sometimes also beneficial for Training a neural network.

6 Conclusion and Future Work

All the challenging issues discussed in the previous section were not been tackled yet by the researchers. Though some successes were achieved by using deep learning techniques. From this survey one observation we can make, many researchers' uses pretrained networks for evaluate their model. ResNet, VGGNet, GoogleNet networks are the top listed architectures for the researchers. Even though it is not clear that these models will work in all the domains. Recently some good results were achieved by making a fused model of different networks. Though there are some limitations of deep learning techniques, still it is widely used for solving real-time problems. Convolutional neural networks, RNN, LSTM networks create a benchmark performance in computer vision, robotics, speech recognition, and all the domains. In this literature, we also try to introduce the capabilities of different deep learning techniques. As this research field is new, there is a massive gap for improvement. In conclusion, we can say deep learning techniques are the current state of the learning algorithms. We expect that, in feature by using deep learning techniques researchers can solve many unsolved problems. Our work still in progress, in recent features we are trying to detect different chest diseases by using deep learning techniques.

Acknowledgements The author would like to thank unacquainted reviewers for their valuable comments. Author would also like to thank National Institute of Technology, Manipur and also department of Computer Science and Engineering for providing Lab and required infrastructure.

References

Abràmoff MD, Lou Y, Erginay A, Clarida W, Amelon R, Folk JC, Niemeijer M (2016) Improved automated detection of diabetic retinopathy on a publicly available dataset through integration of deep learning. Invest Ophthalmol Vis Sci 57(13):5200–5206. https://doi.org/10.1167/iovs.16-19964

Alexe B, Deselaers T, Ferrari V (2012) Measuring the objectness of image windows. In: PAMI

Anavi Y, Kogan I, Gelbart E, Geva O, Greenspan H (2015) A comparative study for chest radiograph image retrieval using binary texture and deep learning classification. In: 2015 37th annual international conference of the IEEE engineering in medicine and biology society (EMBC), Milan, pp 2940–2943. https://doi.org/10.1109/EMBC.2015.7319008

Anavi Y, Kogan I, Gelbart E, Geva O, Greenspan H (2016) Visualizing and enhancing a deep learning framework using patients age and gender for chest x-ray image retrieval. In: Proceedings of the SPIE 9785, medical imaging 2016: computer-aided diagnosis, vol 978510, 7 July 2016. https://doi.org/10.1117/12.2217587

Andermatt S, Pezold S, Cattin P (2016) Multi-dimensional gated recurrent units for the segmentation of biomedical 3D-data. In: Proceedings of the deep learning in medical image analysis (DLMIA). Lecture Notes in Computer Science, vol 10008, pp 142–151

Anthimopoulos M, Christodoulidis S, Ebner L, Christe A, Mougiakakou S (2016) Lung pattern classification for interstitial lung diseases using a deep convolutional neural network. IEEE Trans Med Imaging 35(5):1207–1216. https://doi.org/10.1109/TMI.2016.2535865

Antony J, McGuinness K, Connor NEO, Moran K (2016) Quantifying radiographic knee osteoarthritis severity using deep convolutional neural networks. arXiv:1609.02469

Avendi M, Kheradvar A, Jafarkhani H (2016) A combined deep-learning and deformable-model approach to fully automatic segmentation of the left ventricle in cardiac MRI. Med Image Anal 30:108–119

Bengio Y, Courville A, Vincent P (2013) Representation learning: a review and new perspectives. IEEE Trans Pattern Anal Mach Intell (2013)

Birenbaum A, Greenspan H (2016) Longitudinal multiple sclerosis lesion segmentation using multi-view convolutional neural networks. In: Proceedings of the deep learning in medical image analysis (DLMIA). Lecture notes in computer science, vol 10008, pp 58–67. https://doi.org/10.1007/978-3-319-46976-8_7

Brosch T, Tang LY, Yoo Y, Li DK, Traboulsee A, Tam R (2016) Deep 3D convolutional encoder networks with shortcuts for multiscale feature integration applied to multiple sclerosis lesion segmentation. IEEE Trans Med Imaging 35(5):1229–1239. https://doi.org/10.1109/TMI.2016.2528821

Buza K, Nagy GI, Nanopoulos A (2014) Storageoptimizing clustering algorithms for high-dimensional tick data. Expert Syst Appl 41:4148–4157

Carreira J, Sminchisescu C (2012) CPMC: automatic object segmentation using constrained parametric min-cuts. In: PAMI

Center Berkeley (2016) Caffe. http://caffe.berkeleyvision.org/

Chang H, Han J, Zhong C, Snijders A, Mao J-H (2017) Unsupervised transfer learning via multi-scale convolutional sparse coding for biomedical applications. IEEE Trans Pattern Anal Mach Intell https://doi.org/10.1109/TPAMI.2017.2656884

Charbonnier J-P, van Rikxoort EM, Setio AAA, Schaefer-Prokop CM, van Ginneken B, Ciompi F (2017) Improving airway segmentation in computed tomography using leak detection with

convolutional networks. Med Image Anal 36:52–60. ISSN 1361-8415. https://doi.org/10.1016/j.media.2016.11.001
Chen X (2010) Google big table
Chen W, Corso JJ (2015) Action detection by implicit intentional motion clustering. In: Proceedings of the IEEE international conference on computer vision, pp 3298–3306
Chen J, Chen Y, Xiaoyong D, Li C, Jiaheng L, Zhao S, Zhou X (2013) Big data challenge: a data management perspective. Front Comput Sci 7(2):157–164
Chen H, Dou Q, Yu L, Heng P-A (2016) Voxresnet: deep voxelwise residual networks for volumetric brain segmentation. arXiv:1608.05895
Chen S, Qin J, Ji X, Lei B, Wang T, Ni D, Cheng J-Z (2017) Automatic scoring of multiple semantic attributes with multi-task feature leverage: a study on pulmonary nodules in CT images. IEEE Trans Med Imaging 36(3):802–804. https://doi.org/10.1109/TMI.2016.2629462
Choi H, Jin KH (2016) Fast and robust segmentation of the striatum using deep convolutional neural networks. J Neurosci Methods 274:146–153. https://doi.org/10.1016/j.jneumeth.2016.10.007
Choudhary P, Hazra A (2019) Chest disease radiography in twofold: using convolutional neural networks and transfer learning. Evol Syst 1–13
Christodoulidis S, Anthimopoulos M, Ebner L, Christe A, Mougiakakou S (2017) Multisource transfer learning with convolutional neural networks for lung pattern analysis. IEEE J Biomed Health Inform 21(1):76–84. https://doi.org/10.1109/JBHI.2016.2636929
Çiçek Ö, Abdulkadir A, Lienkamp SS, Brox T, Ronneberger O (2016) 3D U-Net: learning dense volumetric segmentation from sparse annotation. In: Proceedings of the medical image computing and computer-assisted intervention. Lecture notes in computer science, vol 9901. Springer, pp 424–432. arXiv:1606.06650v1
Dalal N, Triggs B (2005) Histograms of oriented gradients for human detection. In: CVPR
de Brebisson A, Montana G (2015) Deep neural networks for anatomical brain segmentation. In: Proceedings of the computer vision and pattern recognition, pp 20–28
Delakis M, Garcia C (2008) Text detection with convolutional neural networks. In: VISAPP
Deng J, Dong W, Socher R, Li L, Li K, Li F (2009) Imagenet: a large-scale hierarchical image database. In: CVPR
Deng L, Yu D et al (2014) Deep learning: methods and applications. Found Trends R Signal Process 7(3–4):197–387
Dittrich J, Blunschi L, Salles MAV (2011) Movies: indexing moving objects by shooting index images. Geoinformatica 15(4):727–767
dos Santos CN, Gatti M (2014) Deep convolutional neural networks for sentiment analysis for short texts. In: Proceedings of the international conference on computational linguistics (COLING 2014)
Dou Q, Chen H, Yu L, Qin J, Heng PA (2017) Multilevel contextual 3-D CNN's for false positive reduction in pulmonary nodule detection. IEEE Trans Biomed Eng 64(7):1558–1567. https://doi.org/10.1109/tbme.2016.2613502
Emad O, Yassine IA, Fahmy AS (2015) Automatic localization of the left ventricle in cardiac MRI images using deep learning. In: Proceedings of the IEEE engineering in medicine and biology society, pp 683–686. https://doi.org/10.1109/embc.2015.7318454
Everingham M, Eslami SA, Van Gool L, Williams CK, Winn J, Zisserman A (2014) The pascal visual object classes challenge: a retrospective. In: IJCV
Felzenszwalb PF, Girshick RB, McAllester D, Ramanan D (2010) Object detection with discriminatively trained part-based models. In: PAMI
Gao M et al (2016) Segmentation label propagation using deep convolutional neural networks and dense conditional random field. In: 2016 IEEE 13th international symposium on biomedical imaging (ISBI), Prague, pp 1265–1268. https://doi.org/10.1109/ISBI.2016.7493497
Gao X, Lin S, Wong TY (2015) Automatic feature learning to grade nuclear cataracts based on deep learning. IEEE Trans Biomed Eng 62(11):2693–2701. https://doi.org/10.1109/TBME.2015.2444389

Gers FA, Schmidhuber J, Cummins F (2000) Learning to forget: continual prediction with LSTM. In: Neural computation

Ghafoorian M, Karssemeijer N, Heskes T, van Uden I, Sanchez C, Litjens G, de Leeuw F-E, van Ginneken B, Marchiori E, Platel B (2016a) Location sensitive deep convolutional neural networks for segmentation of white matter hyperintensities. arXiv:1610.04834

Ghafoorian M, Karssemeijer N, Heskes T, van Uden IWM, de Leeuw F-E, Marchiori E, van Ginneken B, Platel B (2016b) Non-uniform patch sampling with deep convolutional neural networks for white matter hyperintensity segmentation. In: Proceedings of the IEEE international symposium on biomedical imaging, pp 1414–1417. https://doi.org/10.1109/isbi.2016.7493532

Gidaris S, Komodakis N (2015) Object detection via a multi-region and semantic segmentation-aware CNN model. In: ICCV

Glorot X, Bordes A, Bengio Y (2011) Domain adaption for large-scale sentiment classification: a deep learning approach. In: Proceedings of the international conference on machine learning (ICML 2011)

Goodfellow IJ, Ibarz J, Arnoud S, Shet V (2014) Multi-digit number recognition from street view imagery using deep convolutional neural networks. In: ICLR

Guan Z, Chen L, Zhao W, Zheng Y, Tan S, Cai D (2016) Weakly-supervised deep learning for customer review sentiment classification. In: Proceedings of the international joint conference on artificial intelligence (IJCAI 2016)

Guggilla C, Miller T, Gurevych I (2016) CNN-and LSTM-based claim classification in online user comments. In: Proceedings of the international conference on computational linguistics (COLING 2016)

Gulshan V, Peng L, Coram M, Stumpe MC, Wu D, Narayanaswamy A, Venugopalan S, Widner K, Madams T, Cuadros J, Kim R, Raman R, Nelson PC, Mega JL, Webster DR (2016) Development and validation of a deep learning algorithm for detection of diabetic retinopathy in retinal fundus photographs. J Am Med Assoc 316:2402–2410. https://doi.org/10.1001/jama.2016.17216

Han H, Wen Y, Chua T-S, Li X (2014) Toward scalable systems for big data analytics: a technology tutorial. IEEE Access 2:652–687

He K, Zhang X, Ren S, Sun J (2015a) Spatial pyramid pooling in deep convolutional networks for visual recognition. In: PAMI

He P, Huang W, Qiao Y, Loy CC, Tang X (2015b) Reading scene text in deep convolutional sequences. In: CoRR

He K, Zhang X, Ren S, Sun J (2015c) Deep residual learning for image recognition. In: CVPR

Hinton GE, Salakhutdinov RR (2006) Reducing the dimensionality of data with neural networks. Science 313(5786):504–507

Hong S, You T, Kwak S, Han B (2015) Online tracking by learning discriminative saliency map with convolutional neural network. In: ICML

Huang M, Qian Q, Zhu X (2017) Encoding syntactic knowledge in neural networks for sentiment classification. ACM Trans Inf Syst

Hubel DH, Wiesel TN (1968) Receptive fields and functional architecture of monkey striate cortex. J Physiol

Hwang S, Kim H-E, Jeong J, Kim H-J (2016) A novel approach for tuberculosis screening based on deep convolutional neural networks. In: Proceedings of the SPIE, medical imaging 2016: computer-aided diagnosis, vol 9785, p 97852W, 24 Mar 2016. https://doi.org/10.1117/12.2216198

Jaderberg M, Vedaldi A, Zisserman A (2014) Deep features for text spotting. In: ECCV

Jaderberg M, Simonyan K, Vedaldi A, Zisserman A (2015) Reading text in the wild with convolutional neural networks. In: IJCV

Jafari M, Wang J, Qin Y, Gheisari M, Shahabi AS, Tao X (2016) Automatic text summarization using fuzzy inference. In: 2016 22nd international conference on automation and computing (ICAC). IEEE, pp 256–260

Janowczyk A, Basavanhally A, Madabhushi A (2017) Stain normalization using sparse autoencoders (STANOSA): application to digital pathology. Comput Med Imaging Graph 57:50–61

Johnson R, Zhang T (2015) Effective use of word order for text categorization with convolutional neural networks. In: Proceedings of the conference of the North American chapter of the association for computational linguistics: human language technologies (NAACL-HLT 2015)

Kalchbrenner N, Grefenstette E, Blunsom P (2014) A convolutional neural network for modelling sentences. In: Proceedings of the annual meeting of the association for computational linguistics (ACL 2014)

Kim Y (2014) Convolutional neural networks for sentence classification. In: Proceedings of the annual meeting of the association for computational linguistics (ACL 2014)

Kim H, Hwang S (2016) Scale-invariant feature Learning using deconvolutional neural networks for weakly-supervised semantic segmentation. arXiv:1602.04984

Krause J, Jin H, Yang J, Fei-Fei L (2015) Fine-grained recognition without part annotations. In: CVPR

Krizhevsky A, Sutskever I, Hinton GE (2012) Imagenet classification with Deep Convolutional neural networks. In: NIPS

Lawrence S, Giles CL, Tsoi AC, Back AD (1997) Face recognition: a convolutional neural-network approach. In: TNN

Le Cun BB, Denker JS, Henderson D, Howard RE, Hubbard W, Jackel LD (1990) Handwritten digit recognition with a back-propagation network. In: NIPS

Le Q, Mikolov T (2014) Distributed representations of sentences and documents. In: Proceedings of the international conference on machine learning (ICML 2014)

Li G, Ooi BC, Feng J, Wang J, Zhou L (2008) Ease: an effective 3-in-1 keyword search method for unstructured, semi-structured and structured data. In: Proceedings of the 2008 ACM SIGMOD international conference on management of data. ACM, pp 903–914

Li H, Li Y, Porikli F (2014) Deeptrack: learning discriminative feature representations by convolutional neural networks for visual tracking. In: BMVC

Li W, Jia F, Hu Q (2015) Automatic segmentation of liver tumor in CT images with deep convolutional neural networks. J Comput Commun 3(11):146–151

Li Z, Zhang Y, Wei Y, Wu Y, Yang Q (2017) End-to-end adversarial memory network for cross-domain sentiment classification. In: Proceedings of the international joint conference on artificial intelligence (IJCAI 2017)

Lin T-Y, Maire M, Belongie S, Hays J, Perona P, Ramanan D, Dollár P, Zitnick CL (2014) Microsoft COCO: common objects in context. In: ECCV (2014)

Lin T-Y, RoyChowdhury A, Maji S (2015a) Bilinear cnn models for fine-grained visual recognition. arXiv:1504.07889

Lin D, Shen X, Lu C, Jia J (2015b) Deep LAC: Deep localization, alignment and classification for fine-grained recognition. In: CVPR

Liu J, Zhang Y (2017) Attention modeling for targeted sentiment. In: Proceedings of the conference of the European chapter of the association for computational linguistics (EACL 2017)

Liu W, Anguelov D, Erhan D, Szegedy C, Reed S (2015) SSD: single shot multibox detector. In: ECCV

Lo SCB, Lou SLA, Lin J-S, Freedman MT, Chien MV, Mun SK (1995) Artificial convolution neural network techniques and applications for lung nodule detection. IEEE Trans Med Imaging 14(4):711–718. https://doi.org/10.1109/42.476112

Loshchilov I, Hutter F (2016) Online batch selection for faster training of neural networks. In: ICLR workshop

Lu Y, Javidi T, Lazebnik S (2016a) Adaptive object detection using adjacency and zoom prediction. In: CVPR

Lu X, Xu D, Liu D (2016b) Robust 3D organ localization with dual learning architectures and fusion. In: Proceedings of the deep learning in medical image analysis (DLMIA). Lecture notes in computer science, vol 10008, pp 12–20. https://doi.org/10.1007/978-3-319-46976-8_2

Masci J, Meier U, Cireşan D, Schmidhuber J (2011) Stacked convolutional auto-encoders for hierarchical feature extraction. In: Proceedings of the international conference on artificial neural networks, pp 52–59

McAfee A, Brynjolfsson E, Davenport TH, Patil DJ, Barton D (2012) Big data. The management revolution. Harv Bus Rev 90(10):61–67

Michael K, Miller KW (2013) Big data: new opportunities and new challenges [guest editors' introduction]. Computer 46(6):22–24

Microsoft (2016) CNTK. https://github.com/Microsoft/CNTK

Mikolov T, Chen K, Corrado G, Dean J (2013a) Efficient estimation of word representations in vector space. In: Proceedings of international conference on learning representations (ICLR 2013)

Mikolov T, Sutskever I, Chen K, Corrado G, Dean J (2013b) Distributed representations of words and phrases and their compositionality. In: Proceedings of the annual conference on advances in neural information processing systems (NIPS 2013)

Mishra A, Dey K, Bhattacharyya P (2017) Learning cognitive features from gaze data for sentiment and sarcasm classification using convolutional neural network. In: Proceedings of the annual meeting of the association for computational linguistics (ACL 2017)

Mnih A, Kavukcuoglu K (2013) Learning word embeddings efficiently with noise-contrastive estimation. In: Proceedings of the annual conference on advances in neural information processing systems (NIPS 2013)

Moraes R, Valiati JF, Neto WP (2013) Document-level sentiment classification: an empirical comparison between SVM and ANN. Expert Syst Appl

Morin F, Bengio Y (2005) Hierarchical probabilistic neural network language model. In: Proceedings of the international workshop on artificial intelligence and statistics

Ngo TA, Lu Z, Carneiro G (2017) Combining deep learning and level set for the automated segmentation of the left ventricle of the heart from cardiac cine magnetic resonance. Med Image Anal 35:159–171. https://doi.org/10.1016/j.media.2016.05.009

Nilsback M-E, Zisserman A (2008) Automated flower classification over a large number of classes. In: ICVGIP

Philip Chen CL, Zhang C-Y (2014) Data-intensive applications, challenges, techniques and technologies: a survey on big data. Inf Sci 275:314–347

Plis SM, Hjelm DR, Salakhutdinov R, Allen EA, Bockholt HJ, Long JD, Johnson HJ, Paulsen JS, Turner JA, Calhoun VD (2014) Deep learning for neuroimaging: a validation study. Front Neurosci. https://doi.org/10.3389/fnins.2014.00229

Pluim JP, Maintz JA, Viergever M et al (2003) Mutual-information-based registration of medical images: a survey. In: T-MI

Porkar P (2012) Sensor networks challenges. In: 11th international conference on data networks, DNCOCO '12, 7–9 Sept 2012

Poudel RPK, Lamata P, Montana G (2016) Recurrent fully convolutional neural networks for multi-slice MRI cardiac segmentation. arXiv:1608.03974

Poultney C et al (2006) Efficient learning of sparse representations with an energy-based model. In: Proceedings of the advances in neural information processing systems, pp 1137–1144

Prasoon A, Petersen K, Igel C, Lauze F, Dam E, Nielsen M (2013) Deep feature learning for knee cartilage segmentation using a triplanar convolutional neural network. In: Proceedings of the medical image computing and computer-assisted intervention. Lecture notes in computer science, vol 8150, pp 246–253. https://doi.org/10.1007/978-3-642-40763-5_31

Qian Q, Tian B, Huang M, Liu Y, Zhu X, Zhu X (2015) Learning tag embeddings and tag-specific composition functions in the recursive neural network. In: Proceedings of the annual meeting of the association for computational linguistics (ACL 2015)

Qian Q, Huang M, Lei J, Zhu X (2017) Linguistically regularized LSTM for sentiment classification. In: Proceedings of the annual meeting of the association for computational linguistics (ACL 2017)

Rajkumar A, Lingam S, Taylor AG, Blum M, Mongan J (2017) High-throughput classification of radiographs using deep convolutional neural networks. J Digit Imaging 30:95–101. https://doi.org/10.1007/s10278-016-9914-9

Redmon J, Divvala S, Girshick R, Farhadi A (2016) You only look once: unified, real-time object detection. In: CVPR

Riabov A, Liu Z (2006) Scalable planning for distributed stream processing systems. In: ICAPS, pp 31–41

Rifai S, Vincent P, Muller X, Glorot X, Bengio Y (2011) Contractive auto-encoders: explicit invariance during feature extraction. In: Proceedings of the international conference on machine learning, pp 833–840

Rowley HA, Baluja S, Kanade T (1998) Neural network-based face detection. In: PAMI

Salakhutdinov R, Hinton GE (2009) Deep Boltzmann machines. In: Proceedings of the international conference on artificial intelligence and statistics, vol. 1, art no 3

Sarraf S, Tofighi G (2016) Classification of Alzheimer's disease using fMRI data and deep learning convolutional neural networks. arXiv:1603.08631

Sermanet P, Eigen D, Zhang X, Mathieu M, Fergus R, LeCun Y (2013) Overfeat: integrated recognition, localization and detection using convolutional networks

Setio AAA et al (2016) Pulmonary nodule detection in CT images: false positive reduction using multi-view convolutional networks. IEEE Trans Med Imaging 35(5):1160–1169. https://doi.org/10.1109/TMI.2016.2536809

Shen W, Zhou M, Yang F, Yang C, Tian J (2015a) Multi-scale convolutional neural networks for lung nodule classification. In: Ourselin S, Alexander D, Westin CF, Cardoso M (eds) Information processing in medical imaging. IPMI 2015. Lecture notes in computer science, vol 9123. Springer, Cham

Shen W, Yang F, Mu W, Yang C, Yang X, Tian J (2015b) Automatic localization of vertebrae based on convolutional neural networks. In: Proceedings of the SPIE on medical imaging, vol 9413, p 94132E

Shi B, Bai X, Yao C (2015) An end-to-end trainable neural network for image-based sequence recognition and its application to scene text recognition. In: CoRR

Siddiqa A, Hashem IAT, Yaqoob I, Marjani M, Shamshirband S, Gani A, Nasaruddin F (2016) A survey of big data management: taxonomy and state-of-the-art. J Netw Comput Appl 71:151–166

Simard PY, Steinkraus D, Platt JC (2003) Best practices for convolutional neural networks applied to visual document analysis. In: ICDAR

Simonyan K, Zisserman A (2015) Very deep convolutional networks for large-scale image recognition. In: ICLR

Stollenga MF, Byeon W, Liwicki M, Schmidhuber J (2015) Parallel multi-dimensional LSTM, with application to fast biomedical volumetric image segmentation. In: Proceedings of the advances in neural information processing systems, pp 2998–3006

Suk H-I, Shen D (2016) Deep ensemble sparse regression network for Alzheimer's disease diagnosis. In: Proceedings of the medical image computing and computer-assisted intervention. Lecture notes in computer science, vol 10019, pp 113–121. https://doi.org/10.1007/978-3-319-47157-0_14

Sun W, Zheng B, Qian W (2016) Computer-aided lung cancer diagnosis with deep learning algorithms. In: Proceedings of the SPIE, medical imaging 2016: computer-aided diagnosis, vol 9785, p 97850Z, 24 Mar 2016. https://doi.org/10.1117/12.2216307

Suzani A, Rasoulian A, Seitel A, Fels S, Rohling R, Abolmaesumi P (2015) Deep learning for automatic localization, identification, and segmentation of vertebral bodies in volumetric MR images. In: Proceedings of the SPIE medical imaging, vol 9415, p. 941514

Tang D, Qin B, Liu T (2015) Document modelling with gated recurrent neural network for sentiment classification. In: Proceedings of the conference on empirical methods in natural language processing (EMNLP 2015)

Tang D, Qin B, Feng X, Liu T (2016) Effective LSTMs for target-dependent sentiment classification. In: Proceedings of the international conference on computational linguistics (COLING 2016)

Tarando SR, Fetita C, Faccinetto A, Brillet P-Y (2016) Increasing CAD system efficacy for lung texture analysis using a convolutional network. In: Proceedings of the SPIE, medical imaging

2016: computer-aided diagnosis, vol 9785, p 97850Q, 24 Mar 2016. https://doi.org/10.1117/12.2217752

Tran PV (2016) A fully convolutional neural network for cardiac segmentation in short-axis MRI. arXiv:1604.00494

Uijlings JR, van de Sande KE, Gevers T, Smeulders AW (2013) Selective search for object recognition. In: IJCV

Vaillant R, Monrocq C, Le Cun Y (1994) Original approach for the localisation of objects in images. In: IEE proceedings—vision, image and signal processing

van Grinsven MJJP, van Ginneken B, Hoyng CB, Theelen T, Sánchez CI (2016) Fast convolutional neural network training using selective data sampling: application to hemorrhage detection in color fundus images. IEEE Trans Med Imaging 35(5):1273–1284. https://doi.org/10.1109/TMI.2016.2526689

van Tulder G, de Bruijne M (2016) Combining generative and discriminative representation learning for lung CT analysis with convolutional restricted Boltzmann machines. IEEE Trans Med Imaging 35(5):1262–1272. https://doi.org/10.1109/TMI.2016.2526687

Vincent P, Larochelle H, Bengio Y, Manzagol P-A (2008) Extracting and composing robust features with denoising autoencoders. In: Proceedings of the international conference on machine learning, pp 1096–1103

Vivanti R, Ephrat A, Joskowicz L, Karaaslan O, Lev-Cohain N, Sosna J (2015) Automatic liver tumor segmentation in follow-up CT studies using convolutional neural networks. In: Proceedings of the patch-based methods in medical image processing workshop, MICCAI'2015, pp 54–61

Waibel A, Hanazawa T, Hinton G, Shikano K, Lang KJ (1990) Phoneme recognition using time-delay neural networks. In: Readings in speech recognition, pp 393–404

Wang X, Gupta A (2015) Unsupervised learning of visual representations using videos. In: ICCV

Wang Z, Wang X, Wang G (2015a) Learning fine-grained features via a cnn tree for large-scale classification. arXiv:1511.04534

Wang X, Liu Y, Sun C, Wang B, Wang X (2015c) Predicting polarities of tweets by composing word embeddings with long short-term memory. In: Proceedings of the annual meeting of the association for computational linguistics (ACL 2015)

Wang J, Yu L-C, Lai RK, Zhang X (2016b) Dimensional sentiment analysis using a regional CNN-LSTM model. In: Proceedings of the annual meeting of the association for computational linguistics (ACL 2016)

Wang X, Jiang W, Luo Z (2016c) Combination of convolutional and recurrent neural network for sentiment analysis of short texts. In: Proceedings of the international conference on computational linguistics (COLING 2016)

Wang S, Yao J, Xu Z, Huang J (2016e) Subtype cell detection with an accelerated deep convolution neural network. In: Proceedings of the medical image computing and computer-assisted intervention. Lecture notes in computer science, vol 9901, pp 640–648. https://doi.org/10.1007/978-3-319-46723-8_74

Wang C, Elazab A, Wu J, Hu Q (2017) Lung nodule classification using deep feature fusion in chest radiography. Comput Med Imaging and Graph 57:10–18. ISSN 0895-6111. https://doi.org/10.1016/j.compmedimag.2016.11.004

Williams RJ, Zipser D (1989) A learning algorithm for continually running fully recurrent neural networks. Neural Comput 1(2):270–280

Witten IH, Frank E, Hall MA, Pal CJ (2016) Data mining: practical machine learning tools and techniques. Morgan Kaufmann

Xiao T, Zhang J, Yang K, Peng Y, Zhang Z (2014) Error-driven incremental learning in deep convolutional neural network for large-scale image classification. In: ACMMM

Xie W, Noble JA, Zisserman A (2016) Microscopy cell counting and detection with fully convolutional regression networks. Comput Methods Biomech Biomed Eng Imaging Vis 1–10

Xu H, Su F (2015) Robust seed localization and growing with deep convolutional features for scene text detection. In: ICMR

Xu J, Luo X, Wang G, Gilmore H, Madabhushi A (2016a) A deep convolutional neural network for segmenting and classifying epithelial and stromal regions in histopathological images. Neurocomputing 191:214–223. https://doi.org/10.1016/j.neucom.2016.01.034

Xu Y, Li Y, Liu M, Wang Y, Lai M, Chang EI-C (2016b) Gland instance segmentation by deep multichannel side supervision. arXiv:1607.03222

Yan Z, Jagadeesh V, DeCoste D, Di W, Piramuthu R (2014) HD-CNN: hierarchical deep convolutional neural network for image classification. arXiv:1410.0736

Yang L, Luo P, Loy CC, Tang X (2015) A large-scale car dataset for fine-grained categorization and verification. In: CVPR

Yang W, Chen Y, Liu Y, Zhong L, Qin G, Lu Z, Feng Q, Chen W (2017a) Cascade of multi-scale convolutional neural networks for bone suppression of chest radiographs in gradient domain. Med Image Anal 35:421–433. ISSN 1361-8415. https://doi.org/10.1016/j.media.2016.08.004

Yang M, Tu W, Wang J, Xu F, Chen X (2017b) Attention-based LSTM for target-dependent sentiment classification. In: Proceedings of AAAI conference on artificial intelligence (AAAI 2017)

Yoo D, Park S, Lee J-Y, Paek AS, So Kweon I (2015) Attentionnet: Aggregating weak directions for accurate object detection. In: CVPR

Younes L (1999) On the convergence of markovian stochastic algorithms with rapidly decreasing ergodicity rates. Stoch: Int J. Probab Stoch Process 65(3/4):177–228

Yu A, Grauman K (2014) Fine-grained visual comparisons with local learning. In: CVPR

Yu J, Jiang J (2016) Learning sentence embeddings with auxiliary tasks for cross-domain sentiment classification. In: Proceedings of the conference on empirical methods in natural language processing (EMNLP 2016)

Yu L, Yang X, Chen H, Qin J, Heng PA (2017) Volumetric convnets with mixed residual connections for automated prostate segmentation from 3D MR images. In: Proceedings of the thirty-first AAAI conference on artificial intelligence

Zhang W, Li R, Deng H, Wang L, Lin W, Ji S, Shen D (2015a) Deep convolutional neural networks for multi-modality isointense infant brain image segmentation. Neuroimage 108:214–224. https://doi.org/10.1016/j.neuroimage.2014.12.061

Zhang C, Yao C, Shi B, Bai X (2015b) Automatic discrimination of text and non-text natural images. In: ICDAR

Zhao J, Zhang M, Zhou Z, Chu J, Cao F (2016) Automatic detection and classification of leukocytes using convolutional neural networks. Med Biol Eng Comput. https://doi.org/10.1007/s11517-016-1590-x

Zhu Y, Wang L, Liu M, Qian C, Yousuf A, Oto A, Shen D (2017) MRI Based prostate cancer detection with high-level representation and hierarchical classification. Med Phys 44(3):1028–1039. https://doi.org/10.1002/mp.12116

Zilly J, Buhmann JM, Mahapatra D (2017) Glaucoma detection using entropy sampling and ensemble learning for automatic optic cup and disc segmentation. Comput Med Imaging Graph 55:28–41. https://doi.org/10.1016/j.compmedimag.2016.07.012

Validation of a New Method of Pediatric Refraction: Large Aperture Lens Rack

Anupam Sahu, Samrat Chatterjee, Deepshikha Agrawal, and Pradeep Chand Dubey

Abstract *Purpose*: To minimize the problems faced during pediatric retinoscopy, a modification to the standard lens rack was devised at our institute. This study aims to validate the refractive errors determined by large aperture lens rack (LALR) as compared to the results with trial lens method and analyze time taken for trial lenses versus LALR. *Methods*: It was a single center, prospective, pilot study. New patients in the age group of 5 to 13 years with decreased vision, improving to 20/20 with pinhole were included. Cycloplegicretinoscopy was done with both trial lenses and LALR for all patients. Post mydriatictest(PMT) was done at 1 week. Paired T-test was used to compare groups. Correlations were observed using Pearson's 2 tailed correlation. *Results*: Mean refractive error in spherical equivalent was –0.95D with trial lens and –0.65D with LALR ($p = 0.07$). Mean PMT value was –1.0D. Mean cylinder measured with trial lens was –1.48D and –1.26D with LALR ($p = 0.06$). Mean cylinder value at PMT was 1.16D. Trial lens value was significantly different from PMT value ($p < 0.01$). Time taken for LALR was 12 s more than trial lenses ($p = 0.58$). Both spherical equivalent and cylindrical values were significantly correlating with each other for both groups ($p < 0.01$). *Conclusions*: The results obtained by LALR were similar to retinoscopy done with trial lenses for the total refraction. LALR was more accurate in estimating the cylindrical values. We expect time taken for LALR to decrease further with repeated usage.

Keywords Retinoscopy · Lens rack · Refraction · Pediatric

1 Introduction

Refraction among preverbal children has always been difficult. Various attempts have been made in an attempt to quicken the procedure while retaining reliability, including the use of automated refractometers Wood (1987). However manual retinoscopy under cycloplegia remains the gold standard of pediatric refraction Guha (2017),

A. Sahu (✉) · S. Chatterjee · D. Agrawal · P. C. Dubey
MGM Eye Institute, Raipur, Chhattisgarh, India
e-mail: anupam@mgmeye.org

A. A. Rizvanov et al. (eds.), *Advances in Biomedical Engineering and Technology*, Lecture Notes in Bioengineering,
https://doi.org/10.1007/978-981-15-6329-4_11

Erdurmus et al. (2007), Ozdemir et al. (2015). Cycloplegicretinoscopy requires time, patience, and expertise of the examiner along with cooperation of the child.

A retinoscopy rack (ret rack) is an equipment that contains many different spherical lenses in different powers. However the commonly available ret rack has a reduced aperture size which makes it difficult to use in children due to continuous eye movements.

To minimize the problems faced during pediatric retinoscopy, a modification to the standard lens rack was devised at our institute. The idea was to reduce the redundancy of placing trial lenses by developing a product where all/nearly all the lenses (plus or minus) of various powers will be in one lens rack. Also, it should be quick and easy to carry. The lens rack devised consists of rectangular-shaped hard-coated CR-39 lenses (30 × 18 mm) which are easily available in the market, mounted on to a wooden rack (Fig. 1). These are light weight, non-allergic, and easy to handle. There are 4 lens racks for plus powers and 4 lens racks for minus powers which offer a range of ±20 D. Each lens rack has six lenses in ascending order of power.

This study aims to validate the refractive errors determined by large-aperture lens rack (LALR) as compared to the results with trial lens method and analyze time taken for trial lenses versus LALR.

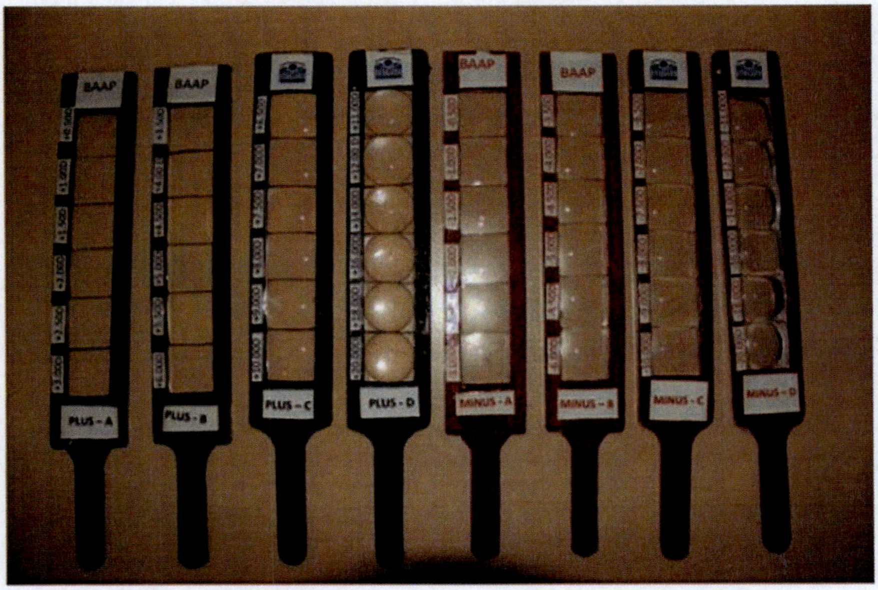

Fig. 1 The large aperture lens rack

2 Methods

It was a single center, prospective, pilot study from 16 February 2017 to 24 March 2017 at MGM Eye Institute, Raipur. Institutional Ethics Committee approval was obtained and the study was conducted in full accord with the tenets of the Declaration ofHelsinki. New patients in the age group of 5 to 13 years with decreased vision (less than 20/20 unaided), improving to 20/20 on LogMAR chart with pinhole were included. Children or eyes with amblyopia or other pathological causes in which vision did not improve to 20/20 were excluded. Also, children who did not cooperate for the post mydriatic test at 1 week were excluded. A written informed consent was taken from all parents/guardians before enrolling. Consent was also obtained for the use of photographs contained in the medical record for the purpose of publication. Once enrolled it was ensured that none of the senior optometrists selected for the refraction did the comprehensive eye examination pre-dilatation. Cycloplegicretinoscopy was done with both trial lenses and LALR by separate senior optometrists who were randomized by lottery (Figs. 2 and 3). Post mydriatic test (PMT) is an assessment of the findings of cycloplegic refraction by subjective means after the effect of cycloplegia is eliminated. It was done at 1 week again by lottery randomization.

The values of refractive error (in diopters) in terms of spherical equivalent (spherical amount + ½ cylindrical amount) and cylindrical value were collected and entered in SPSS (Version 16) for LALR, trial lenses and PMT, and compared using a paired T-test. Time taken for retinoscopy was recorded with a stop watch by a neutral person for both trial lenses and LALR and the average time taken was then compared. These

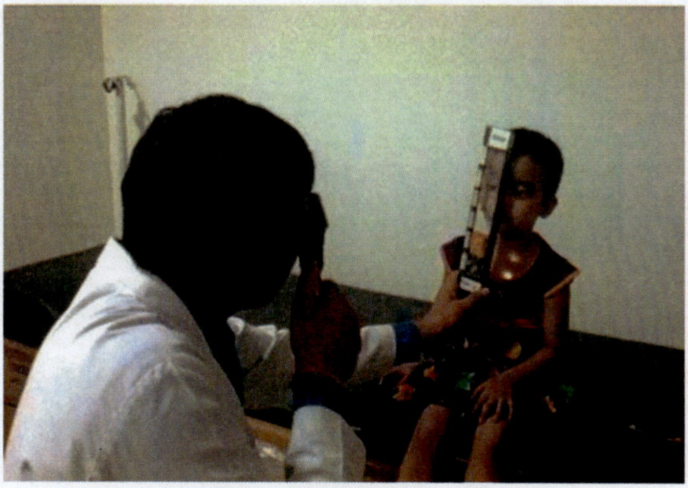

Fig. 2 Retinoscopy performed with LALR

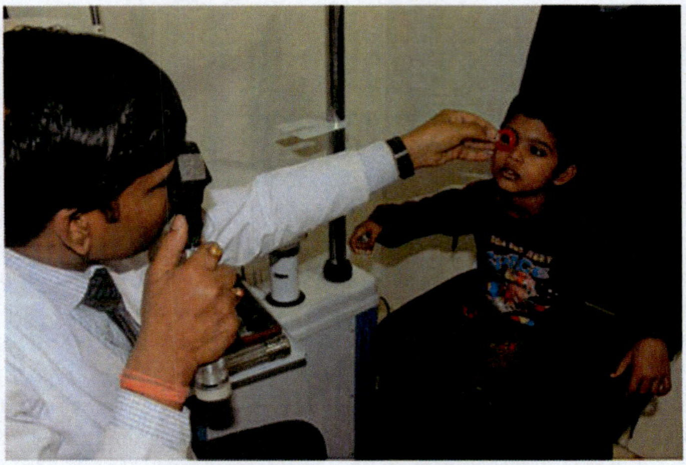

Fig. 3 Retinoscopy performed with trial lenses

were compared using paired T-test with p value < 0.05 considered significant. Correlations were observed using Pearson's 2 tailed correlation and p value < 0.01 was considered significant.

3 Results

28 eyes of 15 patients whose parents consented for the trial and who completed all steps were examined. 2 eyes with amblyopia were excluded. 53.33% (n = 8) were male. The mean age was 8.8 ± 2.5 years.

3.1 Comparison Between Retinoscopy Values

Mean refractive error in spherical equivalent (SE) was –0.95 ± 2.3 Diopter (D) with trial lens and –0.65 ± 2.5D with LALR. Though the mean dioptric difference was 0.3D, paired T-test showed that this difference was not statistically significant (p = 0.07). Mean PMT value in SE was –1.0 ± 1.8D. This was closer to the value achieved with trial lens. But this difference was also statistically insignificant with trial lens having a p value of 0.76 when paired with PMT value and LALR having a p value of 0.11.

Mean cylinder measured with trial lens was –1.48 ± 1.4D, whereas it was –1.26 ± 1.3D with LALR. Both had no statistically significant difference (p = 0.06). The mean cylinder value at PMT was 1.16 ± 1.2D. This was closer to the LALR values (p

Table 1 Comparison of retinoscopy values between groups

Factors assessed	Trial lens group	LALR group	PMT value
Mean refractive error in SE	−0.95 D	−0.65 D	−1.0 D
Mean cylinder measured	−1.48 D	−1.26 D	−1.16 D
Time taken	152 s	164 s	NA

= 0.21) with trial lens retinoscopy overestimating the cylindrical value significantly ($p < 0.01$).

Time taken in retinoscopy by trial lens was 152 ± 63.6 s for each patient whereas time taken by LALR was 164 ± 57.7 s. Time taken for retinoscopy was evaluated in 15 patients. Though LALR took 12 s more for the same set of patients, the difference was not significant (paired T-test; $p = 0.58$) (Table 1).

3.2 *Correlation Between Retinoscopy Values*

To evaluate whether the values were correlating through the entire spectrum of refractive errors studied, Pearson's correlation was used. Both spherical equivalent and cylindrical values showed strong linear correlation with each other for both groups and their correlation was similar with PMT values ($p < 0.01$).

4 Discussion

Various attempts have been made to get reproducible refractive error calculations in children. A significant focus is on automated refractometers as they are quick and can be done by non-clinical persons. However, even after multiple instruments, it has found acceptability only as a screening device for the field where non-clinical persons can carry out the retinoscopy within a short span of time over a large number of childrenGuha (2017); Erdurmus et al. 2007; Ozdemir et al. 2015). However, cycloplegic refraction is still required in those preverbal children who fail the screening testMorgan et al. 2015).

In our short trial, the large aperture lens rack provided comparable results to trial lens retinoscopy. In terms of astigmatism, it was significantly closer to the final PMT values, indicating that the large aperture may provide a better window to evaluate the retinoscope reflex. While it was expected to shorten the time of retinoscopy, LALR in fact took slightly longer. We presume it could be because of a learning curve for the new instrument, though our optometrists were using it for a few weeks before the study. We aim to try it out with our newer trainees for feedback.

4.1 Limitations

Our study has several limitations. First, this was a pilot study to assess the comfort of our optometrists with the device and these results should not be taken as clinically relevant at this point of the study. Second, we may not have been able to nullify the inter and intraobserver differences as noticed for cycloplegicretinoscopy (McCullough et al. 2017).

5 Conclusion

The results obtained by LALR were similar to retinoscopy done with trial lenses for the total refraction. Cylinder values for astigmatism measured by LALR were significantly closer to the PMT values. We expect time taken for the procedure to decrease further with repeated usage. A larger study, taking into account inter and an intra observer difference has been planned.

Overall, the LALR provides a lightweight device which can be carried even to more remote locations, unlike the whole retinoscopy trial set. It also provides a large aperture for better appreciation of astigmatism. It is a much cheaper alternative to automated refractors and can be made with locally available materials.

References

Erdurmus M, Yagci R, Karadag R, Durmus M (2007) A comparison of photorefraction and retinoscopy in children. J Am Assoc Pediatr Ophthalmol Strabismus 11:606–611

Guha S et al (2017) A comparison of cycloplegic autorefraction and retinoscopy in Indian children. Clin Exp Optom 100:73–78

Morgan IG, Iribarren R, Fotouhi A, Grzybowski A (2015) Cycloplegic refraction is the gold standard for epidemiological studies. Acta Ophthalmol (Copenh) 93:581–585

McCullough SJ, Doyle L, Saunders KJ (2017) Intra- and inter- examiner repeatability of cycloplegic retinoscopy among young children. Ophthalmic Physiol Opt J Br Coll Ophthalmic Opt Optom 37:16–23

Ozdemir O, Özen Tunay Z, Petriçli IS, Ergintürk Acar D, Erol MK (2015) Comparison of non-cycloplegic photorefraction, cycloplegic photorefraction and cycloplegic retinoscopy in children. Int J Ophthalmol 8:128–131

Wood ICJ (1987) A review of autorefractors. Eye 1:529–535

Comparative Evaluation of in Vitro Antioxidant, Amylase Inhibition and Cytotoxic Activity of Cur-Pip Dual Drug Loaded Nanoparticles

Trilochan Satapathy, Prasanna Kumar Panda, and Gitanjali Mishra

Abstract Objective: To comparatively evaluate in vitro antioxidant, amylase inhibition, and cytotoxic potential of Cur-Pip dual drug incorporated nanoparticles developed by nano co-precipitation method. Methods: Cur-nanoparticles, Pip-nanoparticles, and Cur-Pip dual drug-loaded nanoparticles were developed by nano co-precipitation method. The amylase inhibition assays of formulated nanoparticles were performed by using the chromogenic DNSA method. The total antioxidant activity was performed by using Butylated hydroxyanisole (BHA) as reference standard. The cytotoxicity potential of nanoparticles formulations at different concentrations such as 25, 50, 100, 250, 500, and 1000 μg were evaluated by agar diffusion method using bacteria E. coli AB 1157. Dimethylsulfoxide (DMSO) was used as solvent. Results: The amylase inhibition assay of three different formulations were performed at concentrations 1000, 500, and 100 μg/ml. Out of these three formulations curcumin nanoparticles exhibited maximum amylase inhibition i.e. –404.92%. Total antioxidant capacity of different nanoformulations were evaluated at concentrations 10, 50, and 100 μg. Antioxidant potentials are expressed as equivalents of ascorbic acid. The Cur + Pip nanoparticles showed maximum antioxidant activity, i.e., 34.6. In cytotoxicity study, none of the formulations exhibited any zone of inhibition. Conclusions: The Cur + Pip nanoparticles showed maximum antioxidant activity in comparison with cur-nanoparticles as well as Pip-nanoparticles whereas cur-nanoparticles exhibited better amylase inhibition over other two nanoformulations. None of the developed formulation showed any cytotoxic activity.

Keywords Nanoparticles · Antioxidant · Amylase inhibition · Cytotoxic

T. Satapathy (✉) · G. Mishra
Berhampur University, Bhanjabihar, Berhampur, Ganjam Odisha-760007, India
e-mail: drtsatapathy@gmail.com

P. K. Panda
University Department of Pharmaceutical Sciences, Utkal University, Vanivihar, Bhubaneswar Odisha-751004, India

A. A. Rizvanov et al. (eds.), *Advances in Biomedical Engineering and Technology*, Lecture Notes in Bioengineering,
https://doi.org/10.1007/978-981-15-6329-4_12

1 Introduction

Curcumin is a yellow polyphenol isolated from the rhizome of herb turmeric (Curcuma longa). Curcumin is not only used in India but also used in several countries like China and Southeast Asia as a spice. Exhaustive study has been carried out on curcumin by various researchers indicated that, Curcumin possess distinct pharmacological activities (Anand et al. 2008) and used in the treatment of various ailments such as anorexia, biliary disorder, hepatic disorder, wounds. (Aggarwal and Harikumar 2009) Further research demonstrated that, Curcumin alone or in combination with other antineoplastic agents exhibited potent anticancer potential which was experimentally evidenced by its inhibitory effects observed on the growth and multiplication of in vitro cell lines as well as In-vivo melanoma, pancreatic carcinomas, Ovarian carcinoma, etc. (Aggarwal et al. 2003) Previous research by various researchers indicated that, Curcumin also possess exciting properties as a signaling molecule which down-regulates gene expression associated with angiogenesis, set-off apoptotic mechanism, and cell cycle arrest. (Gururaj et al. 2002; Belakavadi and Salimath 2005; Shishodia et al. 2005) Recently Jia et al. (2018) investigated the effects of curcumin nanoparticles in inhibition of diabetic neuropathic pain (DNP) and found that, the nanoparticles containing curcumin able to decrease the up-regulated P2Y12 expression in the dorsal root ganglia (DRG), reduced the up-regulation of IL-1β and Cx43, and decreased phosphorylated-Akt (p-Akt) in the dorsal root ganglia of rats with diabetes mellitus. Abdel-Mageid et al. (2018) also investigated the effectiveness of curcumin nanoparticles against complications associated with experimentally induced diabetes in rats, curcumin nanoparticles has greater therapeutic effectiveness in the treatment of Diabetic cardiomyopathy, This is due to by decrease the cardiac inflammation, myocardial fibrosis, as well as programmed myocardial cell deaths. Now-a-days Curcuminoids also gaining special attention in various cosmeceutical products. (Joshi and Pawar 2015) Despite having all these pharmacological properties, curcumin is not approved by the regulatory authorities as a therapeutic agent because of its stability issues (SundarDhilip Kumar et al. 2018). To overcome the problems several strategies have been encouraged such as encapsulation of curcumin in nanoparticles, microspheres, Liposomes, etc., to enhance the curcumin delivery. Extensive review of literature has been done and found that, the bioavailability of curcumin can be enhanced by the addition of piperine (bioenhancer) and by formulate as a nanoparticle. Development of formulations containing controlled drug delivery of therapeutic agents with improved stability and bioavailability is the main aspects of today's research on which many research can be done. Polymeric nanoparticles are considered as a promising drug delivery system that can helpful to overcome the problems associated with Curcumin delivery with stability and enhanced bioavailability. Among the polymers used to formulate the nanoparticles, Polycaprolactone (PCL), Poly (lactic-co-glycolic acid) or PLGA, Poly(n-butyl cyanoacrylate) or PBCA, etc., are considered as ideal polymers because of their low toxicity, biodegradability. Hence, in the present research, our efforts have been devoted to formulate three different nanoformulations by using nanoprecipitation method and to screen these formulations for total antioxidant, amylase inhibition as well as bacterial-based cytotoxicity study.

2 Materials and Methods

2.1 Materials

DNS solution: Onegm of DNS was dissolved in 2 N NaOH, and to it thirty gm of potassium sodium tartarate was added and the whole volume was made up to 100 ml. Amylase (Diastase procured from HiMedia, Mumbai, Cat No. RM-638), Butylated hydroxyl anisole (BHA), and other chemicals used in this study were pure analytical grade and obtained from commercial sources. These chemicals were used in experiment without any further purification. Curcumin and piperine were kind gift sample from Sunpure, New Delhi, India.

2.2 Methods

Three different nanoparticle formulations were developed by nano co-precipitation method. The amylase inhibition potential of these nanoformulations was evaluated by chromogenic DNSA (3, 5-Dinitrosalicylic acid) method. The total antioxidant activity was performed by using Butylated hydroxyl anisole (BHA) as reference standard and cytotoxicity potential of different nanoparticle formulations at different concentrations were evaluated by agar diffusion method using bacteria E. coli AB 1157.

2.3 Preparation of Nanoparticles

In this research, nanoparticles were developed by nanoprecipitation method. Three different formulations such as Cur-NP, Pip-NP, and Cur-Pip dual drug-loaded nanoparticles were developed by using polymer Chitosan/Polycaprolactone (PCL). Briefly, the pure drug (Curcumin/Piperine), polymer (Chitosan), and copolymer (PCL) were taken at their appropriate ratios and dissolved into 5 ml of (90%) acetic acid solution. The homogenous solution was formed due to continuous stirring which then added drop wise into a beaker containing distilled water (50 ml). Finally the nanoparticles formed which was subjected to centrifuge at 13,000 rpm/min for 30 min. After completion of the centrifugation process, the supernatant solution was discarded and the remaining content re-suspended with 10 ml of fresh distilled water. The nanoparticles were then subjected for further analysis.

2.4 Amylase Inhibition Assay

The amylase inhibition assay was performed by using the chromogenic DNSA method. (Miller 1959; Sudha et al. 2011) The total assay mixture consists of one thousand four hundred (1400) μl of 0.05 M sodium phosphate buffer (pH 6.9), Fifty (50) μl of amylase (Diastase procured from HiMedia, Mumbai, Cat No. RM-638) and samples at concentration thousand (1000), Five hundred (500) and hundred (100) μg/ml were incubated at 37 °C for ten (10) min. After pre-incubation, Five hundred (500) μl of 1% (w/v) starch solution was added in the above buffer to each tube and allowed to incubate at thirty seven degree centigrade (37 °C) for 15 min. The reaction was terminated with 1.0 ml DNSA reagent, placed in a boiling water bath for five min, then allowed to cool to room temperature and the absorbance was recorded at 540 nm. The control amylase exhibited 100% enzyme activity and did not contain any sample of analysis. To eliminate the absorbance by sample, the appropriate extract controls with the extract in the reaction mixture in which the enzyme was added after adding DNS (Fig. 1).

The liberated sugar was then determined by the help of standard maltose curve and activities were calculated according to the following formula given below:

$$\text{Activity} = \frac{\text{Conc.of Maltose libereted} \times \text{ml of enzyme used}}{\text{Mol.wt of Maltose} \times \text{incubation time (min)}} \times \text{dilution factor} \tag{1}$$

One unit of enzyme activity is defined as the amount of enzyme required to release one micromole of maltose from starch per min under the assay conditions. The inhibitory/induction property shown by the sample was compared with that of control and expressed as percent induction/inhibition. This was calculated according to the following formula:

$$\%\ \text{inhibition/induction} = \frac{\text{Activity in presence of compound}}{\text{Control Activity}} \times 100 \tag{2}$$

2.5 Analysis of Acarbose as Standard Inhibitor

Acarbose was used as a standard inhibitor and it was assayed at above-mentioned test sample concentrations. The assay method was similar to the above-mentioned procedure, instead of test samples, acarbose was added. The results were compared to that of test sample.

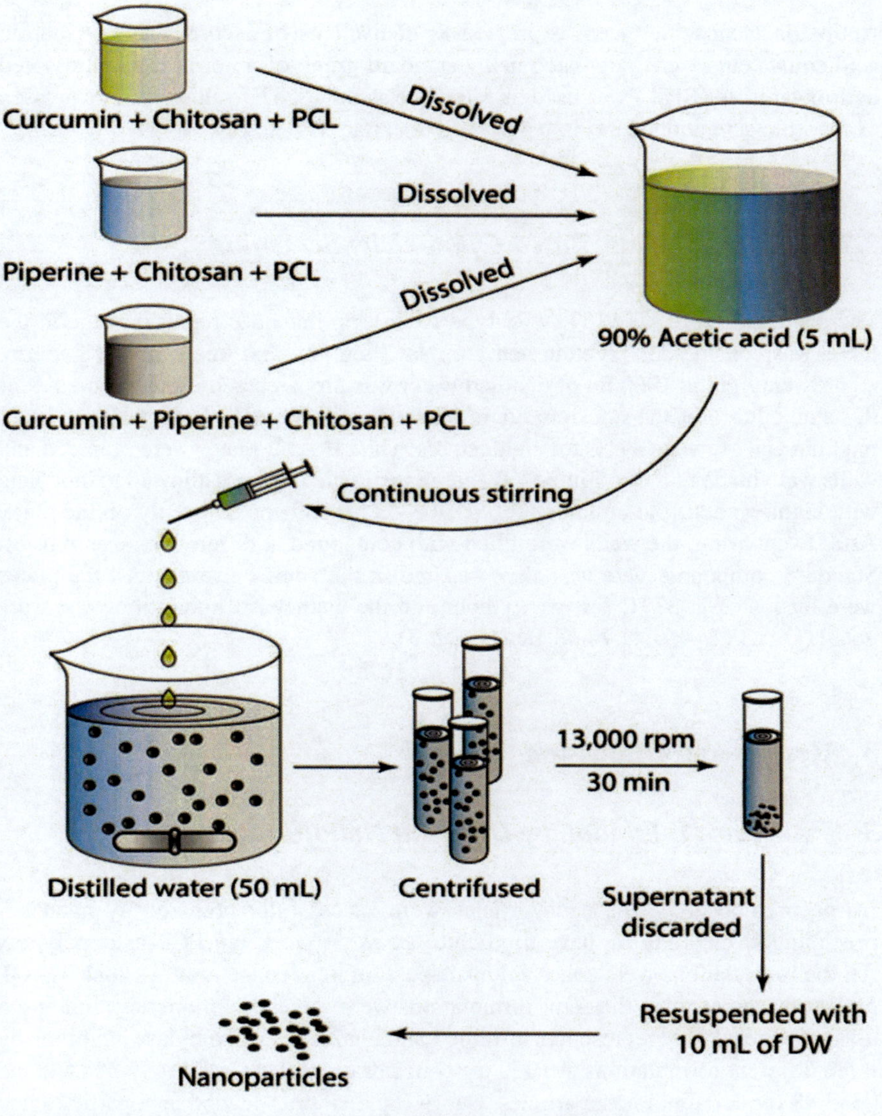

Fig. 1 Schematic representation of preparation of nanoparticle formulations

2.6 *Determination of Total Antioxidant Capacity*

Using a series of test tubes samples were taken in different concentrations (10, 50, and 100 μg). To this, 1.9 mL of reagent solution (0.6 M sulfuric acid, 28 mM sodium phosphate, and 4 mM ammonium molybdate) was added. Then the tubes were allowed to incubate at 95 °C for ninety (90) min after that allowed to cool. The absorbance of the solution of each was recorded at 695 nm against a blank.

Antioxidant capacities were expressed as equivalents of ascorbic acid. Ascorbic acid equivalents were calculated using standard graph of ascorbic acid. Butylated hydroxyanisole (BHA) was used as reference standard. The values were expressed as ascorbic acid equivalents in μg per mg of extract (Prieto et al. 1999).

2.7 *Bacterial Strain-Based Cytotoxicity Screening:*

The bacteria E.coli AB 1157 (wild-type strain), proficient to repair damage in the DNA was considered. Tryptone ten gm, NaCl ten gm, and Yeast extract five gm, Agar twenty gm in 1000 ml of distilled water was prepared and used as a media for this study. Initially, the stock culture of bacteria was revived by inoculating in broth medium and grown at 37 °C for eighteen hrs. The LB Agar plates were prepared and wells were made in the solidified LB agar plate. Each plate was allowed to inoculate with eighteen hour old cultures (100 μl, 10–4 cfu) and spread evenly on the plate. After twenty min, the wells were filled with compound at different concentrations. Standard compound plate was also prepared in the similar manner. All the plates were incubated at 37 °C for twenty hour and the diameter of inhibition zone were noted (Figs. 2, 3, 4, 5, 6, 7 and Tables 1, 2, 3).

3 Results and Discussion

3.1 *Amylase Inhibition by Different Nanoformulation*

In the present study, the nanoparticles were successfully prepared by nano co-precipitation method. We have used chitosan as polymer and PCL as copolymer. All the formulations were spherical in shape. Different concentrations such as 100, 500, 1000 mg of three different formulations were screened for amylase inhibition assay. The results are represented in Table 1 and Fig 2. The % of amylase inhibition by three different formulations at 1000 μg concentrations were –404.92, –358.54, and –393.88 for curcumin nanoparticles, Cur + Pip dual drug-loaded nanoparticles and Piperine nanoparticles, respectively, but in case of total antioxidant activity, Cur + Pip dual drug-loaded nanoparticles at 100 μg showed better antioxidant activity i.e., 34.6 where other two formulations pip-NP and Cur-NP exhibited 30.5 and 28.2, respectively. This result for total antioxidant activity indicated that curcumin and piperine in combination possess better antioxidant activity than the other two formulations. In cytotoxicity study, none of the formulations exhibited any zone of inhibition hence no cytotoxic in nature.

4 Conflict of Interest

The authors declare no conflict of interest.

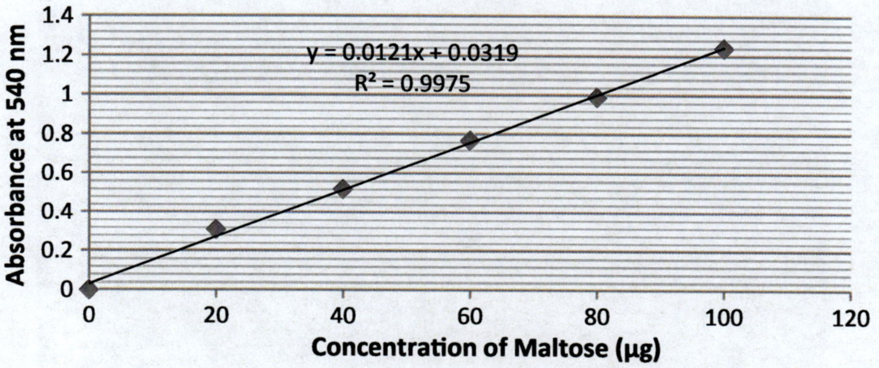

Fig. 2 Standard maltose curve

Fig. 3 Amylase inhibition assay by different nanoformulations

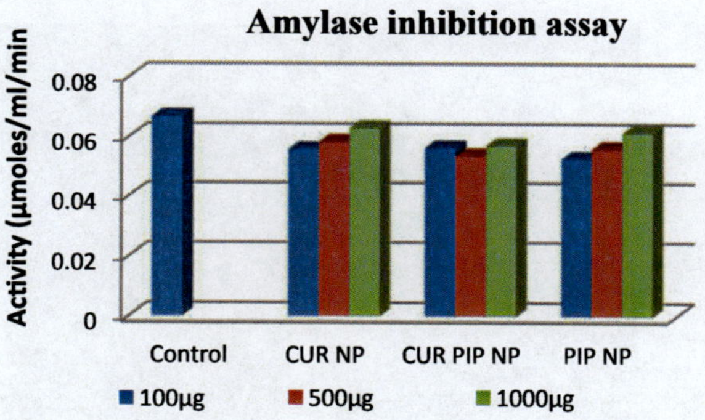

Fig. 4 Total antioxidant activity by different nanoformulations

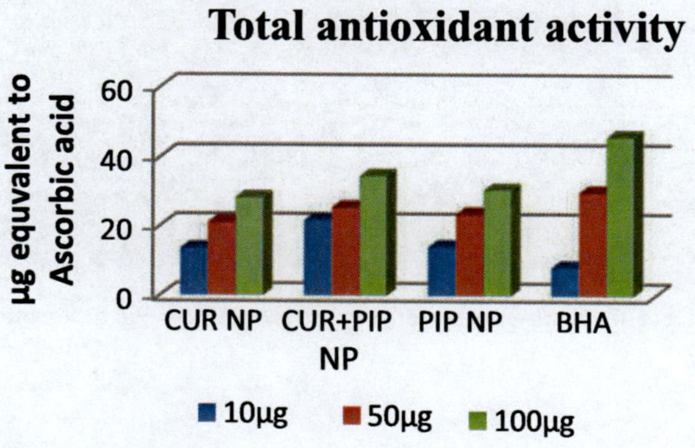

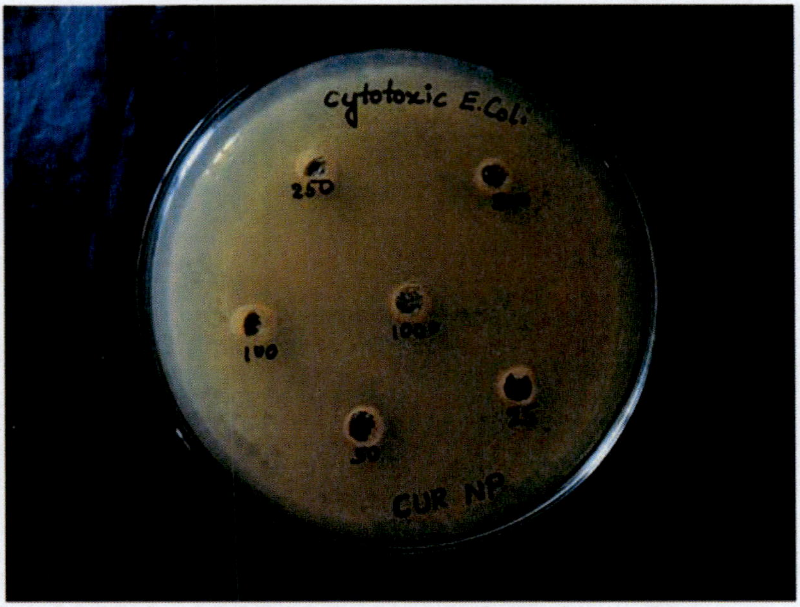

Fig. 5 *E.coli*-based cytotoxic activity of Curcumin nanoparticles

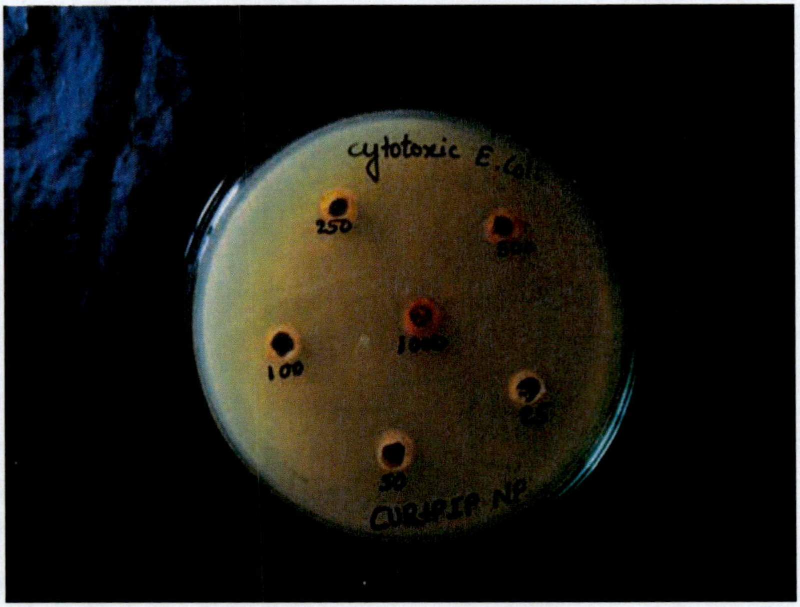

Fig. 6 *E.coli*-based cytotoxic activity of Cur + Pip nanoparticles

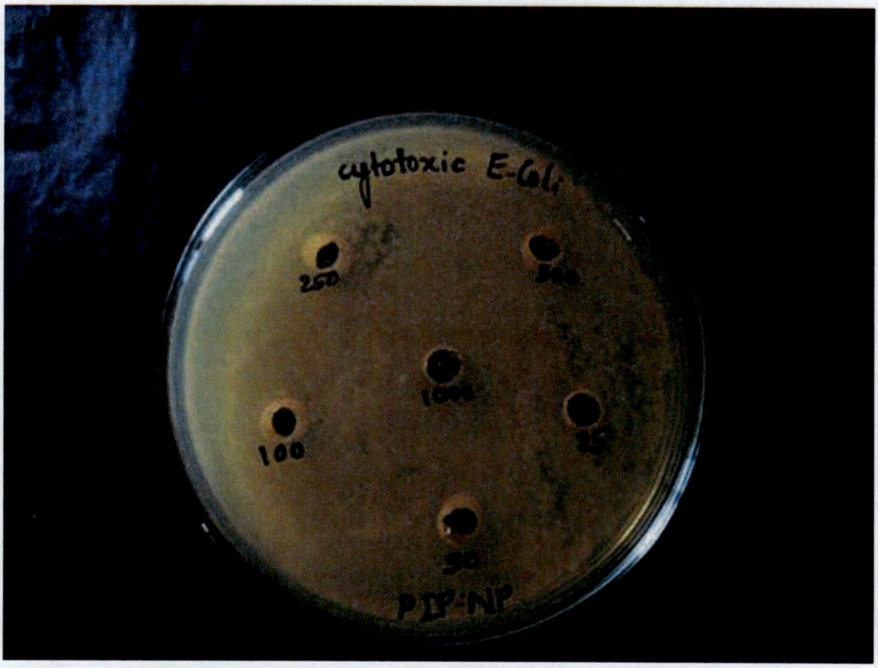

Fig. 7 *E.coli*-based cytotoxic activity of Piperine nanoparticles

Table 1 Composition of different nano particle formulations

Formulation	Curcumin(mg)	Piperine(mg)	Chitosan	PCL
Cur-NP	100	–	50	950
Pip-NP	–	100	50	950
Cur + Pip-NP	50	50	50	950

Table 2 Effect of different nanoformulations on In vitro amylase inhibition

Sample details	Concentrations	OD at 540 nm	Maltose liberated (μg)	Activity (μmoles/ml/min)	Activity (%)	Inhibition (%)
CUR-NP	Control	2.92	240.75	0.066817	536.00	−436.00
	100 μg	2.4415	200.88	0.055751	447.22	−347.22
	500 μg	2.5535	210.21	0.058341	468.00	−368.00
	1000 μg	2.7525	226.79	0.062943	504.92	−404.92
CUR-PIP-NP	100 μg	2.4675	203.04	0.056352	452.05	−352.05
	500 μg	2.3545	193.63	0.053738	431.08	−331.08
	1000 μg	2.5025	205.96	0.057161	458.54	−358.54
PIP-NP	100 μg	2.311	190.00	0.052732	423.01	−323.01
	500 μg	2.4585	202.29	0.056144	450.38	−350.38
	1000 μg	2.693	221.83	0.061567	493.88	−393.88

Table 3 Effect of different nanoformulations on total antioxidant activity

Concentration	Total antioxidant activity			
	CUR-NP	CUR + PIP-NP	PIP-NP	BHA
10 μg	13.6	21.65	14	8.18
50 μg	21.05	25.4	23.5	29.9
100 μg	28.2	34.6	30.5	45.73

Acknowledgements Authors are thankful to Sunpure extract Pvt Ltd for providing Curcumin and Piperine as a gift sample for this research work.

References

Abdel-Mageid AD, Abou-Salem MES, Salaam NMHA, El-Garhy HAS (2018) The potential effect of garlic extract and curcumin nanoparticles against complication accompanied with experimentally induced diabetes in rats. Phytomedicine 43:126–134

Aggarwal BB, Harikumar KB (2009) Potential therapeutic effects of Curcumin, the anti-inflammatory agent, against neurodegenerative, cardiovascular, pulmonary, metabolic, autoimmune and neoplastic diseases. Int J Biochem Cell Biol 1:40–59

Aggarwal BB, Kumar A, Bharti AC (2003) Anticancer potential of curcumin: pre-clinical and Clinical studies. Anticancer Res 23:363–398

Anand P, Thomas SG, Kunnumakkara AB, Sundaram C, Harikumar KB, Sung B (2008) Biological activities of curcumin and its analogues (Congeners) made by man and mother nature. BiochemPharmacol 76:1590−1611

Belakavadi M, Salimath BP (2005) Mechanism of inhibition of ascites tumor growth in mice by curcumin is mediated by NF-kB and caspase activated Dnase. Mol Cell Biochem 273:57–67

Gururaj AE, Belakavadi M, Venkatesh DA, Marme D, Salimath BP (2002) Molecular mechanisms of anti-angiogenic effect of curcumin. Biochem Biophys Res Commun 297:934–942
Jia T, Rao J, Zou L, Zhao S, Yi Z, Wu B, Li L, Yuan H, Shi L, Zhang C, Gao Y, Liu S, Xu H, Liu H, Liang S, Li G (2018) Nanoparticle-Encapsulated Curcumin Inhibits Diabetic Neuropathic Pain Involving the P2Y12 Receptor in the Dorsal Root Ganglia. 11:01–12
Joshi LS, Pawar AH (2015) Herbal cosmetics and cosmeceuticals: an overview. Nat Prod Chem Res 3:170
Miller GL (1959) Use of dinitrosalicylic acid reagent for determination of reducing sugar. Anal Chem 31:426–428
Prieto P, Pineda M, Aguilar M (1999) Spectrophotometric quantitation of antioxidant capacity through the formation of a phosphomolybdenum complex: specific application to the determination of vitamin E1. Anal Biochem 269:337–341
Shishodia S, Amin HM, Lai R, Aggarwal BB (2005) Curcumin (diferuloylmethane) inhibits constitutive NF-kappa B activation, induces G1/S arrest, suppresses proliferation, and induces apoptosis in mantle cell lymphoma. Biochem Pharmacol 70:700–713
Sudha P, Smita SZ, Shobha Y, Bhargava ARK (2011) Potent a-amylase inhibitory activity of Indian Ayurvedic medicinal plants. BMC complementary and alternative medicine 11:5
SundarDhilip Kumar S, Houreld N, Abrahamse H (2018) Therapeutic potential and recent advances of curcumin in the treatment of aging-associated diseases. Molecules 23(4):835

Improved ERP Classification Algorithm for Brain–Computer Interface of ALS Patient

Vyom Raj, Shreya Sharma, Mridu Sahu, and Samrudhi Mohdiwale

Abstract The study on Amyotrophic Lateral Sclerosis (ALS) patient to identify the non-target or target stimulus based on event-related potential provide a way to improve P300 speller based Brain–Computer Interface (BCI). In the current work channel wise EEG data taken for the research. Feature extraction and Feature selection techniques based on Fourier and Wavelet transform and Statistics have been implemented to get the required features among the redundant one. By classifying the features categorized in 3 labels stated above by using support vector machine (SVM). The study reveals that the classification accuracy is improved in Morlet wavelet-based feature than statistical features for different channels taken in consideration.

Keywords Brain computer interface · Amyotrophic lateral sclerosis · Wavelet ransform

1 Introduction

Muscle movements such as speaking, walking, etc., are caused by motor neurons. The loss of functioning of motor neuron or decay of nerve cell causes a neurological disease named Amyotrophic Lateral Sclerosis (ALS). The person affected by this disease loses its control over the muscle movement due to death of motor neuron. The signal transmission process of muscle movement starts from brain's upper motor neuron called cerebral cortex and move toward spinal cord, i.e., lower motor neuron, from the spinal cord the signal reached to specific muscle and perform functioning. As the process starts from upper motor neuron so, the degenerative disease brings gradual impairment starts from upper motor neuron to the spinal cord. ALS patient suffer from cognitive impairments, become quadriplegic and loses ability of communication even gestural communication due to complete failure of limb functioning (PietroCipresso et al. 2012; Mitsumoto and Rabkin 2007). Such critical cases attract the researchers to work on ALS patient to make their life easier. No physical movement needs special

V. Raj · S. Sharma · M. Sahu (✉) · S. Mohdiwale
National Institute of Technology Raipur, Raipur, India
e-mail: mrisahu.it@nitrr.ac.in

A. A. Rizvanov et al. (eds.), *Advances in Biomedical Engineering and Technology*, Lecture Notes in Bioengineering,
https://doi.org/10.1007/978-981-15-6329-4_13

care and support in all the aspect of life such as dietary, rehabilitation, respiration, and communication via devices. Brain–computer interface (BCI) device is the recent advancement for ALS patient to talk with their dear ones and communicate. The main concept of BCI is to transform the brain signal, i.e., EEG signal to respective word which enable a patient to talk via thinking. Deciphering the EEG data itself is a challenging task. Various methods are proposed to improve the performance of BCI. In contribution toward improvement of BCI we proposed a hybrid algorithm for feature extraction and selection. Performance evaluation of different wavelet has been done via Support Vector Machine (SVM).

The rest of the paper structured as follows. Section 2 describes the literature review related to current work. Section 3 provides description of dataset. Section 4 provides the detail description of proposed method for performance improvement. Section 5 provides the result and analysis. Conclusion of the current work given in Sect. 6.

2 Literature Review

Today's era is digital era of communication, with this regard BCI plays very important role in communication for patient suffered from ALS disease. Nicolas F Ramsey et al. presented a fMRI-based method of BCI implant of new generation BCI (Ramsey 2014). The complete flow of BCI-based EEG signal processing shown in Fig. 1

Various researches is going for making the life easy of ALS patient based on EEG signal processing. Hao-Teng Hsu et al. explores amplitude-frequency characteristics of steady-state visual evoked potential (SSVEP) for obtaining the implementation feasibility of frontal SSVEP on BCI system. In this paper authors found frontal SSVEP on no hair-bearing area provides more comfort for BCI-based application due to lower amplitude on frontal SSVEP-based region than occipital SSVEP-based region (Hsu and I-Hui Lee 2016). Rosario Sorbello et al. improves the BCI performance by utilizing biofeedback factor obtained by considering the mental state of

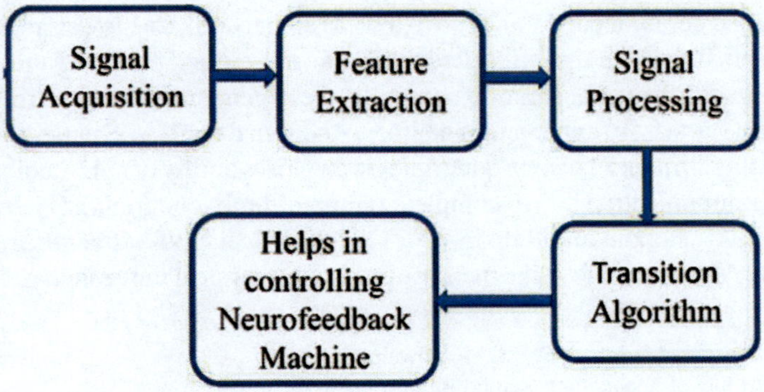

Fig. 1 Complete work flow of BCI using EEG signal

human, futher it shows performance of the robotic feedback by post hoc analysis for maintaining the focus on expected task (Sorbello et al. 2018). The research is going on to evaluate the current potential for BCI-based P300 speller. This potential can be used further for online BCI-based applications (SadafIqbal et al. 2017). Yazıcı et al. proposed statistical feature-based EEG signal classification without pre-processing. The result obtained is higher efficient than the results uses pre-processing steps (Yazıcı and Mustafa Ulutas 2015). Faraz Akram et al. proposed an algorithm for word typing using P300 speller system with smart dictionary of suggesting word (FarazAkram and Metwally 2013).

Yoon et al. investigates the spatial and temporal features of ERP and the result shows that the spatial features are more important than temporal features (Yoon et al. 2018).Thomas P. et al. presents spectral feature-based classification for biomedical signal. Multi wavelet transform, Independent component analysis are the extracted features for signal classification in that paper and found accuracy of 92% with SVM but on different dataset (Thomas and Moni 2016). Hong, K. Set al. presents the feature extraction and classification method for EEG-BCI, in which power spectral density, logarithmic band power are extracted features and linear discriminant analysis is most widely used classifier (Hong et al. 2018).

Various literature discussed above motivate us to research in the area of BCI for improvement in the devices designed for ALS patient. In the current work, we focused on ERP classification using wavelet-based feature extraction and selection technique and find the effect of number of channels selected while evaluation of technique.

3 Dataset Description

BCI2000 dataset has taken for the research in the current study. The dataset consists of recordings of P300 signals obtained from BCI2000 with the help of technique discussed by Farewell and Donchin. 8 patients suffering from ALS disease had to gaze one character out of 36 different character and the task is to predict the correct character while recording of EEG. The recordings are characterized in 3 different states, i.e., no stimulus, non-target stimulus, target stimulus.

4 Methodology

The process of improving the feature extraction and selection method involve definite steps, such as Data acquisition, Feature extraction, Feature selection, and Classification. The Flowchart shown below represents the steps wise procedure followed for enhancement of existing techniques (Fig. 2).

The details of each step provided in subsequent subsection.

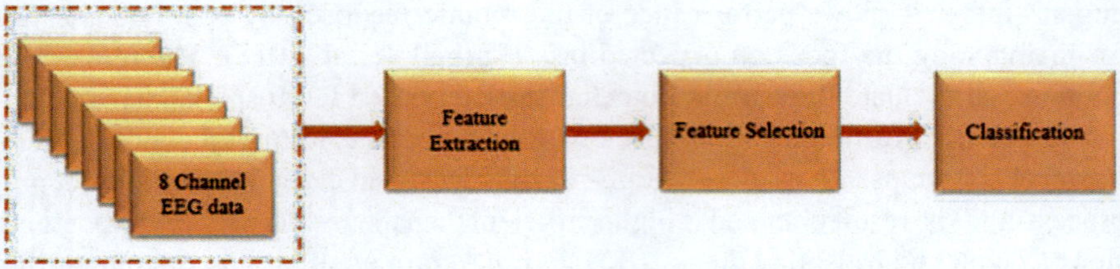

Fig. 2 Flowchart of proposed method

4.1 Data Acquisition

BCI 2000 dataset is used for performing the research on ALS patient. The dataset has been obtained on 8 channel EEG acquisition device. To perform the experiment on each channel EEG data, channel wise data has been imported. Eight channels that were used for recording EEG based on the 10–10 standard (Fz, Cz, Pz, Oz, P3, P4, PO7, and PO8) (Riccio et al. 2013). The obtained EEG signal sampled at 256 Hz. Bandpass filter with band frequency 0.1 to 30 Hz has been used to get the desired frequency range of EEG signal.

4.2 Feature Extraction

The large amount of data always may or may not be important. Introducing large amount of redundant data always maximizes the complexity of the analysis. Features are the important characteristics of data represented in less dimensionality than original data to identify data belongs to particular class. In the current work, features of EEG signal extracted by EEG Analysis package of R. Windowing of original data, double windowing, Spectrum of windowed data, Continuous wavelet transform are the feature extraction methods taken to obtain the features.

Fast Fourier Transform. Fast Fourier Transform (FFT) is an efficient algorithm for finding the Discrete Fourier Transform (DFT) of a sequence with reduced time complexity. FFT divides the signal into small frames over a period of time and differentiate it according to frequency component present in the signal (Welch 1967). For N components, FFT can be calculated by using formula:

$$X^{[k]} = \sum_{n=0}^{N/2-1} x[2n]W_N^{2kn} + W_{Nk} \sum_{n=0}^{N/2-1} x[2n+1]W_{Nkn}^{2} \quad (1)$$

where $x(n)$ is signal of length n, ${W_N}^{kn}$ is twiddle factor.

FFT is calculated by dividing the signal into even and odd components. The above two terms in the Eq. 1 are those even and odd components of EEG signal.

Significance of calculating FFT for ALS patient data is to obtain the different frequency components of EEG data in discretized manner to evaluate them according to frequency bands of EEG signals.

Wavelet Transform. Wavelet transform offers time–frequency representation of data/signal for localization, but not only in the sin or cosine, i.e., Fourier domain but also in real space. In the current work HAAR, Gaussian derivatives, and Morlet wavelet has been implemented for feature extraction from the dataset. The general function for calculating wavelet transform is given in Eq. 2 (SulimanBelal et al. 2018).

$$f(a,b) = \int_{-\infty}^{\infty} f(x)\psi_{a,b}^{*}(x)d(x) \tag{2}$$

where $\psi_{a,b}(x)$ is wavelet function chosen.

The HAAR wavelet is a square-shaped function which allows the target function to be represented in orthonormal basis. As this wavelet function is not continuous hence can-not be differentiable which allows to analyze the signals with sudden changes. Gausian and Morlet wavelets are less sensitive and allow trade between spatial and frequency domain resolution. Thus, they are more popular.

4.3 Feature Selection

Acquiring relevant feature from the set of features for simplification, less training time with improved efficiency and generalization is the main objective of feature selection. In the current work False Discovery Rate (FDR), Analysis of Variance (ANOVA) has been implemented for feature selection problem. FDR provides the number of false observations that could be rejected and ANOVA provides the statistical test among different 8 channel's data to analyze the differences among different channels. FDR is the method to adjust the p-value of test such that the total false positive errors are less (Benjamini and Hochberg 1995). For V number of false positives and R number of rejected null hypothesis FDR can be calculated by

$$Z = \frac{V}{R} \tag{3}$$

$$FDR = E(Z) \tag{4}$$

ANOVA is the technique used to analyze the differences among the mean of a group in a given sample of population. This is useful for comparing and choosing one sample from the set of samples (Sokal and Rohlf 1969). This value also useful to rank the features or variable which tells more discrimination among different classes.

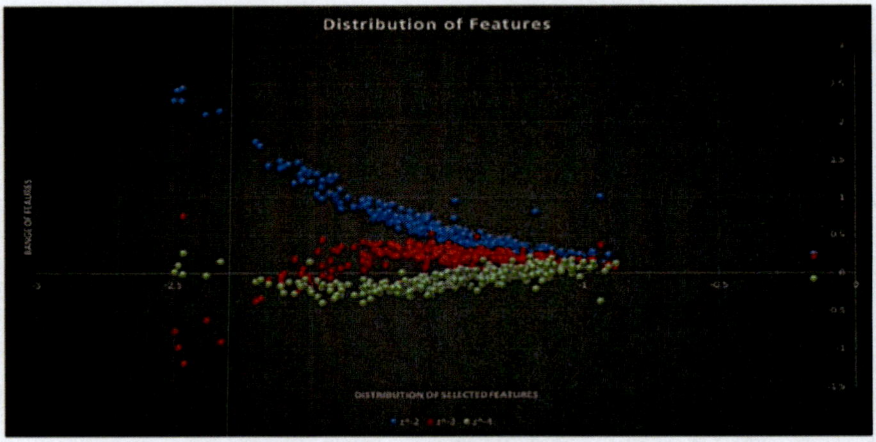

Fig. 3 Distribution of features after feature selection

In the current work, after feature selection a feature vector of length 304*5 has been obtained.

After feature selection the distribution of feature shown in Fig. 3.

After feature selection statistical operations are evaluated channel wise to get optimum feature. Mean, median, product, geometric and harmonic mean are calculated. On the same set of features obtained after selection procedure HAAR, Gaussian, Morlet wavelet is applied for another set of optimum features. These sets of features are now classified to check performance of proposed method.

4.4 *Classification*

After accomplishment of data acquisition channel wise, feature extraction and feature selection, Support Vector Machine(SVM) is applied for classification of acquired features from different statistical and Wavelet Transforms. SVM transforms the data into high dimensional feature space and creates the decision boundary by maximizing the margin on obtained support vector. Various kernel functions can be used as per the data distribution for accurate classification. In the current work polynomial kernel is used as it is efficient as per data distribution.

5 Results

In Table 1, classification accuracy after the application of SVM is presented. It is evident that the best classification accuracy is obtained when we have all the 8 channels taken into consideration and the conditional measure chosen is variance.

Table 1 Channel Wise Accuracy on Statistical Parameters

Channel	Mean	Median	Variance	Product	Geometric	Harmonic
Fz	63.02	78.21	79.43	74.50	63.27	71.63
Fz,Cz	62.71	81.01	72.53	71.77	82.50	73.90
Fz,Cz,Pz	64.09	83.29	73.89	74.42	83.10	74.62
Fz,Cz,Pz,Oz	66.72	84.11	74.21	75.77	84.77	77.25
Fz,Cz,Pz,Oz,P3	71.41	85.22	77.23	75.91	86.55	78.21
Fz,Cz,Pz,Oz,P3,P4	76.83	87.11	88.92	77.64	81.2	89.27
Fz,Cz,Pz,Oz,P3,P4,PO7	82.42	85.11	95.33	78.91	82.34	84.50
Fz,Cz,Pz,Oz,P3,P4,PO7,PO8	88.92	89.11	97.83	80.91	90.61	93.24

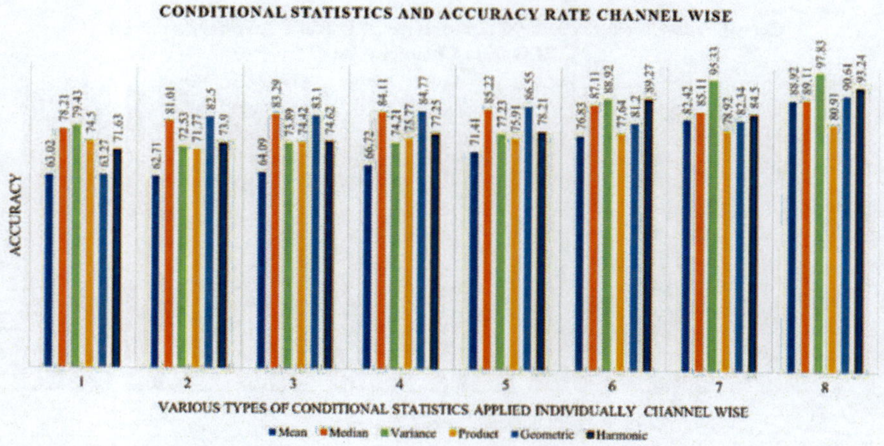

Fig. 4 Conditional Statistics and Accuracy Channel Wise

When all 8 channels are chosen along with the conditional measure variance, we get the classification accuracy of 97.83%. The bar graph in figure also represent the channel wise conditional statistics and accuracy (Fig. 4).

Further we went on to choose the best types of wavelet transformation that could increase classification accuracy. The results of the same have been mapped in Table 2. It has been established that the Morlet method of wavelet transformation gives the best classification accuracy. Combined with variance conditional measure it increases the classification accuracy to 98.28% (Fig. 5).

6 Conclusion

The classification of dataset after feature extraction and selection while varying certain parameters have been accomplished in the current work. The main idea was to obtain the best conditional measures and the optimum number of channels. From

Table 2 Channel wise accuracy on different wavelet transforms

Channels	Haar	Gaussian1	Gaussian2	Morlet
Fz	80.41	82.93	80.43	72.42
Fz,Cz	74.52	76.21	77.64	77.61
Fz,Cz,Pz	75.62	76.54	70.33	82.45
Fz,Cz,Pz,Oz	76.51	77.89	75.31	83.89
Fz,Cz,Pz,Oz,P3	78.11	79.03	80.41	87.61
Fz,Cz,Pz,Oz,P3,P4	89.31	90.21	92.22	93.22
Fz,Cz,Pz,Oz,P3,P4,PO7	96.24	96.72	93.42	96.72
Fz,Cz,Pz,Oz,P3,P4,PO7,PO8	97.91	97.94	96.21	98.28

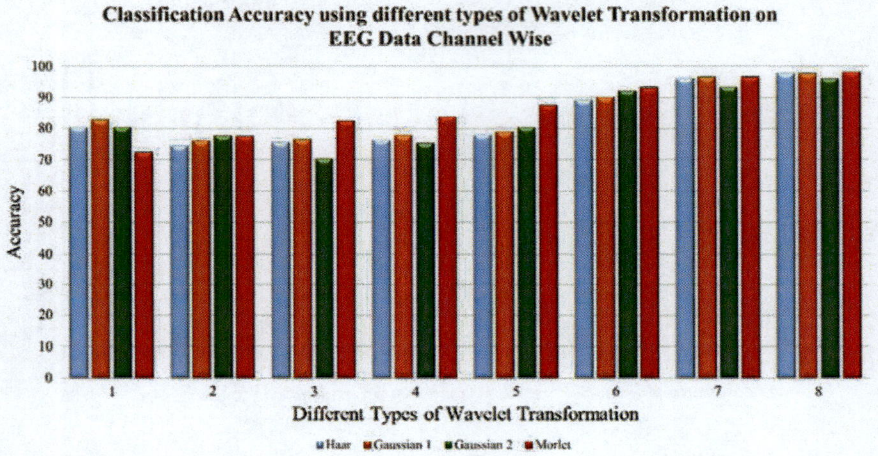

Fig. 5 Wavelet Transformation and Accuracy Channel Wise

the experiment it is clear that if the number of channels is incremented it increases the classification accuracy. On the other hand morlet wavelet transform also improves the performance of classification.

References

Akram F, Metwally MK, Han H-S, Jeon H-J, Kim T-S (2013) A novel p300-based bci system for words typing. In Brain-Computer Interface (BCI), 2013 International Winter Workshop on, ppp 24–25. IEEE

Belal S, Cousins J, El-Deredy W, Parkes L, Schneider J, Tsujimura H, Zoumpoulaki A, Perapoch M, Santamaria L, Lewis P (2018) Identification of memory reactivation during sleep by eeg classification. NeuroImage 176:203–214

Benjamini Y, Hochberg Y (1995) Controlling the false discovery rate: a practical and powerful approach to multiple testing. J Royal Statist Soc Series B (Methodological), 289–300

Cipresso P, Carelli L, Solca F, Meazzi D, Meriggi P, Poletti B, Lul´ e D, Ludolph AC, Silani V, Riva G (2012) The use of p300-based bcis´ in amyotrophic lateral sclerosis: from augmentative and alternative communication to cognitive assessment. Brain and behavior 2(4):479–498

Hong KS, Khan MJ, Hong MJ (2018) Feature extraction and classification methods for hybrid fNIRS-EEG brain-computer interfaces. Frontiers in Human Neuroscience, 12

Hsu H-T, Lee I-H, Tsai H-T, Chang H-C, Shyu K-K, Hsu C-C, Chang H-H, Yeh T-K, Chang C-Y, Lee P-L (2016) Evaluate the feasibility of using frontal ssvep to implement an ssvep-based bci in young, elderly and als groups. IEEE Trans Neural Sys Rehabil Eng 24(5):603–615

Iqbal S, Rizvi BA, Muhammed Shanir PP, Khan YU, Farooq O (2017) Detecting p300 potential for speller bci. In: Communication and Signal Processing (ICCSP), 2017 International Conference on, pp 0295–0298. IEEE.

Mitsumoto H, Rabkin JG (2007) Palliative care for patients with amyotrophic lateral sclerosis:prepare for the worst and hope for the best. Jama 298(2):207–216

Ramsey NF (2014) Exploration of the brain for optimal placement of bci implants in paralyzed people. In Brain-Computer Interface (BCI), 2014 International Winter Workshop on, pp 1–3. IEEE

Riccio A, Simione L, Schettini F, Pizzimenti A, Inghilleri M, Belardinelli MO, Mattia D, Cincotti F (2013) Attention and p300-based bci performance in people with amyotrophic lateral sclerosis. Frontiers in human neuroscience 7:732

Sorbello R, Tramonte S, Emanuele Giardina M, La Bella V, Spataro R, Allison B, Guger C, Chella A (2018) A human–humanoid interaction through the use of bci for locked-in als patients using neuro-biological feedback fusion. IEEE Trans Neural Sys Rehabil Eng 26(2):487–497

Sokal RR, Rohlf FJ (1969) The principles and practice of statistics in biological research. WH Freeman and company, San Francisco, pp 399–400

Thomas P, Moni R (2016) Methods for improving the classification accuracy of biomedical signals based on spectral features. Technology 7(1):105–116

Welch P (1967) The use of fast fourier transform for the estimation of power spectra: a method based on time averaging over short, modified periodograms. IEEE Trans Audio Electroacoust 15(2):70–73

Yazıcı M, Ulutas M (2015) Classification of eeg signals using time domain features. In 2015 23nd Signal Processing and Communications Applications Conference (SIU) pp 2358–2361

Yoon J, Lee J, Whang M (2018) Spatial and time domain feature of ERP speller system extracted via convolutional neural network. Computational intelligence and neuroscience

Size Reduction in Multiband Planar Antenna for Wireless Applications Using Current Distribution Technique

Pravin Tajane and P. L. Zade

Abstract The proposed planar antenna with U shaped slots on patch and asymmetrically rectangular slots on ground plane for WLAN/Bluetooth/WiMAX/HYPERLAN applications. The proposed antenna is reduced the size by using the surface current distribution technique for the multiband applications. The surface current shows that which part is responsible to create the resonance frequencies. The maximum current is concentrated in the rectangular slots for 2.42, 3.41, and 5.4 GHz frequencies. Those portions do not have any surface current that part is not responsible to produce the resonance frequencies. The designed antenna 1 doesn't have any surface current at the upper side of patch means that part does not take part actively participation to create the resonance frequencies so those portion is removed without affecting anything on other parameter. The planar antenna 2 is applicable for multiband applications after size reduction in antenna 1 using surface current distribution technique. The Proposed antenna with U shaped slot on patch and asymmetrically rectangular slots on ground plane which resonates 2.42 GHz, resonates 3.41 and 5.4 GHz covering ISM (Industrial, Scietific, and Medical) and WLAN, Bluetooth, Zigbee, WiMAX, and HYPERLAN. The length is reduced 40 to 30 mm with the help of current spread technique. An overall dimension of antenna is $30 \times 26\text{mm}^2$ from $40 \times 26\ \text{mm}^2$. After reducing the size of antenna is also applicable for the multiband operation.

Keywords Multiband · Surface current · Rectangular slots

P. Tajane (✉)
YCCE, Nagpur, India
e-mail: ptajane8@gmail.com

P. L. Zade
DMIETR, Wardha, India
e-mail: zadepl@yahoo.com

A. A. Rizvanov et al. (eds.), *Advances in Biomedical Engineering and Technology*, Lecture Notes in Bioengineering,
https://doi.org/10.1007/978-981-15-6329-4_14

1 Introduction

1.1 Background

Now a day's microstrip patch antenna became very popular due to low size, easily fabricated, low weight, portable, and easy to install with electronic circuits. So the patch antenna is easily embedded with any type of devices which may portable or non portable. There are different band in different wireless application which is the alternative of cable so we can easily transmit the data from source to destination without disturbing of any information. There are many wireless applications such as industrial, medical, and scientific application (ISM band) use in various portable or non-portable devices for cordless head phone, baby monitoring, and Bluetooth earpiece, etc. Earlier antenna is used the single band for single application by using simple formulae. From the last decade instead of using single band for single antenna, the multiband antenna became very popular by using slits, slots, and partial ground plane technique. The size of antenna is reduced by taking the help of slits, slots, surface current distribution, and defected ground structure. The multiband antenna is applicable to different band in several portable advance devices such as laptop, cellular phone sets, stylish phones, etc. A multiple of antenna designs such as series capacitor is presented to reduce the size of antenna and double circular slot ring resonator is produced multiple band (Singh and Mahesh 2018), the square shape slot at ground plane reduce size and produce multiband (Melkeri and Hunagund 2017), combination of E and T shape slots on the patch and some ground plane adjustment is responsible to shifted frequency 3 to 0.9 GHz which reduce the size of size of antenna (Ambh and Singhal 2016), Couple of circular rings is added in ground plane to reduce the size of antenna for maintaining reference band (Lastname et al. 2015), the pair of mirrored L-shaped strips for multiband operation is used in Tshaped monopole antenna which reduce size of antenna (Jui-Han 2015), Meander slots is responsible to reduce the size of antenna (Brocker et al. 2014), Spiral planar inverted F is used to reduce the size of patch antenna (Achmad Munir 2014), the defected microstrip structure is used to the minimize size of antenna (Singh 2011), Koch shaped fractal defect is used in patch antenna to reduce the size of antenna (Kordzadeh and Hojat Kashani 2009) instead of Split-Ring Resonator, the Complementary Split-Ring Resonator (CSRR) is used to reduce the size of antenna (karimzadeh Baeel et al. 2007), G strip on front side and U strips on back side of substrate is used to reduce the size of antenna as well as to bring several band (Prabhu et al. 2018), circular cross slots are used on patch and ground plane to reduce the size of antenna (Nada and Tawfeeq 2017), have been proposed different technique to reduce the size of antenna for several applications like GSM, WiMAX, WLAN, Bluetooth, Zigbeeand Hyperlan, etc.

1.2 Proposed Technique

Easily we can designed single band by using predefined formulae for the different individuals application like WLAN, WiMAX, HYPERLAN, Zigbee, and Bluetooth. The same single band antenna is designed to produce multiple bands with the help of fractal shape, circular ring, slits, slots, metamaterials, etc. The size reduction of antenna play a very important role in portable devices such as cellular phone. The multiband and size reduction of antenna is very important instead of using single band. The proposed antenna has surface current distribution technique to reduce the size of antenna by maintaining the multiband operations. The Surface current distribution technique is one good technique to reduce the size of antenna. In which area have more surface current and which area does not have surface current. So those portions of antenna do not covers with the surface current meaning of that part does not take in activity to radiate the electromagnetic wave. We can remove that part to reduce the size of antenna. The planar monopole antennas have less gain due to the small size of antenna because the electrical length of patch antenna is directly proportional to gain of antenna. The current is carried from port to end of copper part on patch or ground plane but disturbance is created in between end of copper part on patch or ground plane so the length of electric current is increased which improve the gain of antenna. The capacitive and inductive effect is responsible for the shifting center frequency from higher to lower side or lower to higher side.

Fc = $1/2\,\pi\sqrt{LC}$ L-inductance.

Fc-center frequency C-capacitance.

The Bandwidth enhancements are inversely related to quality factor of antenna. The quality factor depends on copper size of antenna. The bandwidth can be improved by adding slot in copper layer so copper area decreases means quality factor decreases.

Bandwidth 1/Q.F.

Q.F.-Quality factor.

The main aim to reduce the size of antenna with the help of surface current distribution on the back and front side of antenna. After reducing the size of proposed antenna is also worked industrial, scientific, and medical (ISM) applications. It can be assist the upcoming designers to reduce the size of antenna using the surface current distribution technique.

2 Antenna Design and Simulation Approach

Initially designed of planar antenna is shown in Fig. 1 without size reduction. The FR4 substrate is used to designed antenna with relative permittivity 4.3 and height 1.6 mm.

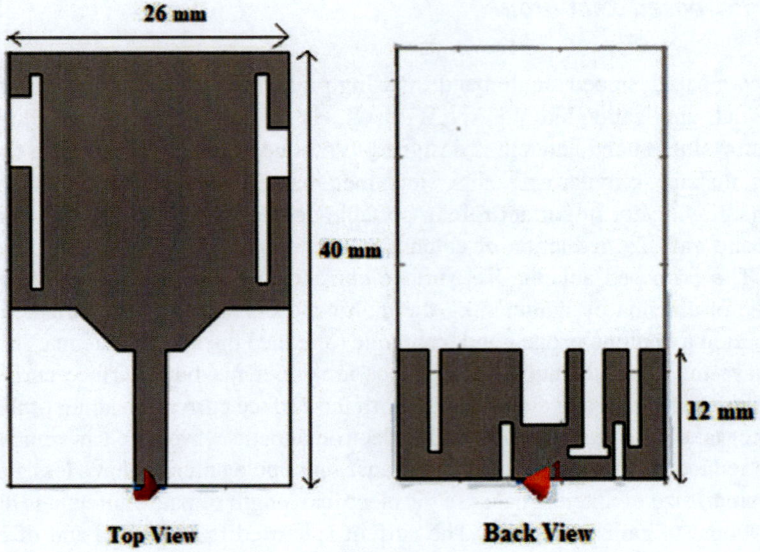

Fig. 1 Antenna 1

Figure 1 shows top view and back view of planar antenna which size is 40 × 26 mm^2. Figure 2a shows that where the maximum current is concentrated so that part is responsible to produce 2.41 GHz resonance frequency.

Figure 2b shows that where the maximum current is concentred so that part is responsible to produce 3.5 GHz resonance frequency. Figure 2c shows that where the maximum is current concentrated so that part is responsible to produce 5.33 GHz resonance frequency.

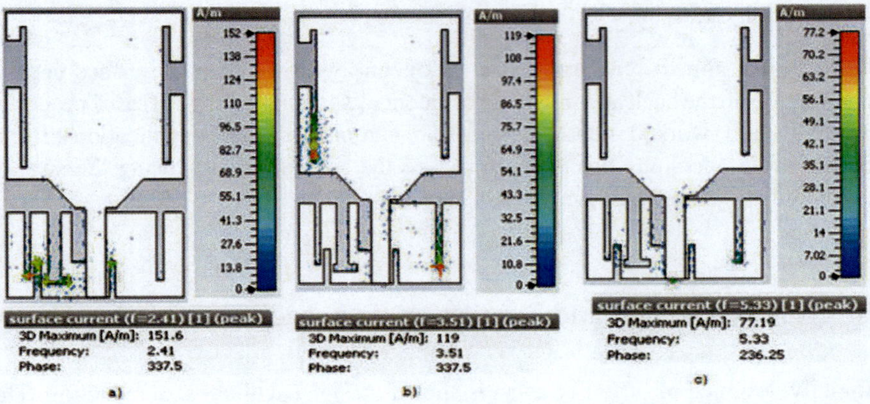

Fig. 2 Surface current distribution for **a** 2.41 GHz, **b** 3.5 GHz, and **c** 5.3 GHz

Figure 3 shows presented antenna after reduction with the help of surface current distribution on the planar antenna. Antenna 1 is 40 mm length but above 30 mm to 40 mm that part does not responsible to produce 2.41, 3.5, and 5.3 GHz center frequencies. So those portions removed from the antenna 1 and reduced size of antenna 2 is 30 × 26 mm^2 by maintaining the same resonance frequencies with useful bandwidth. There is 25% size reduction by using surface current distribution technique.

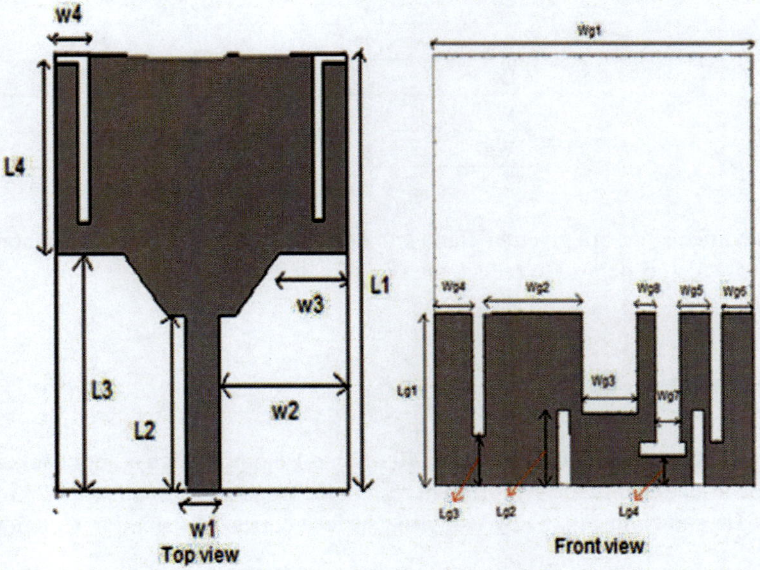

Fig. 3 Geometry of the proposed antenna after reduction (Antenna 2)

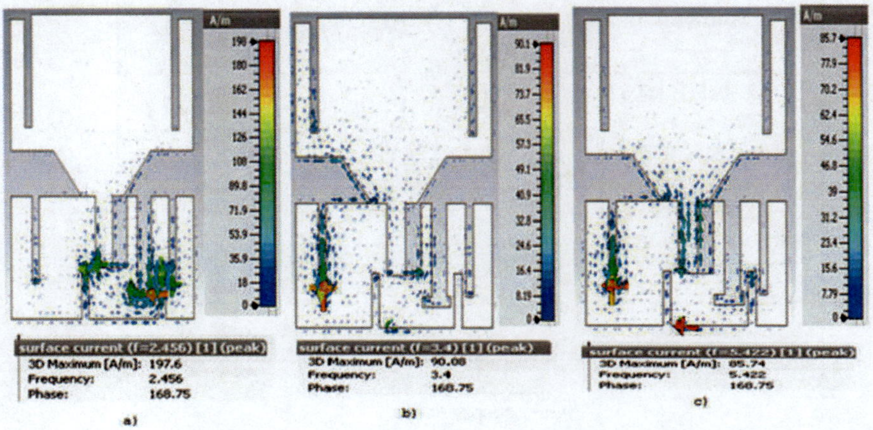

Fig. 4 Surface current distribution for the frequencies **a** 2.45 GHz, **b** 3.4 GHz, and **c** 5.422 GHz

Table 1 Antenna designed parameters with dimensions (units in mm)

Designing parameters	Dimensions (mm)	Designing parameters	Dimensions (mm)
W1	3	Wg3	4.5
W2	11.5	Wg4	3
W3	6	Wg5	2.5
W4	3	Wg6	2.5
L1	30	Wg7	2
L2	12	Wg8	1.5
L3	16.5	Lg1	12
L4	13	Lg2	5.25
Wg1	26	Lg3	3.45
Wg2	8	Lg4	2

The dimensions are given in Table 1 to designed the multiband planar monopole antenna which is helpful to reduce the size of antenna (Fig. 4).

3 Simulation Result

The designed antenna 1 dimensions is $40 \times 26 \times 1.6\ mm^3$ which is applicable for the multiband operations for the different resonance frequencies such as 2.42, 3.5, and 5.43 GHz shown in Fig. 5. By using the surface current distribution technique the

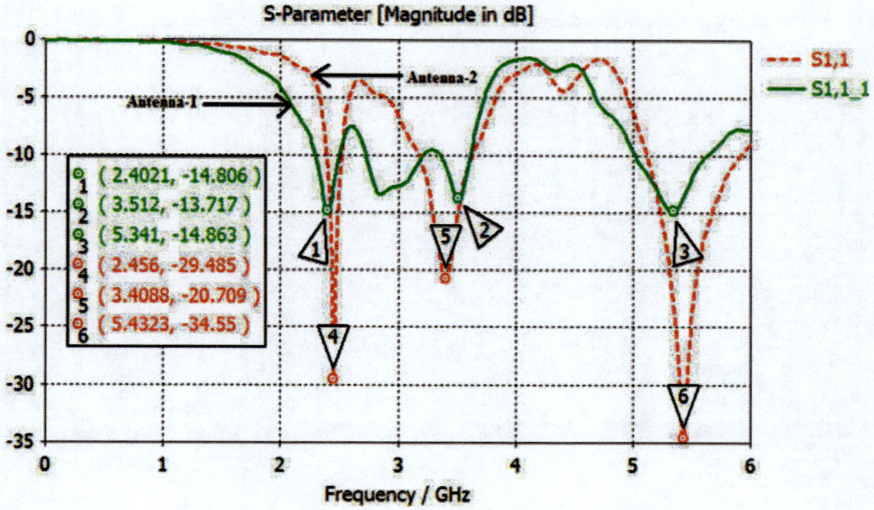

Fig. 5 Return Loss characteristics of designed antenna 1 and antenna 2

size is reduced in designed antenna 1 because those part does not having any surface current i.e. that portion does not take part to produce radiations shown in Fig. 2 for the resonance frequencies 2.42, 3.5, and 5.43 GHz. The dimensions of designed antenna 2 is $30 \times 26 \times 1.6\ mm^3$ which is reduced from designed antenna 1 using surface current distribution technique. There is 25% size reduction without disturbing required bands which is present in designed antenna 1. The standard frequency range of Zigbee, Bluetooth, WLAN, and ISM is 2.4 to 2.5 GHz. The proposed antenna is covered from 2.39 to 2.51 GHz with bandwidth of 120 MHz. The standard frequency of WiMAX is 3.4 to 3.6 GHz range which is covered in proposed antenna with 3.2 to 3.63 GHz with bandwidth of 430 MHz. The HYPERLAN 1 AND HYPERLAN 2 is covered from 5.15 to 5.85 GHz frequency range but in presented antenna is covered from 5.12 to 5.91 GHz with bandwidth of 490 MHz. The proposed antenna is useful to required bandwidth for multiband operation. All above simulation is done with the help of computer simulation technology (CST) software. CST is electromagnetic simulator software which gives the behavior of electric and magnetic of designed antenna in time or frequency domain (Fig. 6).

4 Experimental Results

The Vector Network Analyser (VNA) is used to find the results of fabricated planar monopole antenna. The actual physical view of fabricated planar monopole antenna is as shown in Fig. 7 (Table 2).

In Fig. 7 shows the SMA connector is used to interface between fabricated antenna and Vector Network Analyser to find out the measure results of planar antenna.

Figure 8 shows the measured results of proposed antenna for the reflection coefficient which is nearly same simulated results of designed antenna so the proposed planar antenna is applicable to the multiband oprations.

5 Conclusions

In this paper, the size of planar monopole antenna is reduced by using surface current distribution technique without disturbing any required bands. The portion of antenna which does not have surface current that part does not radiates the electromagnetic waves at outward side. Whereas the portion having more current are responsible to produced the resonance frequencies. The simulated results of antenna 1 shows that, it is applicable to multiband operation by achieving required band with resonance frequencies shown in Fig. 5. After size reduction in antenna 1 using surface current distribution technique is also applicable for multiband operation shown in Figs. 5 and 8.

The fabricated result of proposed antenna is nearly same with simulated results. This surface current distribution technique is very easy method to reduce the size of

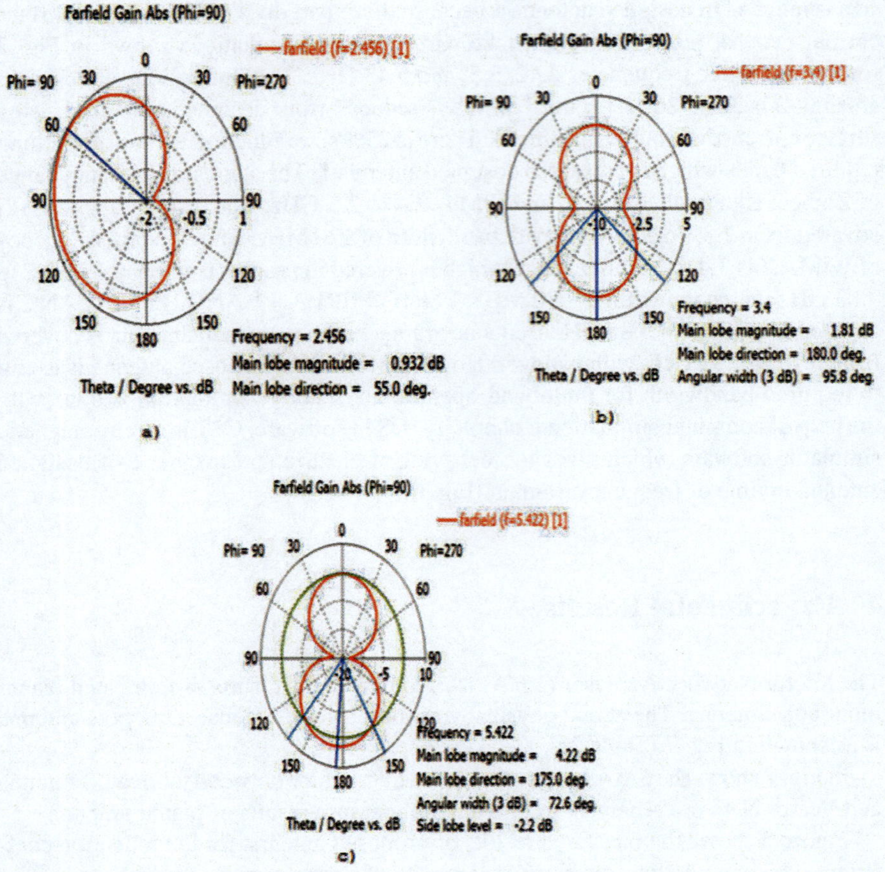

Fig. 6 Pattern of polar type radiation for **a** 2.54 GHz **b** 3.57 GHz **c** 5.43 GHz

antenna because we can easily see which part of antenna have much more surface current or which portion does not have surface current. Hence this technique is helpfulto reduce the size of antenna without affecting anything forwireless applications to required resonance frequencies.

The ISM band ranges 2.4–2.5 GHz which is internationally reserved for industrial, scientific, and medical purpose excluding telecommunications. It is unlicensed band which do not require permission from government authorities. This band is useful in medical field like diathermy machines and hyperthermia therapy. These devices are created electromagnetic radiation which disturbs communication by using matching frequency. The diathermy machines use electromagnetic waves in the ISM bands to concern profound heating to body for recreation and curing. The hyperthermia therapy is used microwaves to warm tissue to destroy cancer cells.

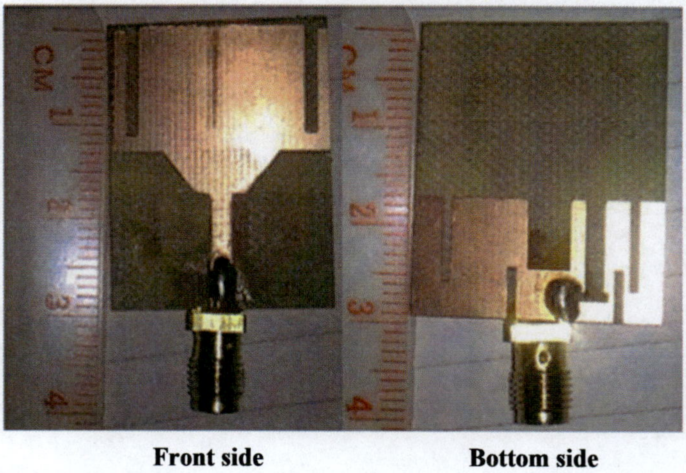

Fig. 7 Fabricated of the proposed planar antenna

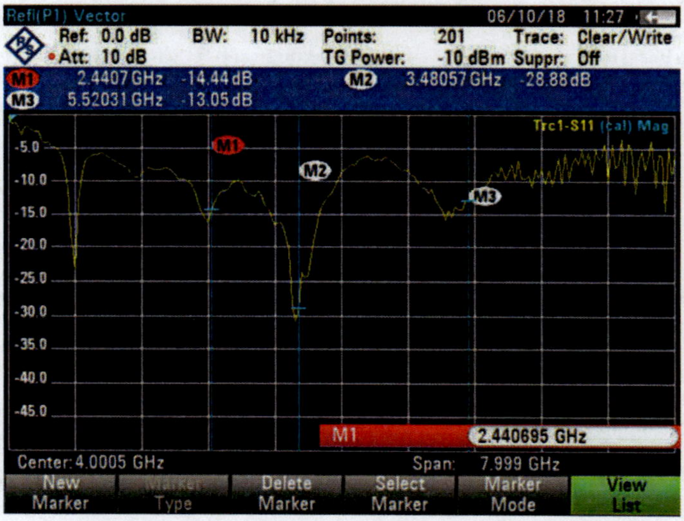

Fig. 8 Measured Characteristics of S11

Table 2 Comparison with proposed antenna and other authors

References	Designed technique	Size reduction in %
Singh et al. (2018)	Modified double circular slot ring resonator(MDCSRR) on ground plane and inter digital capacitor(IDC) on patch	68.83
Melkeri and Hunagund (2017)	Square shaped slot at the ground plane	90
Ambh and Singhal (2016)	E and T shaped slots on ground plane	86.73
Ripin et al. (2015)	Circular ring on ground plane	42
Lu et al. (2015)	pair of mirrored L-shaped strips	20
Proposed	Surface current distribution technique	25

References

Ambh A, Singhal PK (2016) Size reduction of rectangular microstrip patch antenna for gsm application. IRJET 3:255–258

Brocker DE, Werner DH, Werner PL (2014) Dual-band shorted patch antenna with significant size reduction using a meander slot. IEEE Antennas and Propagation Society International Symposium (APSURSI), USA

karimzadeh Baeel R, Dadashzadehl G, Geran Kharakhilil F (2007) Using of CSRR and its equivalent circuit model in size reduction of microstrip antenna. IEEE

Kordzadeh A, HojatKashani F (2009) A new reduced size microstrip patch antenna with fractal shaped defects. Prog Electromagnet Res 11:29–37

Kushwah VS, Tomar GS (2011) Size reduction of microstrip patch antenna using defected microstrip structures. International Conference on Communication Systems and Network Technologies, IEEE

Lu J-H, Zeng B-R, Li Y-H (2015) Planar multi-band monopole antenna for WLAN/WiMAX applications. ISANP, conference January, Taiwan

Melkeri VS, Hunagund PV (2017) Study and analysis of sqmsa with square shaped defected ground structure. IAIM IEEE, India

Munir A, Harish A, Chairunnisa C (2014) Size reduction of UHF planar inverted-F antenna with patch geometry modification. IEEE, ISAP December

Prabhu P, Manikandaswamy S, Saminathan T, Muthu Kumaran B (2018) A compact G and U strip folded planar multiband antenna for wireless applications. Int J Pure Appl Mathemat 118:77–83

Ripin N et.al. (2015) Size miniaturization and bandwidth enhancement in microstrip antenna on a couple circular rings Dgs. Int J Latest Res Sci Technol ISSN (Online): 2278–5299 4(4):27–30, July-August

Singh AK, Abegaonkar MP, Koul SK (2018) Miniaturized multiband microstrip patch antenna using metamaterial loading for wireless application. Progress Electromagnet Res C, 83:71–82

Tawfeeq NN (2017) Size reduction and gain enhancement of a microstrip antenna using partially defected ground structure and circular/cross slots. Int J Elect Comput Eng 7(2):894~898, April 2017, Iraq

Classification of Hepatic Disease Using Machine Learning Algorithms

Lokesh Singh, Rekh Ram Janghel, and Satya Prakash Sahu

Abstract Liver is one of the biggest gland and only organ of the human body which can restore harmed cells. Disorders in liver function affects food digestion and releasing of harmful toxic substances from the body which may results in severe liver diseases like jaundice, abdominal pain abdominal swelling, etc. These diseases require clinical care by experts. Early diagnosis is the only demand that remains to prevent this speedy loss of liver function. Only few of the specific algorithms are employed to operate medical instruments (CT-scan, MRI, EEG, ECG, etc.) to diminish time and the expenditure in diagnosing the liver disease. This work demonstrates the use of numerous classification methods (which were not experimented over ILPD dataset) in designing the classification model. The performance of classification methods are assessed over several measures. ILPD dataset sourced from UCI machine learning repository is used in performing the experiments. Nine classifiers of two classification models are employed in experimentation, namely, SVM and Ensemble. Classifiers are analyzed and compared on various validation measures. Results prove that Coarse Gaussian SVM achieves highest accuracy of 71.4% with less training time achieved of 0.61 s.

Keywords SVM · Ensemble · Coarse gaussian SVM · Machine learning · ILPD

L. Singh (✉) · R. R. Janghel · S. P. Sahu
National Institute of Technology, G.E Road, Raipur 492001, India
e-mail: lsingh.phd2017.it@nitrr.ac.in

R. R. Janghel
e-mail: rrjanghel.it@nitrr.ac.in

S. P. Sahu
e-mail: spsahu.it@nitrr.ac.in

A. A. Rizvanov et al. (eds.), *Advances in Biomedical Engineering and Technology*, Lecture Notes in Bioengineering,
https://doi.org/10.1007/978-981-15-6329-4_15

1 Introduction

Liver is the biggest meaty organ of the body which fulfills various vital functions including detoxification and releasing of harmful chemicals from the body increases the metabolism. Among its various functions, the main function of liver is the blood purification before allowing it to flow in rest of parts of the body. It is helpful in blood clotting by producing proteins (Vijayarani and Dhayanand 2015).

In the field of computer science machine learning classification algorithms are appropriate for the diagnosis of various diseases and can be employed for designing automated medical diagnosis device. There may be certain factors like obesity, intake of drugs on regular basis, mental stress, consumption of alcohol, contaminated water, etc., which may be the cause of liver disorders like liver cancer, hepatitis—A, B, C, D, and E, etc. Indications of liver disease includes—abdominal ache and swelling, intensify itching, nausea, exhaustion, back-ache, etc. These and many more symptoms of liver disorders are hard to diagnose at an early stage as liver continues functioning when it is partially injured. Early prediction at an early stage is the only remedy to reduce the mortality rate due to liver disorders (Ghosh and Waheed 2017).

Factors Affecting Lever (Rajeswari 2010) (Vijayarani and Dhayanand 2015):

Various factors that might affect vital functions of liver if not focused like: Obesity, Viruses, Excessive use of alcohol, Hereditary, and Diabetes.

Liver Disorders (Priya 2018)**:** Common Liver disorders are discussed in this section.

1. Hepatitis: Lever usually gets infected by viruses like hepatitis A, B, and, C. There are some other causes of Hepatitis due to excessive intake of alcohol, medicines, obesity, etc.
2. Cirrhosis: Any factor which harms the liver can lead to Cirrhosis which makes the liver unfit to function in a proper manner.
3. Liver cancer: Cirrhosis may lead to lever cancer like Hepatocellular Carcinoma.
4. Liver failure: Several factors are responsible for the failure of liver including contaminated food, excessive usage of alcohol, obesity, and many more.
5. Hemochromatosis: one of the disorder responsible for damage of liver is the iron which silently sets in the liver due to hemochromatosis which leads to multiple health problems.

Symptoms of Liver Disease (Gregorio Maldini): The following are the symptoms which sign liver disease are—Severe Jaundice, Intensify abdominal pain, Abnormal swelling, More confusion, Acute bleeding, Nausea, Pale stools, Increase in fatigue, and Sudden weight loss.

Building Blocks of Liver: One of the major building block of liver is Phosphatidylcholine (PC) (Kidd 1996). Other building blocks are:

1. Vitamin B12—to manage nerve cells
2. Folic Acid—synthesis of DNA, RNA
3. Iron—for generation of red blood cells,

4. Vitamin A—for healthy vision, teeth
5. Vitamin D—promote bone growth
6. Vitamin K—blood clotting.

The main purpose of designing this prediction model is the early diagnosis of liver diseases using machine learning classification algorithms which were never been implemented for classification of liver patient which may play an important role in reducing the mortality rate due to liver disease disorders. The results achieved are hopefully fruitful comparatively previously used algorithms in terms of various measures like accuracy and training time, etc.

The rest of the research work is organized in following manner: Section-II describes Related Work regarding Liver Disease Prediction, Section-III defines the System Architecture, Classification Algorithms and Dataset used, Section-IV determines evaluated Results, and Section-V concludes the results of the research work through Conclusion.

2 Related Work

Dinu et al. in (2017) focused on A.I-based techniques in designing a system for predicting and diagnosing various diseases which can help doctors or experts in analyzing the disease correctly. The strength of deep learning in designing such a system can raise up the true positive rate while reducing the false positive rates and improves the decision-making ability. The author has proved in this research that the employed A.I algorithm gives significant accuracy in diagnosing diseases. Details on Liver Disorder Diagnosis using Artificial Neural Network can also be found in (Janghel 2016). Ghosh et al. in (2017) performed experiments over UCLA and AP dataset (statistics course dataset). Several machine learning algorithms like K* algorithm, NBC, Bagging, Logistic, and Rep Tree are employed for diagnosing of diseases. All these algorithms are evaluated on various measures like accuracy, precision, sensitivity, and specificity. Results from the experiments show that K* algorithm outperforms comparatively other algorithms as it provides highest accuracy with minimum error rate and thus can diagnose liver disorders accurately. Kumar et al. in (2017) employed various machine learning classification algorithms like C4.5, Random Forest, CART, Random Tree, and REP tree for prediction of liver diseases. 80% of data is employed as training data and 20% of data is used for testing. The database results obtained from the experiment shows the highest accuracy of 79.22% from Random Forest algorithm. Eshraghi et al. in (2013) designed a model for the prediction of liver diseases. Eleven classification algorithms are analyzed and compared over the dataset on different factors like accuracy, precision, and recall. After performing various investigations Bayesian boosting is used in the experiment to achieve better results. Vijayarani et al. in (2015) used two machine learning algorithms named Naïve Bayes and Support Vector Machine for diagnosing liver disease. These two approaches are analyzed and compared over accuracy and execution time

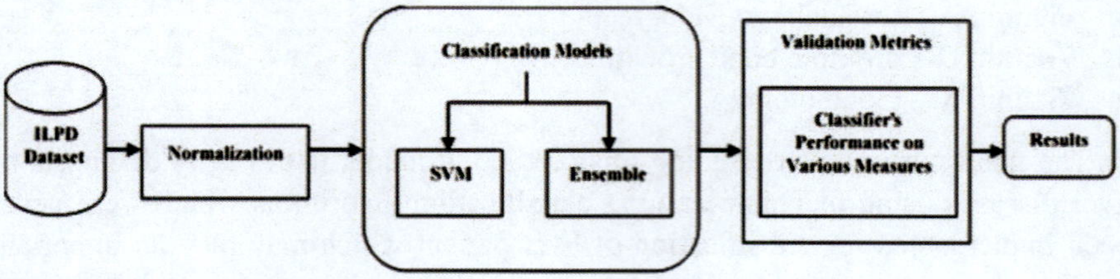

Fig. 1 Proposed model diagram

measures. During experimentation SVM has achieved highest accuracy while Naïve Bayes has achieved minimum execution time.

3 Classification Experiments

3.1 System Architecture

Proposed model diagram is depicted in Fig. 1. The diagram shows the process undergone in the research.

3.2 Normalization

Normalization can be defined as a scaling approach or can be referred as mapping method or a preprocessing stage. Through Normalization we can obtain novel range from the current range. It tends to be useful in approaches like prediction (Patro and Kumar).

3.3 Classification Algorithms

For experimentation, classification models are categorized into two parts, namely, SVM and Ensemble, which comprises of various classifiers. Below is the brief description of algorithms used.

Support Vector Machine. Support Vector Machine is a supervised learning approach firstly introduced by Cortes and Vapnik in 1995. Their objective behind designing the SVM is transforming binary class data into feature space or higher dimensional space and find an optimum separating hyperplane among set of hyperplanes which provides maximum margin between two different classes (Sisodia 2010). SVM is a binary classifier which has the ability to classify both linear and non-linear separable data. SVM locates the decision boundary by very small subset of training samples called as the support vectors (Sisodia 2014).

The class y is defined as follows:

$$y^{(i)} = \begin{cases} -1 \; if \; w^T x^{(i)} + b \leq -1 \\ 1 \; if \; w^T x^{(i)} + b \leq 1 \end{cases} \tag{1}$$

Can also be written as $y^{(i)}$ $(w^T x^{(i)} + b) \geq 1$. The objective of SVM is to fulfill two requirements:

a. There should be maximum possible distance between the two decision boundaries. The equation of hyperplane can be defined as $w^T x + b = 1$ and the distance between the hyperplane can be defined mathematically as $w^T x + b = 1$ which should be maximize. The distance between the margin is defined as $\frac{2}{||w||}$. If we want to maximize the distance i.e. max $\frac{2}{||w||}$, equivalently we have to minimize min $\frac{||w||}{2}$.
b. Support Vector Machine should also correctly classify all $X^{(i)}$, which is the ith sample in the dataset, i.e.,

$$y^{(i)}(w^T x^{(i)} + b) \geq 1, \qquad \forall i \in \{1, ..., N\} \tag{2}$$

Linear SVM, Cubic SVM, Quadratic SVM, Fine Gaussian, Medium Gaussian SVM, and Coarse Gaussian SVM are the classifiers employed in SVM classification model (Alpaydin 2010). Comparative description of different SVM Classifiers are described in Table 1 [https://www.mathworks.com/help/stats/choose-a-classifier.html].

Ensemble. In supervised learning Ensemble classifiers are defined as set of individual classifiers which are trained on a database in a supervised fashion. The concept behind the ensemble classifier is the recompense of errors (Rahman and Tasnim 2014). It works in three steps: in first step it creates multiple datasets from the original training data, in second step it multiple classifiers, in third step it combines all the classifiers to obtain more accurate predictions using weighted majority voting. Three methodologies are employed in designing an effective ensemble system. Following three methodologies are known as basic pillars of ensembles. (1) Sampling of data, (2) training of member classifiers, and (3) combining the classifiers. Ensemble methods not only tends to be useful in reducing the variance but also helpful in increasing the generality very effectively. Ensemble-based systems gain their strength in managing the large volume of data in an incremental manner (Cha Zhang and Yunqian Ma 2012) Boosted Trees, Bagged Trees, and RUS Boosted Trees are the classifiers employed in Ensemble classification model. Table 2 represents comparative description of all ensemble classifiers [https://www.mathworks.com/help/stats/choose-a-classifier.html].

Table 1 Comparative description of all SVM classifiers

Classifier type	Prediction speed	Memory usage	Interpretability	Model flexibility
Linear SVM 	Binary: Fast Multiclass: Medium	Medium	Easy	Low, Makes a simple linear separation between classes
Quadratic SVM 	Binary: Fast Multiclass: Medium	Binary: Medium Multiclass: Large	Hard	Medium
Cubic SVM 	Binary: Fast Multiclass: Medium	Binary: Medium Multiclass: Large	Hard	Medium
Fine Gaussian SVM 	Binary: Fast Multiclass: Medium	Binary: Medium Multiclass: Large	Hard	High, decreases with kernel scale setting. Makes finely detailed distinctions between classes, with kernel scale set to sqrt (P)/4
Medium Gaussian SVM 	Binary: Fast Multiclass: Medium	Binary: Medium Multiclass: Large	Hard	Medium, Medium distinctions, with kernel scale set to sqrt (P)
Coarse Gaussian SVM 	Binary: Fast Multiclass: Medium	Binary: Medium Multiclass: Large	Hard	Low, Makes coarse distinctions between classes, with kernel scale set to sqrt (P)*4, where P is the number of Predictors

Table 2 Comparative description of ensemble classifiers

Classifier type	Prediction speed	Memory usage	Interpretability	Model flexibility
Boosted trees	Fast	Low	Hard	Medium to High
Bagged trees	Medium	High	Hard	High
RUS boosted trees	Fast	Low	Hard	Medium

Table 3 Attribute information of the dataset

S.No	Features	Description	Type
1	Age	Age of the patient	Real number
2	Gender	Gender of the patient	Categorical
3	TB	Total bilirubin	Real number
4	DB	Direct bilirubin	Real number
5	Alkphos	Alkaline phosphotase	Integer
6	Sgpt	Alanine aminotransferase	Integer
7	Sgot	Aspartate aminotransferase	Integer
8	TP	Total protiens	Real number
9	ALB	Albumin	Real number
10	A/G Ratio	Albumin and globulin ratio	Real number
11	Class	Selector field used as class label	Binomial class

3.4 Dataset Used

For experimental purpose MATLAB tool is used to perform the experiment. In this study ILPD (Indian Lever Patient Dataset) is chosen for building and testing classification models, which is taken from UCI repository. Table 3 shows attribute information of the dataset. The information in the dataset is collected from north east of Andhra Pradesh, India. The dataset comprises of 10 attributes and 1 selector field (class) which classifies the records as liver patient and non-liver patient. The dataset has a total 583 instances, out of which 416 records classifies liver patient and 167 records classifies non-liver patient. Out of all 583 instances 441 records are male patient's records and 142 records are female patient records (Bendi Venkata Ramana, M. Surendra Prasad Babu 2018).

Brief Description Of Attributes (Sumedh Sontakke, Jay Lohokare 2017):

1. **Age**: This attribute represents age of the patient.
2. **Gender**: This attribute represents gender of the patient.
3. **TB**: Bilirubin is a substance which is constructed during the break down of old red blood cells. It helps the liver in digesting the food.
4. **DB**: Also known as conjugated bilirubin, it is usually passed in small amount through kidneys and excreted via urine.
5. **Alkphos**: It is a type of enzyme found in our body in various tissues. It is used for the diagnosis of damage of liver and disorders in bone.
6. **Alanine Aminotransferase**: Alanine Aminotransferase is a kind of enzyme found in the cells of our liver and kidney. It is considered as the initial screening test for diagnosing liver disease.

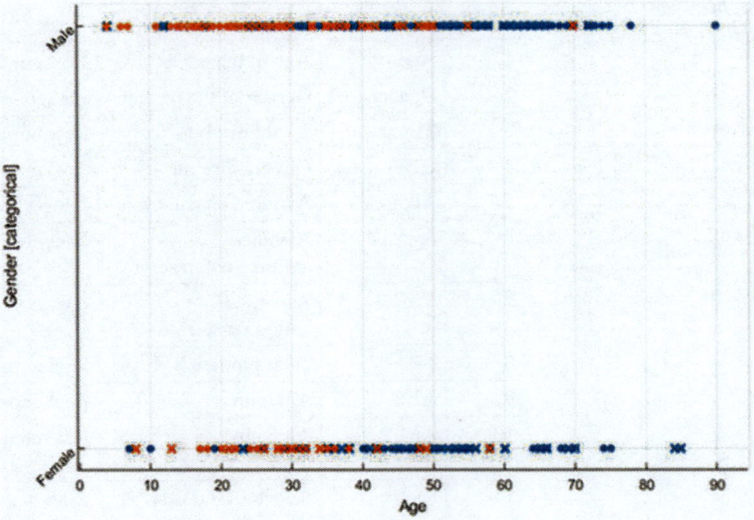

Fig. 2 Scatter Plot of original dataset: Indian Patient Liver Dataset (ILPD)

7. **Aspartate Aminotransferase**: It is a type of enzyme which when released shows the indication of damage of liver or muscles.
8. **Total Protiens**: Total protein test measures the total amount of albumin and globulin in our body.
9. **Albumin**: It is one of the main protein induced in the lever to maintain intravascular colloid osmotic pressure (COP).
10. **A/G Ratio**: The albumin/globulin (A/G) ratio is used to evaluate liver and kidney disease.
11. **Class**: It is a selector field used for binary classification.

Scatter plot diagram of ILPD original dataset is shown in Fig. 2.

Validation Metrics. Following are the details of the validation measures used in the experimentation.

- **Accuracy**: Statistical measure of correctly predicted instances to the total no of instances. It can be defined as:

$$A = \frac{TP + TN}{Total\ no\ of\ samples}$$

- **Precision**: Statistical measure of correctly predicted positive instances to the total no of predicted positive instances. It can be defined as:

$$P = \frac{TP}{TP + FP}$$

- **Recall**: Measure of proportion of retrieved relevant observation among the total relevant observations. It can be defined as:

$$R = \frac{TP}{TP + FN}$$

- **F-1 Score**: Measure of the weighted average of Precision and Recall. It can be defined as:

$$F1 = \frac{2TP}{(2TP + FP + FN)}$$

- **ROC**: Receiver Operating Curve is Measure to plot true positive rate against the false-positive rate.

4 Results

This section illustrates the results obtained in the experimentation using various classification methods which are deemed as superior data mining approaches employed in health-care for diagnosing the diseases. MATLAB tool is being used in experimentation. In total nine machine learning classifiers are employed in performing the experiment over ILPD dataset. tenfold cross-validation method is used for the validation of different classifiers. Table 4 represents comparative performance of different classifiers calculated over measures like Sensitivity, Specificity, Precision, and F-1 score.

As seen from Table 4 there is substantial difference between the sensitivity, specificity, precision, and F1-score of SVM and ensemble methods. Out of six SVM classifiers highest sensitivity is achieved by Cubic SVM 77.52% while in ensemble classifiers the highest sensitivity of 88.01% is achieved by RUS Boosted Trees. Sensitivity is deemed as the diagnostic ability where a person is diagnosed with disease

Table 4 Classification results of classifiers on various measures

Model Type	Classifiers	Sensitivity %	Specificity %	Precision %	F1-Score %
SVM	Linear SVM	71.36	0.0	100.0	83.28
	Quadratic SVM	75.20	48.42	88.22	81.19
	Cubic SVM	77.52	45.51	79.57	78.53
	Fine gaussian SVM	73.35	48.15	93.27	82.12
	Medium gaussian SVM	71.36	0.0	100.0	83.28
	Coarse gaussian SVM	71.36	0.0	100.0	83.28
Ensemble	Boosted trees	73.96	38.10	81.25	77.43
	Bagged trees	75.05	47.87	88.22	26.95
	RUS boosted treess	88.01	45.36	61.78	72.60

as positive. High sensitivity indicates few false negatives which illustrates few cases of disease are missed. The specificity in diagnosis indicates the person having no disease as negative.

Likewise, the highest 48.42% specificity is achieved by Quadratic SVM. The percentage of specificity achieved over ILPD dataset is quiet low Linear SVM along with Medium and Coarse Gaussian SVM has achieved 100% precision and highest F-1 Score of 83.28%. Figure 3 shows graphical representation of Table 4, a comparative performance of 9 classifiers over various measures.

The performance values of different machine learning classifiers calculated over Accuracy and Training-time measures. Table 5 represents the accuracy achieved and the execution time taken by various classifiers for the prediction of liver disease.

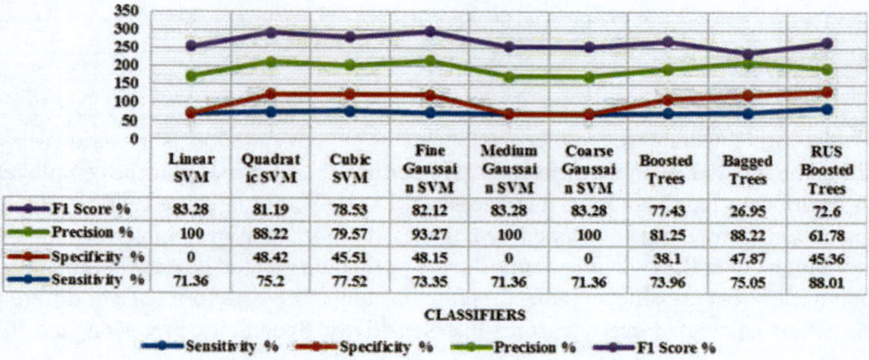

Fig. 3 Classifier's performance on various measures

Table 5 Classification results of classifiers on accuracy and training time

Model type	Classifiers	Accuracy %	Training time (sec)
SVM	Linear SVM	71.4	1.76
	Quadratic SVM	70.8	9.59
	Cubic SVM	69.0	51.1
	Fine gaussian SVM	71.0	0.80
	Medium gaussian SVM	71.4	0.63
	Coarse gaussian SVM	72.2	0.61
Ensemble	Boosted trees	66.2	7.49
	Bagged trees	70.7	4.98
	RUS boosted trees	66.7	5.32

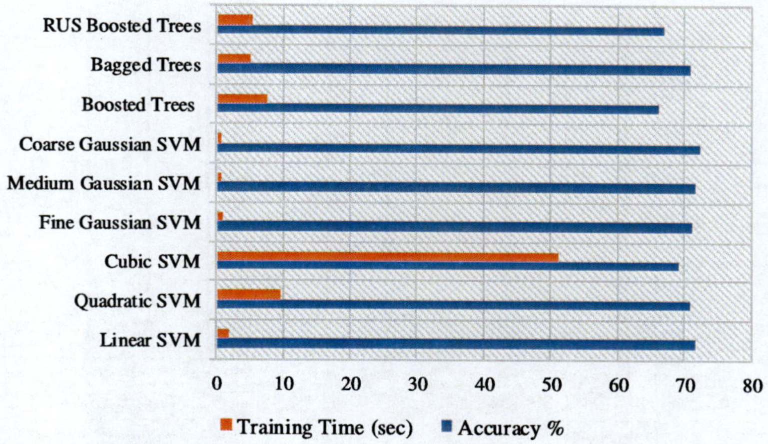

Fig. 4 Classifier's performance on accuracy and training time

Figure 4 shows graphical presentation of comparative performance of 9 classifiers over accuracy and training time.

It is clearly visible from the Table 5 that Coarse Gaussian SVM achieves the highest accuracy of 72.2% among all classifiers with minimum training time of 0.61 s. Confusion Matrix and ROC (Receiver Operating Characteristic) Area of Coarse Gaussian SVM is depicted in Fig. 5. Parallel coordinate plot is used to visualize high dimensional data, where each observation is represented by the sequence of its coordinate values plotted against their coordinate indices. For Coarse Gaussian SVM it is depicted in Fig. 6.

After performing the experiment, Table 5 concludes that Coarse Gaussian SVM outperforms among all classifiers using the ILPD dataset by achieving an accuracy of 72.2% using minimum training time of 0.61 s.

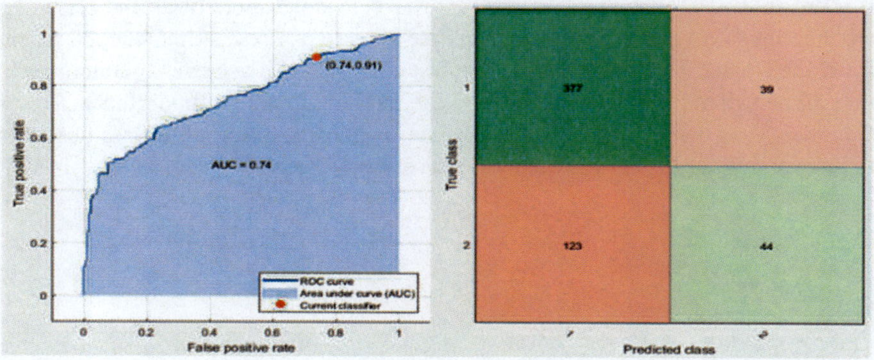

(a) (b)

Fig. 5 ROC and confusion matrix of coarse gaussian SVM

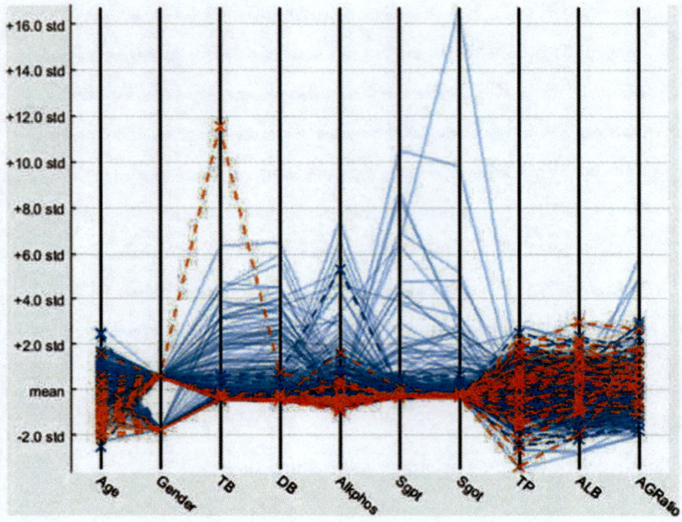

Fig. 6 Parallel coordinate plot of coarse gaussian SVM

5 Conclusion

The main strength of the designed model is the use of Machine learning algorithms used for classification in this study which were never been used to predict Liver disease. Though Liver diseases are hard to predict so this research work focuses on the key role of classification algorithms in predicting liver diseases. Classification algorithms are being chosen as they gain their strength in classifying the instances correctly once the system is trained while requiring less human efforts. Two classification models with their corresponding classifiers are used in conducting the research. For experimentation purpose Indian Liver Patient Database is chosen which is taken from UCI machine learning repository. Performances of all the classifiers are measured on various factors to fulfill the objective of research. Obtained results conclude Coarse Gaussian SVM as the best classifier as it achieves optimum accuracy among all the employed classifiers with minimum training time. Hence, Coarse Gaussian SVM outperforms in Liver Disease Diagnostic. Some more classifiers need to be explored for further research with the same objective.

References

Alpaydin E (2010) Introduction to Machine Learning Second Edition

Bahramirad S, Mustapha A, Eshraghi M (2013) Classification of liver disease diagnosis : a comparative study. IEEE, pp 42–46

Bendi Venkata Ramana, M. Surendra Prasad Babu NBV (2018) ILPD (Indian Liver Patient Dataset) Data Set. 1–2

Zhang C, Ma Y (2012) Ensemble machine learning. Springer International Publishing, New York Dordrecht Heidelberg London

Dinu AJ, Ganesan R, Joseph F, Balaji V (2017) A study on deep machine learning algorithms for diagnosis of diseases. 12:6338–6346

Ghosh SR, Waheed S (2017) Analysis of classification algorithms for liver disease diagnosis. J Sci Technol Environ Inf 5:361–370

Janghel RR, Shukla A, Verma K (2016) Soft computing based expert system for Hepatitis and liver disorders. In: 2016 IEEE International Conference on Engineering and Technology (ICETECH). pp 740–744

Kidd PM (1996) Phosphatidylcholine, a superior protectant against liver damage. Altern Med Rev 1:258–274

Kumar A, Sahu N (2017) Categorization of Liver Disease Using Classification. 5:826–828

Maldini G (2014) What are the Signs of Liver Disease. [Online]. Available: https://www.hawaiipacifichealth.org/media/3429/liver-disease-flyerr1.pdf

Rajeswari P GSR (2010) Analysis of Liver Disorder Using Data mining algorithm. Glob J Comput Sci Technol 10:48–52

Patro S, Sahu KK (2015). Normalization: a preprocessing stage. arXiv preprint arXiv:1503.06462

Priya MB, Juliet PL, Tamilselvi PR (2018) Performance analysis of liver disease prediction using machine learning algorithms. Int Res J Eng Technol 5:206–211

Rahman A, Tasnim S (2014) Ensemble classifiers and their applications. A Review Abstract 10:31–35

Sisodia D, Shrivastava SK, Jain RC (2010) ISVM for face recognition. In: Proceedings—2010 International Conference on Computational Intelligence and Communication Networks, CICN 2010. IEEE, pp 554–559

Sisodia D, Singh L, Sisodia S (2014) Fast and accurate face recognition using SVM and DCT. IEEE, pp 1027–1038

Sontakke S, Jay Lohokare RD (2017) Diagnosis of liver diseases using machine learning. International Conference on Emerging Trends & Innovation in ICT (ICEI). IEEE, Pune, pp 129–133

Vijayarani S, Dhayanand S (2015) Liver disease prediction using SVM and Naïve Bayes algorithms. Int J Sci Eng Technol Res 4:816–820

Anti-hyperlipidemic and Antioxidant Activities of a Combination of Terminalia Arjuna and Commiphora Mukul on Experimental Animals

Jhakeshwar Prasad, Ashish Kumar Netam, Trilochan Satapathy, S. Prakash Rao, and Parag Jain

Abstract The present study has been undertaken to evaluate antihyperlipidemic and antioxidant activities of Terminalia Arjuna (TA) and Commiphoramukul (CM) standardized extracts in combination at their predetermined doses. The hyperlipidemia in animals (rats) was induced by high fat diet by mixing Indian vanaspati ghee and coconut oil in the ratio of 3:1 (v/v). Acute toxicity was performed according to Organization of Economic Cooperation and Development (OECD) -423 and observed for behavioral changes, hematological and biochemical alteration if any or not for 14 days. The result of toxicity studies did not indicate any major changes in the result in comparison with control group of animals. The combination of plant extracts exhibited significant antihyperlipidemic activity in comparison to control group. The level of triglycerides (TGL), cholesterol (CHO), low-density lipoprotein (LDL), and very-low-density lipoprotein (VLDL) got reduced with increased level of high-density lipoprotein (HDL). The result of test drug and standard drug showed similar value with some minor difference. Hematology data of hyperlipidemic rats showed safety level of blood components after treatment with test drug. The test drug also revealed good antioxidant activity by normalization of superoxide dismutase (SOD) and nitric oxide (NO) levels. Thus, further study required to determine the active constituents from plants extracts required for biological activities.

Keywords Terminalia arjuna · Commiphoramukul · Super oxide dismutase · Nitric oxide

1 Introduction

Higher amount of lipids or fats in the blood is characterized as hyperlipidemia. It is a family disorder in which fatty contents get increased abnormally. However, increased amount of fats increases the risk of coronary heart disease (CHD) and also plays role in body's metabolic processes. Individual's diet also shows impact on

J. Prasad (✉) · A. K. Netam · T. Satapathy · S. Prakash Rao · P. Jain
Department of Pharmacology, Columbia Institute of Pharmacy, Tekari, Raipur, CG, India
e-mail: jhakeshwarprasad03@gmail.com

A. A. Rizvanov et al. (eds.), *Advances in Biomedical Engineering and Technology*, Lecture Notes in Bioengineering,
https://doi.org/10.1007/978-981-15-6329-4_16

hyperlipidemia; high cholesterol diet and food containing more saturated fats lead to increased blood cholesterol, and triglycerides levels. Other disorders, such as diabetes mellitus, kidney disease, and hypothyroidism, may promote hypertriglyceridemia (Fronzo and Ferrannini 1991). Most people who have hyperlipidemia also having relevant other complications such as diabetes, high cholesterol and have difficulty in managing all three conditions at a time. Hence combination therapies of more than two or three drugs are prescribed by physicians or clinicians those in turn produce severe adverse effects. Calcium channel blockers (CCBs) are one of the most potentially lethal prescriptions, which may worsen hyperlipidemia if administered excessively (Saeed and Larik 2017).

Frequent administration of CCBs may cause rapid fall in blood pressure, decreased heart rate, and cardiac arrest. However, overdoses of sustained-release formulations result in delayed onset of dysrhythmias, shock, sudden cardiac collapse, and bowel ischemia. Among the anti diabetic agents, Di-Peptidyl Peptidase-IV (DPP-IV) inhibitors are drug of choice for treatment of Type-II diabetes and recent research revealed that, long term administration of DPP-IV inhibitors at their therapeutic doses also causes pancreatic cancers. Herbal drug had been used since ancient times for welfare of the mankind and several research have been done to identify the active compound responsible for therapeutic activity. The active components of plants when taken together may give synergistic effect,when they have co-administered for the treatment of multifactorial disorders such as diabetes associated with hypertension and dyslipidemia. Terminalia Arjuna(TA) is a wild herb containing various chemical constituents. Among these arjunetin and arjunosides acts as a major constituent already been reported for having affinity for Na+ - K+ ATPase Pump. (Urizar and Moore 2003) Commiphoramukul (CM) also reported for having antihyperlipidemic activity. The objective of this present research work is to evaluate the affinity of Terminalia arjuna for Na+ - K+ ATPase Pump blocking effect which in turn may be useful as an antihypertensive agent. When more than two drugs are administered at a time there may be a chance of drug interactions. Hence, toxicity studies need to be carried out for these combination therapies and to achieve better therapeutic response of Terminalia arjuna and Commiphora mukul standardized extract at their predetermined ratio for their synergistic effect in the treatment of high fat diet induced hyperlipidemia using experimental animals in Rats (Dobrian et al. (2000).

2 Materials and Methods

2.1 Drug and Chemical Reagents

Terminalia Arjuna and Commiphoramukuldried extracts were received as a gift sample from SUNPURE Pvt. Ltd New Delhi (India). Carboxy methyl cellulose (0.5–5%) was purchased from LOBA Chemie Pvt. Ltd. Mumbai. Atorvastatin was procured from Sun Pharmaceuticals Pvt. Ltd. Mumbai, Maharashtra,

(India). Halothane was purchased from Korten Pharmaceuical Pvt. Ltd. Shanti-Sthal, Shirgaon-Palghar, Thane-Mumbai (India), and Formaldehyde was purchased from Merck Life Science Pvt. Ltd., Vikroli East, Mumbai, Maharashtra. All other chemicals used was of highest analytical grade-commercially available.

Experimental Animals. Healthy adult Male Albino Wistar rats weighing 180–200 gm were obtained from the Animal House Facility of Columbia Institute of Pharmacy, Raipur, Chhattisgarh, (India) having certificate number CIP/IAEC/2017/103 and Regd. No.1321/PO/ReBi/S/10/CPCSEA. The animals were kept and maintained under controlled environmental conditions with temperature (23 ± 2 °C), relative humidity (40–50%), and 12/12 h light/dark cycle. The animals received a standard pellet diet (Hindustan lever limited, India) and water ad libitum. The animals used in the present study were cared as per the principles and guidelines of Institutional Animal Ethics Committee (IAEC), and in accordance with the CPCSEA, New Delhi, India. The animals were acclimatized to laboratory conditions for at least seven days before initiation of the experiment.

Acute Toxicity. The acute toxicity was evaluated as per OECD guideline-423. Animals were received dose of Terminalia arjuna along with Commiphora mukul 250 mg/kg body weight orally administered by using an oral feeding needle after short fasting period. The general behavior of the animals was continuously monitored for 30 min, 1, 2, and 3 h after dosing, periodically during the first 24 h (with special attention given during the first 4 h) and then daily observed for 14 days.

Experimental Study. The experiment was carried out on animals (albino Wistar rats) to determine therapeutic effectiveness of combination study. In this experiment rats of either gender were randomly divided into four groups. Each group consists of five animals either gender (n = 5). All the animals were administered high fat diet for induction of hyperlipidemia (Table 1).

Induction of hyperlipidemia in rats. Hyperlipidemia was induced by feeding rats on diet rich in fats. It was prepared by mixing India Vanaspati ghee and coconut oil (3:1, v/v). This diet was given per-oral to rats at a dose of 3 ml/kg body weight daily (Munshi et al. 2014).

Table 1 Allocation of animals into various groups for therapeutic effectiveness study

Groups	Treatment	Doses
1	Control group	Drinking water (Oral)
2	Toxic group	High fat diet (3 ml/kg)
3	Standard drug (Atorvastatin)	10 mg/kg
4	Test group [CM+ TA (50:50)]	500/kg/body weight (Oral)

Hematological Study. The blood was collected with EDTA anticoagulant through retro orbital puncture for biochemical estimation. The evaluated blood parameters were red blood cell count (RBC), blood hemoglobin concentration, basophil, eosinophil and neutrophil granulocytes, lymphocytes, and monocytes, hematocrit value, mean corpuscular volume (MCV), mean corpuscular hemoglobin (MCH), mean corpuscular hemoglobin concentration (MCHC), white blood cell (WBC), and platelet counts.

2.2 *Biochemical Estimation of Antioxidants*

Superoxide Dismutase (SOD) Assay. The assay was performed by the production of superoxide from oxygen molecule using reduced b-nicotinamide adenine dinucleotide (NADH) as a reductant and phenazine methosulphate (PMS) as a catalyst. Nitrobluetetrazolium (NBT) was used as an indicator that turned blue when reduced by superoxide. Change in color was monitored spectrophotometrically in the visible range at 560 nm. While adding test drug to the reaction; the antioxidants (superoxide scavengers) competed with NBT to react with superoxide. The percent inhibition of NBT reduction was used to quantify superoxide-scavenging.

Procedure. 10% w/v tissue homogenate in 0.15 M TrisHCl or, 0.1 M phosphate buffer was prepared and centrifuged at 15,000 rpm for 15 min at 4 °C. The supernatant (0.1 ml) was taken and considered it as sample. Then 0.1 ml sample + 1.2 ml sodium pyrophosphate buffer (pH 8.3, 0.052 M) + 0.1 ml phenazinemethosulphate (186 μM) + 0.3 ml of 300 μM Nitroblutetrazolium + 0.2 ml NADH (750 μM) were mixed and incubated at 30 °C for 90 s followed by addition of 0.1 ml glacial acetic acid. This was then stirred with 4.0 ml n-butanol and allowed to stand for 10 min followed by centrifugation, and butanol layer was separate. The Optical Density (OD) of the rest of the sample was measured at 560 nm by taking butanol as blank (Paoletti et al. 1986; Sapakal et al. 2008).

Nitric Oxide Estimation. Nitric oxide is produced due to oxidative stress occurring in the brain. The assay was performed by taking 100 μl of serum sample in a test tube and added 400 μl of carbonate buffer (pH 9.0) followed by addition of copper cadmium alloy fillings (0.15 g). The reaction was stopped by addition of sodium hydroxide (100 μl of 0.35 M) and zinc sulphate solution (400 μl of 120 mM) under vortex mixing. Then the solution was allowed to stand for 10 min and centrifuged at 4000 rpm for 10 min. The clear supernatant solution (500 μl) was transferred to another test tube in which 500 μl of Griess reagent was added. The absorbance was noted spectrophotometrically at 548 nm. A standard curve (1–100 μM) was plotted using sodium nitrite to calculate the concentration of nitrite (Giustarini et al. 2008; Sastry et al. 2002).

Procedure. Mix the following in a spectrophotometer cuvette (1 cm pathlength) i.e., 100 μL of Griess reagent, 300 μL of the nitrite-containing sample and 2.6 mL of deionized water. Then incubated the mixture for 30 min at room temperature. A photometric reference sample was prepared by mixing 100 μL of Griess reagent and 2.9 ml of deionized water. Measured the absorbance of the nitrite-containing sample at 548 nm relative to the reference sample. Absorbance readings were converted to nitrite concentrations as described in calibration.

Histopathological Examination. The animals were anaesthetized with halothane and blood was collected by retro orbital puncture for biochemical estimation. The animals were again anaesthetized by using excess halothane and sacrificed by cervical dislocation method. The abdominal portions were cut opened and heart was dissected out. The Heart was removed immediately and transferred into 10% formalin solution for routine histopathological examination. The samples were taken from the sections of rat heart tissue with highest macroscopic damage. The heart tissue specimen from each animal was removed and fixed in 10% formalin solution then cut into 5 μm thickness, stained using hematoxylin eosin for the histopathological examination. They were made using a rotary microtome, 5 μm thickness sections were cut from the tissue samples embedded in paraffin and placed on standard glass slides. The paraffin was melted with a period of approx 12 h in an incubator at 58 °C. The samples were then stained with haematoxylene and eosin (H&E) according to the protocol. Qualitative analyses were performed on 400× magnified images.

3 Results

3.1 The Effect of Terminalia Arjuna Along with Commiphora Mukul on Behavioral Changes

The results of oral acute toxicity study indicated minor behavioral changes and no mortality observed in animals through the 3-days period following single oral administration at all selected dose levels of the Terminalia arjuna along with Commiphora mukul (Table 2 and Fig. 1).

3.2 The Effect of Terminalia Arjuna Along with Commiphora Mukul on Hematological Changes

See Table 3.

Table 2 Effect of Terminalia arjuna along with Commiphora mukul on lipid profile level in Albino Wistar rats

Parameters	Duration	Sex	Control	Toxic group	Standard	Test
TGL (mg/dL)	0 Day	M	62.5 ± 0.26	78.4 ± 0.22	63 ± 0.34	60.6 ± 0.33
		F	63.4 ± 0.36	77.3 ± 0.20	64 ± 0.32	59.4 ± 0.37
	10th Day	M	75.5 ± 0.23	89 ± 0.21	87.9 ± 0.24	73 ± 0.25
		F	76.6 ± 0.34	88 ± 0.19	88.7 ± 0.27	72 ± 0.28
CHO (mg/dL)	0 Day	M	53 ± 0.23	74 ± 0.25	57.2 ± 0.34	51.5 ± 0.36
		F	54 ± 0.25	73 ± 0.24	58.5 ± 0.32	50.4 ± 0.34
	10th Day	M	56.2 ± 0.34	71 ± 0.31	70.1 ± 0.33	54.5 ± 0.23
		F	57.2 ± 0.37	70 ± 0.30	71.3 ± 0.35	53.3 ± 0.25
HDL (mg/dL)	0 Day	M	11.4 ± 0.25	3 ± 0.27	11.5 ± 0.22	12.3 ± 0.33
		F	12.3 ± 0.34	2 ± 0.25	12.2 ± 0.20	12.7 ± 0.2
	10th Day	M	13 ± 0.23	4 ± 0.21	18.3 ± 0.31	17.5 ± 0.27
		F	14 ± 0.34	3 ± 0.20	19.4 ± 0.33	18.4 ± 0.25
LDL (mg/dL)	0 Day	M	23.6 ± 0.32	37 ± 0.30	22.5 ± 0.24	23.8 ± 0.21
		F	24.5 ± 0.31	36 ± 0.28	23.3 ± 0.25	24.3 ± 0.23
	10th Day	M	28.9 ± 0.25	26.7 ± 0.23	34.8 ± 0.32	44.8 ± 0.34
		F	29.7 ± 0.27	25.5 ± 0.21	35.5 ± 0.33	45.7 ± 0.32
VLDL(mg/dL)	0 Day	M	15.2 ± 0.31	28 ± 0.29	15.7 ± 0.25	14.4 ± 0.31
		F	14.5 ± 0.33	27 ± 0.27	16.5 ± 0.23	13.2 ± 0.33
	10th Day	M	15.8 ± 0.21	13.7 ± 0.19	17.9 ± 0.33	18.5 ± 0.34
		F	16.6 ± 0.24	12.6 ± 0.17	18.7 ± 0.31	17.7 ± 0.32

Mean ± SEM (n = 5)
Triglycerides (TGL), Cholesterol (CHO), High density lipoprotein (HDL) Low density lipoprotein(LDL), Very-low-density lipoprotein (VLDL)

3.3 *Biochemical Parameters Studies*

Superoxide Dismutase Assay

See Table 4 and Fig. 2.

Nitric Oxide (NO) Assay

See Table 5 and Figs. 3 and 4.

Histopathological Examination of Heart. Animal organ (Heart) histopathology report is shown below (Plates 1, 2, 3, 4, 5 and 6).

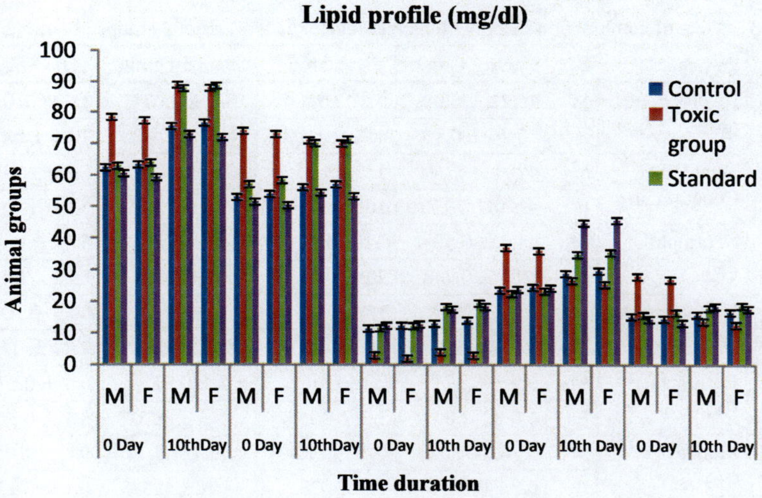

Fig. 1 Graph showing the effect of various treatments on lipid profile in different group of animals. All values are reported as Mean ± SEM (n = 5)

4 Discussion

Hyperlipidemia is a multifactorial disorder involving interactions among environmental, vascular, neuroendocrine, and genetic factors. The prevalence of hyperlipidemia is increasing in India as well as all over the world. Apart from these, the other cause include is more complex i.e., association of type-2 diabetes mellitus as well as obesity. Those are polygenic factors. This complexity makes it difficult to diagnose the disorder properly that make the researchers to look major contributions toward the developments of new drug/new entity for effective treatment. As the drugs available in the market for the treatment of hyperlipidemia associated with diabetes are limited, many patients need the combination therapy of anti-lipidemics that in turn causes various side effects. Hence the herbal therapy has come into existence.

Terminalia arjuna has traditionally been used for the treatment of various heart disorders for more than centuries. It improves cardiac muscle function subsequently improving pumping activity of the heart. Among the active constituent present in the Terminalia arjuna the saponin glycoside thought to be responsible for the ionotropic effect while flavonoids and oligomeric proanthocyanidins (OPCs) provide free radicals antioxidant activity. In other way Commiphora mukul an oleo gum-resin has been used as medications since Vedic period for the effective treatment of number of vascular disorders such as atherosclerosis, hypercholesterolemia, obesity, etc., but the scientific evidence for the combination of these two (Terminalia arjuna and Commiphora mukul) has not been established till yet. So this present study has been undertaken to evaluate the safety and effectiveness of both the drug at their

Table 3 Effect of combination therapy on hematological data of various groups of animal

S. no.	Particulars	Sex	Control Group	Toxic Group	Standard group	Test Group
1	Heamoglobin (gm%)	M	16.22 ± 0.08	5.21 ± 0.04	15.3 ± 0.07*	11.28 ± 0.10*
		F	15.26 ± 0.107	4.20 ± 0.03	14.46 ± 0.120*	10.4 ± 0.141
2	Total WBC Count (cmm)	M	4280 ± 37.41	1120 ± 31.21	4030 ± 50.99	5060 ± 143*
		F	4100 ± 70.710	1090 ± 29.19	3880 ± 135.64	5010 ± 86.023*
3	Neutrophils (%)	M	61.4 ± 0.50	24 ± 0.39	58.8 ± 0.8	57.2 ± 0.8
		F	59.4 ± 0.748	23 ± 0.37	57.4 ± 0.927	56.2 ± 0.8
4	Lymphocytes (%)	M	33.8 ± 0.37	15 ± 0.35	31.6 ± 0.50	30 ± 0.70*
		F	32 ± 0.707	14 ± 0.32	30.6 ± 0.927	28.2 ± 0.860
5	Eosinophils (%)	M	6.2 ± 0.37	2 ± 0.31	4.4 ± 0.50	3.4 ± 0.87
		F	4.6 ± 0.509	1 ± 0.30	3 ± 0.707	3.1 ± 0.860
6	Monocytes (%)	M	03 ± 00	0.3 ± 0.25	02 ± 0.70	02. ± 0.45
		F	02 ± 00	0.2 ± 023	01 ± 0.583	0.1 ± 0.43
7	Basophiles (%)	M	00 ± 00	00 ± 00	00 ± 00	00 ± 00
		F	00 ± 00	00 ± 00	00 ± 00	00 ± 00
8	RBC Count (%)	M	8.302 ± 0.00	1.3 ± 0.27	6.766 ± 0.00	5.694 ± 0.03
		F	7.286 ± 0.012	1.1 ± 0.24	5.73 ± 0.010	4.75 ± 0.014
9	Platelet Count (%)	M	3.728 ± 0.00	0.2 ± 0.00	2.354 ± 0.00	2.174 ± 0.01
		F	2.726 ± 0.009	0.1 ± 0.01	1.33 ± 0.010	1.34 ± 0.018
10	Mean Platelet Value (Million/cmm)	M	10.28 ± 0.09	1.65 ± 0.07	8.722 ± 0.15	8.502 ± 0.14
		F	9.38 ± 0.106	1.35 ± 0.05	7.56 ± 0.107	7.46 ± 0.145
11	Packed Cell Volume (Million/cmm)	M	41.74 ± 0.05	21 ± 0.03	39.24 ± 0.09*	37.36 ± 0.10
		F	40.42 ± 0.106	20 ± 0.01	38.5 ± 0.141*	36.6 ± 0.114
12	Mean Corpuscular Volume (Cu micron)	M	50.548 ± 0.00	14 ± 0.04	48.57 ± 0.00	57.584 ± 0.00
		F	47.76 ± 1.788	13 ± 0.02	47.5 ± 0.141	56.55 ± 0.014
13	Mean Corpuscular Hemoglobin (Pictograms)	M	19.28 ± 0.06	2.65 ± 0.07	17.602 ± 0.00*	16.742 ± 0.00*
		F	18.42 ± 0.106	1.58 ± 0.05	16.54 ± 0.012	15.74 ± 0.014
14	Mean Corpuscular Hemoglobin Con. (mg/dl)	M	38.174 ± 0.00	14 ± 0.09	36.4 ± 0.50	35.29 ± 0.00*
		F	37.18 ± 0.012	13 ± 0.07	35.2 ± 0.860	34.75 ± 0.018
15	Red Cell Distribution Width (%)	M	15.62 ± 0.05	5 ± 0.04	13.64 ± 0.05	11.36 ± 0.10*
		F	14.36 ± 0.107	4 ± 0.02	12.5 ± 0.141*	10.5 ± 0.141

Mean ± SEM (n = 5), P = < 0.005 (*)

Table 4 Superoxide dismutase levels in different groups of animal

S. no.	Groups	SOD level
1	Control	42.00 ± 0.690
2	Toxic group	11.25 ± 1.24
3	Standard	39.41 ± 0.597
4	Test group	27.47 ± 1.194

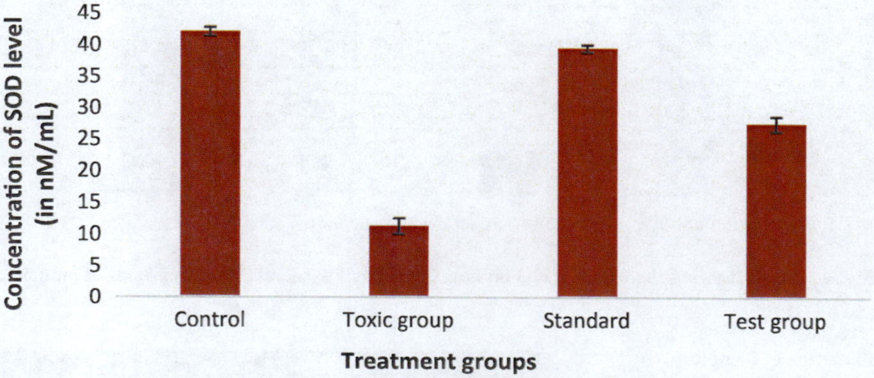

Fig. 2 Graph showing the SOD levels in homogenized heart tissue of different groups of animal

Table 5 Nitric oxide levels in different groups of animal

S. no.	Groups	NO level
1	Control	13.42 ± 0.144
2	Toxic group	2.12 ± 0.683
3	Standard	11.67 ± 0.382
4	Test group	10.17 ± 0.289

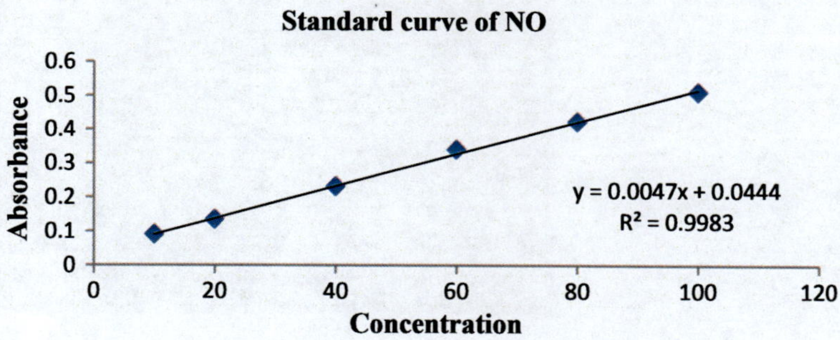

Fig. 3 Graphical representation of standard curve of NO (Nitric Oxide)

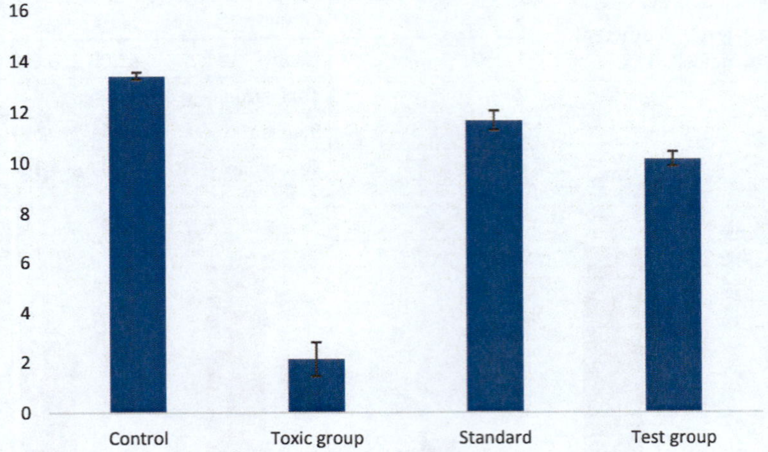

Fig. 4 Graph showing the levels of NO homogenized heart tissue of different groups of animal

Plate no. 1 Sample preparation of heart control group

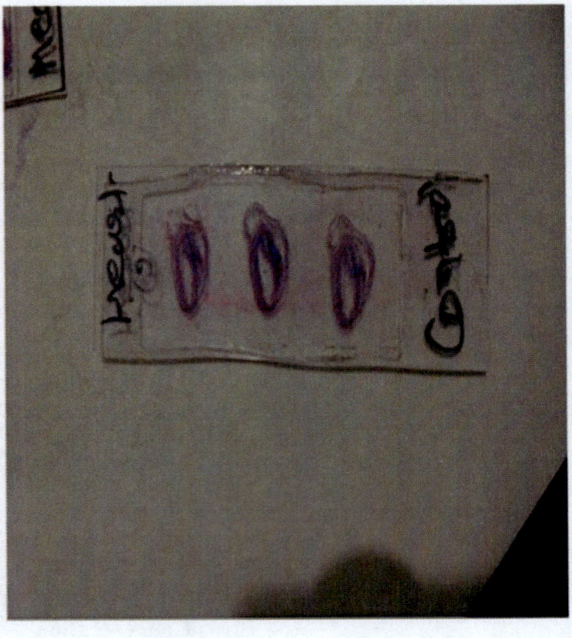

predetermined dose level ratio by using high cholesterol diet hyperlipidemia in rat model.

The oral acute toxicity study for combination of both drugs in rats was carried out. The results for the acute toxicity study indicated that, there were no morbidity and mortality in animals of all the groups. The combination of drugs exhibited decreased level of TGL, CHO, LDL, and VLDL but increased level of HDL. Thus, representing

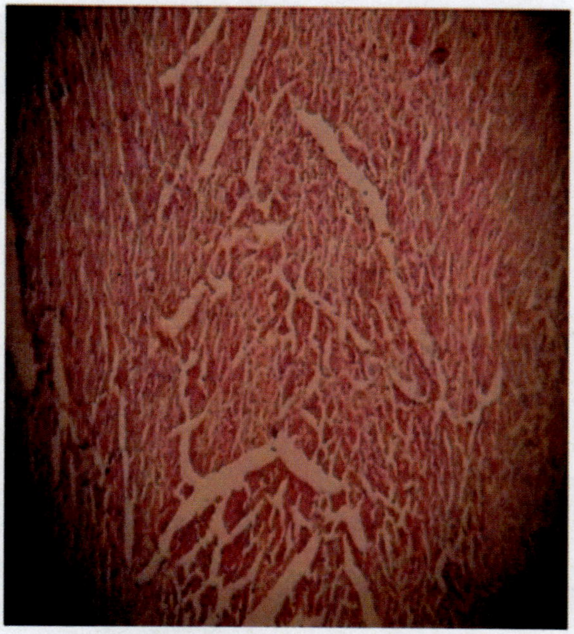

Plate no. 2 Effect of vehicle on histopathological changes of heart tissue in control group

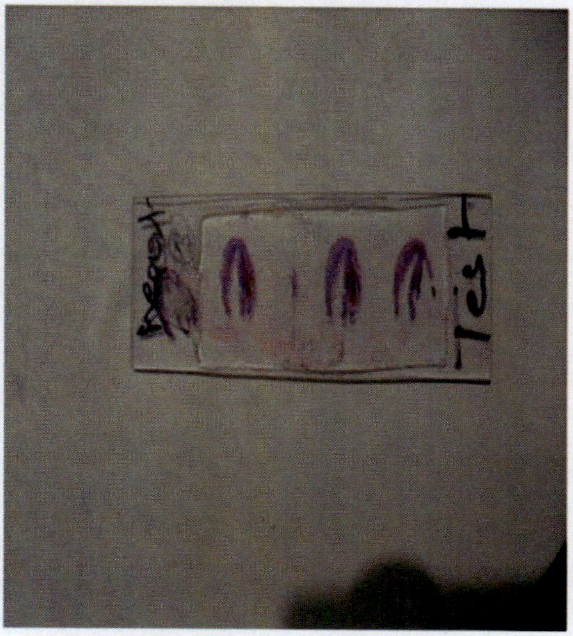

Plate no. 3 Sample preparation of heart test group

Plate no. 4 Effect of Terminalia arjuna along with Commiphora mukul histopathology report of heart test group

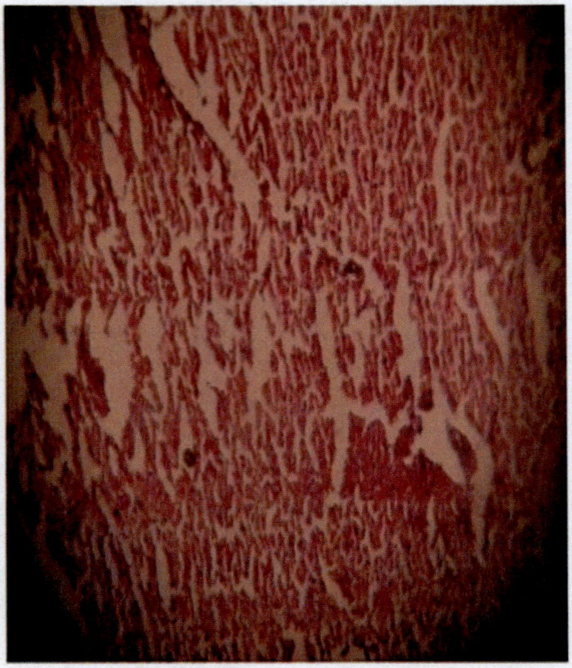

Plate no. 5 Sample preparation of heart standard Group

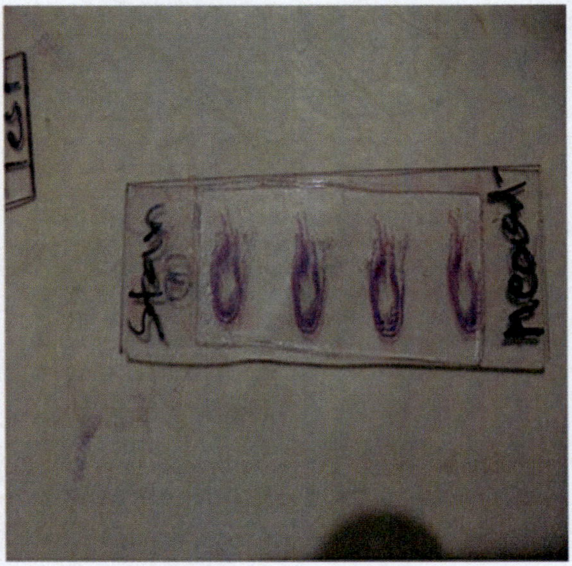

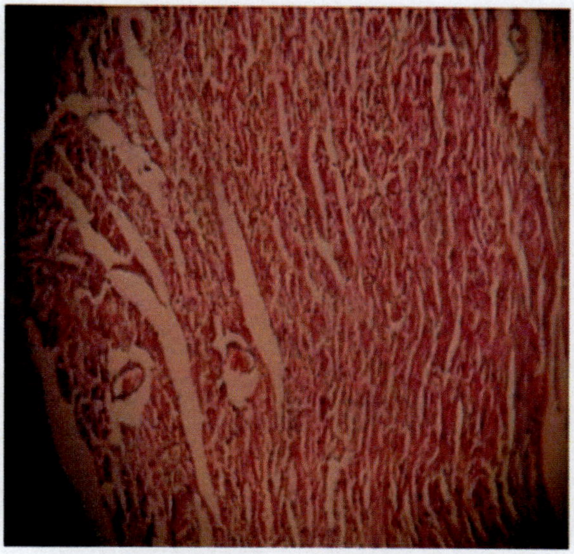

Plate no. 6 Effect of Atorvastatin histopathology report of heart standard group

antihyperlipidemic effect in comparison to control group. The results of antioxidant activity (SOD and NO level) revealed that the combination therapy showed good antioxidant activity on 10th day. Further, exhaustive study is required to determine active constituents and establish the exact mechanism responsible for biological activities.

5 Conclusions

In this present study, various parameters were evaluated for establishment of safety and effectiveness of combination therapy containing Terminalia arjuna and Commiphora mukul. Both the drugs in combination with their predetermined ratios exhibited significant antihyperlipidemic and antioxidant properties. The result of oral acute toxicity study did not show any behavioral changes and mortality.

References

De Fronzo RA, Ferrannini E (1991) Insulin resistance: a multifaceted syndrome responsible for NIDDM, obesity, hypertension, dyslipidemia, and atherosclerotic cardiovascular disease. Diabetes Care 1;14(3):173–194; Horowitz BZ (2001) Toxicity, calcium channel blocker. E-Medicine: Instant Access to the Minds of Medicine, September 6, 2001

Dobrian AD, Davies MJ, Prewitt RL, Lauterio TJ (2000) Development of hypertension in a rat model of diet-induced obesity. Hypertension 35(4):1009–1015

Giustarini D, Rossi R, Milzani A, Dalle Donne (2008) IBT-M in E. Nitrite and nitrate measurement by Griess reagent in human plasma: evaluation of interferences and standardization. In: Nitric Oxide, Part F [Internet]. Academic Press, p 361

Munshi RP, Joshi SG, Rane BN (2014) Development of an experimental diet model in rats to study hyperlipidemia and insulin resistance, markers for coronary heart disease. Indian J Pharmacol 46(3):270

Paoletti F, Aldinucci D, Mocali A, and Caparrini A, (1986) A sensitive spectrophotometric method for the determination of superoxide dismutase activity in tissue extracts. Anal Biochem 153:536–541

Saeed A, Larik FA (2017) Current Topics in herbal medicine with applications in cardiovascular diseases. Cardiovasc Dis 1(1):257

Sapakal VD, Shikalgar TS, Ghadge RV, Adnaik RS, Naikwade NS, and Magdum CS (2008) In vivo screening of antioxidant profile: a review. J Herb Med Toxicol 2(2):1–8

Sastry KV, Moudgal RP, Mohan J, Tyagi JS, Rao GS (2002) Spectrophotometric determination of serum nitrite and nitrate by copper-cadmium alloy. Anal Biochem 306:79–82

Urizar NL, Moore DD (2003) GUGULIPID: a natural cholesterol-lowering agent. Annu Rev Nutr 23(1):303–313

Epileptic Seizure Detection Using Deep Recurrent Neural Networks in EEG Signals

Archana Verma and Rekh Ram Janghel

Abstract Epilepsy is a neurological ailment that influence around 1% of mankind. Around 10% of the United States population experience at least a single convulsion in their life. Epilepsy is distinguished by the inclination of the brain to generate unexpected bursts of strange electrical action which disrupts the normal functioning of the brain. Generally patients not really recognises their conditions so mostly experts suggests that electroencephalography (EEG) for seizure detection. In this research, we implement Recurrent Neural Networks (RNNs) based on Gated Recurrent Unit (GRU) with and without Wavelet Filter technique for early seizure detection. In this paper work in 5-layer of GRU Recurrent Neural Networks (RNNs) technique is implemented to distinguish healthy and seizure class. The proposed system accomplished an accuracy of with DWT 98.50% and without 97.0%.

Keywords Deep neural networks · Epilepsy · EEG · Electroencephalogram · GRU · RNN · Recurrent neural networks · Seizure detection

1 Introduction

Epilepsy is a association of typical neurological conditions of a mind distinguished by recurrent unprovoked convulsions which are a consequence of sudden bursts of abnormal electrical discharges in the brain (Guo et al. 2012; Orosco et al. 2013; Fayek et al. 2017). According to report by WHO (Orosco et al. 2013) around 50 million people are influenced by epilepsy worldwide (Talathi 2017; Acharya et al. 2013; Cook 2013; Guo et al. 2010a). Nearly one in every 100 persons are influenced by a convulsion in their lifespan (Guo et al. 2011), approximately 2.4 million new cases of epilepsy are reported every year globally (Acharya et al. 2013; Geetha 2012). Then,

A. Verma (✉) · R. R. Janghel
Department of Information Technology, National Institute of Technology, Raipur, India
e-mail: archana1694@gmail.com

R. R. Janghel
e-mail: rrjanghel.it@nitrr.ac.in

A. A. Rizvanov et al. (eds.), *Advances in Biomedical Engineering and Technology*, Lecture Notes in Bioengineering,
https://doi.org/10.1007/978-981-15-6329-4_17

the manifestation of an epileptic is eccentric and its movement of activity has been less understood (Guo et al. 2011), approximately 2.4 million new cases of epilepsy are reported every year globally (Acharya et al. 2013; Iasemidis 2003).

Epileptic patients experience varying symptoms depending on the section and the expanse of the brain that is affected, Epileptic seizures can induce gloomy physical, social consequences, and psychological, which include loss of consciousness, injury, and abrupt death. Epilepsy is of two types depending upon the degree of involvement of the brain tissue, which are, generalized seizures and partial seizures, generalized seizures roughly involve nearly the complete brain, partial seizures develop in a particular section of the brain and remain limited to that section (Guo et al. 2010a). Electroencephalography (EEG) is a famous electrophysiological technique to comprehend the complicated action of human brain (Gandhi et al. 2010). EEG directly gauges and registers the electrical action of the mind. Spontaneous EEG signals are classified into several rhythms based on their frequencies, which are band (3–4 Hz); band (4–8 Hz); band (8–13 Hz); band (13–30 Hz) (Zhou and J. 2004; Parvez and Paul 2016). An EEG is especially useful at times when the brain is at risk by providing a sensitive indication of cerebral functioning. It is long time ranges signals consequently. An extended EEG recording is required early examinations for demonstrated proof of this anomalous movement to be an advantage to detect in epilepsy and cerebral tumors. These days EEG signals are utilized to get data relevant to the diagnosis, prognosis, and treatment of these abnormal conditions. EEG is registered using electrodes placed on the scalp and have tiny amplitudes of the range of 20 V (Selvan and Srinivasan 1999; Ammar and Senouci 2016). The electrodes are placed as per the 10–20 international system which has been shown in the Fig. 1.

Usually EEGs contain massive amounts of information and detection of traces of epilepsy requires a visual inspection of the total span of the EEG by a specialist

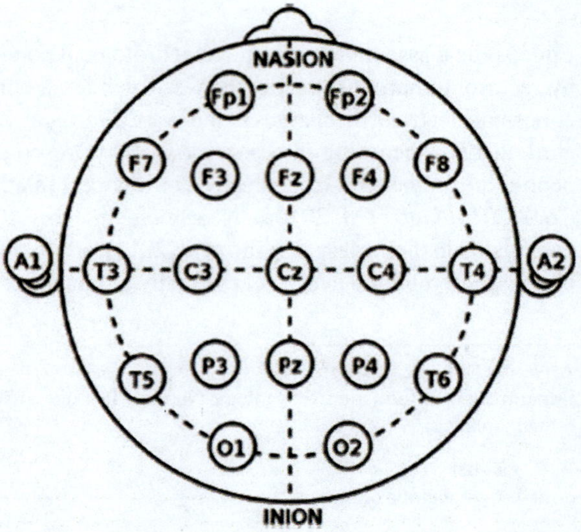

Fig. 1 Standardized electrode placement scheme (Talathi 2017)

which is a cumbersome task (Guo et al. 2010b; Lin et al. 2016). Hence, developing automated epileptic seizure espial system is noteworthy for reviewing EEGs. Over a past two decades, many algorithms have been applied for epileptic seizure detection, which include time–frequency analysis methods, non-linear statistical models, and more present day machine learning methodologies, for example, neural systems and Support Vector Machines (SVM), however inspite of many Progress, current EEG analysis approach are a long way from expert with majority of the methods being considered because of their high false detection rates. Moreover, Sabrina Ammar [34] have work on EEG signals seizer detection in single-channel with the help of extreme learning machine. Also NihalFatmaGler use EEG signals classification for employing Lyapunov exponents using RNN. Tawfik et al. (2016) dose a hybrid automated system use for epileptic seizures detection in help of EEG signals. Kumar et al. (2008) work with Recurrent Neural Network Classifier use for Epilepsy Seizure detection using an automated Wavelet Entropy for feature selection. A novel deep structure was currently presented which attempts to minimize the false alarm rate on EEG signals. This framework incorporates CNNs with RNNs to convey the state of the art performance and use for enhancing accuracy.

2 Methodology

2.1 *Dataset*

A data elucidated by Andrzejak et al. (2001) was employed for the current research. The total data set is comprised of 5 subsets, (marked as Z, O, N, F, S) every subset having 100 single-channel EEG fragments each being on 23:6 s time span sampled at 173:6 Hz. The fragments were chosen from regular multi-channel EEG records after visual examination for artifacts for e.g., because of muscle action or eye movements. Z, O sets contain fragments that have been obtained from EEG recordings conducted on 5 healthy people. People were in an awaken condition and eye open (Z) and eye closed (O). Sets N, F, and S are emerged from an EEG. Fragmentation set F has been registered for the epileptogenic zone, and fragment in set is hippocampal. To do so, set N and F encompassed only activity measured during seizure free spans, set S contains only seizure activity.

2.2 *Discrete Wavelet Transform*

Analyzing Wavelet transform is a spectral estimation procedure in which any general function can be communicated as an endless arrangement of wavelet. Behind of this idea analysis contain communicating a signal as a linear combination of a specific arrangement of set obtained by shifting and dilating one single function called a mother wavelet. The disintegration of the signal prompts an arrangement of coefficient called wavelet coefficient. So that a signal can be reformed a linear combination of the wavelet functions weight by as a wavelet coefficient. The key element of wavelets is the time–frequency limitation (Subasi 2007b). It implies that a large portion of the vitality of the wavelet is confined to a limited time interim. Frequency limitation implies that the Fourier change is band constrained. EEG signal will uncover highlights identified with the transient nature of the signal which is not clear by the Fourier transform. The separated wavelet coefficients give a minimal illustrates that demonstrates the vitality dissemination of the EEG signal in time and frequency (Cho et al. 2014).

2.3 Recurrent Neural Networks (RNNs)

With refined recurrent hidden units like the Gated Recurrent Unit (GRU) and the Long-Short-Term-Memory (LSTM) have turned into a mainstream decision for modeling temporal sequences (Chung et al. 2014; Che et al. 2018), Conventional Feed Forward Neural Networks are trained on labeled data till the prediction error is minimized (Gandhi et al. 2010), whereas the decision of RNN at time t depends on the decision at which the RNN reached at time step t-1.

Inspired by the success of LSTMs Chung et al. (2014), proposed a network architecture which is simpler and efficient compared to LSTMs known as the Gated Recurrent Unit (GRU) (Che et al. 2018) the architecture of a GRU has been depicted, Example of a RNN is shown in Fig. 2.

We used the Keras library for implementation of Gated Recurrent Unit (GRU), 60% of the information was utilized for preparing and 40% of information is utilized for approval.

Lately, Recurrent Neural Networks (RNNs) with sophisticated recurrent hidden units like the GRU and the LSTM have turned to a famous choice for modeling temporal sequences (Talathi 2017; Graves 2013; Fayek et al. 2017; Sathyanarayana et al. 2016), Conventional Feed Forward Neural Networks are trained on labeled data till the prediction error is minimized (Fig. 3).

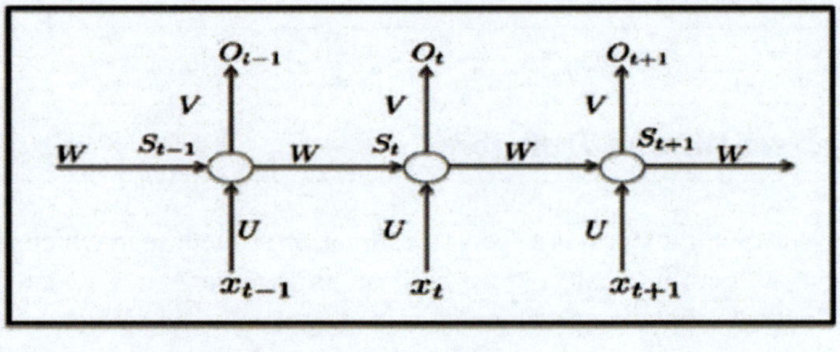

Fig. 2 Recurrent neural networks (Gandhi et al. 2010)

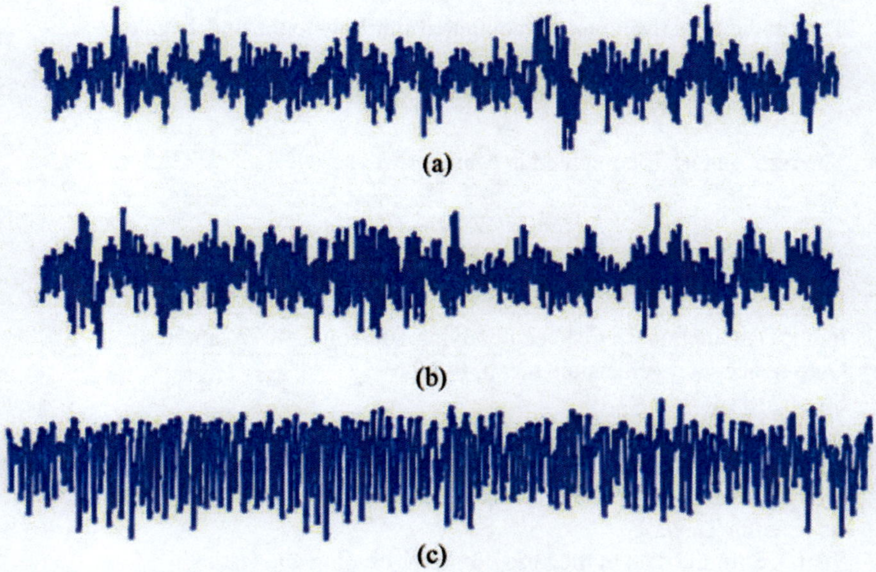

Fig. 3 Raw EEG data belonging to each of subsets **a** Z, **b** O, and **c** S

The decision of RNN at time t depends on the decision at which the RNN reached at time step t-1. In particular we focus on the 2-class classification of a given EEG fragment into either healthy or, ictal states. As subset consists of 100 EEG segments, from each of the subjects 60% of the data is used to prepare training and the remainder is used for validation. Hence the training and testing data consists of 150 segments.

2.4 Gated Recurrent Unit (GRU)

GRU was put forward by Graves et al. (2013) so as to enable every recurrent unit adaptively capture the dependencies among various time range. GRU, similar to LSTM consists of gating units which regulate the flow of data within unit, however it doesn't consist of independent memory cells. The activation h_t^j of the GRU at time t directs initiation between the before activation h_{t-1}^j and the candidate activation $\tilde{h}_{t-1}^j$:

$$h_t^j = (1 - z_t^j)h_{(t-1)}^j + z_t^j \tilde{h}_t^j \tag{1}$$

The update gate z_t^j determines the amount by which the unit updates its activation.

$$z_t^j = \sigma(W^z x_t + U^z h_{(t-1)})^j \tag{2}$$

The candidate activation $\tilde{h}_t^j$ is computed similarly to the update gate:

$$\tilde{h}_t^j = \tanh(\mathrm{W}x_t + \mathrm{U}(r_t * h_{(t-1)}))^j \tag{3}$$

The reset gate r_t^j is computed in a similar way:

$$r_t^j = \sigma(W^r x_t + U^r h_{(t-1)})^j \tag{4}$$

Algorithm 1: Pseudocode for Gated Recurrent Unit (GRU)

Input: Data dictionaries each embodying 100-single 4097 samples
Output: accuracy, precision, recall, F-Score
Step.1. Read Z001 - Z100;
Read O001 - O100;
Read S001 - S100;
Step.2. Create model using Keras library comprising of five layers as listed in Table. 1
Step.3. Split the data in the ratio 60:40 for training and testing
Step.4. Train the model using 60% data using model.fit() function
Step.5. Test the model using 40% data using model.predict() function.
Step.6. Determine the precision, recall, F-Score using the sklearn library.
Step.7. Determine the confusion-matrix with the help of mlxtend
Step.8. Calculate accuracy from the confusion-matrix.
end

In Fig. 4 we firstly take raw EEG signals set of Z, S, and O and we do shuffling then split data as 60% for training and 40% for testing then perform RNN Model (GRU) then done classification as normal and seizure classes. Secondly we use DWT as feature extraction and get better performance and work for RNN Model (Table 1).

Prediction: the following parameters were used to evaluate our proposed method.

$$\text{Precision (P)} = \frac{TP}{(TP + FP)} \tag{5}$$

$$\text{Recall (R)} = \frac{TP}{(TP + FN)} \tag{6}$$

$$\text{F-Score} = \frac{\left(1 + \beta^2\right)(P)(R)}{\left(\beta^2 \cdot P \cdot R\right)} \tag{7}$$

TP = True Positive
FN = False Negative
FP = False Positive

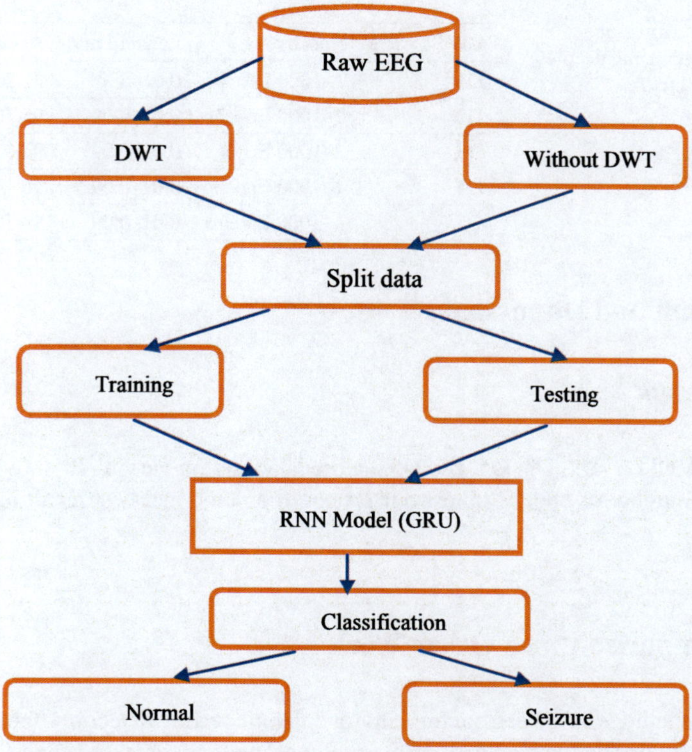

Fig. 4 Flowchart of proposed methodology

Table 1 Parameter Design for GRU

Layer type	Output shape	Parameters
Input	(51, 80, 1)	0
GRU 0	(51,100, 1)	30,600
F_c	(51, 100, 1)	10,100
GRU 1	(51, 100)	60,300
LR	(51, 3)	303

TN = True Negative
P = Precision
R = Recall.

Table 2 Performance measurement using deep RNN with DWT

DWT Levels	Epochs	Leaning rate	Accuracy (%)
D1	1–1000 Epochs	0.01– 0.09	91.20
D2	1–1000 Epochs	0.01–0.09	95.40
D3	1–1000 Epochs	0.01–0.09	97.00
D4	1–1000 Epochs	0.01–0.09	96.10
D5	1–1000 Epochs	0.01–0.09	98.50

3 Result and Discussion

3.1 Result

We check all D1, D2, D3, D4, and D5 featured data set for varying (0–0.09) learning rate and number of epochs to test our proposed method and best result includes in Table 2.

4 Comparison with Other Work

Multiple methods have been put forward for epileptic seizure detection. Table demonstrates a differentiation between various methods, methods assessed on the same data set have been incorporated so that a distinction between the results is realistic.

5 Conclusion and Future Work

EEG signs can be utilized to separate among ordinary and epileptic conditions of the cerebrum. In this work, we put forward an epileptic seizure detection system built on Deep Recurrent Neural Networks using Gated Recurrent Unit (GRU) which can be trained with high accuracy and classify the EEG segments as either healthy or ictal. Our purposed method is capable of achieving an accuracy of up to 97% and with DWT comes 98.50%. The performance of purposed model is better than few reckon in that Table 3. The benefit of this proposed model presented in this paper is large data set. In this proposed method may require more diversity of data to optimum performance. In this paper, we can improve performance of our purposed method by increasing the number of samples. The performance of this method can be improved by increasing the number of samples.

Table 3 Synopsis of erstwhile research for detection of epileptic and normal classes

Authors	Classifiers	Accuracy (%)
Kannathal et al. (2005)	Diagnostic neural Network	97.20
Sadati et al. (2006)	Adaptive Neuro-Fuzzy Inference systems (ANFIS)	92.20
Subasi (2007a)	Adaptive neural fuzzy network	85.90
Guo et al. 2009)	Mixture expert model (a modular neural network)	94.50
Cho et al. (2014)	Feed Forward ANN	95.20
Proposed method	Deep Recurrent Neural Networks (RNN) based on Gated Recurrent Unit (GRU)without DWT	97.00
Proposed method	Deep Recurrent Neural Networks (RNN) based on Gated Recurrent Unit (GRU) with DWT	98.50

References

Acharya UR, Sree SV, Swapna G, Martis RJ, Suri JS (2013) Automated EEG analysis of epilepsy: a review. Knowl-Based Syst 45:147–165

Ammar S, Senouci M (2016) Seizure detection with single-channel EEG using extreme learning machine. In: 2016 17th international conference on sciences and techniques of automatic control and computer engineering (STA), pp 776–779

Andrzejak RG, Lehnertz K, Mormann F, Rieke C, David P, Elger CE (2001) Indications of nonlinear deterministic and finite-dimensional structures in time series of brain electrical activity: dependence on recording region and brain state. Phys Rev E 64(6):61907

Che Z, Purushotham S, Cho K, Sontag D, Liu Y (2018) Recurrent neural networks for multivariate time series with missing values. Sci Rep 8(1): 6085

Cho K, Van Merriënboer B, Bahdanau D, Bengio Y (2014) On the properties of neural machine translation: Encoder-decoder approaches. arXiv1409.1259

Chung J, Gulcehre C, Cho K, Bengio Y (2014) Empirical evaluation of gated recurrent neural networks on sequence modeling. arXiv1412.3555

Cook MJ et al (2013) Prediction of seizure likelihood with a long-term, implanted seizure advisory system in patients with drug-resistant epilepsy: a first-in-man study. Lancet Neurol 12(6):563–571

Fayek HM, Lech M, Cavedon L (2017) Evaluating deep learning architectures for speech emotion recognition. Neural Netw 92:60–68

Gandhi T, Panigrahi BK, Bhatia M, Anand S (2010) Expert model for detection of epileptic activity in EEG signature. Expert Syst Appl 37(4):3513–3520

Geetha G (2012) Detecting epileptic seizures using Electroencephalogram: a novel technique for seizure classification using fast walsh-hadamard transform and hybrid extreme learning machine. In: Proceedings of the second international conference on computational science, engineering and information technology, pp 253–260

Graves A (2013) Generating sequences with recurrent neural networks. arXiv1308.0850

Graves A, Mohamed A, Hinton G (2013) Speech recognition with deep recurrent neural networks. In: 2013 IEEE international conference on acoustics, speech and signal processing (ICASSP), pp 6645–6649

Guo L, Rivero D, Seoane JA, Pazos A (2009) Classification of EEG signals using relative wavelet energy and artificial neural networks. In: Proceedings of the first ACM/SIGEVO summit on genetic and evolutionary computation, pp 177–184

Guo L, Rivero D, Dorado J, Rabunal JR, Pazos A (2010a) Automatic epileptic seizure detection in EEGs based on line length feature and artificial neural networks. J Neurosci Methods 191(1):101–109

Guo L, Rivero D, Pazos A (2010b) Epileptic seizure detection using multiwavelet transform based approximate entropy and artificial neural networks. J Neurosci Methods 193(1):156–163

Guo L, Rivero D, Dorado J, Munteanu CR, Pazos A (2011) Automatic feature extraction using genetic programming: an application to epileptic EEG classification. Expert Syst Appl 38(8):10425–10436

Guo P, Wang J, Gao XZ, Tanskanen JMA (2012) Epileptic EEG signal classification with marching pursuit based on harmony search method. In 2012 IEEE international conference on systems, man, and cybernetics (SMC), pp 283–288

Iasemidis LD et al (2003) Adaptive epileptic seizure prediction system. IEEE Trans Biomed Eng 50(5):616–627

Kannathal N, Lim M, Acharya UR, Sadasivan PK (2005) Entropies for detection of epilepsy in EEG

Kumar SP, Sriraam N, Benakop PG (2008) Automated detection of epileptic seizures using wavelet entropy feature with recurrent neural network classifier. In: TENCON 2008–2008 IEEE region 10 conference, pp 1–5

Lin Q et al. (2016) A novel approach for epileptic EEG signals classification based on biclustering technique. In: 2016 international conference on machine learning and cybernetics (ICMLC), vol 2, pp 756–760

Orosco L, Correa AG, Laciar E (2013) A survey of performance and techniques for automatic epilepsy detection. J Med Biol Eng 33(6):526–537

Parvez MZ, Paul M (2016) Epileptic seizure prediction by exploiting spatiotemporal relationship of EEG signals using phase correlation. IEEE Trans Neural Syst Rehabil Eng 24(1):158–168

Sadati N, Mohseni HR, Maghsoudi A (2006) Epileptic seizure detection using neural fuzzy networks. In: 2006 IEEE international conference on fuzzy systems, pp 596–600

Sathyanarayana A et al (2016) Impact of physical activity on sleep: a deep learning based exploration. arXiv1607.07034

Selvan S, Srinivasan R (1999) Removal of ocular artifacts from EEG using an efficient neural network based adaptive filtering technique, vol 6, no 12, pp 330–332

Subasi A (2007a) EEG signal classification using wavelet feature extraction and a mixture of expert model. Expert Syst Appl 32(4):1084–1093

Subasi A (2007b) EEG signal classification using wavelet feature extraction and a mixture of expert model. Expert Syst Appl 32(4):1084–1093

Talathi SS (2017) Deep Recurrent Neural Networks for seizure detection and early seizure detection systems. arXiv1706.03283

Tawfik NS, Youssef SM, Kholief M (2016) A hybrid automated detection of epileptic seizures in EEG records. Comput Electr Eng 53:177–190

Zhou W, Gotman J (2004) Removal of EMG and ECG artifacts from EEG based on wavelet transform and ICA. In: 26th Annual International Conference of the IEEE Engineering in Medicine and Biology Society, 2004, vol 1, pp 392–395. IEMBS'04

Detection of Disease from Leaf of Vegetables and Fruits Using Deep Learning Technique

Avisha Jaiswal, Saurabh Pathak, Yogesh Kumar Rathore, and Rekh Ram Janghel

Abstract In a country like India where 16% of the total GDP growth is contributed by agriculture alone and majority of the people rely on it for their source of living, it is of utmost importance that the threats to the production of crops should be minimized. Crop leaf disease is major type of diseases suffered by crops; its manual identification is a difficult task leading to wrong treatment and poor production. It gives rise to the need of an accurate automated system to detect plant leaf diseases, and, is made possible by the recent advances in computer vision and deep learning. In this paper we use Convolution Neural Networks to classify and identify vegetable leaf diseases. An open database of Plant Village that contains 54, 306 plant images, with 26 classes of diseases and 14 different crop species is used to curate a new database in reference to our problem domain which contains 5 class of diseases suffered by vegetables having images of diseased and healthy leaves. We have implemented CNN models, namely, Sequential and GoogLeNet with images of leaves as input. The highest success rates achieved are 98.48% and 97.47% for Sequential and GoogLeNet model, respectively. The relatively higher accuracy of the models makes them very useful and eligible to be used to solve the current problem of crop disease in India. It could further be used to make a better vegetable disease identification system.

Keywords GoogleNet · Sequential model · Leaf disease · Plant village dataset · CNN

A. Jaiswal (✉) · S. Pathak · Y. K. Rathore · R. R. Janghel
National Institute of Technology, Raipur, CG, India
e-mail: jaiswal.avisha08@gmail.com

S. Pathak
e-mail: saurabhpathak997@gmail.com

Y. K. Rathore
e-mail: yogeshrathore23@gmail.com

R. R. Janghel
e-mail: rrjanghel.it@nitrr.ac.in

A. A. Rizvanov et al. (eds.), *Advances in Biomedical Engineering and Technology*, Lecture Notes in Bioengineering,
https://doi.org/10.1007/978-981-15-6329-4_18

1 Introduction

India is an agricultural country with 70% of its population relying on agriculture hence, it becomes vital, in a country like India, that cultivation of crops gives maximum yield. Crop health is affected by environmental degradation, pests, chemicals and pesticides and various other factors making it a major challenge in agricultural sector, and the losses to the country's economy and farmers due to it are miserable. Identification of crop disease by optical observation of symptoms is a tedious and complex task leading to mistaken conclusions (Konstantinos and Ferentinos 2018). Faster and accurate predictions of leaf diseases in plants could help to develop an early treatment technique, while considering reducing economic losses (Akila et al. 2018) which could be achieved by the existence of an automated system adding valuable assistance to the task of diagnosis (Konstantinos and Ferentinos 2018)). The simplicity and ease of the system would lead it to be used by the farmers as a valuable tool in the parts of the county with lesser awareness and lacking infrastructure. The recent developments advances in the domain of computer vision backed-up by deep learning has paved way for machine-controlled disease identification (Mohanty and Marcel 2016). Deep learning constitutes a recent modern technique for image processing and data analysis with accurate results (Suk and Shen 2013). The applicability of deep learning is versatile and it has been applied to almost every problem domain including agriculture (Konstantinos and Ferentinos 2018). We have used deep learning for identification and classification of plant leaf disease.

1.1 Deep Learning

Deep learning is a modern branch of machine learning and a subset of artificial intelligence which is inspired from human brain (Suk and Shen 2013) and represents better representational power for future representation. It can be thought as of novel approach in learning representations from data that gives importance on learning of consecutive layers (Chollet 2018). Deep learning neural networks contain many hidden layers which enable them to effectively extract high level features (Suk and Shen 2013). An error of 16.4% was reached by a deep CNN architecture for the classification of pictures into 1,000 probable classes, which was top 5 in 2012 (Krizhevsky et al. (2012). In the following three years, the error rate has been lowered to 3.57% by a variety of architectures in deep convolutional neural networks (Krizhevsky et al. 2012; Zeiler and Fergus 2014; Simonyan and Zisserman 2014). The feasibility of state-of-art deep learning models, which have made computation much easier, have left revolutionary impact on sectors like recognition and processing of (Cunn et al. 1681), voice recognition (Srivastava and Hinton 2014), and alternative equally advanced issues that require working with enormous amount of data, giving a big boom to use of such technologies in areas automation of vehicle driving, artificial intelligence, and inference, etc. The introduction of these deep learning methodologies in the field of agriculture (Carranza-Rojas et al. 2017), and particularly within the field of disease identification, has taken place only since last two extents and the extent is very limited our problem domain specializes in classifying and identifying the disease in leaves of vegetables specifically. Five diseases, namely, Bacterial Spot, Black Rot, Late Blight, Early Blight, and Powdery Mildew, from which various vegetable leaves, suffer are selected and classes for these diseases are made. A database is created having healthy and diseased images of each class of disease, the images were selected from Plant Village dataset that contains large number of images containing different disease (Hughes and Salathé 2015). An accuracy of 98.48% is achieved for sequential model and is maximum so far, GoogLeNet gave an accuracy of 97.47%.

2 Literature Review

The very basic deep learning methodology that has been used throughout this work is Convolution Neural Network (CNN) (Konstantinos and Ferentinos 2018). CNN is a deep learning architecture which is inspired by the visual perception of living beings (Gu et al. 2018). The problems which are complex and are acquainted with large volume of data and pattern recognition, CNN serve as a powerful methodology (Konstantinos and Ferentinos 2018). For the task of identification and classification of plant leaf disease we have used Sequential and GoogLeNet architectures for CNN.

Grinblat et al. (2016) proposed a model which can easily detect three different spices of plant using a powerful neural network (Grinblat et al. 2016). A relatively smaller work was carried out by Sladojevic et al. using same source of data but only for 13 diseases and 5 plants (Sladojevic et al. 2016). The accuracy of these two was in a range of 91–98% depending upon training testing set split. More recently, comparisons were made between some image processing based techniques and deep learning-based techniques which uses different dataset of plant leaves and fruits by Pawara et al. (2017), and the results of CNN beating the performance of those conventional methods were drastic. Finally, models for detection of 9 different tomato diseases and pests were developed by Fuentes et al. (2017) whose performance was satisfactory (Fuentes et al. 2017). "Konstantinos P. Ferentinos compared the performance of various CNN architectures out of which GoogLeNet gave an accuracy of 97.27% high (Konstantinos 2018). Mohanty et al. compared AlexNet and GoogLeNet architectures of CNNs for detection of plant diseases, dataset of leaves images containing 26 plant diseases for 14 different crops was used "in which GoogLeNet achieved a maximum accuracy of 96.21% for a 80 isto 20 train test split in contrast to our work in which GoogLeNet has achieved a success rate of 97.47% for same test train split" (Mohanty et al. 1419). "Sequential architecture so far, has been used for other purposes achieving all time high accuracy of 95.61% for filling of slots in languages which are spoken (Ngoc 2018) whereas our sequential model has achieved a success rate of 98.48%."

3 Methodology

3.1 Data Acquisition

To address the problem of classification and identification of vegetable leaf disease Plant Village (Mohanty et al. 2016) database for images containing 54, 306 images of plant leaves with 26 classes of diseases and 14 different crop species is used to create a new dataset. We have selected 5 classes of diseases for all the vegetable species in Plant Village dataset. The details regarding the classes of diseases and respective number of images in each class are as follows:

Disease
Bacterial Spot: 5123 images
Black Rot: 1801 images
Early Blight: 1966 images
Late Blight: 2909 images
Powdery Mildew: 2887 images
Healthy: 8680 images.

Since the images obtained from this source were already preprocessed, hence all the preprocessing tasks, labeling, and augmentation are ignored to curate the database. To avoid the issue of over fitting complete database was divided into training dataset and testing dataset in a ratio of 80 is to 20. Other such similar divisions like 70 is to 30 should not have any effects on final classification results.

3.2 Methodology

Artificial Intelligence. The study of imitating and automating human thinking is artificial intelligence and the mathematical models via which it can be achieved are artificial neural networks (Konstantinos and Ferentinos 2018). Artificial neural networks have synapse that interconnects their neurons via which they mimic the functioning of brain (Konstantinos and Ferentinos 2018). The main characteristics of artificial neural networks are their ability to learn by themselves through supervised learning and training. They are trained on some dataset that contains certain specific matchings of the input and output data regarding the system which we want our artificial neural network to model.

Convolutional Neural Networks. Convolutional Neural Networks are evolutionary and transformed version of artificial neural networks (LeCun and Bengio 1995), where main feature is that they have a number of layers and reduce the requirement for number of artificial neurons to a very large extent in comparison to normal artificial neural networks (Konstantinos and Ferentinos 2018) and "applies" the same for image recognition purpose. CNN specializes in tasks related to recognition of images hence various models of CNN have been developed for image recognition.

The two architectures that we have used in our work are Sequential and GoogLeNet model.

Sequential Architecture. Sequential model is the most basic type of model for Convolutional Neural Network. The layers in this model are simply stacked and sequentially arranged (see Fig. 1) (Ngoc 2018). The layers have input, output, input_shape, output_shape. Each of the layers of sequential model has a defining configuration. In our implementation of sequential model there are a total of 8 layers in which 4 are convolutional and 4 are max-pooling and these are alternatively arranged. To build a sequential model add layers to it one by one and compile the model with a loss function, an optimizer, and optional evaluation metrics. Then use the dataset to fit the model (Ngoc 2018).

GoogLeNet Architecture. GoogLeNet architecture is a very deep and wide architecture with 22 layers. The specialty of GoogLeNet is that even after having a wide and deep architecture the number of parameters used in it is considerably low (Hughes and Salathé 2015). The peculiar feature of GoogLeNet architecture is that it's a type of network architecture (see Fig. 2) (LeCun et al. 2015) which are in the form of inception module. 1X1, 3X3, and 5X5 convolutional and a max-pooling layer are used in parallel by inception module which helps it to capture a range of characteristics in parallel. Output of all these parallel layers is concatenated by a filter concatenation. GoogLeNet is made by 9 such inception modules (Going et al. 2015).

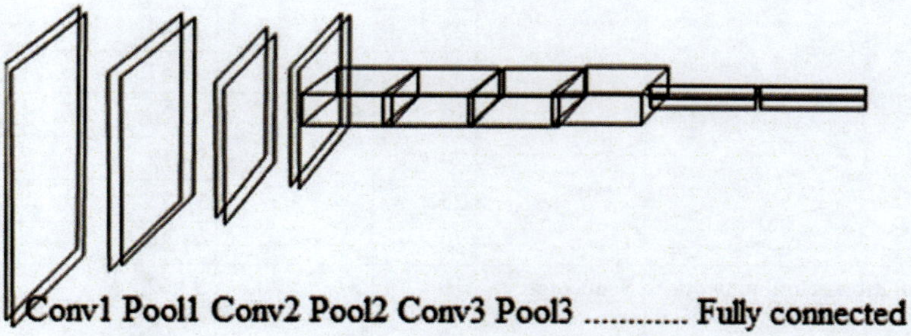

Fig. 1 Architecture of sequential model (Ngoc 2018)

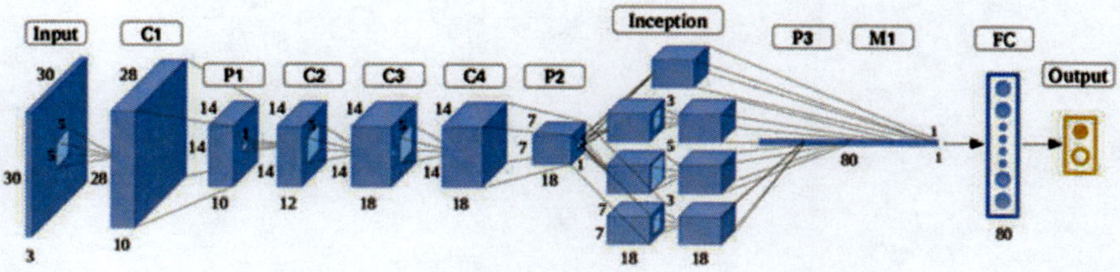

Fig. 2 GoogLeNetArchitecture (LeCun et al. 2015)

4 Result

Performance of both the models is assessed by training the model on the dataset curated from Plant Village database.The parameters that have been varied in order to assess the performance of architectures are batch size and number of epochs, as can be seen in Table 1. In GoogLeNet for batch size 32 success rate kept on increasing as we increased the number of epochs and touched an all time high when number of epochs were 30, then decreased exponentially. Similar phenomenon occurred for Sequential model at when the batch size was 32 and numbers of epochs were 30. So the maximum success rates achieved are 98.48% and 97.47% in the case of sequential and GoogLeNet architectures, respectively, for batch size 32 and number of epochs 30 in both cases (see Table 1).

Here, x-axis shows the number of epochs (i.e. 10, 20, 30, 40) varying with two different batch size of input data i.e., 32 and 64 and y-axis shows the accuracy of particular model (see Fig. 3).

Table 1 Table showing success rates by varying parameters

Neural network	Architecture	Epoch	Batch sisze	Accuracy (Diseased vs. healthy) (%)
Convolution neural network	GoogLeNet	10	32	93.15
		20		94.26
		25		96.84
		30		97.47
		10	64	92.26
		20		94.12
		25		95.73
		30		91.39
Convolution neural network	Sequential	10	32	93.62
		20		91.65
		25		96.36
		30		98.48
		10	64	93.45
		20		89.74
		25		95.45
		30		95.87

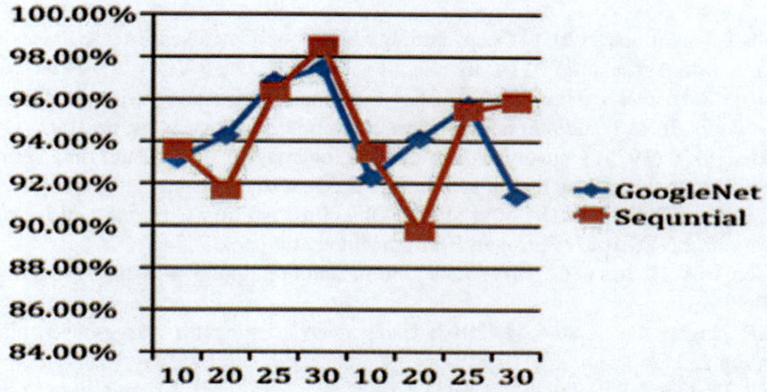

Fig. 3 Result comparison

5 Conclusion

In this work we used convolutional neural networks to develop and assess systems for classification and identification of vegetable leaf diseases "accurately, so as to solve the problem of misleading treatments which arise due to wrong identification of disease that further leads to huge losses in crop production." The relatively higher accuracy of the models makes them very useful and eligible to be used to solve the current problem of crop disease in India. It could further be used to make a better vegetable disease identification system. The maximum success rates achieved are 98.48% and 97.47% in the case of sequential and GoogLeNet architectures, respectively.

References

AkilaM, Deepan P (2018) Detection and classification of plant leaf diseases by using deep learning algorithm. In: International journal of engineering research & technology, ICONNECT conference proceedings, vol 6, issue 7

Carranza-Rojas J, Joly AAJ, Bonnet P, Hervé HG (2017) Goëau, Erick Mata-Montero. Automated Herbarium Specimen Identification using Deep Learning, In: Proceedings of TDWG

Chollet F (2018) Title of deep learning with python

Fuentes A, Yoon S, Kim SC, Park DS (2017) A robust deep-learning-based detector for real-time tomato plant diseases and pest recognition. Sensors 17:2022

Grinblat GL, Uzal LC, Larese MG, Granitto PM (2016) Deep learning for plant identification using vein morphological patterns. Comput Electron Agric 127:418–424

Gu J, Wang Z, Kuen J (2018) Recent advances in convolutional neural networks, pattern recognition, vol 77. Elsevier, pp 354–377

Hughes DP, Salathé M (2015) An open access repository of images on plant health to enable the development of mobile disease diagnosis

Konstantinos P (2018) Ferentinos Deep learning models for plant disease detection and diagnosis. Comput Electron Agricul 145(2018):311–318

Konstantinos T, Ferentinos P (2018) Deep learning models for plant disease detection and diagnosis. Comput Electron Agricul 145:311–318

Krizhevsky A, Sutskever I, Hinton GE (2012) Imagenet classification with deep convolutional neural networks. In: Advances in neural information processing systems, pp 1097–1105

LeCun Y, Bengio Y (1995) Convolutional networks for images, speech, and time series. In: The handbook of brain theory and neural networks, vol 3361(10)

LeCunn Y, Haffner P, Forsyth DA et al (1999) Object recognition with gradient based learning: shape, contour. LNCS 1681. Springer, Berlin, Heidelberg, pp 319–345

LeCun Y, Bengio Y, Hinton G (2015) Deep learning. Nature 521:436–444. https://doi.org/10.1038/nature14539

Mohanty SP, Hughes D, Salathé M (2016) Using deep learning for image-based plant disease detection, pp 233–240

Mohanty SP, Hughes DP, Salathé M (2016) Using deep learning for image-based plant disease detection. Front Plant Sci 1419

Pawara P, Okafor E, Surinta O, Schomaker L, Wiering M (2017) Comparing local descriptors and bags of visual words to deep convolutional neural networks for plant recognition. In: 6th international conference on pattern recognition applications and methods (ICPRAM 2017)

Simonyan K, Zisserman A (2014) Very deep convolutional networks for large-scale image recognition, pp 1409–1556

Sladojevic S, Arsenovic M, Anderla A, Culibrk D, Stefanovic D (2016) Deep neural networks based recognition of plant diseases by leaf image classification. In: Computational intelligence and neuroscience

Srivastava N, Hinton N (2014) Dropout: a simple way to prevent neural networks from overfitting. J Mach Learn Res 15:1929–1958

Suk H-I, Shen D (2013) Deep learning-based feature representation for AD/MCI classification. Med Image Comput Comput Assist Interv 16(2):583–590

Szegedy C et al (2015) Going deeper with convolutions. In: Proceedings of the IEEE conference on computer vision and pattern recognition, pp 1–9

Vu NT (2018) Sequential standing. Deep learning with Python (2018)

Zeiler MD, Fergus R (2014) Visualizing and understanding convolutional networks. In: Computer vision–ECCV, pp 818–833

Evaluation for Toxicity and Improved Therapeutic Effectiveness of Natural Polymer Co-administered Along with Venocin in Acetic Acid-Induced Colitis Using Rat Model

Ashish Kumar Netam, Jhakeshwar Prasad, Trilochan Satapathy, and Parag Jain

Abstract In these present study, two different series of experiments such as evaluation for toxicity and therapeutic of effectiveness of Aesculus hippocastanum co-administered low viscosity of sodium alginate has been carried out by using rat model (Acetic acid-induced ulcerative colitis). Acute and subacute toxicities were performed according to OECD guideline-423 and 407, respectively. In vitro antioxidant study for the combination therapy was studied. Histopathological examinations were carried out to determined organ level toxicity on long-term use of such combination. The results for In vitro antioxidant study suggested the free radical scavenging activity of the combination therapy. There were no behavioral changes or any morbidity and mortality were observed during the oral acute toxicity study food intake, water intake, and body weight variation in all the group of animals were within the similar pattern that means no much changes has been observed. Among the animals between control tests and standard drug-treated groups. The result of histopathological data indicated that some changes were observed for hemoglobin content, red blood cell count, etc., in some test group which is negligible in comparison with control group. The microscopic feature of histopathological study for the different organs such as kidney, liver, heart, etc., indicated that some degeneration and necrosis were observed in all the test groups of animals. These alterations of histopathological changes may be due to the stress, infection, and administration of test compound in empty stomach. Further study is suggested for determination of appropriate dose and ratios (for combination) to reduce the long-term toxicity and to improve the therapeutic effectiveness for the benefit of the entire society.

Keywords Aesculus hippocastanum · Low viscosity sodium alginate · Acetic acid antioxidant · Ulcerative colitis · Lipidperoxidase · Tumor Necrosis Factor (TNF-α)

A. K. Netam (✉) · J. Prasad · T. Satapathy · P. Jain
Department of Pharmacology, Columbia Institute of Pharmacy, Tekari, Raipur, CG, India
e-mail: ashish.netam52@gmail.com

A. A. Rizvanov et al. (eds.), *Advances in Biomedical Engineering and Technology*, Lecture Notes in Bioengineering,
https://doi.org/10.1007/978-981-15-6329-4_19

1 Introduction

Inflammatory bowel disease (IBD) is an intestinal disorder leads to inflamed and ulcerative intestine. It includes Crohn's disease and ulcerative colitis. Ulcerative colitis affected rectal and colonic mucosa, (Carter et al. 2004.) This is including Crohn's disease and ulcerative colitis which is an idiopathic inflammatory bowel disease of the rectal and colonic mucosa (Morris et al. 1989). It is characterized by colonic inflammation, resulting most probably from the infiltration of polymorphonuclear cells, lymphocytes, monocytes, and plasma cells, accompanied by the overproduction of oxygen free radicals, ultimately leading to mucosal alteration and ulceration (Almenier et al. 2012). Ulcerative colitis is a chronic disease cause disturbance in homeostasis in the gastrointestinal tract and intestinal inflammation (Baumgart and Carding 2007). It is also affecting the mucosal layer of the distal colon and rectum. The main symptoms of ulcerative colitis include diarrhea, abdominal cramps, and recurrent blood in the stools caused by mucosal ulcers. Ulcerative colitis about 50 lakh people have affected by inflammatory bowel disease across the world and India. Annually, 12 lakh cases of IBD have been reported in India but unfortunately only few people are aware about the disease (Lennard-Jones 1989). The recent pharmacological therapy for patients with ulcerative colitis includes nonselective anti-inflammatory drugs and corticosteroids or immunosuppressants, as well as anti-TNF-α agents (Cho et al. 2007). Several abovementioned drugs are used for the treatment of ulcerative colitis but they possess several adverse effects. Hence herbal remedies came into existence as an alternative therapy to overcome the disadvantages of such drug (Kane et al. 2003). These drugs are used to maintain continuing long-term remission, reduction of abnormal colonic inflammation, and control of clinical symptoms, such as diarrhea, rectal bleeding, and abdominal pain (Mowat et al. 2011). Though, the continuous use of these medications can cause serious side effects to patients. Thus, a great effort has been made to develop new drugs to treat ulcerative colitis.

It has been reported that herbal drugs are the best alternative for the treatment of ulcerative colitis. Herbal drugs are safe in comparison to other existing drugs, these can be used to prevent long-term remission, reduction of colonic inflammation, controlling clinical symptoms including rectal bleeding, diarrhea, and abdominal pain. Therefore, to avoid serious adverse effects of allopathic drugs, greater efforts have been paying to develop new drugs to treat ulcerative colitis. The natural herbal drugs Aesculus hippocastanum is used by several researchers for the effective treatment of deep vein thrombosis and other venous disorders which is generally seen in ulcerative colitis by considering the therapeutic effectiveness of Aesculus hippocastanum. Hence we have decided to use Aesculus hippocastanum as a test substance to evaluate its potency. The Sodium alginate is well known as biocompatible, degradable, and nontoxic. It forms a gel without the need of heat. They are also widely used as protective reparative effects. So, in this research we have proposed to administer the natural polymer sodium alginate alone and in combination with Aesculus hippocastanum in their appropriate ratio to determine the toxicity and therapeutic effectiveness against Acetic acid-induced ulcerative colitis in rat model.

2 Materials and Methods

2.1 Drug and Chemical Reagents

Aesculus hippocastanum was received as a gift sample from SUNPURE Pvt. Ltd New Delhi (India). Sodium Alginate was obtained from SD FINE Chem Ltd. Mumbai (India). Glacial acetic acid (AA) 99.8% was purchased from LOBA Chemie Pvt. Ltd. Mumbai (India). Sulfasalazine was procured from WALLACE Pharmaceuticals Pvt. Ltd. Ponda, Goa Maharashtra, (India). The RayBio® Enzyme-linked immunosorbent assay (ELISA) kits for rat TNF-α was obtained from Norcross USA. Lignocaine HCl Gel was purchased from ALVES Healthcare Pvt. Ltd. Mumbai, (India). All other chemicals used were of highest analytical grade commercially available.

2.2 Experimental Animals

Healthy adult Male Albino Wistar rats weighing about 180–200 gm were obtained from the Animal House Facility of Columbia Institute of Pharmacy, Raipur, Chhattisgarh, (India) having certificate number CIP/IAEC/2017/102 and Regd. No. 1321/PO/ReBi/S/10/CPCSEA. The animals were kept maintained under controlled environmental conditions with temperature (23 ± 2 °C), relative humidity (40–50%), and 12/12 h light/dark cycle with unlimited access to standard pelleted diet (chow, food) and water ad libitum, as per CPCSEA guideline. The animals were acclimatized to laboratory conditions for at least seven days before initiation of the experiment.

2.3 Experimental Design

The animals were randomly separated into five groups each containing six animals. Group I (Negative control) was pretreated with vehicle every 12 h, per oral; Group II (Toxic control) acetic acid, 2 ml; Group III (Test group-1) received AesculusHippocastanum 5 mg/kg oral; Group IV (Test group-2) received LVA 5gm/kg intrarectally; Group V (Test group-3) received 5 mg/kg mixture of Aesculus hippocastanum and LVA, intrarectally; Group VI (Reference group) was treated with Sulfasalazine 24 h before acetic acid instillation and for the subsequent five days.

2.4 Acute Toxicity

The acute toxicity was evaluated as per OECD guideline-423. The animals were randomly divided into five groups. They were received a dose of low viscosity sodium alginate along with Aesculus hippocastanum in Wistar rat of 50 mg/kg, body weight orally administered by using oral gavages after short fasting period. The general behavior of the animals was continuously monitored for 30 min, 1, 2, and 3 h after dosing, periodically during the first 24 h and the same treatment was followed for seven days.

2.5 Induction of Colitis

Inductions of ulcerative colitis were used according to the method of Ghasemi-Pirbaluti et al. (2017), with slight modification. The animals have fasted overnight with free access to water. The animals were light anesthetized with halothane. The inducing agent; acetic acid (2 ml, 1%, v/v) was instilled into the anus verge by inserting a medical grade polyurethane tube with 2 mm diameter through the rectum into the colon to a distance of 8 cm. The tube was kept in vertical position during instillation and after instillation to avoid leakage of acetic acid solution. Following the enema, After that, animals were kept in cages with continuous supply of feed and water till 8th day. Halothane was used to anesthetize animals and biochemical estimation was performed by collecting blood by retro-orbital puncture for biochemical estimation. The animals were again anaesthetized by using excess halothane and sacrificed by cervical dislocation. The abdominal portions were cut opened and colon was dissected out. Colon was flushed gently with saline and weighed. It was used for macroscopic scoring and histopathological estimations.

2.6 Hematological Study

The blood was collected with EDTA anticoagulant through retro-orbital puncture for biochemical estimation. The evaluated blood parameters were red blood cell count, blood hemoglobin concentration, basophil, eosinophil and neutrophil granulocytes, lymphocytes, and monocytes, hematocrit, mean corpuscular volume (MCV), mean corpuscular hemoglobin (MCH), mean corpuscular hemoglobin concentration (MCHC), white blood cell, and platelet counts (Byelinska et al. 2018).

2.7 *Antioxidant Activity Lipid Peroxidase/Malonaldehyde (LPO/MDA)*

A colonic tissue sample was homogenized in potassium phosphate buffer (50 mMol/L, pH 7.4, 1 g tissue/5 mL buffer). The total amount of protein in each sample was measured using the Bradford method. The tissue homogenate (10% w/v) was prepared in 0.15 M Tris-HCl buffer (PH 7.4). Then to it 0.2 ml of 8.1% sodium dodecyl sulphate (SDS) + 1.5 ml 20% acetic acid + 1.5 ml 8% Thiobarbutric acid (TBA) were added and volume was made up to 4 ml with distilled water. The above solution was subjected to heat on water bath for 60 min using glass ball as condenser. Then the solution allowed cooling and volume was made up to 5 ml. Then 5 ml of butanol: pyridine (15:1) was added and vortexed for a period of 2 min followed by centrifuge at 3000 rpm for 10 min. The upper organic layer was taken and measured optical density was measured at 532 nm. The absorbance was considered as total malondialdehyde (MDA) formed (Bose et al. 1989; Hagar et al. 2007; Alam et al. 2013).

2.8 *Measurement of TNF-α*

Colon was removed and homogenized in PBS then the amount of protein in each sample was measured via the Bradford method. The results were expressed in pg of cytokine/mg of protein. Assessment of cytokines (TNF- α) was carried out using ELISA kit; in clonic tissue, strips were minced with scissors for 15 s, suspended in 2 ml of 10 mm PBS (7.4 pH) and incubated in a shaking water bath 37 °C for 20 min. The sample was centrifuged and the supernatants were kept at −70 °C. The TNF-α assay using ELISA kit was performed (Wallace et al. 1989; Bose et al. 1989).

2.9 *Histopathological Evaluation*

The samples of highest macroscopic damage were selected from the sections of rat colon tissue. A two cm portion of the colonic tissue specimen from each animal was removed and fixed in 10% formalin solution then cut into 5 μm thickness, stained using hematoxylin eosin for the histopathological examination. They were made using a rotary microtome, 5 μm thickness sections were cut from the tissue samples embedded in paraffin and placed on standard glass slides. The paraffin was melted with a period of approx 12 h in an incubator at 58 °C. The samples were then stained with haematoxylene and eosin (H&E) according to the protocol. Qualitative analyses were performed on 400× magnified images.

Table 1 Observation table of animals behavioral

Test	Gender	Control	Control Reversal	Test-1	Test-2	Test-3	Test-3 reversal
Tremor	M	–	–	–	–	–	–
	F	–	–	–	–	–	–
Convulsion	M	–	–	–	–	–	–
	F	–	–	–	–	–	–
Salivation	M	–	–	–	–	–	–
	F	–	–	–	–	–	–
Diarrhea	M	–	–	–	–	–	–
	F	–	–	–	–	–	–
Sleep	M	–	–	–	–	–	–
	F	–	–	–	–	–	–

3 Result

3.1 Behavioral Changes

The results of oral acute toxicity study indicated that behavioral changes were no mortality and morbidity observed in animals through the 3-days period following single oral administration at all selected dose levels of the low viscosity sodium alginate along with Aesculus hippocastanum (Table 1).

3.2 Body Weight Loss

The result of body weight in different groups of animals reveled that there were no much changes have been observed (Table 2 and Fig. 1).

3.3 Hematological Study

See Tables 3 and 4.

3.4 MDA Activity

See Table 5 and Fig. 2.

Table 2 Effect of Low viscosity sodium alginate along with Aesculus hippocastanum on body weight in Albino Wistar Rats

Time period	Gender	Control	Control reversal	Test-1	Test-2	Test-3	Test-3 reversal
0 Days	M	182.4 ± 2.97	182.6 ± 2.57	182.4 ± 2.97	182.5 ± 2.51	182.6 ± 2.57	182.8 ± 2.98
	F	181.2 ± 1.43	181.5 ± 1.33	181.6 ± 1.54	181.5 ± 1.33	181.4 ± 1.25	181.4 ± 1.32
7 Days	M	183.6 ± 2.13	183.8 ± 2.15	183.5 ± 2.11	183.6 ± 2.12	183.7 ± 2.15	183.8 ± 2.18
	F	182.1 ± 2.12	182.3 ± 2.13	182.5 ± 2.11	182.5 ± 2.13	182.6 ± 2.11	182.9 ± 3.1
14 Days	M	184.5 ± 3.15	184.8 ± 3.16	184.2 ± 3.12	184.6 ± 3.15	183.8 ± 3.18	183.9 ± 3.19
	F	183.2 ± 3.03	183.5 ± 3.05	183.2 ± 3.04	183.3 ± 3.06	183.5 ± 3.07	183.7 ± 3.09
21 Days	M	185.3 ± 2.92	185.4 ± 2.94	185.1 ± 2.93	185.2 ± 2.94	185.5 ± 2.96	185.7 ± 2.97
	F	183.4 ± 2.15	183.6 ± 2.16	183.1 ± 2.07	183.8 ± 2.16	183.6 ± 2.12	185.8 ± 2.99
28 Days	M	185.8 ± 2.98	185.9 ± 2.96	185.5 ± 2.96	185.4 ± 2.94	185.6 ± 2.97	185.8 ± 2.99
	F	185.2 ± 2.94	185.5 ± 2.96	185.4 ± 2.29	185.2 ± 2.94	185.7 ± 2.97	185.9 ± 2.99

Mean ± SEM (n = 6)

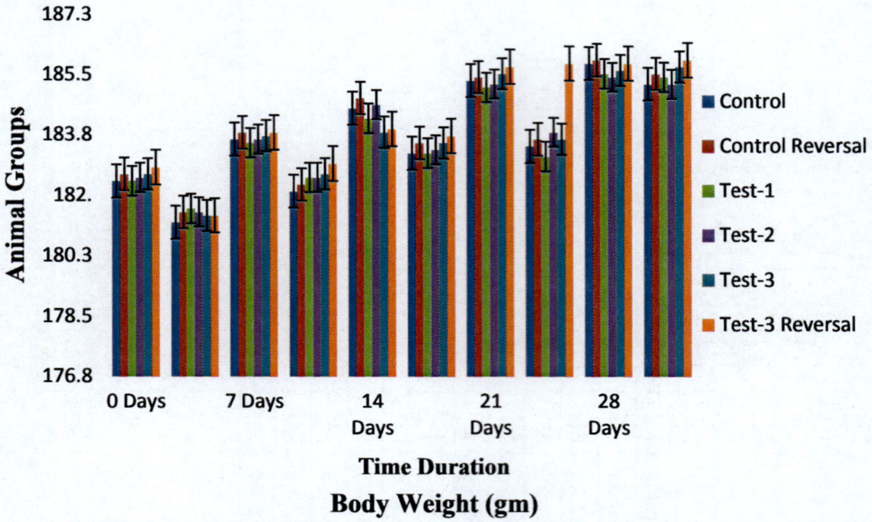

Fig. 1 Graphical representation of animals mean body weight during dosing, number of animals per group n = 10, each group (Five—Male and Five—Female). All value is reported as mean ± SEM (n = 6)

Table 3 Hematological data of various groups of male animals

S. no.	Parameters	Control group	Test-1 group	Test-2 group	Test-3 group
1.	Hb (gm%)	16.48 ± 0.298	16.46 ± 0.102	16.26 ± 0.102	16.2 ± 0.2
2.	WBC (cmm)	2560 ± 40	3420 ± 152.9	4360 ± 112.2	4480 ± 106.77
3.	Neu (%)	42.4 ± 0.244	40.4 ± 0.812	43.8 ± 0.969	44 ± 00
4.	Lym (%)	52 ± 0.004*	53.8 ± 0.969	47.8 ± 0.734	47.2 ± 0.244
5.	Eos (%)	4.2 ± 0.374	5 ± 0.316	4.6 ± 0.244	4.4 ± 0.509
6.	Mon (%)	01 ± 00	01 ± 00	01 ± 00	01 ± 00
7.	Bas (%)	00 ± 00	00 ± 00	00 ± 00	00 ± 00
8.	RBC (%)	7.61 ± 0.002*	7.53 ± 0.046	7.54 ± 0.011	7.53 ± 0.009
9.	Platelet (%)	2.75 ± 0.003*	2.91 ± 0.019	2.94 ± 0.007	2.95 ± 0.013
10.	MPV	9.62 ± 0.058	9.46 ± 0.097	9.4 ± 0.054	9.5 ± 0.004*
11.	PCV	33.38 ± 0.165	33.40 ± 0.329	33.42 ± 0.631	33.46 ± 0.082
12.	MCV	51 ± 0.196	52.69 ± 0.087	52.72 ± 0.12	52.75 ± 0.01
13.	MCHb (Pictograms)	19.72 ± 0.009	19.68 ± 0.026	19.92 ± 0.002*	19.68 ± 0.058
14.	MCHb (mg/dl)	36.47 ± 0.006	37.35 ± 0.011	37.63 ± 0.022	37.66 ± 0.009
15.	RCDW (%)	15.22 ± 0.037	15.28 ± 0.086	15.5 ± 0.094	15.6 ± 0.005*

Mean ± SEM (n = 5), P value = < 0.005(*)

Table 4 Hematological data of various groups of female animals

S. no.	Parameters	Control group	Test-1 group	Test-2 group	Test-3 group
1.	Hb (gm%)	15.48 ± 0.298	14.46 ± 0.102*	14.26 ± 0.102	14.2 ± 0.2
2.	WBC (cmm)	2350 ± 41	2352 ± 42.3	2355 ± 42.2	2355 ± 43.7
3.	Neu (%)	42.4 ± 0.24	40.4 ± 0.23	43.8 ± 0.29	44 ± 00
4.	Lym (%)	51 ± 0.70	51.4 ± 0.69	52.4 ± 0.73	52.6 ± 0.74
5.	Eos (%)	4.2 ± 0.34	4.5 ± 0.36	4.6 ± 0.34	4.6 ± 0.39
6.	Mon (%)	01 ± 00	01 ± 00	01 ± 00	01 ± 00
7.	Bas (%)	00 ± 00	00 ± 00	00 ± 00	00 ± 00
8.	RBC (%)	7.61 ± 0.002	7.63 ± 0.046	6. ± 0.011	7.60 ± 0.009*
9.	Platelet (%)	3.75 ± 0.003	3.91 ± 0.019*	2.74 ± 0.007	3.14 ± 0.013
10.	MPV	9.62 ± 0.058	9.46 ± 0.097	10.3 ± 0.054	10.6 ± 0.050*
11.	PCV	47.38 ± 0.165	33.84 ± 0.329	38.22 ± 0.631	42.46 ± 0.082
12.	MCV	55.41 ± 0.196	51.22 ± 0.087	52.724 ± 0.12	49.938 ± 0.01
13.	MCHb (Pictograms)	19.72 ± 0.009	20.68 ± 0.026	19.92 ± 0.02*	18.68 ± 0.058
14.	MCHb (mg/dl)	35.47 ± 0.006	40.35 ± 0.011	37.73 ± 0.022	37.66 ± 0.009
15.	RCDW (%)	15.72 ± 0.037	16.28 ± 0.086*	16.5 ± 0.094*	16.36 ± 0.05*

Table 5 Serum MDA levels in different groups of animals

S. no.	Groups	MDA level
1	Control	8.22 ± 0.509
2	Test-1	5.23 ± 0.152
3	Test-2	4.57 ± 0.052
4	Standard	4.78 ± 0.072

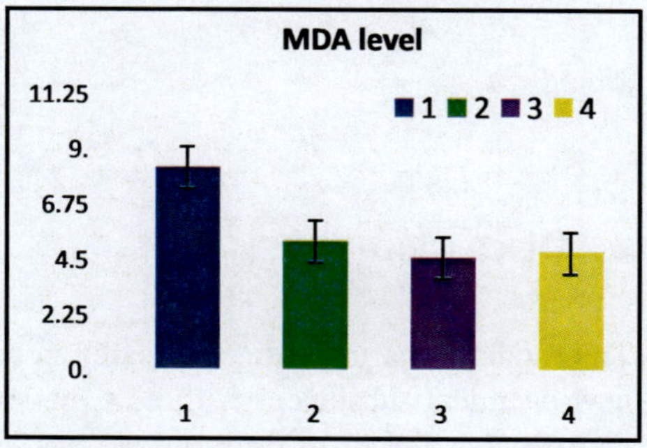

Fig. 2 Graph showing the level of serum MDA in homogenized colon tissue of different group of animals

Table 6 TNF-α level in different groups

S. No.	Groups	TNF-α level
1	Control	5.53 ± 0.672
2	Test-1	4.84 ± 0.252
3	Test-2	4.59 ± 0.239
4	Standard	5.13 ± 0.584

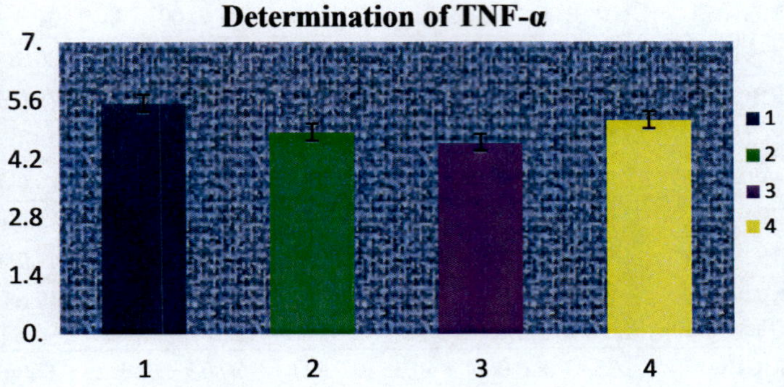

Fig. 3 Graph showing the level of TNF-α in homogenized colon tissue of different group of animals

3.5 *TNF-α Activity*

See Table 6 and Fig. 3.

3.6 *Histopathology*

See Fig. 4.

4 Discussion

The modern pharmaceutical research is concerned with all aspects of identifying new chemical substances with new modes of action. In particular, the economics of treatment linked to drug dosage has led compound to new drug development technologies. As a result, treatments are now becoming more reasonable for wide sections of society, including the financially challenged. Few marketed products are available for the effective treatment of IBD such as Sulfasalazine, Mesalazine, Balsalazide, Antileukotriene like Montelukast, etc. Natural polymers are evaluated for their wound-healing effect in case of IBD. Steroidal drugs such as prednisone,

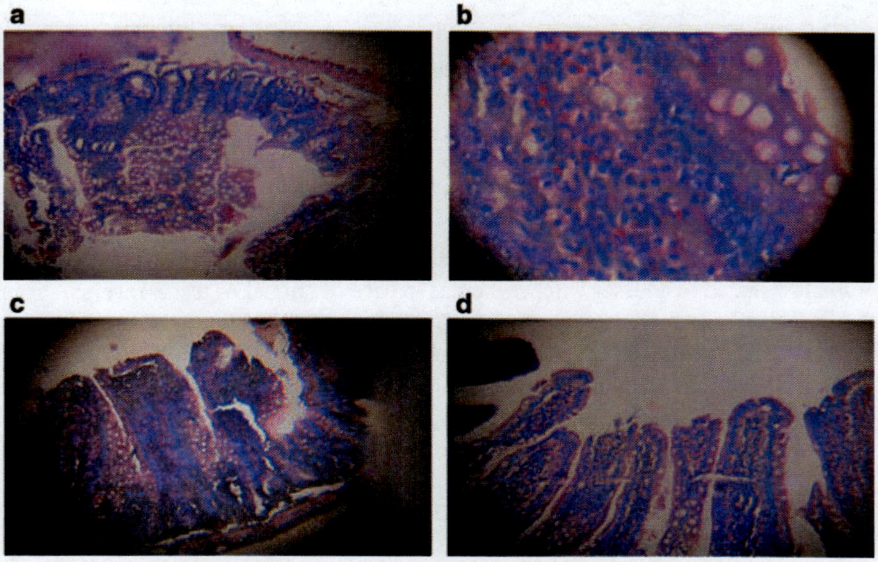

Fig. 4 **a** Control group. **b** Test group-1. **c** Test group-2. **d** Standard group

etc., on long-term use cause severe side effect like Cushing syndrome. Many herbal products possess anti-inflammatory, antioxidant activity has been evaluated for their anti-IBD effect. Venocin from Horse chestnut seed extract has been shown to have Antioxidant, Anti-inflammatory, Wound healing, and Supports circulation property.

Several abovementioned allopathic drugs are used for the treatment of ulcerative colitis but they possess several adverse effects. Hence herbal remedies came into existence as an alternative therapy to overcome the disadvantages of such drugs. Among the natural herbal drugs Aesculus hippocastanum is used by several researchers for the effective treatment of deep vein thrombosis and other venous disorder which is generally seen in ulcerative colitis by considering the therapeutic effectiveness of Aesculus hippocastanum. We have decided to use Aesculus hippocastanum as a test substance to evaluate its potency against acetic acid-induced ulcerative colitis using rat model. Further it has been evidenced from the literature that sodium alginate is a natural polymer used for various gastrointestinal disorders though sodium alginate is obtained from natural source and devoid of any side effects/adverse effects. The Sodium alginate is well known as biocompatible, degradable, and nontoxic. It forms a gel without the need of heat. They are also widely used as protective reparative effects. Though we have taken the combination of sodium alginate which is co-administered with Venocin, there is a need to evaluate the toxicity and safety of the combination.

Two series of studies have been carried out for the evaluation of toxicity and therapeutics effectiveness for the combination of Aesculus hippocastanum along with low viscosity sodium alginate at their predetermined dose.

The oral toxicity study has been carried out as per OECD-423. The results of the oral acute toxicity study indicated that there were no mortality and morbidity observed in animal of all the groups. The result of body weight in different groups of animals reveled that there were no much changes has been observed similarly, food and water consumption all the group of animals showing similar pattern of result during the entire course of experiment the effect of combination therapy on Hematological data of all the groups of animal when depicted in Table no. 15 and Fig no. 6–18. The hemoglobin content except control and control reversal group of animals reduced to a lesser extent whereas in case of animal of test-3 reversal group the value has been increased up to 17.3. The WBC count in all groups of animals has been increased. To some extent in comparison to control and control reversal group neutrophils lymphocytes and eosinophils value did not indicated much change in all groups of animals. RBC count has been decreased in Test-1, Test-2 as well as Test-3 reversal group whereas Test-3 group of animals having similar RBC count value in comparison to control reversal group platelet count and mean platelet value shown no much change in all group of animals. The value for packed cell volume has been reduced in Test-1, Test-2, and Test-3 reversal group of animals. The results for Mean corpuscular volume having similar pattern of results in all groups of animals. The results of mean corpuscular Hemoglobin content indicated that it has been increased in all the test groups of animal in comparison to control and control reversal group.

The antioxidant activity for the combination therapy has been determined by the estimation of MDA and TNF-α. The results for MDA have been depicted in table no 17, 18, and graphical represented in graph no. 20 and 21. The results indicated that Test compound showing decreased in concentration of MDA and TNF-α. Hence the combination therapy possesses free radical scavenging activity.

On the termination of experiment the animal was sacrificed as per CPCSEA guidelines and Subjected organ where isolated and subjected for histopathological examination to determine the toxicity of combination therapy at different organ level. The histopathological indicated that on long-term use of combination therapy some degeneration and necrosis have been observed in Test group of animals whereas no changes in microscopic features were pointed out in control and control reversal group of animals. These may be due to the excessive stress less food intake, etc. From the above finding it has been observed that the combination of Aesculus hippocastanum and Low viscosity sodium alginate at predetermined dosed possess very good free radical scavenging activity and anti-inflammatory activity and TNF-α inhibiting activity. To reduce the long-term use of organ level toxicity, further study is suggested to adjust the dose level and duration which produce better therapeutic effectiveness which in turn pave the way for the development of new drug.

5 Conclusion

The present study has been undertaken to establish the improved therapeutic effectiveness of Aesculus hippocastanum co-administered with LVA. From the previously published scientific data, Aesculus hippocastanum is used for the treatment of various venous disorders such as deep vein thrombosis, etc., and Sodium alginate is a natural polymer having good mucosal protective activity. Hence, the present study has been proposed to determine the toxicity and synergistic effect of both the drugs with their appropriate ratios. The study has been carried out using Wistar rats and acetic acid was used as inducing agent for ulcerative colitis. Various In vitro studies such as Malondialdehyde (MDA) carried out and the result revealed that the combination possess good level of antioxidant activity. Hematological and biochemical findings also support our hypothesis but long-term administration of combinations (AesculusHippocastanum along with low viscosity sodium alginate) alters the mucosal integrity which has been observed from histopathological findings of some vital organs. So, our findings suggest that further detailed study is required to establish the exact mechanism of mucosal degeneration which in turn reduces the long-term organ level toxicity that will help the researchers to decide for further new drug development for the benefit of the society.

References

Alam MN, Bristi NJ, Rafiquzzaman M (2013) Review on in vivo and in vitro methods evaluation of antioxidant activity. Saudi Pharm J 21(2):143–152

Almenier HA, Al Menshawy HH, Maher MM, Al Gamal S (2012) Oxidative stress and inflammatory bowel disease. Front Biosci (Elite Ed) 1(4):1335–1344

Baumgart DC, Carding SR (2007) Inflammatory bowel disease: cause and immunobiology. Lancet 369(9573):1627–1640

Bose R, Sutherland GR, Pinsky C (1989) Biological and methodological implications of prostaglandin involvement in mouse brain lipid peroxidation measurements. Neurochem Res 14(3):217–220

Byelinska IV, Kuznietsova HM, Dziubenko NV, Lynchak OV, Rybalchenko TV, Prylutskyy YI, Kyzyma OA, Ivankov O, Rybalchenko VK, Ritter U (2018) Effect of C60 fullerenes on the intensity of colon damage and hematological signs of ulcerative colitis in rats. Mater Sci Eng C 1(93):505–517

Carter MJ, Lobo AJ, Travis SP (2004) Guidelines for the management of inflammatory bowel disease in adults. Gut 53(suppl 5):v1–6

Cho JY, Chang HJ, Lee SK, Kim HJ, Hwang JK, Chun HS (2007) Amelioration of dextran sulfate sodium-induced colitis in mice by oral administration of β-caryophyllene, a sesquiterpene. Life Sci 80(10):932–939

Ghasemi-Pirbaluti M, Motaghi E, Najafi A, Hosseini MJ (2017) The effect of theophylline on acetic acid induced ulcerative colitis in rats. Biomed Pharmacother 1(90):153–159

Hagar HH, El-Medany A, El-Eter E, Arafa M (2007) Ameliorative effect of pyrrolidine-dithiocarbamate on acetic acid-induced colitis in rats. Eur J Pharmacol 554(1):69–77

Kane S, Huo D, Aikens J, Hanauer S (2003) Medication nonadherence and the outcomes of patients with quiescent ulcerative colitis. Am J Med 114(1):39–43

Lennard-Jones JE (1989) Classification of inflammatory bowel disease. Scand J Gastroenterol 24(sup170):2–6

Morris GP, Beck PL, Herridge MS, Depew WT, Szewczuk MR, Wallace JL (1989) Hapten-induced model of chronic inflammation and ulceration in the rat colon. Gastroenterology 96(3):795–803

Mowat C, Cole A, Windsor AL, Ahmad T, Arnott I, Driscoll R, Mitton S, Orchard T, Rutter M, Younge L, Lees C (2011) Guidelines for the management of inflammatory bowel disease in adults. Gut 60(5):571–607

Wallace JL, MacNaughton WK, Morris GP, Beck PL (1989) Inhibition of leukotriene synthesis markedly accelerates healing in a rat model of inflammatory bowel disease. Gastroenterology 96(1):29–36

Finite Element Analysis of Traumatic Brain Injury Due to Blunt Impact of Different Durations

Tanu Khanuja and Harikrishnan Narayanan Unni

Abstract This study proposes a detailed biomechanical model of the human head to study the effect of non-penetrating (blunt) head impacts of different durations on intracranial organs. A patient-specific high biofidelity three-dimensional finite element human head model is developed from the magnetic resonance imaging (MRI) data and segmented into five volumes namely, skull, cerebrospinal fluid (CSF) with ventricular system, cerebrum, cerebellum, and brain stem and each segment is assigned with appropriate material properties. The model validated against the impact experiment based on human cadaver is used to perform simulation with a range of blunt impact durations and biomechanical analysis is performed by investigating the maximum intracranial pressure (ICP) and von Mises stress distribution across the brain. The probability of loss of consciousness and tissue damage is studied based on the ICP and von Mises stress values. The coup and contrecoup phenomena is also studied with the localization, extension and, intensity of tissue damage based on the injury tolerance criteria present in the literature.

Keywords Blunt impact · Finite element human head model · Intracranial pressure · von Mises stress · Coup and ContreCoup phenomena

1 Introduction

Traumatic brain injury (TBI) is the injury to the head organs due to sudden impact or jolt to the head which can result in temporary or permanent damage to the brain (Maas et al. 2008). Most of these injuries are invisible from the outside and current imaging modalities may or may not be able to detect the physical damage to the brain (Parikh et al. 2007). Typically, head impact durations causing TBI are in the range

T. Khanuja (✉) · H. N. Unni
Department of Biomedical Engineering, Indian Institute of Technology, Hyderabad, India
e-mail: bm12m14p000002@iith.ac.in

H. N. Unni
e-mail: harikrishnan@iith.ac.in

A. A. Rizvanov et al. (eds.), *Advances in Biomedical Engineering and Technology*, Lecture Notes in Bioengineering,
https://doi.org/10.1007/978-981-15-6329-4_20

of few milliseconds and can cause closed or penetrating head injuries (Maas et al. 2008). However, it is not always possible to perform cadaveric experiments on the human brain to understand the injury phenomena. Consequently, the finite element method is used as a pronounced tool to understand the brain injury mechanism.

The finite element method is used as a noninvasive approach to replicate injury phenomena to the head and to study the tolerance criteria for different types of injuries. In last three decades, a large number of numerical models of the human head are developed to study the computational mechanics of the skull fracture and TBI with linear material properties and simplistic geometries (Zhao et al. 2015; Siswanto and Hua 2012; Yan and Pangestu 2011; Yoganandan and Pintar 2003; Kleiven 2003). Recently, a few detailed geometrical structures modeled with non-linear material properties are used for the computational study. However, these models have segmented the brain similar to a sphere or ellipsoid without the appropriate representation of sulcus and gyrus structures on the brain surface (Zhao et al. 2015; Yan and Pangestu 2011; Kleiven 2003). In addition, most of the experimental and computational studies focus on static and dynamic impact analysis with impact duration ranging from 3 to 20 ms (Nahum et al. 1977; Trosseille et al. 1992; El et al. 2008; Kleiven 2006). The short duration blunt impacts produce non-penetrating closed brain injury where the brain is not directly exposed to the impact (Hannay et al. 2004). These non-penetrating blunt impacts can be catastrophic and may not be detectable just after injury using imaging modalities. The computational study on patient-specific finite element head models can be used to give a clear picture of such blunt injury mechanisms.

The present study focuses on the generation of a new three-dimensional finite element human head model from a patient-specific MRI data with a clear representation of sulci and gyri folds on the brain surface. In addition, the validated model is used for investigating the head injury mechanism with blunt impacts of durations ranging from 0.5 to 2.5 ms. The broad injury prediction and soft tissue damage criteria are discussed on the basis of threshold intracranial pressure (ICP) and von Mises stress summarized in literature.

2 Material and Methods

2.1 MRI Predicated 3-D Human Head Model and Mesh Generation

At present, the magnetic resonance imaging (MRI) is the most sensitive imaging modality to accurately capture the soft tissue configurations of the human body (Lee and Newberg 2005). Hence, to specifically capture the sulci and gyri folding on the cerebrum surface the present finite element model is approximated from the axial MRI images obtained from "The whole brain atlas" of the Harvard medical school (Harvard Medical School). The image-based meshing platform AMIRA 5.6

(Institute academic license) is used to generate the three-dimensional finite element human head model from the MRI image data. The image slices are labeled on the basis of region of interest (ROI) as represented in Fig. 1. The process of segmentation includes semi-automated tools such as thresholding, magic wand, contour fitting, and interpolation. In addition, the "remove islands" tool, "smooth labels" tool, and image processing filters are utilized to smoothen the segmented surface within the mask. The segmented surfaces are concatenated and converted to the coupled volumes. The segmented head model comprises of five rudimentary structures namely, skull, cerebrospinal fluid with ventricular system, cerebrum, cerebellum, and brain stem as illustrated in Fig. 2.

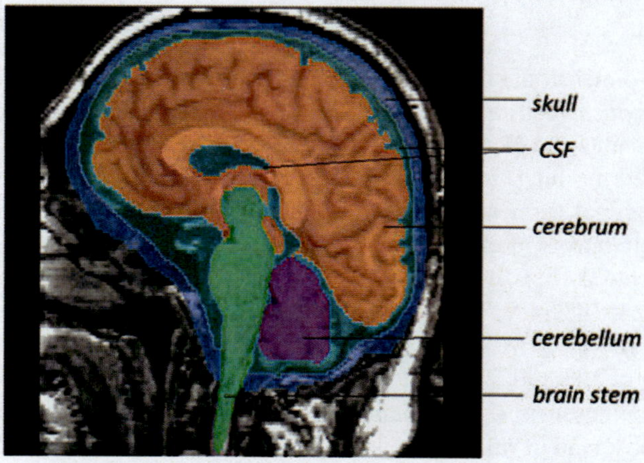

Fig. 1 Illustration of segmentation of ROI in one of the sagittal MRI slice

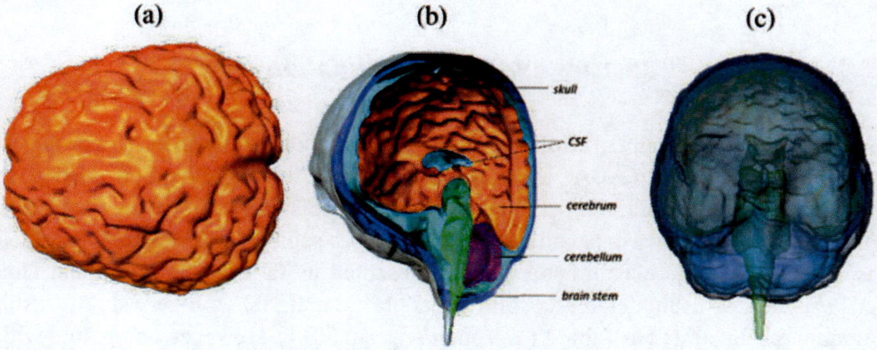

Fig. 2 Illustration of segmented volumes in AMIRA 5.6 **a** top view of cerebrum with sulci and gyri structures; **b** sagittal cross-section of full head surface model representing segmented layers; **c** front view of full transparent head model

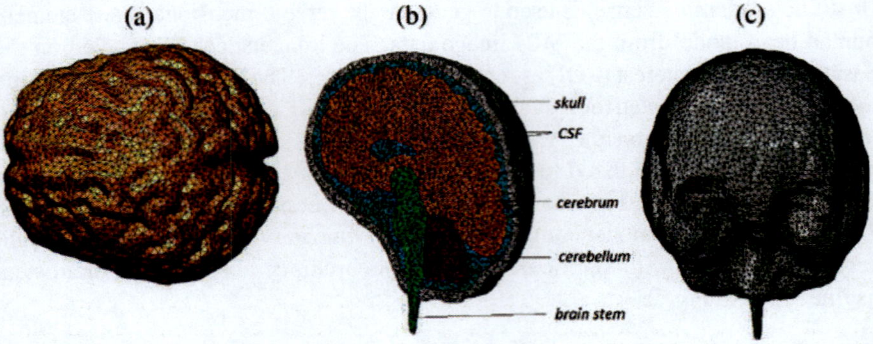

Fig. 3 Illustration of the meshed model in AMIRA 5.6 with C3D4 type elements **a** cerebrum meshed with 188,742 number of elements; **b** sagittal cross-section illustrating different mesh segments; **c** full head meshed model with 345,597 number of elements

The three-dimensional finite element head model is meshed with 345,597 C3D4 (four nodes, constant strain, linear, solid tetrahedral elements with three degrees of freedom) elements and 62,568 nodes in AMIRA 5.6 using marching cube algorithm and grid-based meshing. In addition, the volumes are coupled using shared nodes between individual segments in order to avoid non-convergence of model simulation which is resulted from model complexity and contacts between structures. The individual segments skull, CSF, cerebrum, cerebellum, and brain stem are meshed with 60,462, 67,624, 188,742, 20,948, and 7821 number of C3D4 elements, respectively as illustrated in Fig. 3. Moreover, the mesh optimization is performed with varying mesh density of 103,890, 288,962, 345,597, and 579,352 elements in order to select the mesh model insensitive to the number of mesh elements. The maximum variation of 1.98% is observed in validation results in between models with 288,962, 345,597, and 579,352 elements. Therefore, to maintain the balance between accuracy and time cost, the model with 345,597 elements is utilized for further simulations.

2.2 *Material Properties and Model Validation*

After meshing the model in AMIRA 5.6, the finite element head model is imported to the finite element solver ABAQUS 6.9/Explicit. Ideally, the skull is composed of three layers of different material properties and their own anisotropy. However due to lack of mechanical property data and complexity in layer segmentation, skull is modeled as a linear isotropic elastic material as represented in Table 1 (Siswanto and Hua 2012). The CSF is modeled as a solid linear elastic material with very small elastic modulus as illustrated in Table 1 (Yan and Pangestu 2011). However, being fluid with consistency similar to water, CSF should be modeled as Newtonian fluid but in order to avoid solid–fluid coupling difficulty during simulation CSF is modeled as a solid layer. The higher Poisson's ratio is considered to represent the incompressibility of

Table 1 Material properties for skull, CSF, and soft tissues

Segment	Density (kg/m^3)	Elastic modulus (MPa)	Poisson's ratio	Viscoelastic coefficients
Skull	3000	7300	0.22	–
CSF	1000	0.15	0.499	–
Brain (cerebrum, cerebellum, brain stem)	1040	0.0199	0.499	$g_1 = 0.5837$, $\tau_1 =$ 0.02571 $g_2 = 0.2387$, $\tau_2 =$ 0.0257

Table 2 Maximum ICP and von Mises stress during different impact pulses

Impact pulse duration (ms)	Maximum ICP (kPa)	Maximum von Mises stress (kPa)
2.5	183	69.75
2	206.185	64.4
1.5	226	55.8
1	237.96	43.2
0.5	476.5	22.9

the material. The brain tissue comprising of cerebrum, cerebellum, and brain stem are modeled as an incompressible linear viscoelastic material using two-term Prony series model as collated in Table 1 (Rashid et al. 2012; Zhang et al. 2001). The viscoelastic Prony series is given as:

$$G(t) = 1 - \sum_{i=1}^{n} g_i \left(1 - e^{-\frac{t}{\tau_i}}\right) \tag{1}$$

where, $G(t)$ = Dynamic shear modulus,

g_i = relaxation coefficient,

τ_i = characteristics relaxation time.

After assigning the material properties model is validated against the Nahum's experimental data presented in literature (Nahum et al. 1977). The head-neck junction is considered to be freely moving and the semi-sinusoidal impact pulse of 6 ms duration similar to experimental contact force is applied to the frontal side of the skull. The computed ICPs at anterior and posterior sides of the model are compared with the experimental ICPs and a good correlation between model and experimental data is observed. The validation graph is presented by the author in the IEEE BIBE 2018 conference proceeding (Khanuja and Unni 2018).

3 Results and Discussion

Subsequent to the model validation the blunt impact analysis is performed with impacts on frontal-top side of the head as shown in Fig. 4 with five impact pulses of semi-sinusoid shape ranging from 0.5 to 2.5 ms with peak magnitude of 9 kN as shown in Fig. 5. The magnitude of force is chosen based on the force range presented in literature to induce traumatic brain injury (Yoganandan et al. 1995).

Fig. 4 Illustration of impact location on the frontal-top side of the head

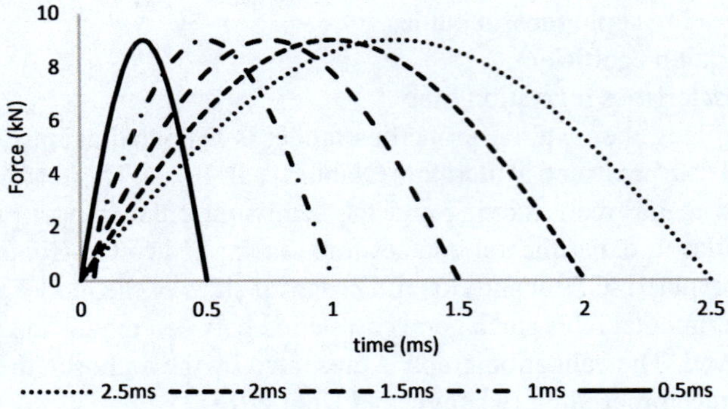

Fig. 5 Illustration of impact force pulses applied on the head

The analysis is performed in ABAQUS 6.9/Explicit with a minimum time increment of 70.9 ns for the 10 ms duration to analyze the immediate effect of impacts on the head. The maximum ICP and von Mises stress values for varying pulse durations are compared as presented in Table 2. It can be reported from the computed results that the maximum value of ICP is increasing with decreasing impact duration which is similar to the coup pressure trend presented in the literature with varying pulse durations (Pearce and Young 2014). In addition to this, our model predicts that the maximum von Mises stress is decreasing with decreasing impact duration. Based on the tolerance criteria existing in the literature the ICP exceeding 300 kPa causes mild traumatic brain injury while the von Mises stress exceeding 38 kPa generates 50% probability of severe neurological damage (Newman et al. 2000; Baumgartner and Willinger 2005). The ICP is used as an indicator of loss of consciousness and concussion and the von Mises stress is used as an indicator of neurological damage to the brain. Based on the threshold value of 300 kPa, the probability of loss of consciousness is observed to be maximum during the shortest impact pulse (0.5 ms) while the probability of maximum neurological damage is observed during longest impact pulse (2.5 ms) as illustrated in Fig. 6. Moreover, the maximum ICP is obtained near the peak of impact pulse while the maximum von Mises stress is obtained near 9 ms in all the cases.

During impact to the frontal-top side of the head, a positive peak pressure is obtained on frontal lobe of the brain and on the other side, the negative pressure of similar magnitude range is observed in contrecoup side, covering occipital lobe of brain and cerebellum as illustrated in Fig. 7. The negative pressure may result in severe transient cavitation in the contrecoup side of the brain (El et al. 2008; Brennen 2003). In addition, after the removal of the impact, the negative pressure is also observed on the coup side which represents the probability of brain cavitation injury on coup side as well however of lesser severity.

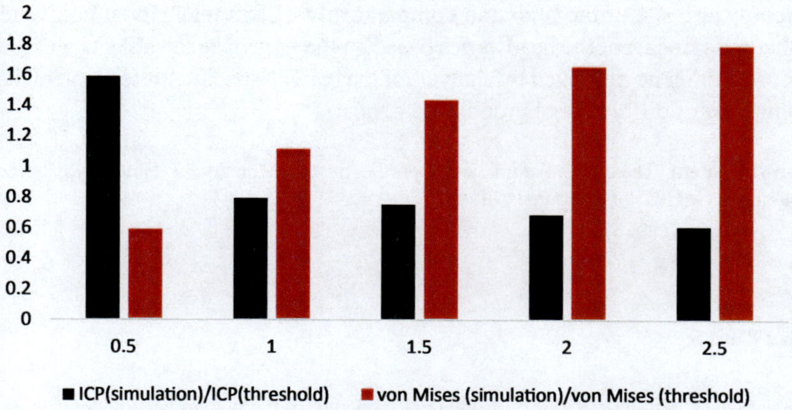

Fig. 6 Probability of loss of consciousness (ICP ratio) and neurological damage (von Mises stress ratio) at 0.5, 1, 1.5, 2, and 2.5 ms duration impact pulses

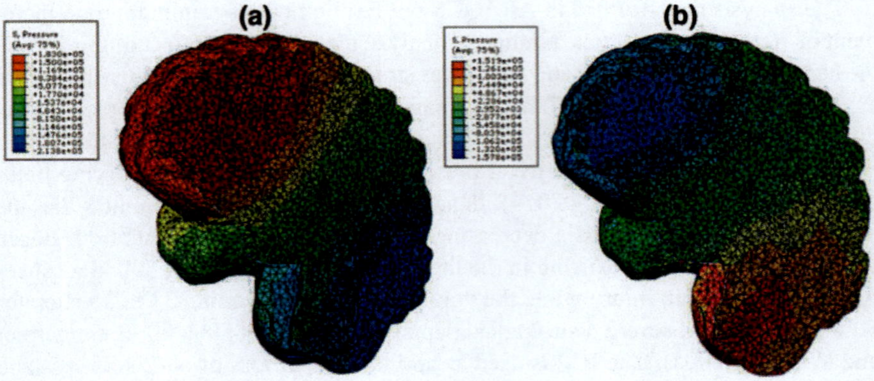

Fig. 7 Illustration of coup and contrecoup ICPs during 2.5 ms duration impact pulse **a** positive pressure on coup site at 1.75 ms, **b** negative pressure on coup site at 4 ms

4 Conclusion

The work presented the ICP and von Mises stress response of brain during short time pulse impacts of varying duration and same magnitude. For short duration impacts, the ICP has been seen as increasing with decreasing time width of the impact pulse whereas the von Mises stress is found to be decreasing with decreasing impact duration. The study concludes that the short duration impacts have higher probability to cause concussion and loss of consciousness without going through huge physical tissue damage. However, the increase in contact force duration will cause more damage to the intracranial organs with severe traumatic brain injuries. Furthermore, the coup and contrecoup pressure phenomena is studied for blunt impacts which indicate the probability of brain tissue cavitation injury on the coup site along with the contrecoup site of the brain but comparatively of less intensity at coup site. As the anatomical scale of the head varies based on the human head subjects, the present work can further be improved by analysis of the patient-specific finite element human head models of different age groups and genders.

Acknowledgment The present work is supported by the Ministry of Human Resource and Development (MHRD), Government of India.

References

Baumgartner D, Willinger R (2005) Numerical modeling of the human head under impact: new injury mechanisms and tolerance limits. In: IUTAM symposium on impact biomechanics: from fundamental insights to applications, pp 195–203

Brennen C (2003) Cavitation in biological and bioengineering contexts. In: Proceedings of the 5th international symposium on cavitation. Osaka, Japan

El ST, Mota A, Fraternali F, Ortiz M (2008) Biomechanics of traumatic brain injury. Comput Methods Appl Mech Eng 197:4692–4701

Hannay HJ, Howieson DB, Loring DW, Fischer JS, Lezak MD (2004) Neuropathology for neuropsychologists. In: Lezak MD, Howieson DB, Loring DW (eds) Neuropsychological assessment. Oxford University Press, Oxford [Oxfordshire], pp 158–62. ISBN 978-0-19-511121-7

Johnson KA, Becker JA (2020) Normal Anatomy in 3-D with MRI/PET (Javascript), the whole brain atlas, Harvard Medical School. http://www.med.harvard.edu/AANLIB/cases/caseNA/pb9.htm. Accessed 31 July 2020

Khanuja T, Unni HN (2018) [Regular Paper] Computational modeling of traumatic brain injury due to impact on different sides of human head. In: IEEE 18th international conference on bioinformatics and bioengineering (BIBE), Taichung, Taiwan, pp 364–370

Kleiven S (2003) Influence of impact direction on the human head in prediction of subdural hematoma. J Neurotrauma 20(4):365–379

Kleiven S (2006) Evaluation of head injury criteria using a finite element model validated against experiments on localized brain motion, intracerebral acceleration, and intracranial pressure. Int J Crashworthiness 11:65–79

Lee B, Newberg A (2005) Neuroimaging in traumatic brain imaging. NeuroRx 2(2):372–383

Maas AI, Stocchetti N, Bullock R (2008) Moderate and severe traumatic brain injury in adults. Lancet Neurol 7(8):728–741

Nahum AM, Smith RW, Ward CC (1977) Intracranial pressure dynamics during head impact. In: Proceedings 21st stapp car crash conference, pp 339–366

Newman J, Barr C, Beusenberg M, Fournier E, Shewchenko N, Welbourne E, Withnall C (2000) A new biomechanical assessment of mild traumatic brain injury, II: results and conclusions. In: Proceedings of international IRCOBI conference on the biomechanics of impact, pp 223–233

Parikh S, Koch M, Narayan RK (2007) Traumatic brain injury. Int Anesthesiol Clin 45(3):119–135

Pearce CW, Young PG (2014) On the pressure response in the brain due to short duration blunt impacts. PLoS ONE 9(12):e114292

Rashid B, Destrade M, Gilchrist M (2012) Hyperelastic and viscoelastic properties of brain tissue in tension. In: Proceedings of the ASME 2012 international mechanical engineering congress and exposition, November 9–15, 2012, Houston, Texas, USA

Siswanto WA, Hua CS (2012) Strength analysis of human skull on high speed impact. Int Rev Mech Eng 6(7):1508–1514

Trosseille X, Tarriere C, Lavaste F, Guillon F, Domont A (1992) Development of a F.E.M. of the human head according to a specific test protocol. In: Proceedings on 36th stapp car crash conference, SAE Paper No. 922527, Society of Automotive Engineers, Warrendale, PA

Yan W, Pangestu OD (2011) A modified human head model for the study of impact head injury. Comput Methods Biomech Biomed 14(12)

Yoganandan N, Pintar FA (2003) Biomechanics of temporo-parietal skull fracture. Journal of Clinical Biomechanics 19(3):225–239

Yoganandan N, Pintar FA, Sances A, Walsh PR, Ewing CL, Thomas DJ, Snyder RG (1995) Biomechanics of skull fracture. J Neurotrauma 12(4):658–668

Zhang L, Yang KH, King AI (2001) Comparison of brain responses between frontal and lateral impacts by finite element modeling. J Neurotrauma 18(1):21–30

Zhao W, Ruan S, Ji S (2015) Brain pressure responses in translational head impact: a dimensional analysis and a further computational study. Biomech Model Mechanobiol 14(4):753–766

Data Dissemination Using Social-Based Attributes in Delay-Tolerant Networks

Sanjay Kumar, Prasoon Shukla, and Sudhakar Pandey

Abstract In rural areas an ambulance carrying a patient may require to connect with experts in a city. In rural areas it is not possible that network connectivity to the ambulance available all the time. In this type of situation concept of delay-tolerant networks may be utilized for the opportunistic connectivity with a person at distance. Delay-Tolerant Network (DTN) is an approach to a wireless network, which possesses an uncertainty in the connection. Due to intermittent connectivity and long delays, routing in DTN turns out to be quite challenging. In recent years, many social-based routing reflects the capabilities of social attributes for disseminating data. In this paper social attributes namely, community, similarity, betweenness, and degree centrality are being used for outlining a routing technique. Using this approach we can improve the quality of service provided by the ambulances while passing through rural areas. A comparison is made with already proposed routing technique, for example, BUBBLE Rap, Epidemic, and experiences a noteworthy improvement in delivery ratio and buffer utilization.

Keywords Delay-tolerant network · Community · Degree centrality · Betweenness centrality · Similarity

1 Introduction

Rural areas do not provide network connectivity to the vehicles passing through these areas due to lack of infrastructure. Vehicles carrying patients may require to establish connection with the doctors available in nearby cities. But this facility can be added to the ambulances only when there is some mechanism to provide network connectivity in the areas where end to end network connectivity is not available all the time. Delay-tolerant networks can be one of the solutions in this type of scenario. Delay-tolerant network (DTN) (Fall 2003; Delay Tolerant Networking Research Group 2004) is an opportunistic network where the connection is intermittent, i.e., network is sparse in

S. Kumar (✉) · P. Shukla · S. Pandey
Department of Information Technology, National Institute of Technology, Raipur, India
e-mail: skumar.it@nitrr.ac.in

A. A. Rizvanov et al. (eds.), *Advances in Biomedical Engineering and Technology*, Lecture Notes in Bioengineering,
https://doi.org/10.1007/978-981-15-6329-4_21

nature. Lack of infrastructure, partitioning of the network, very long delays are the features of delay-tolerant network. Due to these features routing in such network is a quite challenging task and numerous researches are going on. Routing in delay-tolerant network concerns itself with the ability to route or transport the data packets (contains message) from a source to destination It is a generally a difficult task to design an efficient routing protocol for opportunistic networks due to the lack of knowledge about the infrastructure of the network and dedicated connection is not present always present. Routing (Balaganesh et al. 2014; Zhu et al. 2013) in the delay-tolerant network is mainly characterized into two types, namely Flooding- and Forwarding-based strategy. In Flooding-based routing (Balaganesh et al. 2014; Zhu et al. 2013), the multiple copies of the same message will be created and these copies will be delivered to all set of nodes that are encountered by the node containing the data packet. The process goes on till the message is delivered to the destination or time to live (TTL) ends. Some Flooding-based routing techniques are Epidemic routing, Spray, Wait, Single and Two Hop routing techniques, etc. Whereas Forwarding-based (Balaganesh et al. 2014; Zhu et al. 2013) routing uses the knowledge of the network (such as network topology) for determining the relevant path to deliver a message to the destination. Some forwarding-based routing techniques are Source routing, Per-Hop, and Per-contact routing. One special category of the routing techniques exists, namely, Social-based routing (Balaganesh et al. 2014; Zhu et al. 2013) where social features of an opportunistic network are taken to route the message. Some of the social-based routing techniques are Location-Based routing, Label-Based routing, BUBBLE RAP routing, etc.

In this paper, the main intention is to exploit social behaviors of the delay-tolerant network with the help of social attributes: community, similarity, betweenness, and degree centrality. A community is a social unit (a group of people) formed on the basis of some common features such as profession, needs, place (the geographical area where they reside). Within a community (social institutions like family, government, society, etc.) everyone has some role associated with them due to which some interact with more number of peoples while some don't even interact. On this basis popularity of a particular human being is determined, termed as centrality. Every individual belongs to one or more communities and has some likeness with other people. The degree of likeness is measured by similarity which describes how often people interact and pattern they follow to interact with others. In this paper, community, similarity, betweenness, and degree centrality are being exploited for forwarding data packets in DTN. Methodologically, community detection helps in understanding local community structure which further used in designing efficient routing techniques for disseminating data. Newman (2004) has given many centralities to approximate the importance of a node in a network. Betweenness centrality is counted as the frequency a given node encountered on the shortest path followed between a pair of nodes. Degree centrality of a particular node measures the number of nodes encountered to that given node. Similarity measures the likeness between two given nodes by estimating the number of common neighbors between these given nodes. The combination of these three helps in selecting the best possible node to forward the data packets.

Improvement in the delivery probability and buffer utilization with the help of these social attributes are contributions of the paper we are presenting. The rest of the paper is structured as follows: Sect. 2, 3, 4, and 5 explains the social attributes and their significance in routing the data packets. Section 6 introduces the proposed routing technique containing these social attributes in selecting the route and their roles in proposed routing technique. Section 7 deals with performance evaluation of the proposed routing technique in terms of delivery probability and buffer utilization. Section 8 presents the conclusion.

2 Community

A community (George 1955; Gondaliya et al. 2015; Danon et al. 2005; Gor and Dhamecha 2014) is a subgraph where intra-cluster density is greater than inter-cluster density when the density of edges is taken into an account. In a generalized way, when things, individuals, creatures are gathered together in view of some vicinity (e.g., space, time, intrigue, and so on) then they frame a community (e.g., Hindu community and Muslim community, IT Professional community, and CIVIL professional community portrayed in light of religion and calling individually). It is one of the prime attributes of delay-tolerant networks (DTN) as mobile devices are carried by a human being who belongs to certain communities.

In BUBBLE RAP routing technique, it was expected that every last node is at least in one community whether it is global or subcommunity. The global community is that in which each and every node exist that are present in the network. Whereas, in subcommunity, every one of those nodes is available which exist in the global community and are grouped together with a view of some social elements. In this proposed routing technique, the same suspicion is additionally taken.

The various community detection algorithms have been proposed to identify communities in a network such as Weighted Network Analysis (WNA) by Newman (2005) and K-CLIQUE proposed by Palla et al. (2005) which helps in finding communities in a network. In this paper for community detection (CONGA), Cluster-Overlap Newman–Girvan Algorithm (Gregory 2007; Jora and Chira 2016; Kim et al. 2013; Khatoon and Banu 2015; Jancura 2012) is being used, which is an enhanced version of Grivan and Newman (GN) algorithm (Gregory 2007). GN algorithm is first divisive detection algorithm which espouses betweenness to extricate communities from a number of nodes but suffers from poor scalability. CONGA came up with a new concept of "splitting" betweenness of vertices, which enables it to split vertices among various communities. In order to split a vertex V into V1 and V2, firstly a new "imaginary" edge is being added. The cost of the imaginary edge is taken zero. Moreover, no possible paths originating from vertex V traverse this edge hence; the path lengths traversing V will remain untouched and no new shortest paths will be created. Then betweenness CB{V1, V2} needs to be calculated for imaginary edge. V can be splited into two parts choosing one from the 2d (v)-1-1 ways.

3 Betweenness Centrality

In graph theory, centrality (Kim et al. 2014; Freeman 1977; Wan et al. 2016) is a quantitative measure of the relative significance of a vertex within a graph. In the social graph, the centrality of a node delineates its social significance within the social network (e.g., how important a person is within a social network and reflects his activeness). In DTN, the sociological centrality is being used for selecting a node as a relay for forwarding the messages, which means nodes with higher centrality are the best nodes to be selected as a relay.

Betweenness centrality (Freeman 1977; Newman 2005; Manam et al. 2014) of a node "*x*" is defined as the number of times node "*x*" is encountered as a relay in all possible shortest paths between two other nodes (Fig. 1). It is one of the important social attributes as it describes the amount of data disseminated through it. Let us take a graph $G = (V, E)$, where V and E represent the set of nodes and edges in sparse network respectively. "*a*" and "*b*" are the numbered nodes and edges, respectively. It is assumed that each edge $e \in E$ has a positive integer weight $w(e) = 1$ which shows that there exists a dedicated connection between the nodes. A path from vertex source (s) to destination (d) is defined as a sequence of vertices v_i, $0 \leq i < 1$, where $v_0 = s$ and $v_l = d$. A path length is defined as the sum of the weights of edges.

$$w = \sum_{i=0}^{l} w(e_i) \quad \forall\ v_i\ , 0 < i < l \tag{1}$$

$\sigma(s,d)$ is the total shortest path between sand d, and $\sigma(s,v,d)$ represents the number of paths passes through v. $\tau(s,v,d)$ we take as the fraction of shortest paths between s and d passing vertex v, i.e.,

$$\tau(s, v, d) = \sigma(s, v, d) \big/ \sigma(s, d) \tag{2}$$

Betweenness centrality $C_B(v)$ of a vertex v is defined as

Fig. 1 Between centrality of a node taking 1 as source and 100 as destination in 100 node sparse network

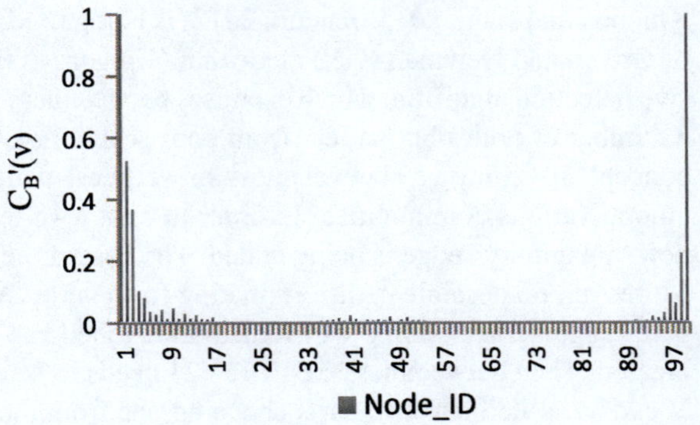

$$C_B(v) = \sum_{s,d \in G(V,E)} \tau(s, v, d) \tag{3}$$

As the networks are exceedingly large so calculating the betweenness indices needs polynomial running time complexities of O(n^2) or O(n^3) and are still prohibitive for networks as it contains a large number of nodes. Moreover, the networks are updated continuously, adding or removing node being able to track the changes in a network and the accompanying centrality is quite tough and infeasible. The local approach uses only the vertices directly adjacent to a target vertex to derive an approximation of the true centrality measure.

In order to obtain a local approximation for betweenness centrality, the numerator in (2) can be decomposed as

$$\sigma(s, v, d) = \sigma(s, v) \cdot \sigma(x, d) \tag{4}$$

Putting the value of σ in (2) and then combining it with (3) we move toward the approximation of CB. After repeatedly applying the decomposition identity as done in (2) lead to those fractions which only includes a predecessor and a successor of v_0, v_1 represents predecessor and successor of the node v. Thus,

$$C'_B(v) = \sum_{a=1}^{N} \sum_{b=1}^{N} \frac{\tau(v_0, v, v_1)}{\tau(v_0, v_1)} \tag{5}$$

4 Degree Centrality

Degree centrality (Kim et al. 2014; Wu et al. 2016; Wan et al. 2016) of a node is determined by the number of links present in that node means a number of nodes connected to it. A node having more number of connections possesses more possibility to deliver a message successfully and can be considered as a popular node.

Let the network contains N number of edges then degree centrality $C_D(x)$ of node "x" is calculated by:

$$C_D(x) = \sum_{i=1}^{N} \sigma(x, i) \tag{6}$$

where $\sigma(x, i)$ represents the status of the connection between node "x" and "i." The value of $\sigma(x, i)$ will be either 1 or 0. "1" indicates that there exists a connection between the nodes whereas the opposite holds true for "0." Now approximating the degree centrality with respect to the maximum possible degree of a node.

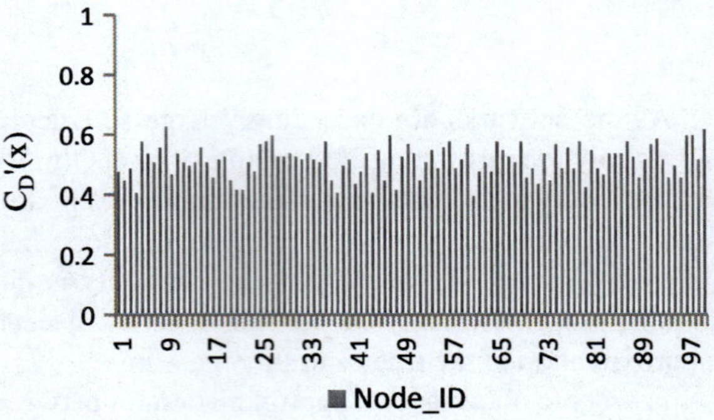

Fig. 2 Degree centrality of a node taking 1 as source and 100 as destination in 100 node sparse network

$$C'_D(x) = \sum_{i=1}^{N} \frac{\sigma(x,i)}{\rho(x)} \tag{7}$$

$\rho(x)$ represents the maximum possible degree of the node in the network. If the network contains N number of nodes then the maximum possible degree of a node will be $\rho(x) = N - 1$. So.

Putting The Value of $\rho(x)$ in (7), we get

$$C'_D(x) = \sum_{i=1}^{N} \frac{\sigma(x,i)}{N-1} \tag{8}$$

In this proposed routing technique, both degree and betweenness centrality are being used. They are used to define the significance of a node in a given network which further used to select the best possible node for forwarding the data in that network. The graph below in Fig. 2 shows the degree centrality of various nodes at a particular instance of time determined from the sample network of 100 nodes taken below.

5 Similarity

Similarity (Zhu et al. 2013; Daly and Haahr 2007; Patel and Gondaliya 2015) is a sort of social attributes which is used to express the relation among the nodes in terms of location, interest, background, etc. It is measured based on common neighbors present between the pair of nodes in a given community whether it is global or subcommunity. Similarity came from the observation that individuals often befriend others who have a similar appeal and perform a similar activity. In this manner, the high similarity between the nodes infers the presence of a good social connection between them. The probability of a future collaboration *P*(*a*,*b*) can be anticipated by the result of similarity and its strength between two nodes "*a*" and "*b*" is calculated by:

$$P(a, b) = \rho \times S(a, b) \tag{9}$$

$$S(a, b) = \sum_{i=1}^{n}\sum_{j=1} |M(a) \cap M(b)| \tag{10}$$

$$S(a, b) = \sum_{i=1}^{n}\sum_{j=1} |M(i, j)| \tag{11}$$

where $M(a)$ and $M(b)$ are the set of neighbors of node "*a*" and "*b*", respectively. This *S*(*a*,*b*) reveals the "similarity" between nodes "*a*" and "*b*," relative to the network topology. The strength of a similarity is defined as the number of times a pair of nodes interacts with each other in a given period of time. Let us suppose in a network 8 numbers of nodes are present and at a particular instance of time then the connection matrix be:

	a	B	c	d	e	f	g	h
a	0	0	1	0	1	1	0	1
b	0	0	1	1	0	1	0	1
c	1	1	0	0	0	1	0	1
d	0	1	0	0	1	0	0	0
e	1	0	0	1	0	1	1	0
f	1	1	1	0	1	0	1	0
g	0	0	0	0	1	1	0	1
h	1	1	1	0	0	0	1	0

If in a network "*N*" number of nodes are present then the size of connection matrix be "$N \times N$." Here in above the size of connection matrix "*N*" is 8, which represent a number of nodes and connection between them. The value of connection matrix changes with time as each is associated with its mobility pattern which changes with time. Then the value of $M(a)$ and $M(b)$ can be calculated from connection matrix.

$M(a)$ matrix is given by:

0	0	1	0	1	1	0	1

M(b) matrix is given by:

0	0	1	1	0	1	0	1

Resultant matrix M is given by:

0	0	1	0	0	1	0	1

Therefore similarity between the nodes a and b will be: $S(a,b) = 3$.

The strength of the similarity (ρ) between the particular pair of nodes will be calculated by the contact history of the pair over a period of time. In the social graph, strength can be represented as the number of times two human beings (belonging to the same community) interact with each other over a specific period. Let us suppose $x(t)$ be a function which represents the interaction pattern between a given pair of nodes. Mathematically, it is calculated as the integration of the total area of all instances of function $x(t)$ over a period "T" (when it is active with another node) divided by area of one instance of a function $x(t)$ calculated over the same time period. Then the value of ρ is given by:

$$\rho = \frac{\int_{t=0}^{T} x(t)}{\int_{t=0}^{t_1} x(t)} \tag{12}$$

where the value of $x(t)$ will be 1 for a traverse of time when two nodes are in contact with each other else it will be 0.

6 Proposed Algorithm

The selection procedure of the node follows in two major steps. These steps are as follows:

Step 1: Each node has its centrality (both betweenness and degree) associated with it. The source node will select the best possible node among its all encountered node that is having the highest betweenness centrality. Then it will be selected as a relay node. On the off chance that more than one node is having the same value of betweenness centrality then next level of selection of a node is done in order to select best possible node. Now degree centrality is mulled over for choosing the most appropriate node among those nodes which is having same betweenness centrality. This procedure of node selection is done till message reaches the destination community. As soon as it reaches the destination community step 2 is followed to perform further forwarding of the message.

Step 2: As the message reaches the destination community then a blend of two social attributes, to be specific degree centrality and the similarity are utilized to select the node for forwarding the message. These attributes are utilized within the destination community as opposed to in a global community because intra-community density is higher than that of inter-community density. Since these social attributes are determined from quite different criteria so a utility function is shaped to consolidating these social attributes, which will be utilized for the decision-making process.

$$U(s, x, d) = \alpha \times P(s, x) + \beta \times C'_D(x) \tag{13}$$

Now,

$$U(s, x, d) = \alpha \times \rho \times \sum_{i=1}^{n} \sum_{j=1}^{n} |M(i, j)| + \beta \times \sum_{j=1}^{n} \frac{\sigma(x, j)}{n - 1} \tag{14}$$

The value of α and β is chosen such a way that relative significance of the two social-based attributes can be perfectly utilized and $\alpha + \beta = 1$. Among all encountered nodes that node will be selected which is having the highest value of the utility function and this process continues until the destination node is reached. Since the buffer size is fixed and if not legitimately oversaw then leads to overhead in the network. So for proficient utilization of buffer various buffer management techniques were proposed (Le et al. 2016; Rashid et al. 2015). Here a message having most minimal TTL will be dropped out when the buffer of the node selected for forwarding the message is full. The message with least TTL is dropped with a view that the probability of successful delivery of that message is less than that of another message.

7 Performance Evaluation

For evaluation-proposed routing technique is evaluated and compared against three existing DTN routing techniques, namely Epidemic routing, BUBBLE Rap, Direct Delivery. The evaluation of these routing techniques is done in a scenario that node generates messages in a specific interval of time during the simulation. Parameters for simulation setups are shown in Table 1.

7.1 Buffer Utilization

Space (memory) assumes a critical part in the routing technique as buffer size of the node is limited and can't be expanded because of different issues, for example, administration issue and power utilizations. Additionally, if irrelevant messages are

Table 1 Simulation Setup for evaluation of routing techniques

Simulation	Setup
Number of nodes	80
Message TTL	300 min
Message sizes	10–100 KB
Buffer sizes	1–10 MB

Fig. 3 Number of data packets used for the successful delivery of message

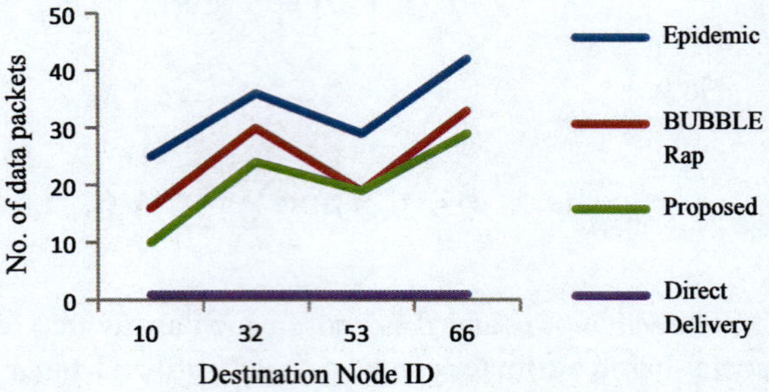

put away then it leads to the dropping of numerous message bundles. So messages should be stored in a productive way and oversee them.

In various multi copies routing techniques such as the Epidemic routing extensive number of the message, imitations are being made, and a number of copies rely on a number of encountered nodes. In the proposed routing technique, two levels of selection are done to choose one most appropriate node for forwarding the message. Therefore, it leads to a significant decrement in the usage of the buffer. The space complexity of a routing technique is defined as the number of replicas of a message created for successful delivery of the single message. As in BUBBLE Rap and Epidemic routing techniques, multiple paths are chosen for forwarding the message whereas in the proposed routing technique single most relevant path is chosen.

So, a number of replicas will be less which is shown in Fig. 3 where distinctive destination nodes are chosen randomly with fixed source node "1" in a dynamic network. From Fig. 3 it can be presumed that for same destination number of replicas utilized by proposed routing is relatively not as much as that of existing routing techniques.

For the fixed buffer size there is a significant dropping of messages in Epidemic routing and BUBBLE Rap. So for efficient utilization of buffer message having most minimal TTL will be dropped out when the buffer of the node selected for forwarding the message is full as explained in Sect. 6.

In Fig. 4, each result in this section is the average result of 50 simulations for 1 MB buffer size scenario, proposed routing technique has higher Buffer Utilization Ratio (BUR) than BUBBLE Rap by 27.3% and Epidemic routing by 37.5% because of optimal node and path selection.

7.2 *Delivery Ratio*

The delivery ratio is defined as the ratio of a number of messages received by the destination node to those generated by the source node. Suppose a number of messages generated by the source node "s" destined to destination node "*d*" be "M_S" and

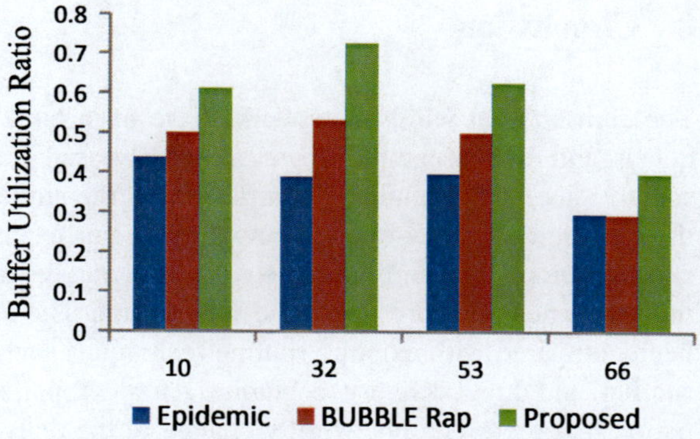

Fig. 4 Relative buffer uutilization ratio of overall network for various destination nodes

messages received by the destination node be "M_D." Mathematically, it can be defined as:

$$Delivery\ Ratio = M_D / M_S \tag{15}$$

In Fig. 5 routing technique is compared with proposed technique and for various times to live of a message generated by the source node. The delivery ratio is been approximated by the simulation run of 50 times. From Fig. 5, it can be concluded that the proposed routing technique possess delivery ratio greater than BUBBLE Rap and Directly delivery but lower than Epidemic routing technique.

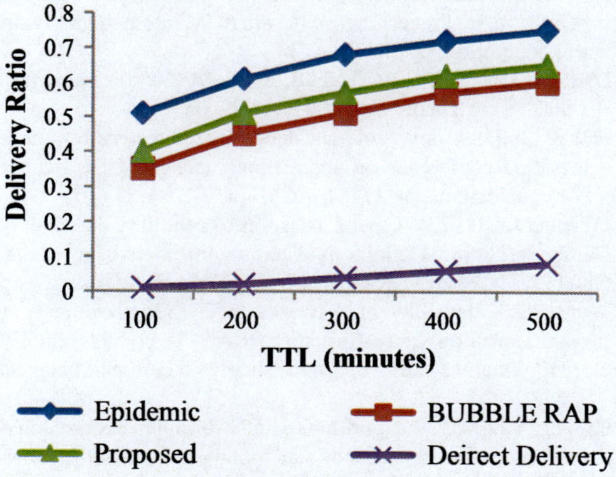

Fig. 5 Buffer utilization ratio of overall network for various destination nodes

8 Conclusion

The conventional wireless networks have been quite questionable due to its long delays and sparse nature. However, social-based routing supplemented with the social traits like community, centrality, and the similarity is a superior answer for the conventional delay-tolerant networks. In this paper by utilizing social attributes, namely community, similarity, betweenness, and degree centrality an optimal selection of the node for forwarding the data packets. The proposed routing technique has been compared with existing routing techniques such as BUBBLE Rap, Epidemic routing, and direct delivery technique. After comparisons, we thought of a determination that there is a noteworthy change in the delivery ratio and buffer has been efficiently utilized. This prompts significant improvement in routing more and more data packets at once. Finally this approach may be implemented in real-world situation like a ambulance carrying a patient through rural areas and communication with doctor in hospital is not regular. The proposed technique will be helpful in improving the quality of service of communication between a doctor and ambulance. Proposal has not been tested in real-world scenario which limits its applicability.

References

Balaganesh M, Sathiya P, Balagowri D (2014) A survey of misbehavior detection scheme in DTN. Int J Comput Appl 107(10)

Daly EM, Haahr M (2007) Social network analysis for routing in disconnected delay-tolerant MANETs. In: Proceedings of the 8th ACM international symposium on mobile ad hoc networking and computing, pp 32–40. ACM

Danon L, Diaz-Guilera A, Duch J, Arenas A (2005) Comparing community structure identification. J Stat Mech: Theory Exp 2005(9):P09008

Fall K (2003) A delay-tolerant network architecture for challenged internets. In: Proceedings of the 2003 conference on applications, technologies, architectures, and protocols for computer communications, pp 27–34. ACM

Freeman LC (1977) A set of measures of centrality based on betweenness. Sociometry, 35–41

George Jr A (1955) Definitions of community: areas of agrement rural sociology. Hillery 20:118–118

Gondaliya N, Shah M, Kathiriya D (2015) A node scheduling approach in community based routing in social Delay Tolerant Networks. In: 2015 International conference on advances in computing, communications and informatics (ICACCI), pp 594–600. IEEE.

Gor HR, Dhamecha MV (2014) A survey on community detection in weighted social network. Int J 2(1)

Gregory S (2007) An algorithm to find overlapping community structure in networks. In: European conference on principles of data mining and knowledge discovery, pp 91–102. Springer, Berlin, Heidelberg

Jancura P (2012) Evolutionary analysis in PPI networks and applications. [Sl: sn].

Jain S, Fall K, Patra R (2004) Routing in a delay tolerant network. In: Proceedings of the 2004 conference on applications, technologies, architectures, and protocols for computer communications, pp. 145–158

Jora C, Chira C (2016) Evolutionary community detection in complex and dynamic networks. In: 2016 IEEE 12th international conference on Intelligent computer communication and processing (ICCP), pp 127–134. IEEE

Khatoon M, Banu WA (2015) A survey on community detection methods in social networks. Int J Educ Manage Eng (IJEME) 5(1):8

Kim CM, Kang IS, Han YH, Park CY (2013) A community detection scheme in delay-tolerant networks. In: Ubiquitous information technologies and applications, pp 745–751. Springer, Dordrecht

Kim CM, Kang IS, Han YH, Jeong YS (2014) An efficient routing scheme based on social relations in delay-tolerant networks. In: Ubiquitous information technologies and applications, pp 533–540. Springer, Berlin, Heidelberg

Le, T., Kalantarian, H. and Gerla, M., 2016, June. A joint relay selection and buffer management scheme for delivery rate optimization in dtns. In *World of Wireless, Mobile and Multimedia Networks (WoWMoM), 2016 IEEE 17th International Symposium on A* (pp. 1–9). IEEE.

Manam VC, Mahendran V, Murthy CSR (2014) Performance modeling of DTN routing with heterogeneous and selfish nodes. Wireless Netw 20(1):25–40

Newman ME (2004) Analysis of weighted networks. Phys Rev E 70(5):056131

Newman ME (2005) A measure of betweenness centrality based on random walks. Soc Networ 27(1):39–54

Palla G, Derényi I, Farkas I, Vicsek T (2005) Uncovering the overlapping community structure of complex networks in nature and society. Nature 435(7043):814

Patel CM, Gondaliya N (2015) Enhancement of social based routing protocol in delay tolerant networks. Int J Comput Appl 122(4)

Rashid S, Ayub Q, Abdullah AH (2015) Reactive weight based buffer management policy for DTN routing protocols. Wireless Pers Commun 80(3):993–1010

Wan L, Zhang H, Liu F, Chen Y (2016) Routing in Delay tolerant networks with fine-grained contact characterisation and dynamic message replication. In: 2016 IEEE 17th international symposium on a world of wireless, mobile and multimedia networks (WoWMoM), pp 1–6. IEEE

Wu J, Wang J, Liu L, Tanha M, Pan J (2016) A data forwarding scheme with reachable probability centrality in DTNs. In: 2016 IEEE on Wireless communications and networking conference (WCNC), pp 1–6. IEEE

Zhu Y, Xu B, Shi X, Wang Y (2013) A survey of social-based routing in delay tolerant networks: Positive and negative social effects. IEEE Commun Surv Tutor 15(1):387–401

Hiding Patient Information in Medical Images: A Robust Watermarking Algorithm for Healthcare System

Ritu Agrawal, Manisha Sharma, and Bikesh Kumar Singh

Abstract Electronically transmission of the medical images is an essential prerequisite in a Healthcare system. During transmission, the medical images can be hacked and altered either partially or in totality, unless aptly safeguarded. Digital watermarking schemes are being utilized to protect the integrity of the medical image. Digital watermarking is a technique of embedding watermark in a medical image, to maintain confidentiality and security to the inserted watermark. The watermark used in the medical image can either be an image, text, audio, or video. Electronic Patient Record (EPR) in an image form is used as a watermark. A non-blind watermarking scheme for healthcare system using Discrete Cosine Transform (DCT) is presented in this paper. Block-based Cosine transformation is applied to the host medical image to obtain different frequency coefficients. The middle frequency band of Cosine transformation is considered for embedding watermark, as this frequency band provides an extra robust resistant to watermark. Before hiding the EPR information in the image, binary EPR information is encoded using Convolution Error Correcting Code (ECC) and decoded using Viterbi decoder, to enhance the accuracy of the detection process. To provide additional robust to the coded EPR data, M-ary modulation is further applied before embedding. The imperceptibility and robust performance evaluation of the scheme is tested on standard online Digital Imaging and Communications in Medicine (DICOM) brain image database by changing the watermark embedding factor. The proposed scheme is tested using various performance measures and is observed to be highly robust and imperceptible.

Keywords Medical image watermarking · Electronic patient record · Error control coding and M-ary modulation

---

R. Agrawal (✉) · M. Sharma
Department of Electronics and Telecommunications, Bhilai Institute of Technology, Durg, Chhattisgarh, India
e-mail: ritube_03@yahoo.co.in

B. K. Singh
Department of Biomedical Engineering, National Institute of Technology, Raipur, Chhattisgarh, India

A. A. Rizvanov et al. (eds.), *Advances in Biomedical Engineering and Technology*, Lecture Notes in Bioengineering,
https://doi.org/10.1007/978-981-15-6329-4_22

1 Introduction

Maintaining health is a prerequisite for sustaining the human resources, the most valuable resources available on the earth. Active assistance of physicians and associated personnel are required to achieve better health. In the age of converging technologies, providing information about a patient for correct and accurate diagnosis through electronic means and media plays a vital role. Due to rapid growth in the domain of Information and Communication Technology (ICT), many advanced means like telemedicine, telesurgery, telediagnosis, etc., have been evolved to ease the workings of the physicians with the active usage of internet and multimedia technologies. However, the privacy of medical data in electronic form poses potential threats both external as well as internal to maintain the patient–physician relationship for accurate delivery of healthcare. Internal threats can be overcome to a greater extent through proper management whereas external threats are to be handled by exploiting the existing technologies. For efficient and effective implementation of the patient–physician relationship, healthcare systems, it is necessary to share medical information over open networks. Exchange of information over open network is raising various complex legal and ethical issues, including image retention and fraud, privacy, malpractice liability, etc. Thus the major challenge to maintain a better patient–physician relationship, privacy and secured transaction of medical images is of prime importance.

As the intruders pose a challenge to the privacy of the medical images being exchanged over an open communication channel, it needs to be addressed and resolved. The issue of unwanted access during transmission of the image can be effectively handled by using Hiding technique. Hiding is a technique in which the message signal or the information embed in the host/cover images without any much degradation in the visual quality to the host image. There are two major techniques of hiding the image during transmission, namely, Steganography and Watermarking schemes. In this paper we only address the watermarking scheme in the medical image.

Watermarking of medical images addresses the specific issues of maintaining the integrity of medical images that provides the assurance that the data was not accidentally or deliberately modified during transmission over communication channels. Medical images are as important as other medical data. They play a key role in almost all phases of the healthcare systems like detection, diagnosis, therapy follow-up, and so on. Hence hiding of the image data during transmission is of major concern. The present work deals with the designing of a novel watermarking scheme in the area of medical imaging. In the watermarking scheme, the DCT as a host image is used for insertion and retrieval of the watermark. A combination of ECC and M-ary modulation was applied to the watermark, after that the coded message was inserted in the mid-band DCT coefficient of the cover medical image. Choice of considering mid-frequency Cosine transformation band was to provide good imperceptibility and robust to the inserted watermark.

Related work

A chronological survey on the previous works made by the researchers in the field of medical image watermarking schemes and embedding patients' information in the cover image is presented in the following paragraphs. However looking into the vastness of the field undertaken, the survey has been presented in parts like information hiding and enhancement of robustness of the system using spread spectrum technique.

Information Hiding. Mostafa et al. (2010) formulated a blind EPR hiding scheme using Discrete Wavelet Packets Transform (DWPT) in the application of telemedicine. Before embedding watermark, BCH ECC was applied to the binary watermark, for improving the robust property of the watermarking scheme. Though, the watermark embedding scheme is simple but the time required for extracting the watermark was high due to BCH error correction code.

Nambakhsh et al. (2011) proposed a contextual based dual watermarking scheme for PET images. Two watermarks, text information and ECG signal of the patient were inserted in the selected preferred texture area of the medical image. The robust assessment of the scheme was tested only for three known attacks. The security of the system was greatlhanced as dual watermarks were embedded in the PET image but the system required higher execution time due to higher level of complexity of the system.

Das and Kundu (2011) developed a multiple and non-blind watermarking scheme using contourlet transform and Discrete Cosine transform. The perceptual quality and the robust performance of the retrieved watermark were high as combinations of transform domain were used. The system was found to be complex as two different transformation techniques were used.

Kannammal and Shuba Rani (2014) reported a dual-level security for medical image transaction using encryption and watermarking. The watermarking techniques used were LSB and Wavelet transform. Before embedding the watermark, different types of encryption technique like RSA, AES, and RC4 were applied to provide security to the watermark image. Due to the application of different encryption techniques for embedding, the system became very complex and the execution time was high.

Singh et al. (2015a) reported a robust, non-blind, and multiple watermark embedding technique using Singular value decomposition (SVD) and Discrete Wavelet Transform (DWT) for telemedicine application. Moreover, to provide additional robustness to text watermark four different ECC were applied. Owing to the application of dual watermarks, the security scheme got remarkably improved, however, due to system complexity the execution time requirement also got enhanced.

Singh et al. (2015b) devised a novel spread spectrum based multiple watermark embedding scheme using DWT. To improve the robust performance of the test watermark error control coding BCH was applied. Though the security of the watermarking system got enhanced as spread spectrum technique is applied yet the time required for extracting the watermark was high due to BCH error correction code and the system's complexity increased due to application of two different transformation techniques.

Mehto and Mehra (2016) devised a combined watermarking algorithm using DWT and DCT. EPR watermark was inserted in the host image such that the technique is imperceptible to the Human Visual System (HVS). The watermarking technique is too complex as two different transformation techniques were used for the embedding and extraction process. Moreover, no common processing attacks were applied to evaluate the robustness of the system.

Swaraja (2017) reported a multiple watermark embedding scheme using DCT, DWT, and SVD for authentication and integration in the application of medical images. The security of the watermarks was enhanced by applying lossless arithmetic coding to text watermark and Arnold transform to the image watermark. On the contrary, the system became complex as different transformation techniques were used for the embedding and extraction process. Also, no common processing attacks were applied to evaluate the robustness of the system.

Zear et al. (2018) formulated a combined watermarking scheme in the application of Healthcare using DCT, DWT, and SVD. To enhance the security of the multiple watermarks were applied. The proposed scheme performances were examined using different parameters at varying gain factor. Use of the multiple watermarks the security of the system increased, however, the higher complexity of the system increased to execution time.

Enhancement of robustness of watermarking system using spread spectrum techniques. In digital watermarking the Spread Spectrum (SS) modulation principle has been applied due to its excellent characteristics viz. security protection and robustness. The use of the wide spectrum of the host signal in the message hiding process puts a limit on the data rate subject to a given embedding distortion.

Kutter (1999) reported that the role of M-ary modulation is an effectual way to get a substantial improvement in robust performance. Also the performance did not diminish under noise like distortion, for example, lossy JPEG compression.

Hajjaji et al. (2011) designed a secure based watermarking scheme for transmission of medical images using a hybrid combination of Code Division Multiple Access (CDMA), Discrete Wavelet transform (DWT), and Error Correcting Code (ECC). By applying this combination the perceptual quality of the designed scheme is improved but at the same time the computational complexity of the system is high.

From the previous discussions on various existing techniques, few research gaps have been identified which are discussed through the following paragraphs:

- The EPR watermark needs to be inserted in order to maintain security during transmission to the medical image. It is of utmost importance that the watermark should not deform the medical image of interest. Hence to be on the safe side, in the present work a judicious decision has been taken to use DCT for the embedding process. As its high degree of spectral compaction the DCT property is popular for multidimensional applications; like better perceptual invisibility, adequate robustness, reasonable complexity, and hence less execution time. At the same time DCT domain watermarking is popular for hardware implementation because there are a number of fast algorithms exist (Mohamed et al. 2011).

- As reported by the previous researchers that the data rates are limited to a maximum value for a given embedding distortion if the wide spectrum of the cover data in EPR hiding is applied. Hence there is a scope in improvement using M-ary modulation over binary signaling principle. The M-ary signaling scheme offers an excellent anti-jamming performance and interference rejection properties. Hence the areas of M-ary signaling principle in Spread Spectrum (SS) approach in DCT remain as one of the subject matters.
- In order to detect the watermark properly and to avoid false detection, Error Control Coding (ECC) can be effectively used. Only few of the researchers used ECC for the detection purpose.

1.1 Contribution and Outline of the Paper

We proposed a combined approach for embedding EPR using ECC and M-ary modulation in the middle DCT frequency band of the cover image. Using the combined approach of EPR hiding in the image, not only there is a reduction in the bandwidth requirement but also improvement in the system efficacy.

(a) Security of the image watermark is enhanced by using ECC and M-ary modulation. By using this combination of techniques the system becomes complex but at the cost of robust improvement.
(b) The proposed watermarking scheme performance is evaluated on publically available DICOM dataset.
(c) The proposed approach imperceptibility and robustness performances are evaluated using different attacks and at different embedding factor.

The rest of the article is comprised as. Section 2 represents the Materials and methods. Results and discussion were presented in Sect. 3. The conclusion drawn from this study is discussed in Sect. 4.

2 Materials and Methods

The DICOM brain image database source, the proposed watermarking embedding, and the extraction flow diagram is discussed and enumerated in Fig. 1 and Fig. 2, respectively.

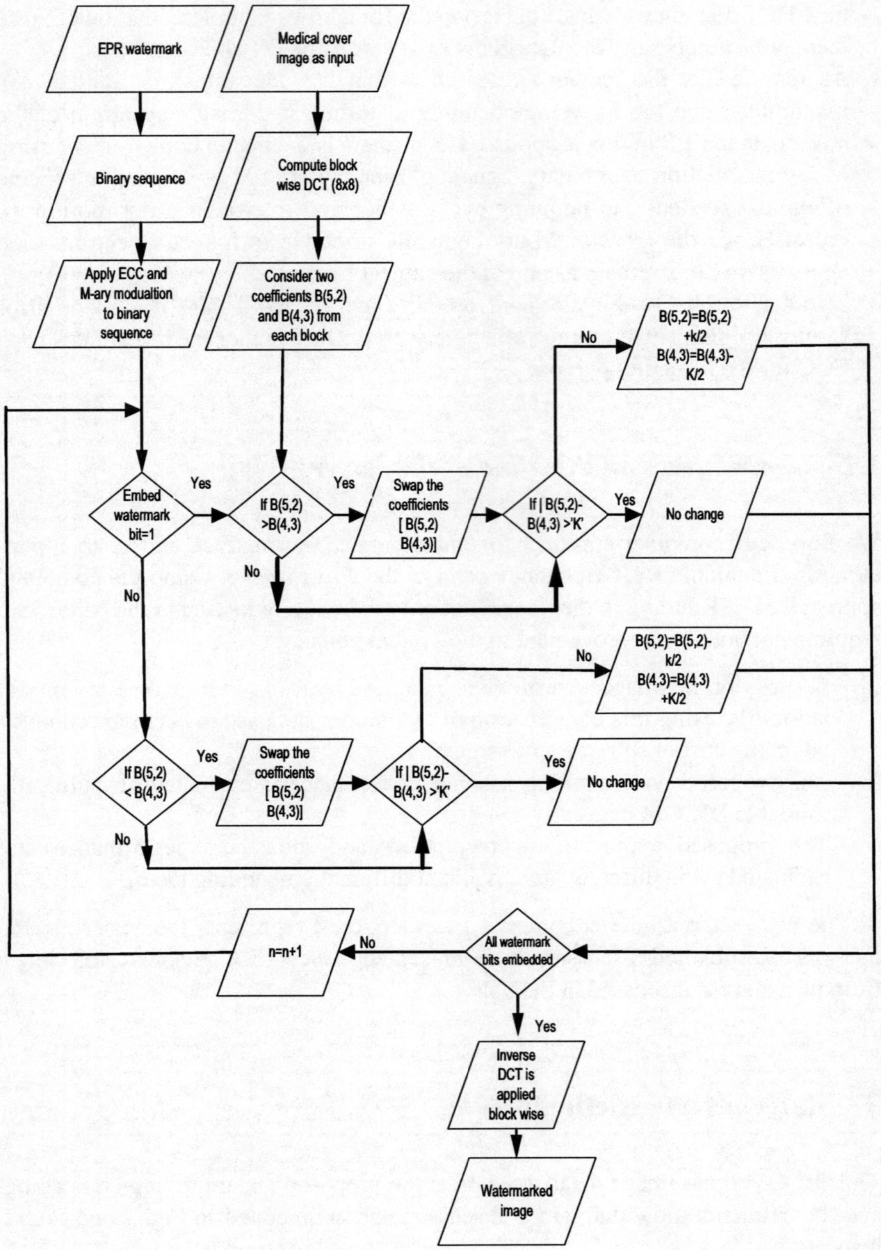

Fig. 1 Flow diagram of watermark embedding

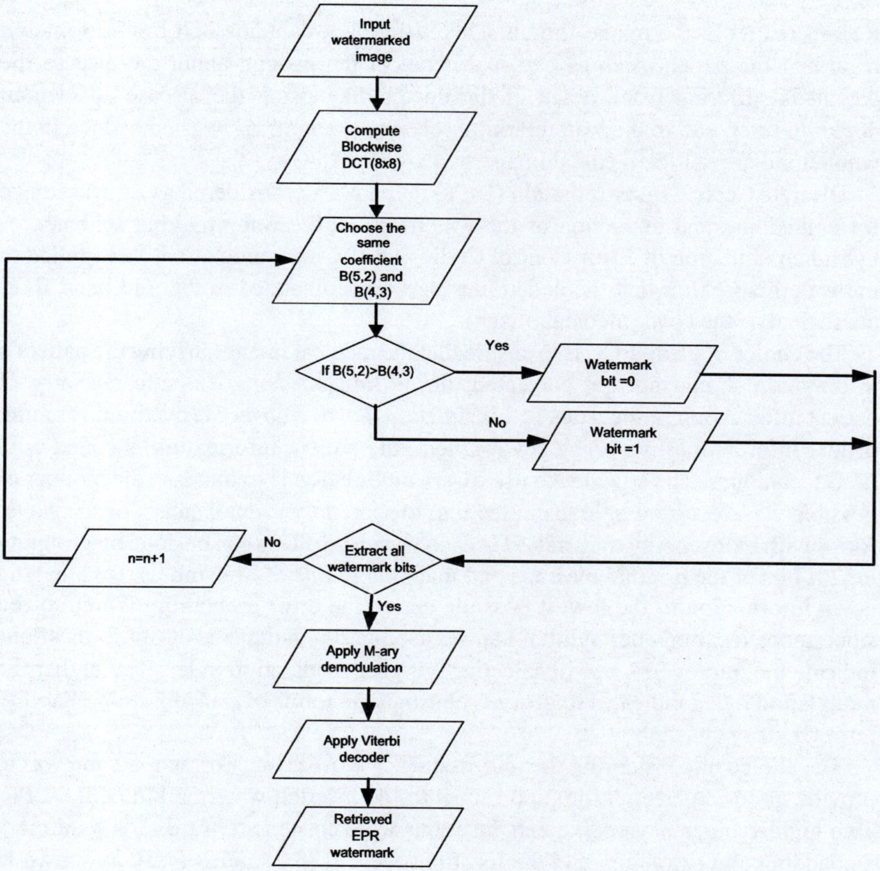

Fig. 2 Flow diagram of extraction algorithm

2.1 Brain Image Dataset

The dataset considered in the experiment was obtained from DICOM brain MR image, a standard online database (Chan and Siu 1992). For the purpose of the watermark embedding and extraction, 22 brain tumor infected images were considered in the analysis. However, these datasets do not have any ground truth images.

2.2 Proposed Watermarking Scheme [Embedding and Extraction]

This section explains the watermark embedding and extraction process for the proposed watermarking scheme. The watermark used was the Electronic Patient

Record (EPR) in the image format. This watermark contained all the information regarding the patient such as personal detail of the patient, about the disease the patient is suffering from, detail of the doctor diagnosing the disease and whom doctor to refer, etc. In the watermarking scheme, a watermark was embedded in the whole/total area of the medical image.

Discrete Cosine Transformation (DCT) domain was considered as a cover image for embedding and extraction of the watermark in the watermarking schemes. A hybrid combination of Error Control Coding and M-ary modulation was applied to the watermark, after that the coded message was embedded in the mid-band DCT coefficient of the cover medical image.

The choice of using ECC is to ensure that the medical image carrying the patient's information should not get corrupted during transmission. The safe recovery of patient information is important in this situation. So, to recover the maximum amount of text information in a noisy environment, the patient information is coded with ECC techniques. The key idea to use M-ary modulation is to increase the number of possible waveforms being transmitted, but to send only a small subset of the waveforms with more energy transmitted in each waveform. This can be done by grouping log2M bits of the original message and mapping to one of M symbols. By breaking down information to the lowest possible entity, the error probability is minimized since more locations per symbol can be used and simultaneously more locations indicate the improved scope of selecting a higher modulation index. The higher the modulation index indicates improved robustness in terms of transmission reliability through noisy channels.

The choice of considering the mid-frequency Cosine transformation band was to provide good imperceptibility and robust to the inserted watermark, i.e., the EPR. Two middle-frequency coefficients locations were chosen arbitrarily for watermark embedding and extraction and the locations were *B* (5,2) and *B* (4,3); as shown in Table 1 with grey colored. Watermark bit "1" was embedded if coefficient *B* (5,2) < *B* (4, 3); otherwise bit "0" was embedded. The relative modification in the preferred coefficient magnitudes was carried out using the watermark embedding factor "*K*."

Table 1 DCT block showing 64 coefficients

B(1,1)	B(1,2)	B(1,3)	B(1,4)	B(1,5)	B(1,6)	B(1,7)	B(1,8)
B(2,1)	B(2,2)	B(2,3)	B(2,4)	B(2,5)	B(2,6)	B(2,7)	B(2,8)
B(3,1)	B(3,2)	B(3,3)	B(3,4)	B(3,5)	B(3,6)	B(3,7)	B(3,8)
B(4,1)	B(4,2)	B(4,3)	B(4,4)	B(4,5)	B(4,6)	B(4,7)	B(4,8)
B(5,1)	B(5,2)	B(5,3)	B(5,4)	B(5,5)	B(5,6)	B(5,7)	B(5,8)
B(6,1)	B(6,2)	B(6,3)	B(6,4)	B(6,5)	B(6,6)	B(6,7)	B(6,8)
B(7,1)	B(7,2)	B(7,3)	B(7,4)	B(7,5)	B(7,6)	B(7,7)	B(7,8)
B(8,1)	B(8,2)	B(8,3)	B(8,4)	B(8,5)	B(8,6)	B(8,7)	B(8,8)

3 Results and Discussion

In this scheme DICOM brain images have been considered. The DICOM images have an extension as DCM and to be converted into JPEG format. All the medical images used in the study are of size 256 × 256 pixels as host/cover image. A binary image size 32 × 32 pixels is considered as the watermark to be embedded instead of using the actual EPR. In this watermarking scheme, the watermark is embedded in the total medical image. All the experiments were conducted on the processor configuration as 2 GHz, 4 GB RAM using the MATLAB R2013 software platform. The DICOM dataset with the original cover medical images (top) along with their watermarked images (bottom) are shown in Fig. 3. The binary watermark in an image form is shown in Fig. 4. From Fig. 3, it is observed that there is no visual difference between the original image and the watermarked image.

3.1 Performance Measures

The robust performance and perceptual/visual quality measurement of the proposed scheme are evaluated using Normalized Cross-Correlation (NCC) (DICOM 2017)

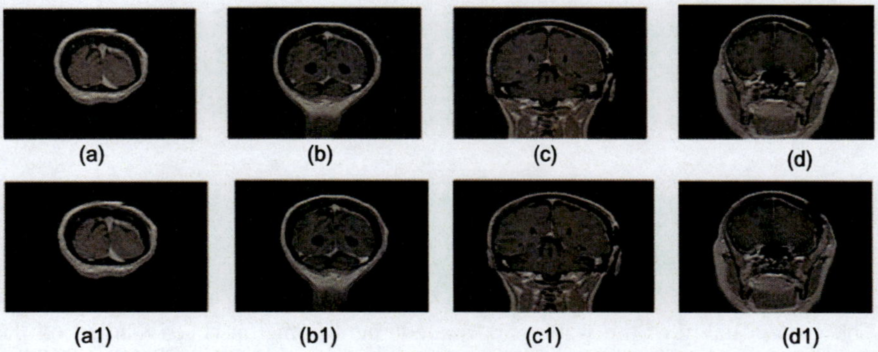

Fig. 3 Original medical image database (top (**a**–**d**)) together with their watermarked version (bottom (**a1**–**d1**))

Fig. 4 Original binary watermark image

and Peak Signal to Noise Ratio (PSNR) (DICOM 2017), respectively. Perceptual quality performance metric (PSNR) is evaluated between the watermarked image and the original host image, where robust performance measure (NCC) is evaluated between the retrieved and the original watermark.

The equations used for evaluating the performance measures are:

a. Peak signal to noise ratio (PSNR):

$$\mathrm{PSNR} = 10\log_{10}\frac{\left(I_{i,j}^{2}\right)_{\max}}{\mathrm{MSE}} \tag{1}$$

where $(I_{i,j})\max$ is the maximum intensity of the original image $I(i, j)$. MSE is mean square error. PSNR is higher for better transformed image. They measure the resemblance between the original and watermarked image. The PSNR value greater than 28 dB is acceptable (Cox and Miller 2002).

b. Normalized Cross-Correlation (NCC): It is the measure of similarity between the original watermark image and the retrieved watermark image. The value lies between "0" and "1." Value "0" indicates no similarity between the two images and value "1" indicates a complete resemblance. It is given by:

$$\mathrm{NCC} = \frac{\sum_{i=1}^{M}\sum_{j=1}^{N}(I(i, j)xF(i, j))}{\sum_{i=1}^{M}\sum_{j=1}^{N}(I(i, j))^{2}} \tag{2}$$

3.2 Visual Quality Evaluation

The visual quality of the watermark embedding scheme has been studied on the basis of qualitative and quantitative analysis. The qualitative analysis of the scheme is evaluated by observing the DICOM images of Fig. 3. It is observed from Fig. 3 that the original cover DICOM images along with their watermarked images are almost similar. Moreover, the visual quality of the scheme is also evaluated using PSNR. The PSNR performances of the watermarking scheme without signal processing attacks are presented in Table 2. From Table, it is observed that the PSNR values decrease as the value of embedding factor K increases. Furthermore, the average PSNR values of the considered images are observed as 57.45 dB and 54.68 dB at embedding factor $K = 10$ and 20, respectively. As the PSNR value is above 28 dB (Kallel et al. 2006). Hence, it is proved that the watermarking scheme is capable of producing good quality watermarked images on the basis of qualitative and quantitative analysis.

Table 2 PSNR verses different embedding factor for different DICOM image

Images	Embedding factor $K = 10$	Embedding factor $K = 20$
	PSNR (dB)	
Image $\neq$ 1	59.05	56.28
Image $\neq$ 2	58.14	55.60
Image $\neq$ 3	56.08	54.26
Image $\neq$ 4	56.56	52.59
Average	57.45	54.68

3.3 Robustness Analysis of DICOM Images

The robust performance of the watermarking scheme is evaluated in terms of NCC for second datasets (DICOM) against different embedding factor ($K = 10$ and $K = 20$). The watermarking scheme robustness validation has been studied by applying different common processing attacks to the watermark image. The various attacks used in the study are: salt and pepper noise (density (D) = 0.01), Gaussian noise (mean (M) = 0 and variance (Var) = 0.01), speckle noise (variance = 0.01), Gaussian Low Pass filter, average filter [3 × 3], median filter [3 × 3], rotation (35 degree), resize (256-128), JPEG (QF = 50), and JPEG 2000 compression attacks. The obtained results in terms of NCC are discussed below in detail:

Robustness Analysis against Different Noise Attacks. Various noise attacks were subjected to the watermarked image. The noise considered in the study are the salt and pepper ($D = 0.01$), Gaussian ($M = 0$, Var = 0.01) and speckle (Var = 0.01). The NCC values for different noise attacks were enumerated in Table 3 with embedding factor $K = 10$ and 20 and the corresponding retrieved watermark from

Table 3 Robustness analysis of DICOM images against different noise attacks

Attacks	NCC values				Average NCC
	Image 1	Image 2	Image 3	Image 4	
Salt and pepper noise (density = 0.01), ($K = 10$)	0.9577	0.9547	0.9597	0.9489	0.9552
Gaussian noise (mean = 0, variance = 0.01), ($K = 10$)	0.9554	0.9617	0.9532	0.9546	0.9562
Speckle noise (variance = 0.01), ($K = 10$)	0.8999	0.9278	0.9131	0.9055	0.9115
Salt and pepper noise (density = 0.01), ($K = 20$)	0.9598	0.9605	0.9632	0.9582	0.9604
Gaussian noise (mean = 0, variance = 0.01), ($K = 20$)	0.9590	0.9621	0.9585	0.9600	0.9599
Speckle noise (variance = 0.01), ($K = 20$)	0.9001	0.9365	0.9055	0.9235	0.9164

the attacked DICOM images for $K = 10$ were shown in Fig. 5. The average NCC of the extracted watermark is 0.9552 and 0.9604 when salt and pepper noise was applied to the same dataset at embedding factor ($K = 10$ and $K = 20$), respectively.

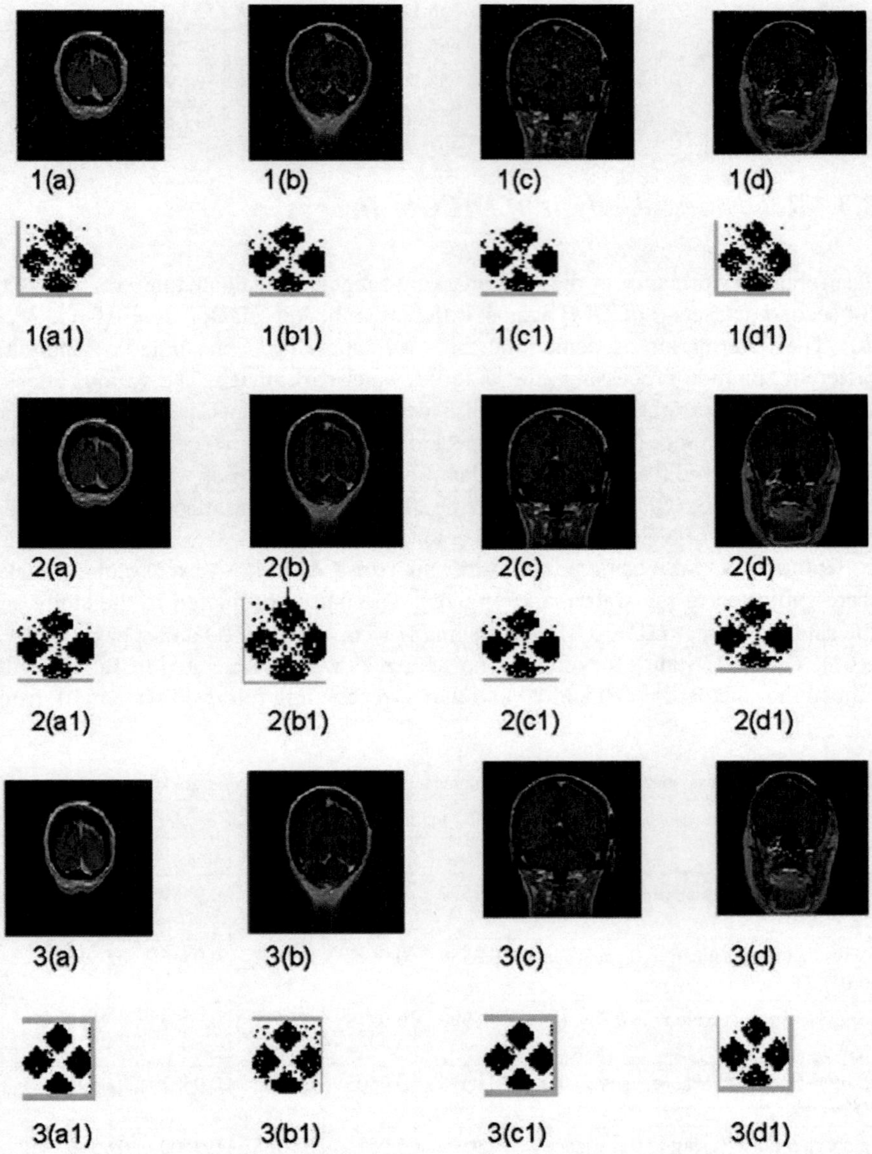

Fig. 5 1 **(a–d)** watermarked image against salt and pepper noise. 1 **(a1–d1)** Extracted watermark against salt and pepper noise. 2 **(a–d)** Watermarked image against Gaussian noise. 2 **(a1–d1)** Extracted watermark against Gaussian noise. 3 **(a–d)** Watermarked image against speckle noise. 3 **(a1–d1)** Extracted watermark

Table 4 Robustness analysis of DICOM images against different filtering attack

Attacks	NCC values				Average NCC
	Image 1	Image 2	Image 3	Image 4	
Gaussian low pass ($K = 10$)	0.9554	0.9669	0.9652	0.9407	0.9570
Averaging [3 × 3] ($K = 10$)	0.9355	0.9466	0.9576	0.9497	0.9473
Median [3 × 3] ($K = 10$)	0.9490	0.9527	0.9534	0.9552	0.9525
Gaussian low pass ($K = 20$)	0.9595	0.9696	0.9652	0.9413	0.9589
Averaging [3 × 3] ($K = 20$)	0.9412	0.9490	0.9578	0.9518	0.9499
Median [3 × 3] ($K = 20$)	0.9512	0.9602	0.9610	0.9554	0.9569

Similarly, the average NCC of the extracted watermark is 0.9562 and 0.9599 when Gaussian noise was applied to the watermarked image to the same dataset for the same embedding factor, respectively. Further, the average NCC value is 0.9115 and 0.9164, respectively, when speckle noise is. It is observed from Table 3 that the watermarking scheme is robust against different noise attacks, as in all the cases the average NCC value is higher than 0.7 (Kallel et al. 2006).

Robustness Analysis against Different Filtering Attacks. Different filtering attacks have been subjected to the watermarked image. The filter considered are the Gaussian Low Pass, average filter [3 × 3] and median filter [3 × 3]. The average NCC of the extracted watermark is 0.9570 and 0.9589 when Gaussian LPF is applied to the watermarked image at embedding factor $K = 10$ and 20, respectively. The average NCC of the extracted watermark were obtained as 0.9473 and 0.9499 when average filter [3 × 3] kernel is applied to the watermarked image for $K = 10$ and 20, respectively. Also, the average NCC value as 0.9525 and 0.9569 when the median filter with [3 × 3] kernel is applied for different values of K, respectively, as already discussed. The NCC values for different filtering attacks has enumerated in Table 4 for different values of K and the retrieved watermark from the same attack at $K = 10$ for the same set of images are shown in Fig. 6.

Robustness Analysis against Different Geometrical Attacks. The watermarked images have been tested for different geometrical attacks, namely, rotation and resizing. The 35 angle rotation and [256-128-256] resizing attacks are applied to the watermarked image, and the results were shown in Table 5 and Fig. 7. The average NCC values of the extracted watermark are obtained as 0.9241 and 0.9255 against rotation attack, respectively, for different values of K (10 and 20), and the average NCC value against resizing attacks for considered values of K was obtained as 0.9301 and 0.9389, respectively, are shown in Table. The extracted watermark for $K = 10$ was shown in Fig. 7. Table and figure reveal that the watermarking scheme is highly robust against geometrical attacks.

Robustness Analysis against Different Compression Attacks. The watermarked images have been tested for different compression attacks, namely, JPEG and JPEG 2000. These attacks were applied to the watermarked image for different K values, and the results were shown in the Table and figure. The average NCC values of the extracted watermark are obtained as 0.9545 and 0.9560 against JPEG

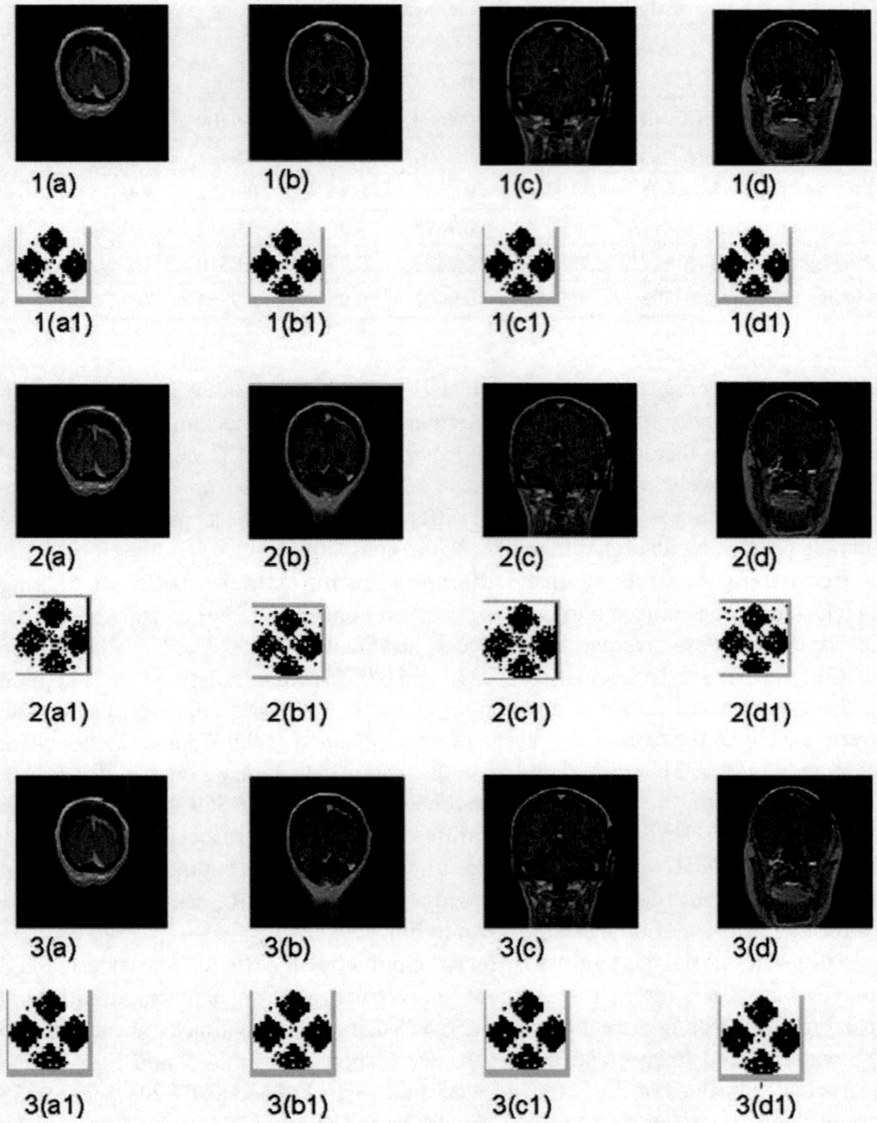

Fig. 6 1 (**a**–**d**) Watermarked image against Gaussian LPF. 1 (**a**1–**d**1) Extracted watermark against Gaussian LPF. 2 (**a**–**d**) Watermarked image against average [3 × 3]. 2 (**a**1–**d**1) Extracted against watermark average [3 × 3]. 3 (**a**–**d**) Watermarked image against median [3 × 3]. 3 (**a**1–**d**1) Extracted watermark against median [3 × 3]

Table 5 Robustness analysis of DICOM images against geometrical attack

Attacks	NCC values				Average NCC
	Image 1	Image 2	Image 3	Image 4	
Rotation (35 degree) ($K = 10$)	0.9349	0.9318	0.901	0.9278	0.9241
Resizing (256-128-256) ($K = 10$)	0.9408	0.9089	0.9364	0.9345	0.9301
Rotation (35 degree) ($K = 20$)	0.9350	0.9328	0.9031	0.9312	0.9255
Resizing (256-128-256) ($K = 20$)	0.9508	0.9112	0.9404	0.9532	0.9389

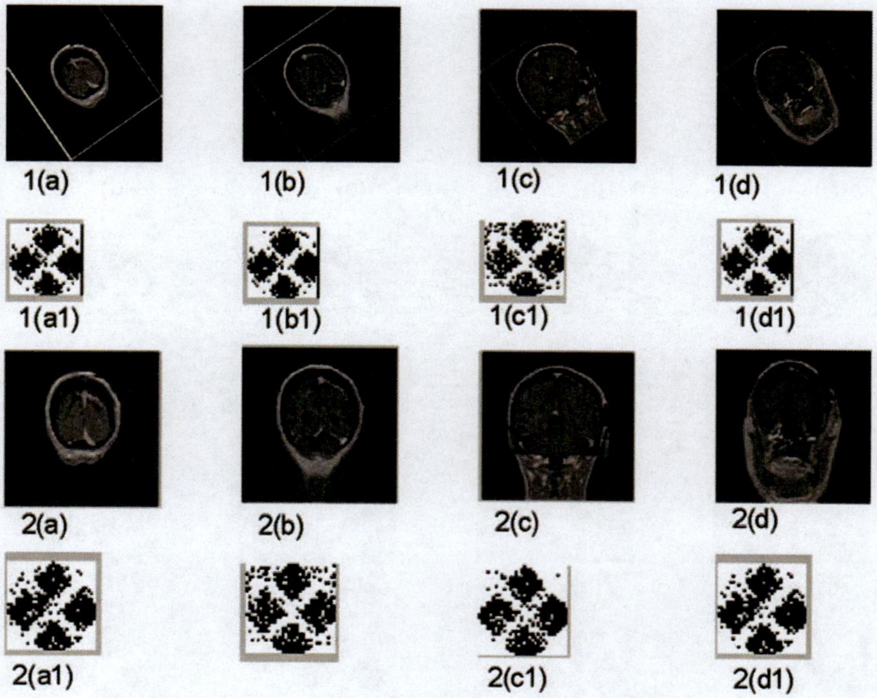

Fig. 7 1 **(a–d)** Watermarked image against rotation (35 degree). 1 **(a1–d1)** Extracted watermark against rotation (35 degree). 2 **(a–d)** Watermarked image against resizing [128-256-128]. 2 **(a1–d1)** Extracted watermark against resizing [128-256-128]

compression attack, respectively, for different values of K, and the average NCC value against JPEG 2000 compression for considering values of K were obtained as 0.9584 and 0.9630, respectively, and are shown in Table 6. The extracted watermark for the same experimental images with $K = 10$ was shown in Fig. 8.

Table 6 Robustness analysis of DICOM images against compression attack

Attacks	NCC values				Average NCC
	Image 1	Image 2	Image 3	Image 4	
JPEG (QF = 50) ($K = 10$)	0.9466	0.9664	0.9586	0.9464	0.9545
JPEG 2000 ($K = 10$)	0.9532	0.9645	0.9640	0.9520	0.9584
JPEG (QF = 50) ($K = 20$)	0.9483	0.9672	0.9601	0.9485	0.9560
JPEG 2000 ($K = 20$)	0.9542	0.9681	0.9697	0.9602	0.9630

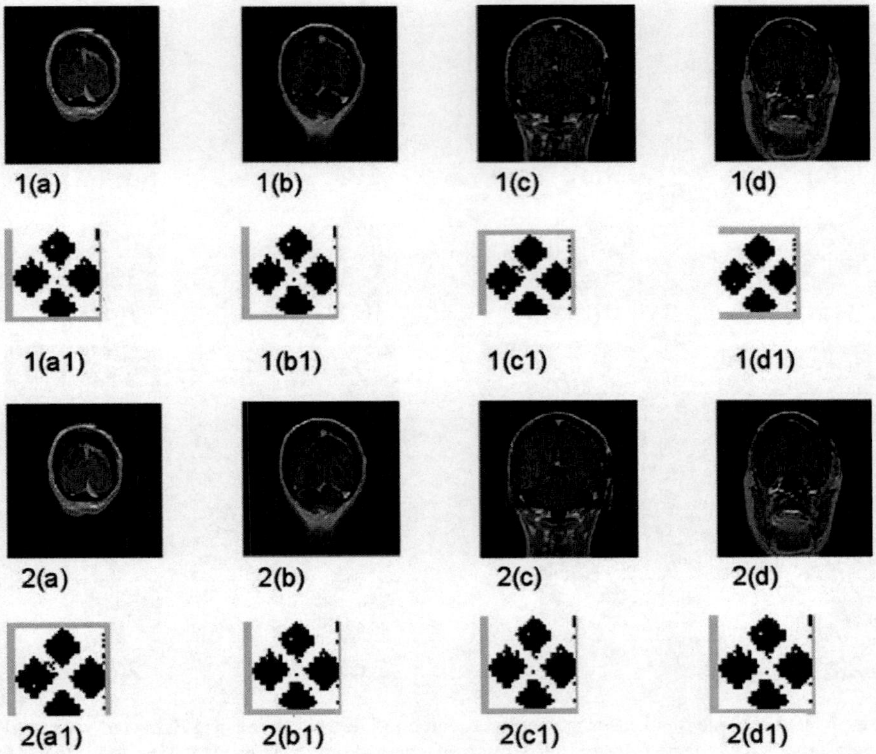

Fig. 8 1 (**a**–**d**) Watermarked image against JPEG (QF = 50). 1 (**a**1–**d**1) Extracted watermark. 2 (**a**–**d**) Watermarked image against JPEG 2000 compression. 2 (**a**1–**d**1) Extracted watermark

4 Conclusion

Now a days, electronic health care system is becoming popular. In this system, the remote area doctors send the patient information interleaving with the medical image to the expert physicians for a second opinion viz open network. To maintain the security to the EPR data medical image watermarking are a potent solution. This paper presents a new watermarking method in the application of medical image where the EPR data is embedded in the host image using DCT transform. To enhance the watermarking scheme robustness, initially the binary EPR information is encoded using ECC and then M-ary modulation is applied.

The attractive features of the proposed work are summarized as follows:

- The watermarking techniques using ECC, M-ary modulation, and DCT exhibit better performance in terms of imperceptibility and robustness.
- Security of the image watermark is enhanced by using ECC and M-ary modulation. By using this combination of techniques the system becomes complex but at the cost of robust improvement. Various techniques were combined to make a balance between the imperceptibility and robustness of the watermarking system. Due to hybrid combination the computational complexity of the system is high and needs to be investigated separately in future communication.

References

Chan YH, Siu WC (1992) On the realization of discrete cosine transform using the distributed arithmetic. IEEE Trans Circ Syst I: Fundam Theory Appl 39(9):705–712

Cox IJ, Miller ML (2002) The first 50 years of electronic watermarking. EURASIP J Adv Sig Process 2002(2):820936

Das S, Kundu MK (2011) Hybrid contourlet-DCT based robust image watermarking technique applied to medical data management. In: International conference on pattern recognition and machine intelligence, pp 286–292. Springer, Berlin, Heidelberg

DICOM Samples Image Sets, http://www.osirix-viewer.com/. Accessed Sept 2017

Hajjaji MA, Mtibaa A, Bourennane E (2011) A watermarking of medical image: method based "LSB". J Emerg Trends Comput Inform Sci 2(12)

Kallel IF, Kallel M, Bouhlel MS (2006) A secure fragile watermarking algorithm for medical image authentication in the DCTdomain. In: Proceedings of IEEE conference on information and communication technology, ICTTA'06 Syria, pp 2024–2029

Kannammal A, Subha Rani S (2014) Two level security for medical images using watermarking/encryption algorithms. Int J Imaging Syst Technol 24(1):111–120

Kutter M (1999) Performance improvement of spread spectrum based image watermarking schemes through M-ary modulation. Lecture notes in computer science, Brisbane, Germany

Mehto A, Mehra N (2016) Adaptive lossless medical image watermarking algorithm based on DCT & DWT. Procedia Comput Sci 78:88–94

Mostafa SA, El-Sheimy N, Tolba AS, Abdelkader FM, Elhindy HM (2010) Wavelet packets-based blind watermarking for medical image management. Open Biomed Eng J 4:93

Nambakhsh MS, Ahmadian A, Zaidi H (2011) A contextual based double watermarking of PET images by patient ID and ECG signal. Comput Methods Programs Biomed 104(3):418–425

Singh AK, Kumar B, Dave M, Mohan A (2015a) Robust and imperceptible dual watermarking for telemedicine applications. Wireless Pers Commun 80(4):1415–1433

Singh AK, Kumar B, Dave M, Mohan A (2015b) Multiple watermarking on medical images using selective discrete wavelet transform coefficients. J Med Imaging Health Inform 5(3):607–614

Swaraja K (2017) A hybrid secure watermarking technique in telemedicine. Int J Eng Technol 9(3):265–270

Zear A, Singh AK, Kumar P (2018) A proposed secure multiple watermarking technique based on DWT, DCT and SVD for application in medicine. Multimedia Tool Appl 77(4):4863–4882

Segmented Lung Boundary Correction in Chest Radiograph Using Context-Aware Adaptive Scan Algorithm

Tej Bahadur Chandra, Kesari Verma, Deepak Jain, and Satyabhuwan Singh Netam

Abstract Chest X-ray (CXR) is the most popular imaging modality used for preliminary diagnosis of pulmonary diseases. In automatic computer-aided diagnosis (CAD), the number of false-positive cases can be reduced by segmenting out the normal anatomical structures. Demarcating lung parenchyma on CXR image is challenging due to complex anatomical structures of the human thoracic cavity. The extracted lungs boundary suffers from undesirable artifacts such as ridges and pits. This paper presents an algorithm to adaptively scan the inner lung boundary to correct the undesirable artifacts. Further, the algorithm is context-aware, it takes care of normal cavity due to the aortic knuckle and diaphragm borders. The algorithm is tested on 138 binary lung mask extracted from digital CXR images from Montgomery dataset. The quantitative, qualitative, statistical results reveal that the proposed algorithm outperforms the existing method. The average increase in segmentation accuracy is 2.561%.

Keywords Chest X-ray · Lung segmentation · Radiography · Anatomical atlas · Thoracic disease · Costophrenic angle

T. B. Chandra (✉) · K. Verma
Department of Computer Applications, National Institute of Technology, Raipur, India
e-mail: tejbahadur1990@gmail.com

K. Verma
e-mail: kverma.mca@nitrr.ac.in

D. Jain · S. S. Netam
Department of Radiodiagnosis, Pt. Jawahar Lal Nehru Memorial Medical College, Raipur, India
e-mail: deepak0808deepak@gmail.com

S. S. Netam
e-mail: sbsnetam@yahoo.com

A. A. Rizvanov et al. (eds.), *Advances in Biomedical Engineering and Technology*, Lecture Notes in Bioengineering,
https://doi.org/10.1007/978-981-15-6329-4_23

1 Introduction

Chest X-ray (CXR) imaging modality is an important, noninvasive clinical adjunct for preliminary diagnosis of interstitial lung diseases. Automatic computer-aided diagnosis (CAD)of ambiguous thoracic disease manifestations on radiographic images is challenging. The elusive normal anatomical structures and vague abnormal opacities in lung fields require localized deep feature analysis to characterize normal and abnormal CXR images. In this context, accurate identification of lung parenchyma regions provides rich structural information like size, shape, lung volume, etc., for subsequent phases of automated diagnosis (Candemir et al. 2014; Jaeger et al. 2014).

The accurate delineation of lung boundary is challenging due to an overlapping complex anatomical structure like rib cage, heart apex, hilum, aortic knuckle, and strong edges of clavicle bones (Chandra and Verma 2020; Chondro et al. 2018) as shown in Fig. 1a, b. Several state of art methods have been proposed over past few years for lung parenchyma segmentation and automated analysis of CXR images. Ginneken et al. (2001) performed a survey on 150 papers and suggested that CAD-based analysis on segmented lungs significantly reduce the false positives and increase the effectiveness of abnormality detection. Shi et al. (2006) used modified SIFT method and patient-specific shape statistics to segment lung fields from a series of chest radiograph of the same patient and achieved segmentation accuracy of 93.8% using the proposed method which is significantly higher than the performance obtained in ASM method (92.6%) (Cootes and Taylor 2004) and Snake method (78.7%) (Kass et al. 1988). Annangi et al. (2010) utilized canny edges features, CP angle corner feature to extract the lung air cavity regions from

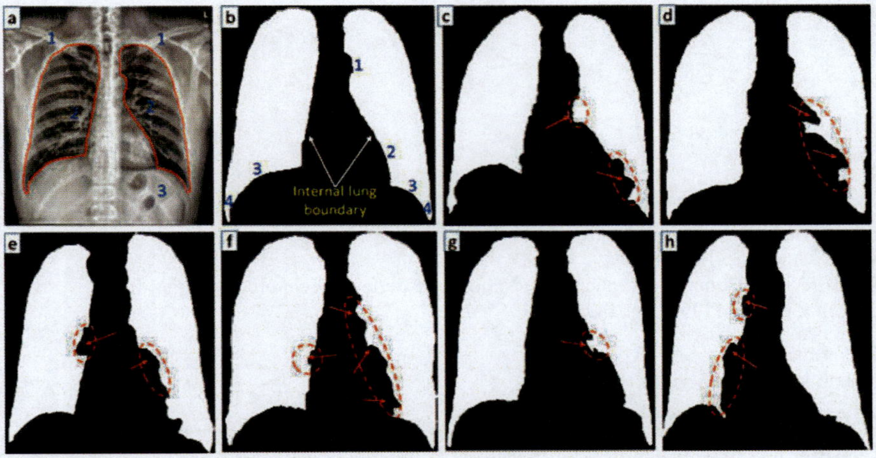

Fig. 1 **a** Normal chest X-ray image with marked lung boundary (1-clavicle bone, 2-left and right hilum, 3-gastric bubble). **b** Corresponding normal binary lung mask (1-aortic knuckle, 2-heart apex, 3-left and right diaphragm border, 4-left and right costophrenic angle). **c–h** Binary lung mask showing undesirable ridges and pits due to strong bone edges or consolidations near inner lung boundary

posterior-anterior (PA) CXR images and attains dice similarity coefficient value of 0.88 $\pm$ 0.07. Jaeger et. al. (2012) used collective lung shape modal, intensity modal, and segmentation mask to perform accurate segmentation. The proposed method provides AUC of 83.12%. Candemir et al. (2012) proposed the lung modal-based graph cut method for lung boundary detection for segmenting lung air cavity and observed 91% of segmentation accuracy on JSRT dataset, further the enhanced version using SIFT flow-based nonrigid deformable registration method to automatically compute the patient-specific binary lung mask is presented in Candemir et al. (2014) attaining higher segmentation accuracy of 95.4% (JSRT dataset), 94.1% (Montgomery dataset), and 91.7% (Maryland dataset). A recent method for lung segmentation using ultrametric contour map and structured edge detector is proposed in Yang et al. (2018) achieving a mean Jaccard index of 95.2%.

Segmented lung regions play an important role in the quantitative estimation of radiographic indices like lung volumes, varying heart dimension, irregular lung shape, cardiothoracic (CT) ratio, costophrenic angle, and other pathologies which can be used as a measure to detect cardiomegaly, pneumothorax, atelectasis, pleural effusion, and other chronic abnormalities. The automatic localization of costophrenic recesses and measurement of costophrenic angle on CXR images are described in Maduskar et al. (2013), Armato et al. (2004), Armato et al. (1998), Hasan et al. (2012), Dallal et al. (2017). A very limited number of methods have been reported in literatures related to lung boundary correction using computed tomography (CT) images. Yim and Hong (2008) proposed curvature-based lung boundary correction method in CT images using 3D volume of interest (VOI) refinement and achieved overlapping area ratio is 96%. Pu et al. (2008) use geometric smoothening of lungs border to include juxtapleural nodules in CT images while reducing the over-segmentation. Liang et al. (2017) proposed boundary tracing algorithm which detects the inflection point more accurately than others. The proposed method shows the 6.46% increase in segmentation precision compared to existing state of art methods. Singadkar et al. (2018) use dominant point marching algorithm to connect concave and convex regions. The obtained average volumetric overlap is 96.97%.

Despite of several improvements in lung segmentation algorithms, boundary correction in CXR images remains an unexplored area. Artifacts like ridges and pits degrade the segmentation accuracy subsequently affecting the performance of CAD-based automated systems. The undesirable artifacts are due to strong edges of the bones or due to disease opacification near lung boundary regions. In this study, a new context-aware adaptive scan correcting algorithm is proposed to adaptively scan the inner lung boundary in patient-specific binary lung mask of CXR image to correct the undesirable ridge and pit as shown in Fig. 1c–h. The working of the algorithm is derived from the core idea presented in paper (Sezaki and Ukena 1973). The algorithm is context-aware, it takes care of the lower diaphragm curve and normal pits due to the aortic knuckle near the upper inner border of the right lung.

The rest of the paper is structured as follows. Section 2, elaborates the detailed implementation of the algorithm. In Sect. 3, the brief description of the materials and method used in this study is presented. The experimental setup, quantitative and qualitative evaluation of results are described in Sect. 4. In Sect. 5, the detailed discussion is presented followed by a conclusion and future direction in Sect. 6.

2 Proposed Algorithm

The proposed context-aware adaptive scan algorithm is an enhancement of method presented in paper (Sezaki and Ukena 1973). Moreover, this proposed algorithm is used to enhance the segmentation accuracy of binary lung mask-based segmentation techniques. In this study, the algorithm uses binary lung mask obtained from the SIFT flow-based nonrigid registration-driven lung segmentation method (Candemir et al. 2014) for correcting ridges and pits on the inner boundary of segmented lung mask.The working principle of the proposed method is illustrated in Fig. 2a, b and a detailed description is presented in Algorithm 1. The algorithm initially set the scan lines at given interval si from top to bottom scan limit (B_{pt}) of binary lung mask (Fig. 2a). Subsequently for each scan line, left and right inner lung border point p_i is retrieved. However, in case of left lung, the scanning starts after skipping the top$_{\mathrm{skip}}$ percent area (usually greater than 30%) from the top to accommodate the normal cavity structure due to the aortic knuckle as shown in Fig. 2a. The algorithm adaptively scan from the point p_i by placing two more points p_{i+1}, and p_{i+2} on the scan line at interval si and $2si$, respectively, as shown in Fig. 2b. If the point p_{i+1} lies outside the range of $(p_i \pm d1)$ and point p_{i+2} lies within the range of $(p_i \pm d2)$ then the point p_{i+1} is considered either pit or ridge and new border point is computed by averaging the first and last value as described in line 14 of the Algorithm 1.

Moreover, if the point p_{i+1} falls in the range of $(p_i \pm d1)$ but the point p_{i+2} lies outside the range of $(p_i \pm d2)$ then the point p_{i+1} is considered as normal border point and the algorithm will simply skip the point. Further, if the points p_{i+1} and p_{i+2} both lies outside the range of $(p_i \pm d1)$ and $(p_i \pm d2)$, respectively, then the algorithm follows the recursive approach with adjusting the learning parameters $\mu_2 + \gamma$ and adapted scan interval $si + \omega$ to deal with large pits or ridges as described in line 16, 17, and 19 of Algorithm 1. Finally, the algorithm stops when the scan interval

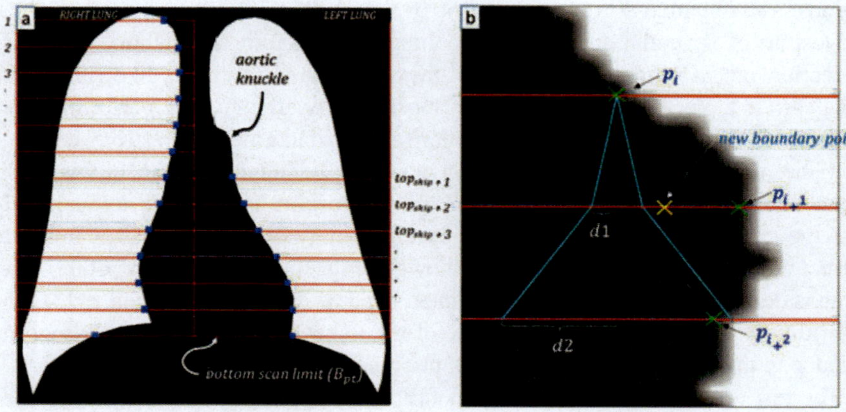

Fig. 2 **a** The binary lung mask showing a scan line on the area of interest. **b** Working principle of the algorithm to correct the ridges and pits in lungs boundary using binary lung mask

$si > si_m$. The values of $d1$ and d2 are adjusted automatically based on the scan interval and the learning parameters $\mu_1$andμ_2 as described in line 8 and 9 of the Algorithm 1. The values of learning parameters are determined empirically after a series of repeated experimentation.

Algorithm 1. Proposed context-aware adaptive scan correcting algorithm

---

Input

1: $I = \{i_1, i_2, i_3 \ldots\ldots\ldots i_n\},\ i_k \in D^N$;//$D^N$is CXR image dataset
2: Initialize $si,\ where, si > 1$;//The starting scan interval
3: Initialize $si_m,\ where, si_m > si$; //The maximum scan interval limit
4: Initialize $\mu_1,\ where, \mu_2 > 0$ // μ_1, μ_2 are learning parameter and is arbitrarily selected
5: Initialize top_{skip}; // To preserve the aortic knuckle pit, should be greater than 30% of image height
6: Set $B_{pt} =$ ***call*** $Locate\_Bottom\_Scan\_Limt\ (I, si_s)$; // $B_{pt}$is bottom scan limit

Output

1: Updated binary lung mask

Algorithm

1: $img \leftarrow Bounded\_Region_{minimal}\ (I)$; //Compute the minimal bounded region of *I*.
2: **Repeat** form $p_i = top_{skip}\ to\ B_{pt}$
3: $p_{i+1} =$ ***call*** $Scan\_Correct\ (img,\ p_i,\ si,\ \mu_1,\ \mu_2,\ si_m)$;
4: Adjust the binary lung mask border according to the new value.
5: **End**
6:
7: **Function**$Scan\_Correct(img,\ p_i,\ si,\ \mu_1,\ \mu_2,\ si_m)$
8: Set $d1 = si \times \mu_1$;//Set pits and ridge range parameters
9: Set $d2 = d1 + (si \times \mu_2)$;//Set pits and ridge range parameters
10:
11: **If** $((p_{i+1} \leq (p_i \pm d1))OR(si > si_m))$**then**
12: **Return** p_{i+1};//stop algorithm when either p_{i+1} or p_{i+1} and p_{i+2} both liesin therange of *d1* and *d2* respectively or scan intervalexceeds the max scan interval limit si_m.
13: **Elseif**$(p_{i+1} \geq (p_i \pm d1))$ AND $(p_{i+2} \leq (p_i \pm d2))$ **then**
14: $p_{i+1} = (p_i + p_{i+2})/2$; //Compute the average point
15: **Elseif** $(p_{i+1} \geq p_i \pm d1)$ AND $(p_{i+2} \geq p_i \pm d2)$ **then**
16: $si = si + \omega$; //increase the scan interval by ω
17: $\mu_2 = \mu_2 + \gamma$; //increase the learning parameter by γ
18: // call with adapted learning parameters
19: $p_{i+1} =$ ***call*** $Scan\_Correct\ (img,\ p_i,\ si,\ \mu_1,\ \mu_2,\ si_m)$;
20: **End**
21: **End**
22:
23: **Function**$Locate\_Bottom\_Scan\_Limt(img,\ si)$
24: **Repeat** from $i = h\ to\ h/2$; // h is the image height
25: $d_i = rb_i - lb_i$; // Compute inner lung border distance d_i
26: $dd_i = d_i - d_{i-1}$;//compute the difference distance from previous scan line to current scan line
27: **End**
28: //assign the location where maximum distance difference lies
29: $B_{pt} = Loc(\max(dd_k));\ k = h,\ h - si,\ h - 2 * si, h - 3 * si \ldots\ldots\ldots h/2$
30: **End**

---

3 Materials and Methods

3.1 Dataset

In this study, the Montgomery Set (Candemir et al. 2014; Jaeger et al. 2014) is used to evaluate the performance of the proposed method. The dataset contains a total 138 digital CXR images comprising 58 images with tuberculosis manifestation and 80 images with normal findings. All the X-ray images are in portable network graphics (PNG) file format of dimension 4020 × 4892 or 4892 × 4020 with a bit-depth of 12 bits. The pixels are linearly spaced in both horizontal and vertical direction with spacing 0.0875 mm. The dataset is created and maintained by National Library of Medicine in collaboration with the Department of Health and Human Services, Montgomery County, Maryland, USA.

3.2 Lung Segmentation

The patient-specific binary lung mask in this study is obtained from nonrigid registration-driven robust lung segmentation method proposed in Candemir et al. (2014). The detailed working of the method is shown in Fig. 3. The method is based on content-based image retrieval (CBIR) technique to get the patient-specific adaptive lung mask. It uses Bhattacharyya shape similarity index to get the most similar reference images to patient X-ray, at the same time partial radon transform is used to handle the little affine variation in input X-rays images. The corresponding lung mask of top *n*-ranked similar reference images are registered using deformable SIFT

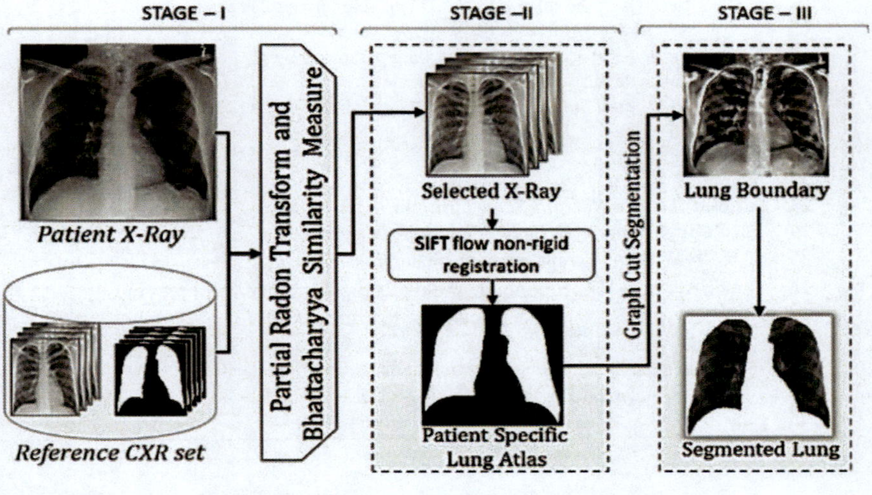

Fig. 3 Three stages of nonrigid registration-driven lung segmentation method

flow registration method to get the patient-specific average lung mask. Finally, the graph cut method is used to cut the original CXR image to get the segmented lungs.

3.3 *Evaluation Measures*

To evaluate and compare the performance of the proposed algorithm qualitative and quantitative evaluation measures are used as described in the following subsections.

Quantitative Evaluation. The Jaccard similarity coefficient (Panigrahi et al. 2019) as described in Eq. 1 is used to measure the overlap between the corrected binary lung mask ($S_{\text{corrected}}$) and the corresponding gold standard ground truth (S_{GT}) mask over all the pixels. Where true positive (TP) represents the correctly identified foreground pixel, false positive (FP) represents the pixels misclassified as foreground and false negative (FN) represent the pixels misclassified as background pixels.

$$\text{JaccardIndex(JI)} = \frac{|S_{\text{corrected}} \cap S_{\text{GT}}|}{|S_{\text{corrected}} \cup S_{\text{GT}}|} = \frac{|\text{TP}|}{|\text{FP}| + |\text{TP}| + |\text{FN}|} \tag{1}$$

$$\text{JaccardDistacne(JD)} = 1 - \frac{|S_{\text{corrected}} \cap S_{\text{GT}}|}{|S_{\text{corrected}} \cup S_{\text{GT}}|} \tag{2}$$

Similarly, *dice's index* (DI) (Dice 1945) as described in Eq. 3 is also computed to measure the area of overlap.

$$\text{Dices Index(DI)} = \frac{|S_{\text{corrected}} \cap S_{\text{GT}}|}{|S_{\text{corrected}}| + |S_{\text{GT}}|} = \frac{2|\text{TP}|}{2|\text{TP}| + |\text{FN}| + |\text{FP}|} \tag{3}$$

Qualitative Evaluation. The visual assessment of the corrected binary lung mask is important since the topography of normal anatomical structure plays an important role in abnormality detection. The lung mask before and after correction is fused (overlapped) with white regions representing the lung mask before correction and pink region representing the lung mask after correction as shown in Fig. 7. These fused lung mask images along with the original CXR images ware provided to an experienced radiologist for visual assessment. The opinion of the radiologist regarding the accuracy of ridges or pits correction using the proposed algorithm is recorded as the level of satisfaction (in %).

Statistical Analysis. Statistical analysis plays an important role in medical image analysis to examine the significance of a method based on the several experimental runs. In this study, Z-test statistics for two samples are used to assess the statistical significance of the proposed algorithm over the existing method (Agnihotri et al. 2017). The Z-score as described in Eq. 4 uses mean and variance of the two populations to accept or reject the null hypothesis (H_0) based on the p-value and critical value at specified significance level (α).

$$Z = \frac{M_1 - M_2 - \Delta}{\sqrt{\frac{\sigma_1^2}{n_1} + \frac{\sigma_2^2}{n_2}}} \tag{4}$$

where, M_1 and M_2 are the means of Jaccard index of existing and proposed method, respectively, Δ is the hypothesized difference between the population means, σ_1 and σ_2 are the standard deviation of two populations and n_1 and n_2 are the sizes of two samples. Where the null hypothesis (H_0) denotes the mean segmentation performance of proposed algorithm and the existing method is equal and the alternate hypothesis denotes the performance of proposed algorithm is significantly higher than the existing method. The assumed null and alternate hypotheses are described in Eq. 5 and Eq. 6, respectively.

$$H_0 : M_1 = M_2 \tag{5}$$

$$H_1 : M_1 < M_2 \tag{6}$$

4 Experimental Setup

This section presents the topological design of an experimental setup to test and compare the performance-proposed algorithm as shown in Fig. 4. The algorithm is tested on 138 binary lung masks extracted using nonrigid registration-based lung segmentation method from Montgomery dataset. Further, the twofold evaluation measures (quantitative and qualitative) are used to evaluate the performance.

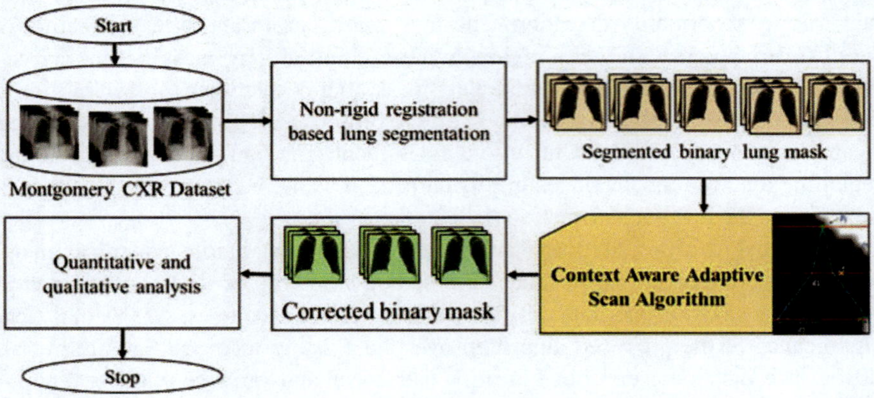

Fig. 4 Topological design of experimental setup

5 Results and Discussion

This section presents the results of the quantitative evaluation and visual assessment of the proposed algorithms. The response of the proposed method while dealing with ridges and pits are illustrated in Fig. 5. It can be observed from Fig. 5a–h that the point p_{i+1} lies outside the range of $p_i \pm d1$ while p_{i+2} lies within the range of $p_i \pm d2$ showing the presence of either ridge or pits.The computed new boundary point is the averaged value of two points p_i and p_{i+2} shown as red-colored cross mark in each image patch. Furthermore, if the size of artifact is large that remains undetected by scan lines at smaller intervals, the algorithm gradually increases the scan interval by ω and widens the range of $d2$ by γ in each iteration as observed from Fig. 5d.

The results of quantitative evaluation measures: average Jaccard index (JI), Jaccard distance (JD), and dice's index (DI) to measure the performance of the algorithm are shown in Table 1. It can be observed from Table 1 that average Jaccard index (overlap measure) between segmented binary lung mask (before correction) and gold standard ground truth (GT) is 0.8960. However, after correcting the lung

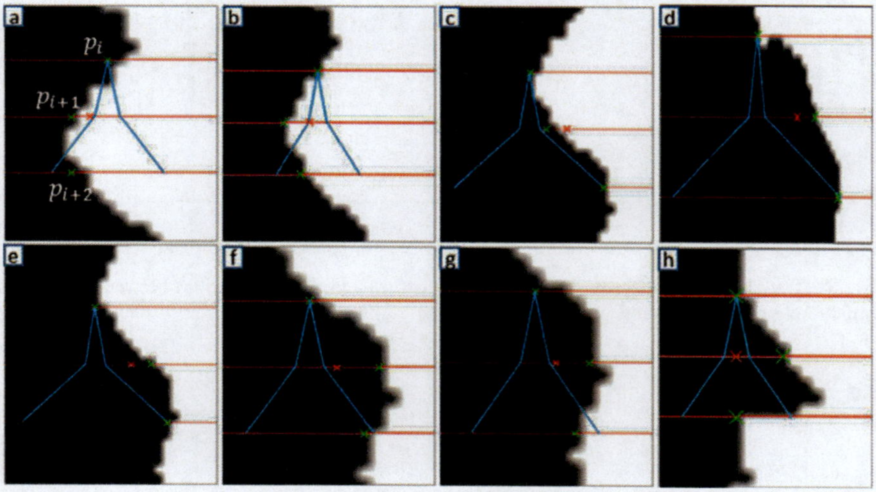

Fig. 5 Scan correcting the ridges and pits of binary lung mask using adaptive scan lines

Table 1 Segmentation performance measure using average Jaccard index (JI), average Jaccard distance (JD), and average dice's index (DI) of binary mask before and after lung mask correction

Evaluation measures	Before correction	After correction
Average Jaccard index	0.8960	0.9216
Average Jaccard distance	0.1040	0.0784
Average Dice's index	0.9451	0.9592

mask the Jaccard index is increased by 0.0256. The improvement in the performance is due to the fact that the ridges and pits due to strong edges of the normal anatomical structure are corrected by the proposed algorithm as shown in Fig. 8. The increase in average dice's index and decreased average Jaccard distance after lung mask correction confirms the promising performance of the proposed algorithm.

Figure 6 shows the plot of the Jaccard similarity index before and after correcting the individual binary lung mask. From the plotted graph it can be observed that the overlap area between corrected lung mask and gold standard ground truth mask is significantly better than that of before correction for each individual lung mask. The performance comparison of the newly proposed and existing method is performed using Receiver Operating Characteristic (ROC) curve as shown in Fig. 7b. From the area under ROC curve analysis it can be observed that the proposed boundary correction method has improved the lung segmentation accuracy attaining AUC of 0.908.

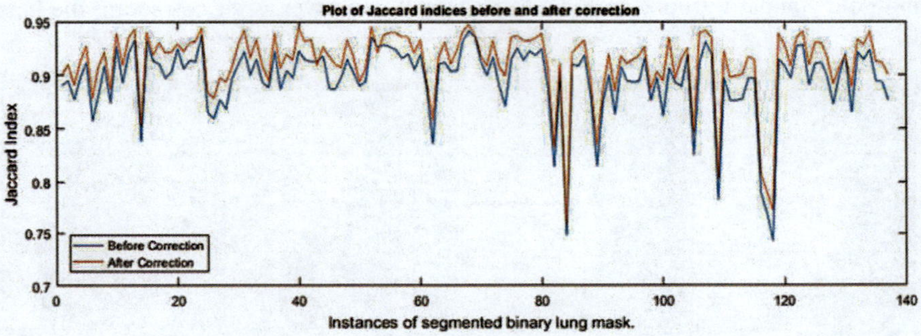

Fig. 6 The plot of the Jaccard similarity index (before and after correction) between individual binary lung mask and gold standard ground truth

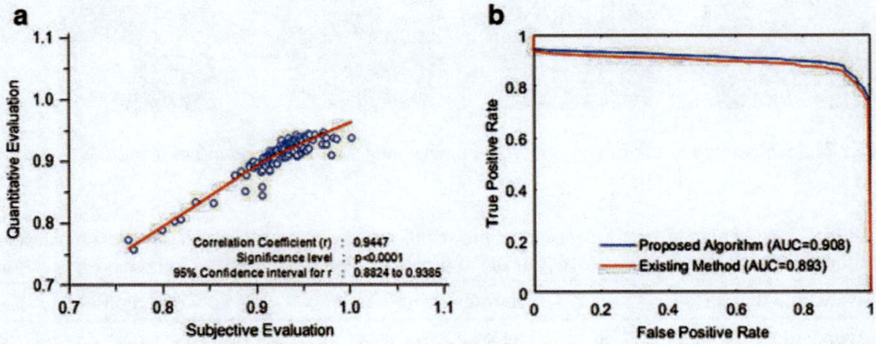

Fig. 7 **a** Correlation coefficient plot between observed quantitative result and result of subjective evaluation by radiologist. **b** The receiver operating characteristic (ROC) curve plot for proposed algorithm and existing method

The Pearson's correlation coefficient (r) plot shown in Fig. 7a shows the strength of linear relationship between observed quantitative results and the results of the subjective evaluation performed by the radiologist. The value of r ranges between −1 to 0 to +1 for negatively correlated, not correlated and, positively correlated, respectively. The obtained value of correlation coefficient $r = 0.9447$ is very close to +1 which denotes that the result of subjective evaluation performed by the radiologist is positively correlated with the results obtained by quantitative evaluation.

The qualitative evaluation of corrected binary lung mask (as shown in Fig. 8) is performed by an expert radiologist by rating (in %) based on visual appearance of the corrected lung mask (as shown in Fig. 8). As per the radiologist suggestion the segmented part of lungs using existing method (anatomical atlas-based segmentation method Candemir et al. 2014) specially in regions with strong bone or disease responses are not properly segmented that affects the computation of features for diseases diagnosis. The proposed method resolves this issue and improve the result based on visual assessment attaining average visual score by 92.53%.The average improvement in visual segmentation result is 2.93% compared to existing method. Moreover, as suggested by the radiologist the algorithm is lacking in some cases, especially when the ridges or pits are large enough that remains undetected or partially detected by the algorithm as shown in Fig. 8b, c, e, h, i, l which can be the future research direction.

The statistical analysis performed using Z-test method for two sample means on obtained quantitative result at 0.05 level of significance (α) is shown in Table 2. From the Table 2, it can be observed that the obtained p-value (0.00006) is significantly smaller than the given significance level ($\alpha = 0.05$) which clearly rejects the null hypothesis. Furthermore the value of calculated Z-score is less than the Z-score at critical point which confirms that acceptance of alternate hypothesis that verifies the statistical significance of the proposed algorithm over the existing method.

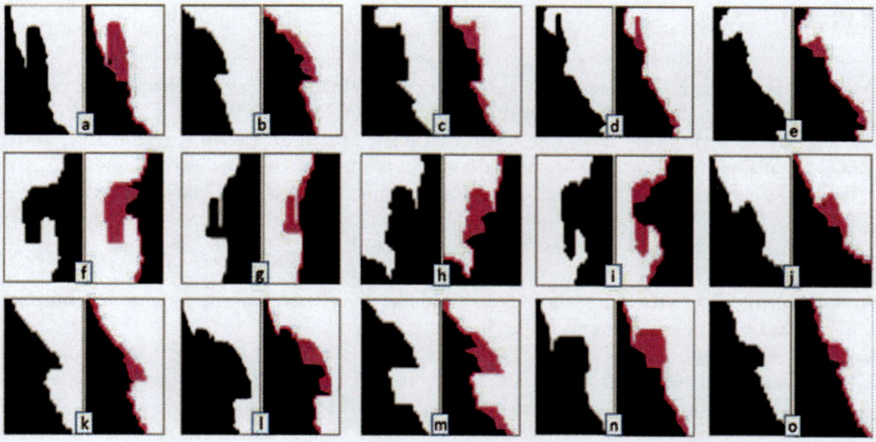

Fig. 8 Corrected binary mask images using the proposed algorithm (white regions show the actual segmentation having ridges and pits while the pink region shows the corrected binary mask)

Table 2 Z-test scores (left tail test) for two sample means at 0.05 level of significance (α) for 138 observations in each samples

	Statistics value
Z-score (calculated)	−3.83981
Z-score (critical)	1.64485
p-value	0.00006

6 Conclusion

In this paper, we have proposed the context-aware adaptive scan algorithm for correcting the artifacts (ridges and pits) in binary lung masks. The proposed method uses adaptive scanning for eliminating the artifacts in the inner lung border improving the segmentation accuracy. The quantitative evaluation measures reveal the better performance of the proposed algorithm attaining an average 2.561% increase in segmentation accuracy compared to existing methods without lung mask correction. Moreover, the results of visual evaluation performed by the radiologist confirm that the corrected binary mask more accurately represents the lung fields. The statistical analysis performed using Z-test score for two sample means confirms the statistical significance of the proposed algorithm.

In future, the proposed algorithm can be enhanced in several ways: (1) Costophrenic recess and diaphragm contours can be corrected intelligently. (2) Artifacts near normal aortic knuckle cavity can be improved. (3) The algorithm can be evaluated on larger datasets. (4) Intra- and inter-operator and observer analysis can also be performed.

References

Agnihotri D, Verma K, Tripathi P (2017) Variable global feature selection scheme for automatic classification of text documents. Expert Syst Appl 81:268–281

Annangi P, Thiruvenkadam S, Raja A, Xu H, Sun XSX, Mao LML (2010) A region based active contour method for x-ray lung segmentation using prior shape and low level features. In: 2010 IEEE International Symposium on Biomedical Imaging: From Nano to Macro, pp 892–895

Armato III SG, Giger ML, MacMahon H (1998) Computerized delineation and analysis of costophrenic angles in digital chest radiographs. Acad Radiol 5(5):329–335. https://doi.org/10.1016/S1076-6332(98)80151-7

Armato III SG, Giger ML, MacMahon H (2004) Method and system for the automated delineation of lung regions and costophrenic angles in chest radiographs

Candemir S, Jaeger S, Palaniappan K, Antani S, Thoma G (2012) Graph cut based automatic lung boundary detection in chest radiographs. In: IEEE healthcare technology cconference: translational engineering in health and medicine, pp 31–34

Candemir S, Jaeger S, Palaniappan K, Musco JP, Singh RK, Xue Z, Karargyris A, Antani S, Thoma G, McDonald CJ (2014) Lung segmentation in chest radiographs using anatomical atlases with nonrigid registration. IEEE Trans Med Imaging 33:577–590

Chandra TB, Verma K (2020) Pneumonia detection on chest X-Ray using machine learning paradigm. In: Proceedings of 3rd international conference on computer vision and image processing. Springer, Singapore, pp 21–33. https://doi.org/10.1007/978-981-32-9088-4_3

Chondro P, Yao CY, Ruan SJ, Chien LC (2018) Low order adaptive region growing for lung segmentation on plain chest radiographs. Neurocomputing. 275:1002–1011

Cootes TF, Taylor CJ (2004) Statistical models of appearance for computer vision. Available on: https://www.face-rec.org/Algorithms/AAM/app_models.pdf. Accessed 24 July 2020

Dallal AH, Agarwal C, Arbabshirani MR, Patel A, Moore G (2017) Automatic estimation of heart boundaries and cardiothoracic ratio from chest x-ray images. In: Medical imaging 2017: computer-aided diagnosis, p 101340K

Dice LR (1945) Measures of the amount of ecologic association between species. Ecology 26:297–302

Hasan MA, Lee S-L, Kim D-H, Lim M-K (2012) Automatic evaluation of cardiac hypertrophy using cardiothoracic area ratio in chest radiograph images. Comput Methods Programs Biomed 105:95–108

Jaeger S, Karargyris A, Antani S, Thoma G (2012) Detecting tuberculosis in radiographs using combined lung masks. In: 2012 Annual international conference of the IEEE on engineering in medicine and biology society (EMBC), pp 4978–4981

Jaeger S, Karargyris A, Candemir S, Folio L, Siegelman J, Callaghan FM, Xue Z, Palaniappan K, Singh RK, Antani SK et al (2014) Automatic tuberculosis screening using chest radiographs. IEEE Trans Med Imaging 33:233–245

Kass M, Witkin A, Terzopoulos D (1988) Snakes: active contour models. Int J Comput Vis 1:321–331

Liang J, Jiang H, Ma L, Liu Y, Toshiya N (2017) A new boundary correction method for lung parenchyma. In: Eighth international conference on graphic and image processing (ICGIP 2016), vol 10225. International Society for Optics and Photonics, p 1022529. https://doi.org/10.1117/12.2266086

Maduskar P, Hogeweg L, Philipsen R, van Ginneken B (2013) Automated localization of costophrenic recesses and costophrenic angle measurement on frontal chest radiographs. In: Medical imaging 2013: computer-aided diagnosis, vol 8670. International Society for Optics and Photonics, p 867038. https://doi.org/10.1117/12.2008239

Panigrahi L, Verma K, Singh BK (2019) Ultrasound image segmentation using a novel multi-scale Gaussian kernel fuzzy clustering and multi-scale vector field convolution. Expert Syst Appl 115:486–498

Pu J, Roos J, Yi CA, Napel S, Rubin GD, Paik DS (2008) Adaptive border marching algorithm: Automatic lung segmentation on chest CT images. Comput Med Imaging Graph 32:452–462

Sezaki N, Ukena K (1973) Automatic computation of the cardiothoracic ratio with application to mass screening. IEEE Trans Biomed Eng:248–253

Shi Y, Qi F, Xue Z, Ito K, Matsuo H, Shen D (2006) Segmenting lung fields in serial chest radiographs using both population and patient-specific shape statistics. In: International conference on medical image computing and computer-assisted intervention. Springer, Berlin, Heidelberg, pp 83–91. https://doi.org/10.1007/11866565_11

Singadkar G, Mahajan A, Thakur M, Talbar S (2018) Automatic lung segmentation for the inclusion of juxtapleural nodules and pulmonary vessels using curvature based border correction. J King Saud Univ - Comput Inf Sci

Van Ginneken B, Ter Haar Romeny BM, Viergever MA (2001) Computer-aided diagnosis in chest radiography: a survey. IEEE Trans Med Imaging 20:1228–1241

Yang W, Liu Y, Lin L, Yun Z, Lu Z, Feng Q, Chen W (2018) Lung field segmentation in chest radiographs from boundary maps by a structured edge detector. IEEE J Biomed Heal Inform 22:842–851

Yim Y, Hong H (2008) Correction of segmented lung boundary for inclusion of pleural nodules and pulmonary vessels in chest CT images. Comput Biol Med 38:845–857

Effect of Temperature and Titania Doping on Structure of Hydroxyapatite

Yash Chopra, Rajesh Kumar, and Howa Begam

Abstract Hydroxyapatite (HAp) doped with titanium dioxide was prepared by wet chemical route. Thus prepared HAp was heat-treated at 800 °C and sintered at 1250 °C. To study the effect of dopant and temperature on crystallographic structure, X-Ray diffraction, and Fourier transform infrared spectroscopy (FTIR) was performed. It was observed that all materials were in single phase and few cases peak shifting were observed. Dopant causes distortion of hexagonal crystal structure, which lessened the lattice parameter and crystallinity. Temperature had also role in changes in unit cell volume and percentage of crystallinity. Similar effect was observed in FTIR result.

Keywords Titanium dioxide · Hydroxyapatite · XRD · FTIR · Unit cell volume · Crystallinity

1 Introduction

Calcium phosphate-based bioceramics are extensively used as implant biomaterial due to its capacity of bone growth and biocompatibility (Kalita et al. 2007). Synthetic hydroxyapatite has engrossed special consideration as bone and teeth implant material. Beside its advantages, there are so many shortcomings as implant material like its inferior osteoinductivity, brittleness (Bodhak et al. 2011). Enhancement of osteoinductivity is still a significant challenge. Other limitations are including high percentage of crystallinity, that causes nondegradability of HAp when it is implanted into the body (Geng et al. 2016). Therefore, there is an emergent concern to make a new generation biomaterial using HAp, that will reduce the shortcomings of synthetic hydroxyapatite and it will help in faster healing after implantation.

Y. Chopra · R. Kumar (✉) · H. Begam
Department of Biomedical Engineering, National Institute of Technology Raipur, Raipur, India
e-mail: rajverma311297@gmail.com

H. Begam
e-mail: howabegam@gmail.com

A. A. Rizvanov et al. (eds.), *Advances in Biomedical Engineering and Technology*, Lecture Notes in Bioengineering,
https://doi.org/10.1007/978-981-15-6329-4_24

The use of metallic ion as substitute element is one approach for betterment of synthetic HAp in terms of mechanical properties, crystallinity, etc. Geng et al. (2016) showed that magnesium (Mg^{2+}) and strontium (Sr^{2+}) ions (Mg/Sr/(Ca + Mg + Sr) concentration: 10/20 mol%) influenced the mineral metabolism at the time of bone remodeling process and its also enhanced osteoblast cell proliferation. Zinc dopant improved new bone formation and it showed more new bone formation compared to undoped HAp when it was implanted in rabbit model (Bhattacharjee et al. 2014). Kalita et al. used zinc and magnesium dopant during the sol–gel synthesis of nano hydroxyapatite and they showed that the zinc and magnesium dopant improved the surface hardness and compression strength of nano-HAp (Kalita and Bhatt 2007). Stanić et al. (2010) used copper and zinc ions as dopant and concluded on the basis of their antimicrobial test results that all metal doped HAp exhibit reduction of viable cell of Escherichia coli, Staphylococcus aureus, and Candida albicans. Ti-doped HAp shows good affinity to organic compound which can be used for antibacterial applications (Tsuruoka et al. 2015). Samudrala et al. (2017) prepared borosilicate glassed doped with titanium dioxide (TiO_2) to prepare compositions with precise degradation rate and superior biological response for bone-tissue engineering domain.

The crystal structure of hydroxyapatite is hexagonal and when dopant ions replace the calcium ions, its crystal structure becomes distorted. The aim of this work was to prepare titania-substituted hydroxyapatite and to analyze their crystallographic behavior due to dopant and temperature.

2 Material and Method

2.1 Materials

For the synthesis of titania doped and undoped hydroxyapatite, the chemicals which were used are: calcium hydroxide, ($Ca(OH)_2$, assay: 96%, Loba Chemie, India), orthophosphoric acid (H_3PO_4, assay: 85%, Loba Chemie, India), and titanium oxide (TiO_2, assay: 94%, Molychem, India).

Synthesis of Pure and Titania Doped Hap. Hydroxyapatite was prepared using wet precipitation method as described elsewhere (Begam et al. 2017). Briefly, 0.6 M orthophosphoric acid was added dropwise for 2 h in 1.008 M calcium hydroxide solution. The chemical reaction was conducted in stirring condition at 80 °C and pH was 11. For doped hydroxyapatite, required amount of titanium oxide (5 wt%) was mixed into calcium hydroxide aqueous solution before addition of orthophosphoric acid solution. The prepared hydroxyapatite slurry was kept for aging for 24 h followed by filtration and washing with distilled water. After that, it was air dried at 80 °C for 24 h and dried cakes were sieved. The thus prepared powder is coded as HAp 80. The powder was calcined at 800 °C with holding time 2 h and heating rate 5 °C per min in a muffle furnace. Calcined powder was coded as HAp 800 °C. The calcined powder was then pressed at pressure of 150 MPa to make pellets using hydraulic press. Pressed pellets were sintered in a muffle furnace at 1250 °C with holding time 2 h and heating rate was 3 °C per min. The sintered samples were coded as HAp 1250 °C.

Characterization. Both undoped and titania doped HAp were studied for phase analysis with X-ray diffractometry (XRD; Bruker diffractometer, D8 advance, Japan). The XRD scan was recorded in the 2θ range of 0–80° with scan speed of 1° per min. For calculation of lattice parameters (*a* axis and *c* axis), two peaks (3 0 0) and (0 0 2) was used. The formula for lattice parameter calculation is given below (Eq. 1) (Bhattacharjee et al. 2014).

$$\frac{1}{d^2} = \frac{4}{3}\left(\frac{h^2 + hk + k^2}{a^2}\right) + \frac{l^2}{c^2} \tag{1}$$

Here, *d* is the distance between adjacent planes in the set of Miller indices (*h k l*). The unit cell volume (*V*) was calculated using the relation $V = 2.589a2c$ (Bhattacharjee et al. 2014). The fraction of crystalline phase, i.e., the crystallinity (X_c) of the samples were also calculated by the following expression:

$$X_c = \left(1 - \frac{V_{112/300}}{I_{300}}\right) \tag{2}$$

Here, $V_{112/300}$ is the intensity of hollow between planes (112) and (300). I_{300} is intensity of the (300) plane (Bhattacharjee et al. 2014). FTIR analysis was performed to study the bonds present in the material using KBr in UV-1800 Shimadzu instrument, Japan. The study was carried out in range of 400–4000 cm^{-1}.

3 Results and Discussion

The XRD patterns of thus prepared powder(both pure and doped) are shown in Fig. 1a. In case of pure HAp 80, it was observed that the peaks were perfectly correlated with standard HAp with hexagonal crystal structure and P63/m space group, described in JCPDS data (card no. 09-0432). Major peaks were found at 2θ angle of 25.89° (0 0 2), 31.79° (2 1 1), 32.09° (3 0 0), 39.7° (3 1 0), 46.6° (2 2 2), and 49.36° (2 1 3). Whereas in doped HAp the peaks were little bit shifted. The characteristic peaks were observed at 2θ angle of 26.04° (0 0 2), 33.05° (3 0 0), 46.87° (2 2 2), and 49.69° (2 1 3).

Both pure and doped calcined powders (Fig. 1b) were phase pure and well-matched with XRD pattern of standard HAp with JCPDS file no 09-0432. The intensities of each peak were reduced in case of doped HAp compared to undoped. Peak shift was observed for doped powder at 300 (2θ angle of 33.03°).

In case of sintered samples, both pure and doped HAp were phase pure and showed peak shift due to sintering at high temperature. Pure for HAp characterstics peaks were observed at 2θ angle of 26.30° (0 0 2), 32.21° (2 1 1), 33.37° (3 0 0), 32.61° (1

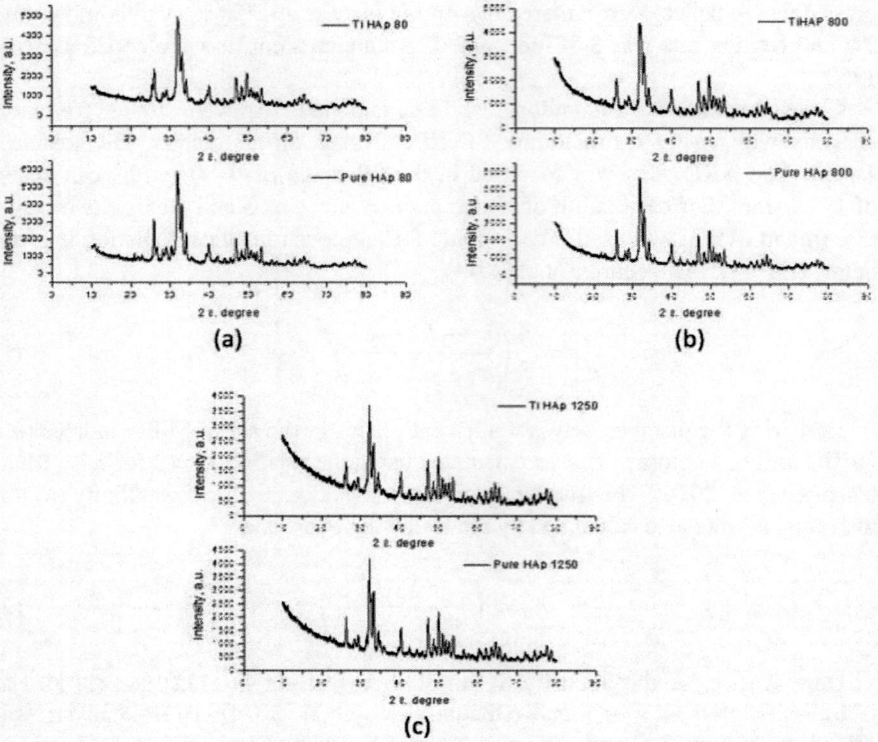

Fig. 1 XRD spectra of pure and doped HAp at various temperature (**a**) at 80 °C, (**b**) at 800 °C, and (**c**) at 1250 °C

1 2), 47.13° (2 2 2), 49.89° (2 1 3). The peaks observed for sintered doped HAp were 2θ angle of 26.23° (0 0 2), 32.19° (2 1 1), 32.55° (1 1 2), 33.32° (3 0 0), 47.10° (2 2 2), and 49.86° (2 1 3). Table 1 shows the peak shifting due to effect of temperature and also due to dopant.

Table 1 Characteristics peak shift due to temperature and doping

Samples	Characteristic peaks					
	0 0 2	2 1 1	1 1 2	3 0 0	2 2 2	2 1 3
Pure HAp 80 °C	×	×	×	×	×	×
Ti HAp 80 °C	√	√	×	√	×	×
Pure HAp 800 °C	×	×	×	×	×	×
Ti HAp 800 °C	×	×	×	√	×	×
Pure HAp 1250 °C	√	√	√	√	√	×
Ti HAp 1250 °C	√	√	√	√	√	×

Table 2 Lattice parameter of hydroxyapatite

Samples	*a* axis, Å	*c* axis, Å
Pure HAp 80 °C	9.650	6.8745
Ti HAp 80 °C	9.4193	6.835
Pure HAp 800 °C	9.4054	6.8563
Ti HAp 800 °C	9.3833	6.8485
Pure HAp 1250 °C	9.3039	6.7869
Ti HAp 1250 °C	9.2903	6.7691

The lattice parameter of hexagonal structure of hydroxyapatite crystal was calculated from XRD spectra as shown in Table 2. The a axis and c axis of doped samples were reduced compared to undoped samples. The size of titanium ion (ion radius 0.605 Å) is small compared to calcium ion (ion radius 0.99 Å). Due to doping, titanium replaces the calcium ions and it caused shrinkage in the crystal structure and therefore the lattice parameters of the crystal were decreased. Similar result was observed in literature in case of zinc doping (Begam et al. 2017). The lattice parameter also went on decrease due to an increase in temperature. As temperature increase, crystal size also decreases and hence also lattice parameters. Both doped and undoped HAp exhibited similar behavior on temperature. Similar observation was reported in literature (Gomes et al. 2012). Pure and zinc doped HAp showed reduced crystal size with temperature. The unit cell volume is presented in Fig. 2a. As the lattice parameter changed with temperature and doping, the unit cell volume was also changed. With increase in temperature there was loss of lattice H_2O from apatite lattice (Miyaji et al. 2005) and hence shrinkage of a and c axis was detected.

The percentage of crystallinity was calculated from the XRD data by taking valley between peak (1 1 2) and peak (300) and the intensity of peak (3 0 0). The crystallinity

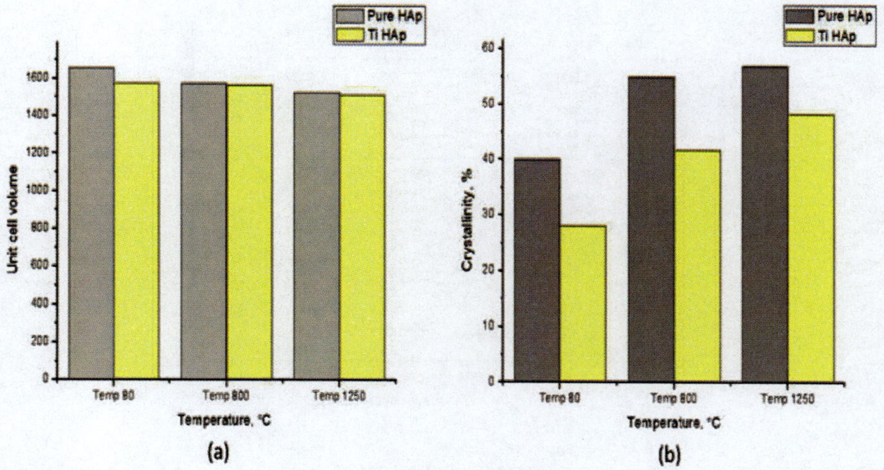

Fig. 2 **a** Unit cell volume and **b** fraction of crystallinity of pure and doped Hap

(X_c) is the amount of crystalline phase within a material. The increase in intensity of the peaks with temperature also attributed the increase in percentage of crystallinity. It was observed that with temperature the percentage of crystallinity was increased as evidenced by the sharp diffraction peaks. There is vital role of dopant on crystallinity of HAp. As shown in Fig. 2b, the doped samples exhibited in decreased crystallinity compared to undoped samples at all temperature. As we seen, dopant causes shortening of lattice parameters and it causes disturbance in HAp crystallization (Begam et al. 2017). Stanić et al. (2010) showed that copper and zinc dopant cause lessening of lattice parameters and also crystallinity.

Figure 3 shows the FTIR spectra of all samples. The graph shows peaks for different functional groups present in the material. There was no peak shifting in doped HAp as depicted in FTIR spectra. The wide peak at 3571 cm^{-1} corresponds to the OH– group. Peaks for phosphate groups were found at 576, 567, 1008, and

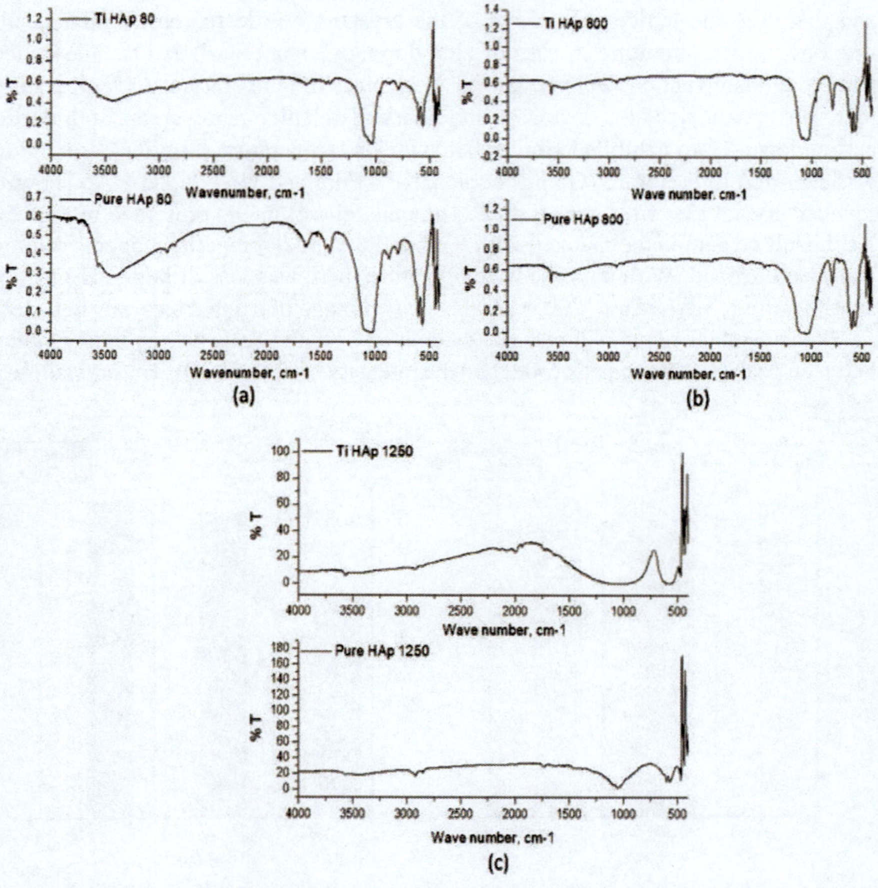

Fig. 3 FTIR spectra of pure and doped HAp at different temperatures (**a**) at 80 °C, (**b**) at 800 °C, and (**c**) at 1250 °C

1055 cm^{-1}. The peaks 1646 and 1420 cm^{-1} correspond to the carbonate groups. It was observed that with temperature the peak for hydroxyl group for calcined powder was sharp and very small. When the samples were sintered at 1250 °C that peaks were absent due to the evaporation of adsorbed OH bonds. This result was agreed with the XRD result as we seen reduction of lattice parameter with temperature.

4 Conclusion

Bioactive calcium phosphate-based bioceramics are extensively used in bone tissue engineering due to their resemblance with the mineral phase of teeth and bones. They have outstanding biocompatibility, osteoconductivity, etc. In this work, a new type of bioceramics was prepared by doping titania in calcium phosphate. The prepared powder was heat treated at 800 °C and 1250 °C and the effect of temperature on crystal structure was studied. The crystallographic behavior of doped HAp was studied using XRD and FTIR. It was observed that the crystallinity was increased with temperature and decreased due to doping. Also, the lattice parameters were reduced due to doping as the titanium replaces by the calcium ions in the structure.

Acknowledgments The authors are sincerely acknowledge the help and facilities provided by the Department of Biomedical Engineering, Department of Biotechnology, Department of Physics, and Department of Metallurgy, NIT Raipur, India for conducting this research work.

References

Begam H, Kundu B, Chanda A, Nandi SK (2017) MG63 osteoblast cell response on Zn doped hydroxyapatite (HAp) with various surface features. Ceram Int 43(4):3752–3760

Bhattacharjee P, Begam H, Chanda A, Nandi SK (2014) Animal trial on zinc doped hydroxyapatite: a case study. J of Asian Ceram Soc 2(1):44–51

Bodhak S, Bose S, Bandyopadhyay A (2011) Bone cell–material interactions on metal-ion doped polarized hydroxyapatite. Mater Sci Eng, C 31(4):755–761

Geng Z, Wang R, Li Z, Cui Z, Zhu S, Liang Y, Liu Z (2016) Synthesis, characterization and biological evaluation of strontium/magnesium-co-substituted hydroxyapatite. J Biomater Appl 31(1):140–151

Gomes S, Nedelec JM, Renaudin G (2012) On the effect of temperature on the insertion of zinc into hydroxyapatite. Actabiomaterialia 8(3):1180–1189

Kalita SJ, Bhatt HA (2007) Nanocrystalline hydroxyapatite doped with magnesium and zinc: synthesis and characterization. Mater Sci Eng, C 27(4):837–848

Kalita SJ, Bhardwaj A, Bhatt HA (2007) Nanocrystalline calcium phosphate ceramics in biomedical engineering. Mater Sci Eng, C 27(3):441–449

Miyaji F, Kono Y, Suyama Y (2005) Formation and structure of zinc-substituted calcium hydroxyapatite. Mater Res Bull 40(2):209–220

Samudrala R, Penugurti V, Manavathi B (2017) Cytocompatibility studies of titania-doped calcium borosilicate bioactive glasses in-vitro. Mater Sci Eng, C 77:772–779

Stanić V, Dimitrijević S, Antić-Stanković J, Mitrić M, Jokić B, Plećaš IB, Raičević S (2010) Synthesis, characterization and antimicrobial activity of copper and zinc-doped hydroxyapatite nanopowders. Appl Surf Sci 256(20):6083–6089

Tsuruoka A, Isobe T, Matsushita S, Wakamura M, Nakajima A (2015) Comparison of photocatalytic activity and surface friction force variation on Ti-doped hydroxyapatite and anatase under UV illumination. J Photochem Photobiol, A 311:160–165

Preparation and Characterization of Cellulose Nano Crystal/PVA/Chitosan Composite Film for Wound Healing Application

Shubham Sen, Rashmi Agrawal, and Howa Begam

Abstract The aim of this study was to develop Cellulose Nano Crystal (CNC)-based sustainable biocomposite film. CNC was extracted from medical absorbent cotton using alkali and acid hydrolysis. The composite films were produced by reinforcing CNC in PVA/Chitosan polymer using solvent casting method. Films were characterized by XRD, contact angle, hemocompatibility, protein adsorption. CNC dispersed in PVA/Chitosan affects the surface properties and it enhanced the hydrophilicity of the composite film. The percentage of hemolysis of both composites was less than 5% which confirmed it as hemocompatible material. The percentage of protein adsorption of CNC-based composite was 39% higher compared to control. This result suggests that CNC is a good reinforcing material for biopolymer composite preparation.

Keywords Cellulose nanocrystal · Biocomposite · Protein adsorption · Hemocompatibility

1 Introduction

Transdermal drug delivery is a potential means to deliver active ingredients into the circulatory system through skin into blood because it is safe, effective, and comfort to patients (Shankar et al. 2018). Transdermal patches deliver drugs at a preset controlled rate to systemic circulation when applied to intact skin. It provides sustained and controlled delivery of molecules to body. Generally high dosage drug is administered into the patch and as the concentration of drug is higher in patch compared to blood, the drug will come into blood through diffusion process through skin for a long period that maintains constant drug concentration into blood (Ravichandiran

S. Sen · R. Agrawal (✉) · H. Begam
Department of Biomedical Engineering, National Institute of Technology Raipur, Raipur, India
e-mail: rashmi22273@gmail.com

H. Begam
e-mail: howabegam@gmail.com

A. A. Rizvanov et al. (eds.), *Advances in Biomedical Engineering and Technology*, Lecture Notes in Bioengineering,
https://doi.org/10.1007/978-981-15-6329-4_25

and Manivannan 2015). There are several biopolymers like gelatin, silk, starch, polyvinyl, alcohol, chitosan which are used to prepare transdermal wound healing patches (Khamrai et al. 2017).

Cellulose is one of the most profuse biopolymer, which is commonly used as reinforcing elements for fabrication of composite biopolymer (Noshirvani et al. 2018). Cellulose is a polymer having glucose monomer, linked by β (1, 5) glycosidic bonds. Cellulose nanocrystal (CNC) is a promising biomaterials exhibiting many advantages like biocompatibility, mechanical properties, high water absorption capacity, low cost, renewability, etc., (Bajpai et al. 2017) and therefore it is attractive to use in bionanocomposites. CNCs has been used to prepare different types of biocomposites using polypropylene, polyvinyl alcohol, polyurethane, etc. Noshirvani et al. (2018) prepared CNC loaded starch/PVA nanocomposite film using solvent casting method for biodegradable packaging application. Bajpai et al. (2017) developed CNC-based chitosan films along with curcumin and silver nano particles for wound dressing application. They observed good results in terms of wound reduction in rat model. Wang et al. (2018) prepared antioxidant films using chitosan and epigallocatechin-3-gallate with bacterial cellulose as reinforcement element. They observed bacterial cellulose improved tensile strength and reduced water solubility and it offered sustained release of drugs for long time. CNC and metallic nano particle-based composite nanomaterials showed good antibacterial properties against pathogenic bacteria (Perumal et al. 2018). Generally acid hydrolysis method is used for preparation of CNC (Singh et al. 2017). CNCs are available in different morphology depending on the source and method of extraction (George and Sabapathi 2015). Thambiraj and Shankaran (2017) extracted CNC from cotton with needle-shaped morphological structure. They did a preliminary experiments on fabrication of CNC-based film using cotton-based CNC. It is evident that there is no dearth of literature on extraction of CNC from cotton using hydrochloric acid, composite preparation, and its detailed characterization.

The aim of this study was to develop CNC-reinforced wound dressing film using chitosan and polyvinyl alcohol. For this, CNC was isolated from medical absorbent cotton using hydrochloric acid hydrolysis. The efficiency of CNC-loaded film was compared with control (film without CNC). The water absorption, swelling properties, contact angles and protein adsorption studies were performed for both film with and without CNC.

2 Materials and Method

2.1 Materials

The chemicals used for preparation of CNC and CNC composite films and for their characterization are given in Table 1

Table 1 List of chemicals used for CNC extraction and film preparation

Chemicals	Chemical formula	Purity	Company name
Hydrochloric acid	HCl	35–38%	Thomas Baker (Chemicals) Pvt. Ltd., India
Sodium hydroxide	$NaOH$	97%	SRL, India
Chitosan	$C_6H_{11}NO_4$	75%	Himedia Laboratories, India
Polyvinyl alcohol	$[CH_2CH(OH)]_n$	–	Loba Chemie, India
Bovine serum albumin (BSA)	–	98%	SRL, India
Ethanol	C_2H_5OH	99.9%	ChangshuHongsheng Fine Chemical Co.Ltd, China
Toluene	C_7H_8	99.5%	SRL, India
Glacial acetic acid	CH_3COOH	99.5%	Loba Chemie, India

2.2 *Preparation of Film*

Cellulose nanocrystals were extracted from medical absorbent cotton using alkali and acid hydrolysis as described in literature (Thambiraj and Shankaran 2017; Abu-Danso et al. 2017). Briefly, cotton was carefully cleaned using distilled water and ethanol to eliminate any type of unwanted residues. After that, it was treated with 2% NaOH and acid hydrolysis was performed using hydrochloric acid. After that, CNC was centrifuged and washed with distilled water. Finally, it was dried at 60 °C for 12 h and the dried powder was used for film preparation.

For the preparation of film solvent casting method was used. For this, 10 w/v% PVA solution was prepared using distilled water at 70 °C. 1 w/v% chitosan solution was prepared in distilled water containing acetic acid (1 v/v%) at 40 °C. After cooling, both solutions were mixed at the ratio of 70:30 in constant stirring followed by ultrasonication for one hour. For CNC loaded film, 3 wt% CNC powder according to dry weight of PVA and chitosan was added before ultrasonication. Both solutions were poured in glass Petri dish followed by drying at 40 °C for 12 h. PVA/Chitosan film was coded as film 1 and CNC/PVA/Chitosan composite was coded as film 2. The optical micrograph of as-prepared film is shown in Fig. 1. The color of both films was white and transparent.

2.3 *Characterization*

Thickness. Thickness of composite films were measured using a micrometer screw gauge to nearest 0.01 mm. Thickness was measured at three random places of composite films and it was observed that thickness of films were same in each point of measurement. There was no difference between film 1 and film 2 and the thickness of each film was 1 mm.

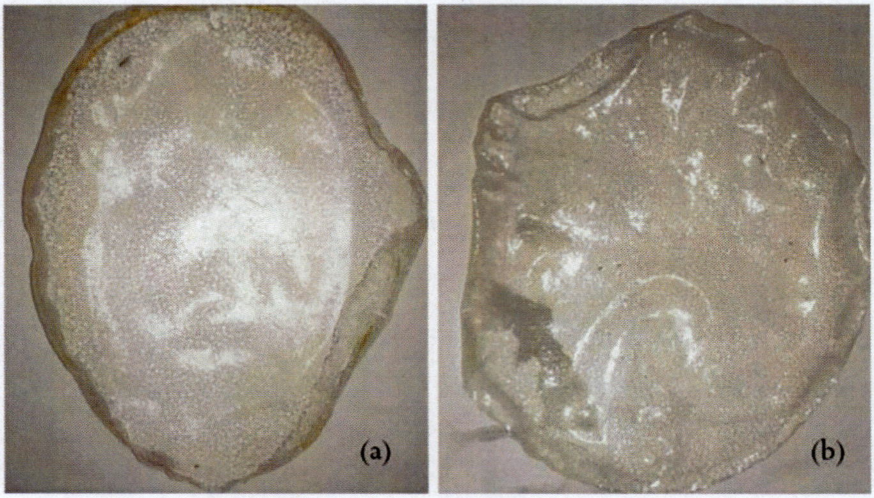

Fig. 1 Optical image of composite **a** film 1 and **b** film 2

X-Ray Diffraction (XRD). XRD of dried specimens was performed by X-ray diffractometry (XRD; Brukerdiffractometer, D8 advance, Japan) using Cu radiation with wavelength $\lambda = 0.154$ nm. The XRD scan was performed in range of 2θ angle of 0–80° and scan speed of 1° per min.

Porosity. Porosity of the CNC films was determined by means of the liquid displacement method. The films were dipped in ethanol (99.9%) for 5–10mins and the wet weight was taken. The percentage of porosity was determined as,

$$\text{porosity} = (W_W - W_D)/\rho \times V$$

where, W_D and W_w are dry weight and wet weight of films, $\rho = 0.789$ g/cm^3 (density of ethanol) and V is volume of films.

Contact Angle. Contact angle was determined by conventional sessile drop technique (phoenix 300 instrument). Distilled water was dropped on the surface of the film and photo of drop and the film was captured by CCD camera.

Protein Adsorption Studies. For protein adsorption studies, specimens of equal weight were immersed in ethanol (99.9%) for one hour and then washed thrice with PBS (pH 7.4) for 30 min. After that, the films were placed in glass test tube containing 2 ml of bovine serum albumin (BSA) (BSA conc. 1 mg/ml) for one hour. After one hour, the films were taken out and the solution was tested for unadsorbed proteins left in it. Proteins present in the solutions were estimated by folin-lowry method using BSA as standard. Absorbance (660 nm) was measured in UV-V is spectrophotometer (Shimazdu, Japan).

Hemolysis Study. In vitro hemocompatibility test was conducted by hemolysis assays described in literature (Samanta and Chanda (2018)). Briefly, human blood was diluted in normal saline in 4:5 ratio. 2–3 g of films were taken in glass test tubes containing 10 ml saline and incubated at 37 °C for 30 min. After that, 0.2 ml diluted blood was mixed in both test tubes and again it was incubated for additional 60 min. Positive control and negative control were prepared by adding 0.2 ml diluted blood in 0.1% sodium bicarbonate (10 ml) and normal saline (10 ml), respectively. Both positive and negative controls were incubated at 37 °C for 60 min.

After that, centrifugation (500 g) was carried for all test tubes for 5minutes, the supernatant was cautiously collected, and optical density was determined using spectrophotometer (Sl 177, ELICO, India) at 545 nm. The hemolysis (%) was determined using the following formula:

$$\%\text{hemolysis} = \frac{\text{O.D. of Test} - \text{O.D. of negative}}{\text{O.D. of positive} - \text{O.D. of negative}} \times 100 \quad (1)$$

2.4 Statistical Analysis

All data were presented as mean ± standard deviation (SD). All specimens were in triplicate ($n = 3$), unless stated otherwise. One way analysis of variance (ANOVA) with a Tukey's post hoc test was carried out by using ORIGIN software. Significant differences were considered as: ($p < 0.001$; $**p < 0.01$; $*p < 0.05$).

3 Results and Discussion

Thickness. The biocomposite films were prepared by using PVA/chitosan and PVA/Chitosan/CNC were uniform without any bubbles. The thickness of both samples measured as 1 mm. Thickness measurement is important because the physical properties of films depend on thickness of the film. There are no significant differences between two films.

Porosity. The porosity of film 1 was calculated as 25.34 ± 17% and it was observed that after addition of CNC the porosity was reduced (17 ± 7.3%). The reduction of porosity of film 2 results in increase in apparent density. The density is beneficial for the mechanical property of scaffold. An optimum porosity is necessary for cell attachments and the movement of nutrients and it ultimately increases the implant's bioactivity. For controlled release of drugs from scaffold, the porosity should be optimum without adversely affecting the structural or mechanical behavior of scaffold (Rambhia and Ma 2015). The release of drugs is generally occurs through diffusion process and after implantation scaffold exhibits strong burst release. Therefore, it is important to reduce initial burst release of drugs and maintain sustained release for longer period.

Contact angle. Contact angle test is generally performed to determine the surface hydrophilicity of any biomaterial. Contact angle depends on the hydrophilic group, which is exposed on the surface of the material. The data of water contact angle of film 1 was $65.23° \pm 0.54°$. When CNC was added in PVA-chitosan-based films, it showed reduction in contact angle ($54.53° \pm 1.41°$). The increase in hydrophilicity of CNC-based films may affect the cell adhesion, cell proliferation (Fig. 2).

XRD study. The XRD spectra of both films are presented in Fig. 3. PVA displayed a semi-crystalline structure for hydrogen bond between the OH^- groups of the PVA chains. Both films exhibited peak at 2θ angle of 20° for superimposed peak of chitosan and PVA (Amaral et al. 2013). Film 2 exhibited peak at 2θ angle of 22.74° for cellulose nanocrystal.

Hemocompatibility study. Toxic materials can destroy erythrocytes and release hemoglobin. Hemocompatibility is considered one of key concerns in tissue engineering, particularly for scaffolds that will come into direct contact with blood. The percentage of hemolysis is important feature in assessment of biocompatibility of materials (Liu et al. 2014). The percentage of hemolysis of film 1 was $2.8 \pm 0.49\%$ and for film 2 it was $3.66 \pm 0.19\%$. As both samples showed hemolysis less than 5%, therefore both films were highly hemocompatible according to the ASTM F756 standard (Samanta and Chanda 2018). The data suggested that CNC-based composite meets the requirement for medical biomaterials.

Protein adsorption study. Protein adsorption is significant in controlling cell reaction to biomaterials (Saravanan et al. 2017). Biomaterials adsorb proteins on their surface to anchor osteoblast cells via integrins (Shankar et al. 2018). Initial protein adsorption is important because it determines the success of any biomaterials (Shankar et al. 2018). When a material is exposed to cell, proteins spontaneously adsorb on the surface of the materials. The adsorption of protein is mainly dependent on the surface chemistry, surface wettability, surface morphology, etc. (Le et al. 2013). In our study, we studied protein adsorptionability for one hour. It was observed that protein adsorption was higher in film 2 compared to the control (Figure 4). Inclusion of CNC in PVA/Chitosan film enhanced protein adsorption by 38%. As we found in contact angle analysis, film 2 showed more hydrophilicity and it can be related to enhanced protein adsorption rate.

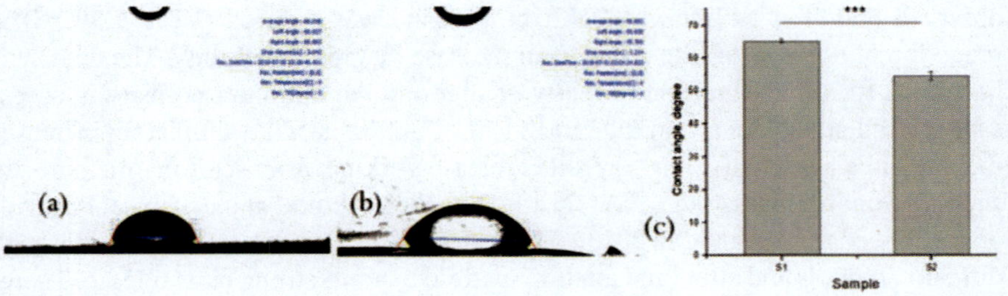

Fig. 2 Water contact angle of **a** film 1 and **b** film 2 **c** Bar diagram showing contact angle difference between two films

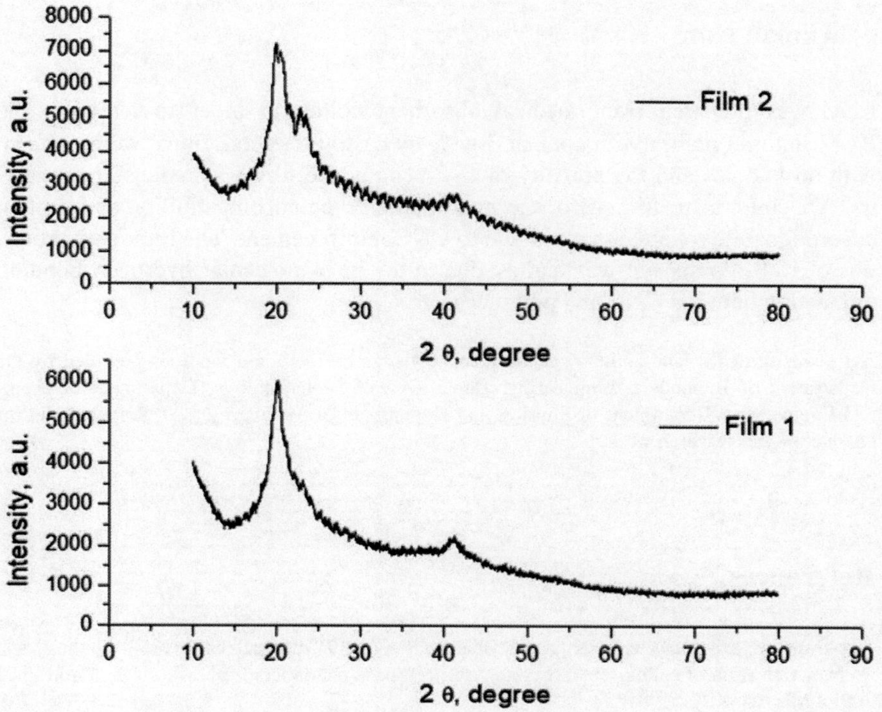

Fig. 3 XRD spectra of composite films

Fig. 4 Bar diagram showing BSA protein adsorption of composite film after 1 h.

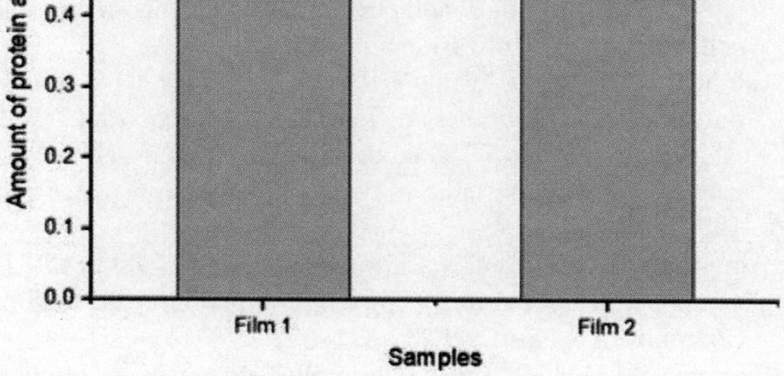

4 Conclusion

CNC was prepared from medical absorbent cotton in order to reinforce the PVA/Chitosan matrix with content 3 wt% by casting method. Films were uniform with no bubbles and the porosity of CNC composite film was reduced compared to PVA/chitosan matrix. Also, the hydrophilicity, hemocompatibility, and protein adsorptionability were enhanced due to CNC reinforcement. The improved proterties of CNC composite was mainly due to the inter-molecular hydrogen bonding interaction between CNC and polymer matrix.

Acknowledgments The authors sincerely acknowledge the help and facilities provided by the Department of Biomedical Engineering, Department of Biotechnology, Department of chemical Engineering, Department of Physics, and Department of Metallurgy, NIT Raipur, India for conducting this research work.

References

Abu-Danso E, Srivastava V, Sillanpää M, Bhatnagar A (2017) Pretreatment assisted synthesis and characterization of cellulose nanocrystals and cellulose nanofibers from absorbent cotton. Int J Biol Macromol 102:248–257

Amaral IF, Sousa SR, Neiva I, Marcos-Silva L, Kirkpatrick CJ, Barbosa MA, Pêgo AP (2013) Kinetics and isotherm of fibronectin adsorption to three-dimensional porous chitosan scaffolds explored by 125I-radiolabelling. Biomatter 3(2):e24791

Bajpai SK, Ahuja S, Chand N, Bajpai M (2017) Nano cellulose dispersed chitosan film with Ag NPs/Curcumin: An in vivo study on Albino Rats for wound dressing. Int J Biol Macromol 104:1012–1019

George J, Sabapathi SN (2015) Cellulose nanocrystals: synthesis, functional properties, and applications. Nanotechnol Sci Appl 8:45

Khamrai M, Banerjee SL, Kundu PP (2017) Modified bacterial cellulose based self-healable polyeloctrolyte film for wound dressing application. Carbohyd Polym 174:580–590

Le X, Poinern GEJ, Ali N, Berry CM, Fawcett D (2013) Engineering a biocompatible scaffold with either micrometre or nanometre scale surface topography for promoting protein adsorption and cellular response. Int J Biomater

Liu Y, Cai D, Yang J, Wang Y, Zhang X, Yin S (2014) In vitro hemocompatibility evaluation of poly (4-hydroxybutyrate) scaffold. Int J Clin Exp Med 7(5):1233

Noshirvani N, Hong W, Ghanbarzadeh B, Fasihi H, Montazami R (2018) Study of cellulose nanocrystal doped starch-polyvinyl alcohol bionanocomposite films. Int J Biol Macromol 107:205–2074

Perumal AB, Sellamuthu PS, Nambiar RB, Sadiku ER (2018) Development of polyvinyl alcohol/chitosan bio-nanocomposite films reinforced with cellulose nanocrystals isolated from rice straw. Appl Surf Sci 449:591–602

Rambhia KJ, Ma PX (2015) Controlled drug release for tissue engineering. J Control Release 219:119–128

Ravichandiran V, Manivannan S (2015) Wound healing potential of transdermal patches containing bioactive fraction from the bark of Ficusracemose. Int J Pharm PharmSci 7(6):326–332

Samanta SK, Chanda A (2018) Study on the structure and properties of crystalline pure and Doped β-tri calcium phosphate ceramics. Mater Today: Proc 5(1):2330–2338

Saravanan S, Chawla A, Vairamani M, Sastry TP, Subramanian KS, Selvamurugan N (2017) Scaffolds containing chitosan, gelatin and graphene oxide for bone tissue regeneration in vitro and in vivo. Int J Biol Macromol 104:1975–1985

Shankar S, Oun AA, Rhim JW (2018) Preparation of antimicrobial hybrid nano-materials using regenerated cellulose and metallic nanoparticles. Int J Biol Macromol 107:17–27

Singh S, Gaikwad KK, Park SI, Lee YS (2017) Microwave-assisted step reduced extraction of seaweed (Gelidiellaaceroso) cellulose nanocrystals. Int J Biol Macromol 99:506–510

Thambiraj S, Shankaran DR (2017) Preparation and physicochemical characterization of cellulose nanocrystals from industrial waste cotton. Appl Surf Sci 412:405–416

Wang X, Xie Y, Ge H, Chen L, Wang J, Zhang S, Feng X (2018) Physical properties and antioxidant capacity of chitosan/epigallocatechin-3-gallate films reinforced with nano-bacterial cellulose. Carbohyd Polym 179:207–220

Automated CAD System for Skin Lesion Diagnosis: A Review

Lokesh Singh, Rekh Ram Janghel, and Satya Prakash Sahu

Abstract Skin cancer is deemed as the lethal type of cancer threatening worldwide with an increase in mortality rate per year. The growing incidences of melanoma skin cancer have introduced numerous treatment options. However, surgical treatment remains the basis for treating skin cancers. Automated skin cancer detection still remains a challenging task in the current scenario. Prior diagnosis of skin cancer requires computer-aided diagnosis. Computerized methods for analyzing images in dermoscopy are a subject of interest as dominant information regarding skin lesions can be retrieved. This paper has discussed the state of art techniques employed in CAD systems by providing domain facets of melanoma accompanied by efficient methods utilized in every phase. The phases comprise image preprocessing techniques, extraction, and selection of significant features, segmentation methods, and classification approaches for the identification of skin lesions. Inapplicability and future trends are discussed in the domain of research.

Keywords Melanoma · CAD system · Segmentation

1 Introduction

Cancer is a severe health issue and deemed as a major cause of an increase in mortality rate worldwide. American Cancer Society conducts a survey every year to evaluate the cancer statistics of the current year and archive the novel cancer instances, death rate, and survival of the United States. Society has estimated the occurrence of 600,920 deaths among 1,688,780 cases due to cancer incidences in 2017. According

L. Singh (✉) · R. R. Janghel · S. P. Sahu
National Institute of Technology, G.E Road, Raipur 492001, India
e-mail: lokesingh@gmail.com

R. R. Janghel
e-mail: janghel1310@gmail.com

S. P. Sahu
e-mail: spsahu.it@nitrr.ac.in

A. A. Rizvanov et al. (eds.), *Advances in Biomedical Engineering and Technology*,
Lecture Notes in Bioengineering,
https://doi.org/10.1007/978-981-15-6329-4_26

Table 1 Estimated new cancer cases and deaths by sex, the United States, 2017 (Siegel et al. 2017)

Cancer cases	Estimated new cases			Estimated deaths		
	Both sexes	Male	Female	Both sexes	Male	Female
Skin (excluding basal and squamous)	95,360	57,140	38,220	13,590	9250	4340
Melanoma of the skin	87,110	52,170	34,940	9730	6380	3350
Other no epithelial skin	8250	4970	3280	3860	2870	990

to the survey, cancer incidence rate and mortality rate are found around 20 and 40%, respectively, higher in men comparatively in women (Siegel et al. 2017). Based on the gender, evaluated cancer statistics of the United States for 2017 is presented in Table 1.

Figure 1 graphically illustrates the estimated new cancer cases and death rate accordingly of three different categories of cancer of both sexes. It is clearly observed from Fig. 1 that the statistics of cancer cases and the death rate found to be higher in men than in women who are subjected to matter, as the statistics is fearsome and horrific.

Skin cancer is one of the lethal forms of cancer considered today. As per the cancer statistics of Europe in 2012, 22,199 cases of deaths are reported among 100,339 cancer cases (Okur and Turkan 2018). Due to the higher death rate and excessive level of metastasis, melanoma is turning as one of the fatal types of cancer in the skin. According to one of the surveys conducted in 2008, melanoma found to be the 19th most mortal cancer worldwide due to its higher incidence rate with a projection of 200,000 new cancer cases in the northern countries (Oliveira et al. 2016a).

Melanoma is a fatal disease that arises in pigment cells referred to as melanocytes (melanin-producing cells). Unlike other skin cancers, it simply dispersed above other

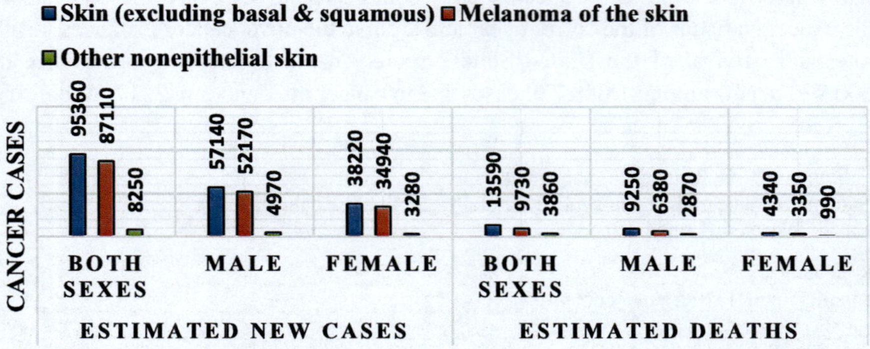

Fig. 1 Evaluation of cancer cases and deaths

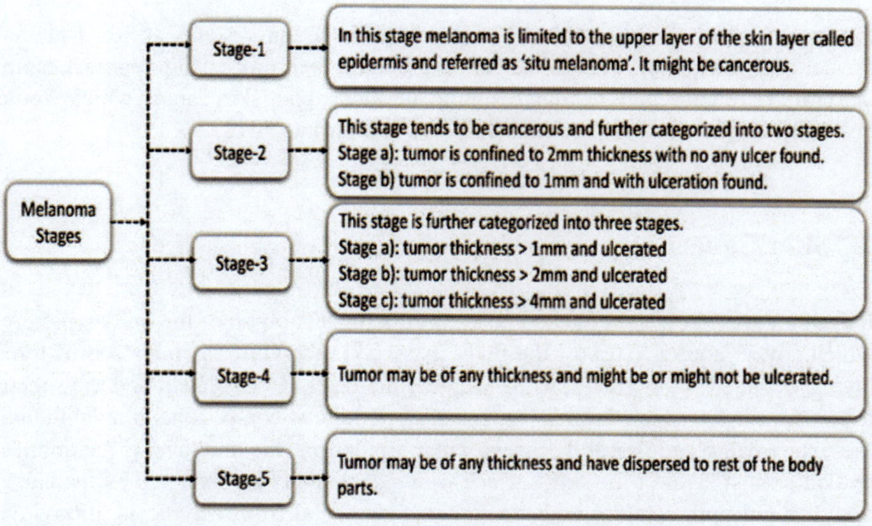

Fig. 2 Melanoma stages (Okur and Turkan 2018)

tissues expeditiously. Cancer might be metastasized by the tissues, lymph system, or by the circulation of the blood. It gradually expands to the closer areas when expanded by tissues, but when it pierces through the lymph system or blood vessels, it expanded over the rest of the body tissues. This usually becomes an ill-posed problem when tissue turns cancerous as it expands through melanoma. Malignant melanoma can be successfully cured if detected at an early stage by a simple visual analysis, as its occurrence is clearly visible on the surface of skin. But the main practical problem that confronts us is that one can define the particular stage of melanoma only when a lesion has undergone surgery. The stage of melanoma can be identified using four primary features—(a) thickness of a tumor, (b) ulceration, (c) dispersion to the lymph nodes, and (d) dispersion to the rest of the body parts. The stages of melanoma are discussed in Fig. 2.

Being aware of early detecting mechanisms of melanoma is the key to prevention from melanoma. Numerous techniques are available that have been developed for better estimation of skin lesions. A country like Turkey has found out the way to cope with this problem by running 19 clinics for investigation of melanoma. Several facilities are now available online for taking appointments in numerous health centers worldwide. 'ABCDE of melanoma' serves as a guideline for increasing the awareness regarding skin lesions, where 'A' stands for Asymmetry, 'B' for Border irregularity, 'C' for distinct Colors, 'D' for Diameter features of lesions, and 'E' represents Evolution which acts as an indicator of the rapid growth of lesions. An individual investigated suspected as per the above-mentioned guidelines accordingly must reach to the experts for clinical examination. This problem can be investigated visually by experts. This phenomenon has been widely observed by investigators with an average inspection accuracy of 65%, which is relatively a low accuracy that has emerged from the imperfection of the examiners. Automated detection systems are thus pivotal for better investigation of skin cancer. Digitized image processing approaches are therefore required which results in significant facts about skin cancer which would be of great importance for the experts (Okur and Turkan 2018).

2 Skin Cancer

It is of great interest to know about skin as the biggest organ of the body structure, which covers approx. 16% of the body mass. Skin structure is composed of three layers, depicted in Fig. 3 are, namely epidermis layer, dermis layer, and the subcutaneous fat layer which further composed of numerous components like epithelial, mesenchymal, glandular, and neurovascular. Epidermis, the first layer is the thinnest and outermost layer of the skin surface which can clearly be observed by the naked eye. It highly protects the body from being damaged from chemicals, ultraviolet rays, etc. This layer serves as a protective shield for the skin surface. Dermis, the second layer is responsible for sweat generation, hair growth, and manages the flow of blood to the skin. Subcutaneous fat layer, the third and bottom layer of the skin, is responsible for connecting the dermis to the body's muscles and bones and manages the body temperature (D'Orazio et al. 2013; Mark Elwood and Jopson 1997; Lowe 2006; Slominski and Wortsman 2013; Borja-Cacho and Matthews 2008; Slominski et al. 2012).

Cancer begins when the growth of cells becomes out of control in the body. Cells can turn cancerous in any part of the body and can spread to the rest of the body areas. Melanoma skin cancer begins in specific types of cell. Figure 3 depicts the three main types of cells in the top layer of the skin (called the epidermis) (Fig. 4).

Squamous cells: These cells are flat and thin by structure and observed in the tissue which constitutes the outer skin layer.

Basal cells: These cells are small in size and round in shape usually found in the lower section of the epidermis. These cells continuously separate themselves for the formation of new cells, so squamous cells can be replaced that constitute the skin's surface.

Melanocytes: These cells can turn into melanoma. They constitute a brown-colored pigment referred to as melanin which turns the skin brownish. This pigment prevents the deep layers of the skin from being harmed by sun exposure. More amount of pigment is generated by the melanocytes when skin comes in contact with the sun turning the skin darker (American Cancer Society 2017).

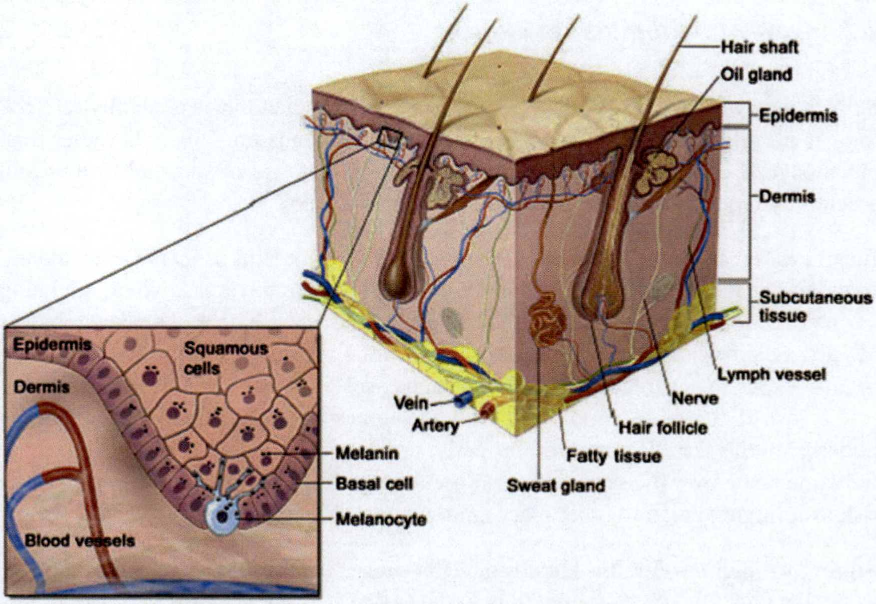

Fig. 3 Types of skin layer and skin cells (Korotkov and Garcia 2012)

2.1 *Melanoma Skin Cancer*

The unusual development of the melanocytes is responsible for the formation of malignant tumors usually referred to as melanoma (Pathan et al. 2018), which is a type of cancer that usually starts in melanocytes. Malignant melanoma is also called cutaneous melanoma (Melanoma et al. 2007). Melanoma tumors are brown or black in color when melanin produced by the melanoma cells, but tumors appear pink or white in color when melanoma cells stop producing melanin.

Melanoma can occur in any part of the skin surface but usually begins on the chest and backside in males and on legs in females. Neck and face are other common parts. Skin with dark pigments are at low risk of having melanoma at these common parts but can be risky at other parts like palm of hands, soles of the feet, and under the nails.

Melanoma occurs more frequently at skin than at the rest of the body parts like eyes, mouth, genitals, etc. Melanoma occurs less commonly compared to the BCC (Basal Cell Cancer) and SCC (Squamous Cell Cancer) but is highly hazardous as it spreads speedily to other body parts if left undetected early (American Cancer Society 2017). Melanoma can't be cured when reached to an advanced stage and possible treatment involves surgery, immunotherapy, chemotherapy, and/or radiation therapy (Society 2011; Kaufman 2005; Jerant et al. 2000).

2.2 Non-Melanoma Skin Cancer

Skin cancer, which doesn't fall under the category of melanoma is generally referred to as non-melanoma skin cancer as their growth depends on skin cells rather than melanocytes. Their nature is entirely different from those of the melanomas and essentially requires different treatment (Kaufman 2005).

Basal cell carcinoma. As per the experts, around eight skin cancers out of ten are found to be BCC. This cancer usually begins to grow on the areas, which are being harmed due to sun exposure in particular to the head and the neck. The development of BCC is very slow and very rarely spread to the rest of the body parts. It can be hazardous when not treated properly, in such cases, it spreads and starts developing in the rest of the body parts and harms the bones and other tissues by attacking underneath the skin. If not cured properly, it might occur again after some time on the same place over the skin surface. Individuals suffering from the BCC are at high risk as it might spread to other places in the body (American Cancer Society 2017).

Squamous cell carcinoma Heading. This cancer is less hazardous than the BCC as per the fact, where two individuals found to be suffering from the SCC out of ten. SCC generally occurs on such areas of the body which are open to sun like face, ears, neck, and lips (American Cancer Society 2017; Kaufman 2005; Jerant et al. 2000).

Keratoacanthomas. Tumors of dome-like shape originated due to sun exposure are termed as keratoacanthomas. They evolve very quickly but later grows slowly. Some of them are not deemed hazardous as they get might be cured automatically with time without any treatment. But, some of them sustain to increase and few of them might spread to the rest of the body parts. The growth of Keratoacanthomas is difficult to examine by the experts which results in improper prediction which is mistakenly considered as SCC by the experts (American Cancer Society 2017).

Merkel cell carcinoma. Merkel cells are neuroendocrine cells that are responsible for hormone making cells. These cells originate over the base of the top skin layer—epidermis. Markel cells are found to be closer to the nerve endings in the skin. They are responsible for sensations of being touched. This cancer begins with the uncontrollable growth of Markel cells. Markel cell carcinoma is also known as neuroendocrine carcinoma as Markel cells are a kind of neuroendocrine cells (American Cancer Society 2017; Jerant et al. 2000).

Kaposi sarcoma. Kaposi sarcoma is a type of cancer in which abnormal cells generate purple, red, or brown color tumors over the skin surface, termed as lesions. Apart from the skin surface, they can also evolve in the rest of the body parts like lymph nodes, lungs, etc. These skin lesions due to Kaposi sarcoma often appear over the legs or the face and cause the affected area to swell painfully. This cancer becomes lethal when lesions grow in lungs, liver, or digestive tract (American Cancer Society 2017).

Lymphoma of the Skin. Lymphoma is a type of cancer that starts growing in white blood cells lies in the immune system of the body. When it starts growing in the skin, it is termed as skin lymphoma or cutaneous lymphoma. Basically, it is of two types:

- Hodgkin lymphoma: it is also termed as Hodgkin's lymphoma or Hodgkin's disease
- Non-Hodgkin lymphoma: it is also termed as non-Hodgkin's lymphoma or NHL (Kaufman 2005; Jerant et al. 2000).

3 Image Acquisition

Cameras and digital videos are the sources to acquire clinical images for the acquisition of the images. Images obtained from mobile distances or acquired at distinct lightning conditions make the imaging inconsistent. The problem arises when the lesion is small in size due to the poor resolution of the image. The presence of artifacts like hairs, reflections, shadows, and skin lines are other hitches in image acquisition which obstruct in analyzing the skin lesions. Indeed, a non-invasive method referred to as ELM (Epiluminescence Microscopy) is employed for acquiring the images. In this method, the lesion is engrossed in oil, and images are obtained with the help of dermatoscopic devices comprises of a camera (Pereira et al. 2016).

A miscellaneous dataset is the key requirement in the development of a robust CAD system for the better detection of melanoma. Table 2 illustrates the dermatological datasets which are publically available.

It is of great interest to know about skin as the biggest organ of the body structure which covers approx. 16% of the body mass. Skin structure is composed of three layers,

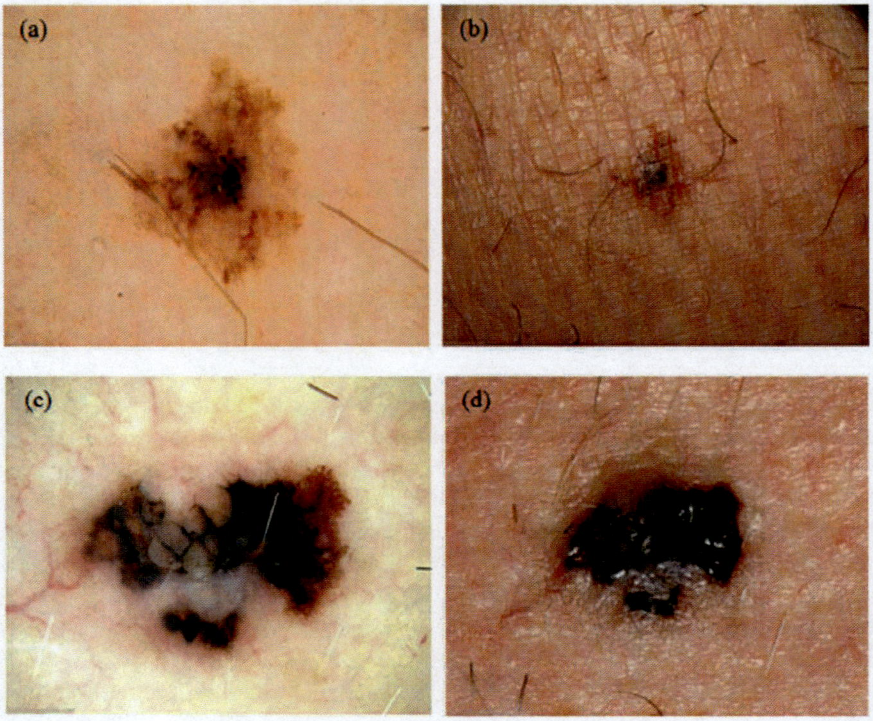

Fig. 4 Examples of dermoscopy (**a** and **c**) and macroscopic (**b** and **d**) images: **a** and **b** are images of melanoma in situ, and **c** and **d** are of invasive melanoma (these images are publicly available in (Oliveira et al. 2016a)

4 CAD System

4.1 Preprocessing

For the betterment of the quality of images, preprocessing is deemed as the initial stage of the diagnosing system where the unwanted noise such as bubbles, hair, etc., is removed as they might result in inaccurate classification. Various causes are responsible for the preprocessing of the source image like (a). low contrast amidst skin lesion and surrounding skin, (b). irrelevant borders, (c). Artifacts such as skin lines, hairs, black frames, etc., which in turn might affect the accuracy. A clear demonstration of the given input image before and after hair removal is illustrated in Fig. 5.

Image enhancement, image restoration, and hair removal are the key steps of image pre-processing and are clearly illustrated in Fig. 6 (Mehta and Shah 2016).

Table 2 Dataset details

Datasets	Number of images	Number of lesion images	Dermoscopic and lesion feature annotation	Reference
ISIC Archive (2016 ISBI challenge)	900	273—ME 627—Non-ME	Globules Streaks	Codella et al. (2018)
ISIC Archive (2016 ISBI challenge)	2000	374—ME 254—Seborrheic keratosis 1372—Benign Nevi	N/W Negative N/W Milla like cysts streaks	Berseth (2017)
PH2	200	80—Common Nevi 40— ME 80—Atypical nevi	Asymmetry Colors Pigment N/W Dots Globules Streaks Regression Areas Blue—White veil	Mendonca et al. (2013)
Dermatology Atlas	8084	80—ME	NO	Dermoscopy Atlas (https://www.dermoscopyatlas.com)
Dermnet Skin Disease Atlas	23,000	190—ME	NO	Atlas of Dermatology (https://www.dermnet.com/)
EDRA Interactive Atlas of Dermoscopy	NO	Invasive melanomas in situ melanoma, Spitz nevi, Clark nevi, Bowen disease (80— ME, 120—Benign)	Pigment N/W Irregular Dots Globules	Argenziano et al. (2000)
Dermofit Image Library	1300	76—ME	NO	Dermofit Image Library—Edinburgh Innovations (https://licensing.eri.ed.ac.uk/i/software/dermofit-image-library.html)
Dermquest	NO	308—ME	Type of ME	Search results (https://www.dermquest.com/results/?q=Malignantmelanoma)
Dermis	NO	300—ME	NO	DermIS.https://www.dermis.net/dermisroot/en/home/index.htm)

*ME = Melanomas, N/W = Network

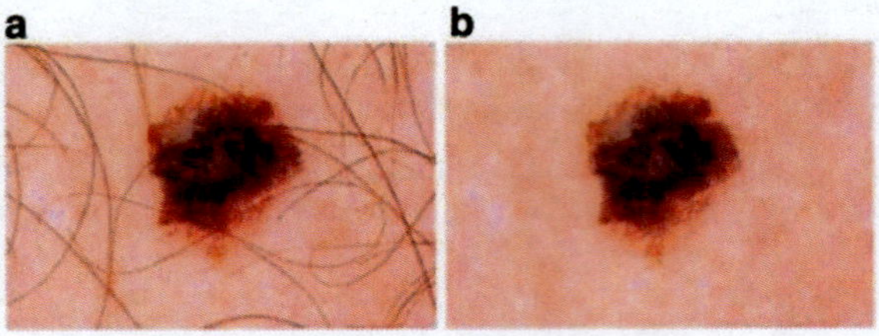

Fig. 5 **a** Original image **b** hair removal.

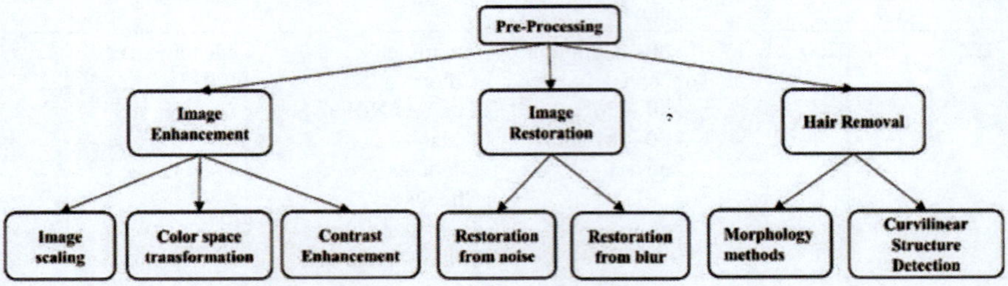

Fig. 6 Preprocessing techniques

4.2 *Lesion Segmentation*

Segmentation is the primary method for the separation of data into a significant region of interest (ROI). The quality of the segmented image is of maximal importance as it directly affects the performance of the CAD system. The lesion is segmented when the boundary of a lesion is delineated from the given input image with the help of automated image segmentation mechanisms (Rodríguez and Sossa 2017).

Figure 7 clearly illustrates the image before and after segmentation. IMAGEJ software can also be employed for the purpose of colored segmentation (Jaleel and Salim 2013) (Figs. 8, 9 and 10).

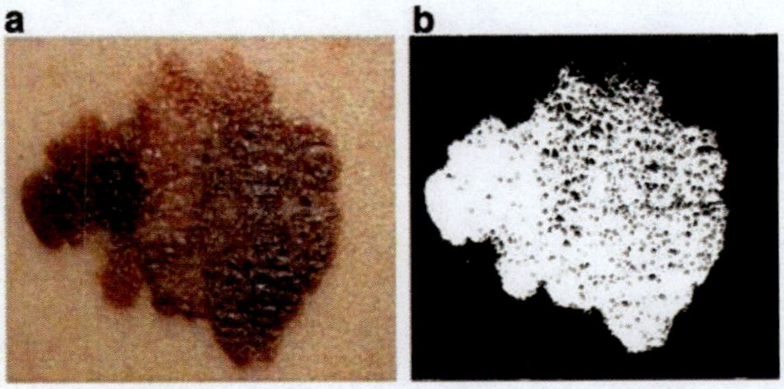

Fig. 7 **a** Original image **b** segmented image (Jaleel and Salim 2013)

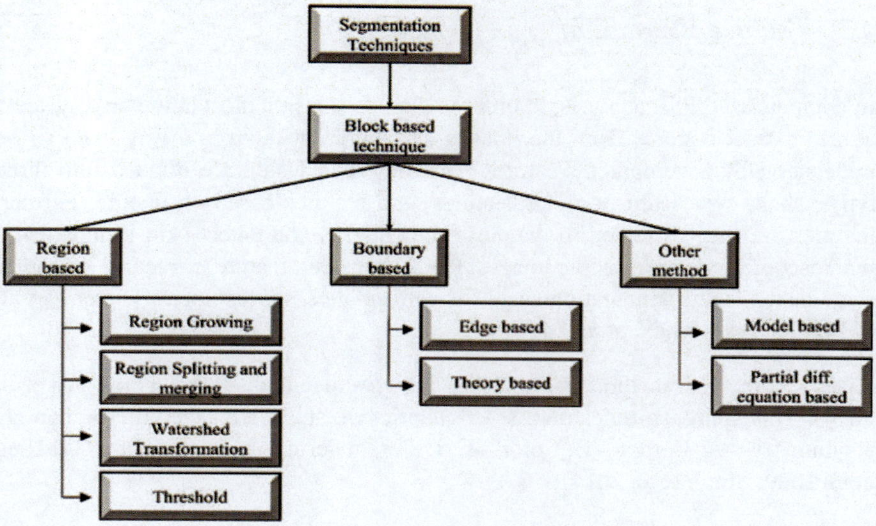

Fig. 8 Image segmentation techniques

Based on several image attributes like histogram, color, gradient, wavelet coefficients, etc., block-based methods separate the image into rectangular blocks. Block-based segmentation methods are further categorized into three types as per the fundamental traits of pixels (Jaglan et al. 2019).

Region-Based Detection Method. In region-based detection methods, regions are generally analyzed by the center of gravity, which is inconvertible corresponding to the rotation, scaling, and skewing and remains constant under random noise and gray-level variation. Segmentation approaches play a key role in diagnosing the region-based features. The accuracy obtained after segmentation notably affects the resulting registration. The image segmentation is performed, respectively, in an iterative manner with the registration wherein at each step segmentation parameters are tuned (Zitová and Flusser 2003).

Boundary-Based Detection Methods. The limitations left uncovered by the region-based approaches are overcome using boundary-based approaches for segmentation and classification. The segmentation of image is performed into regions through unseen changes occurred in the intensity of an image. Boundary based methods are categorized into two, namely ridge detection and edge detection. The ridge detection fulfills the aim by capturing the major axis of symmetry of an enlarged object, while the edge detection fulfills the aim by capturing the borders of the object (Jaglan et al. 2019; Pathan et al. 2018).

4.3 Feature Extraction

In computerized PSL analysis, in order to classify a lesion most automated systems aim to extract features from the images and represent them in a way that can be understood by a computer. Features regarding skin lesion are divided into three types: shape-type features, color features, and features based on texture. Further, features are then extracted to diagnose the patterns and detect skin lesions using macroscopic and dermoscopic images. In CAD system, feature extraction is another stage next to the segmentation stage for performance improvement (Oliveira et al. 2016a; Korotkov and Garcia 2012).

ABCD Rule. This method distinguishes benign and malignant melanoma (Ali et al. 2014). This approach relies on a semi-quantitative analysis of the criteria, namely Asymmetry—A, Border—B, Color—C, Different dermoscopic structures—D (Leo et al. 2009; Singh et al. 2014).

Menzies Method. Menzies's approach basically divides the feature into two categories. The first category is the positive feature group and the second category is the negative feature group. The positive and negative features are discussed in Table 3 (Okur and Turkan 2018).

Seven Point Checklist. This checklist comprises 7 distinct features which when combined act as an indicator of melanoma. Likewise, the ABCD rule, each feature is labeled with a value that is calculated for the analyzed lesion based on existing specific features (Okur and Turkan 2018).

Cash Algorithm. CASH algorithm stands for Color, Architecture, Symmetry, and Homogeneity/Heterogeneity. Feature architecture included by this algorithm is not employed by other scoring algorithms. The architecture indicates the structure of the lesion in terms of dermoscopic structure and colors (Pathan et al. 2018).

Table 3 Menzies's scoring

Negative features	Positivef
• Symmetry of lesion	• Blue-White veil
• Presence of a single color	• Multiple brown dots
	• Pseudopods
	• Radial Streaming
	• Scar-like depigmentation
	• Globules
	• Multiple 5–6 colors
	• Multiple blue-gray dots
	• Broadened network

Shape Features. The organization of an object defines the shape of a particular object which is represented using boundary, region, etc. The investigation of asymmetry features is performed by separating the lesion's region into two sub-regions by an axis of symmetry for analyzing the similarity of the area (Chang et al. 2005).

Color Features. The contrast of the skin lesions is represented by RGB color space. To acquire more information regarding color of the lesion some other color spaces have been utilized, namely RGB, HSV, HVC, CMY, YUV, I1/2/3, L*C*H, CIEXYZ, CIELAB, and CIELUV (Oliveira et al. 2016a).

Texture Features. Basically, texture features can be explained by characteristics like coarseness, contrast, and directionality. For the human visual system, the texture is identical (Lew 2013). Texture feature extraction comprises three main classes, namely Statistical, structural, and transform-based methodologies (Haralick et al. 1973).

4.4 Feature Analysis and Selection

Feature selection is the process of searching the most relevant and significant features among all features. An advantage of employing feature selection methods is the reduced computation time and improved prediction performance in the field of machine learning (Chandrashekar and Sahin 2014). A large number of features, when provided as an input to the classifier, might result in redundancy and increased computational complexity. Therefore, to improve the robustness of the classification model, feature selection is the next step after feature extraction which enhances the classification performance. A significant facet of feature selection is maximal relevancy with less redundancy (Singh et al. 2017).

Filter Approach. No learning methodology is opted by filter-based methods for the selection of the best subset of features. Dominant features are evaluated based on 'scoring' criteria and highly scored features are then chosen for further classification process which makes filter-based methods faster than the wrapper based methods. Following are the filter-based methods described:

(A) **$\chi 2$ Test**

The chi-square test, also represented as $\chi 2$ test or known as Pearson's chi-square test, demonstrates an apparent difference among expected frequencies and observed frequencies. The following equations represents statistical independence between two events x and y.

$$P(xy) = P(x)P(y) \text{ or} \tag{1}$$

$$P(x/y) = P(x) \text{ and } P(y/x) = P(y) \tag{2}$$

(B) **Euclidian Distance**

Euclidean distance can be employed to measure the distance among pair of instances p and q in an N-dimensional feature space. Usually, it is evaluated as the square root of the addition of the squared difference among the related coordinates.

$$d(a, b) = \{\Sigma i(ai - bi)2\}^{1}/_{2} \tag{3}$$

(C) **Information Gain**

Information gain evaluates the amount of information a feature provides regarding the class. It describes the relevancy of an attribute. It indicates the percentage of separation of data by the attribute according to the classification. The entropy can be calculated as follows:

$$\text{Entropy}(s) = \sum_{i=1}^{n} -P_i \log 2 P_i \tag{4}$$

where $n =$ number of classes, $P_i =$ probability of s belongs to class i, (Vanaja 2014).

Wrapper Method. Wrapper-based approaches employ learning methods for the performance evaluation of classification algorithms. Wrapper approaches are better than filter methods as they provide higher classification accuracy but at a cost of higher computational complexity.

- Forward selection: The process begins with a null set and at a time one feature is added, the rest of the features are then appended to the existing subset for the evaluation of the novel subset.
- Backward elimination: The process begins with a featured set comprises of all features and features are removed one at a time.
- Forward stepwise Selection: The process begins with a null set and features are appended and removed one at a time (Cadenas et al. 2013).

Embedded Method. Embedded approaches incorporate variable elimination during the training process which is specific to the employed learning methodologies. This approach interacts directly with the classification methods. It is faster than the wrapper methods. Following are the few embedded feature selection methods:

- Decision Trees.
- Weighted Naïve Bayes.
- Variable elimination using the weighted vector of SVM.

Hybrid Approach. Hybrid method is an amalgamation of filter and wrapper approaches. As a filter-based method candidate, feature subset is chosen from the original feature subset which is then refined by the wrapper based methods. It utilizes the benefits of both of the two methods (Vanaja 2014).

4.5 Classification

Classification comprises the last stage of CAD system to decide the Pigmented Skin Lesion class. A brief description of several classification approaches employed in diagnosing melanoma is explained in this section (Giotis et al. 2015).

SVM. The support vector machine follows the principal of structural risk minimization with an aim to search the classification method which curtails the boundary of the expected error (Gehler and Schölkopf 2009). It searches for the optimal separating hyperplane which gives the maximum margin and seeks for the closest data point of the training dataset amidst 2 classes (Codella 2015).

K-NN. K-NN referred to as K-nearest neighbor is one of the simplest machine learning methods. It works on the principle of feature similarity. A very minimum training phase is required which makes the training process faster. It stores the complete available data points and classifies novel data points using the distance function like Euclidean, Manhattan, and Minkowski functions. Data samples are classified using the majority vote of their neighbors (Rourrur and Rvise 1982; Erickson et al. 2017).

Discriminant Analysis. Discriminant analysis is used for classification predictive problems and is defined by Fisher (1936), where two or more samples or populations are defined a priori, and one or more than one novel sample is then classified into one of the defined samples. It defines the relationship between dependent variables and independent variables (Analysis 2018).

ANN. ANN stands for Artificial Neural Network, also referred to as neural network. It models the complex relationships between input and output. Pattern recognition is the basic task performed by ANN in the field of medical imaging. It is a computational model which relies on the framework of a biological neural network. Information passes from the network directly affects the organization of the ANN as neural network learns on the basis of input and output (Ercal et al. 1994).

Decision Tree. A decision tree is a supervised machine learning algorithm employed in classification problems. A decision tree is a group of questions arranged in a hierarchically structured form and is depicted graphically in tree form. It follows a branching method to demonstrate each possible output of a decision. Decision trees are designed by evaluating a group of training data points for known class labels which are then used for classifying prior unseen samples (Yu et al. 2016).

Naïve Bayes. Naïve Bayes is one of the most efficacious classification algorithms of machine learning. Naïve Bayes classifiers are a collection of probabilistic classifiers that rely on Bayes theorem with an assumption among features. It is not an individual algorithm but a collection of several methods works on the same principle i.e. each classified feature does not depend on each other. It is easy to construct with a simple evaluation and thus useful for big datasets (Lee et al. 2011).

Fig. 9 Linear classifier in two-dimensional space

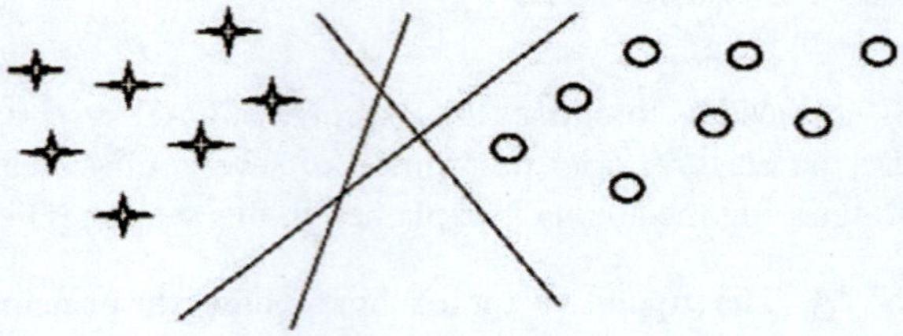

Fig. 10 Illustration of a 3-NN rule in a three-dimensional pattern space

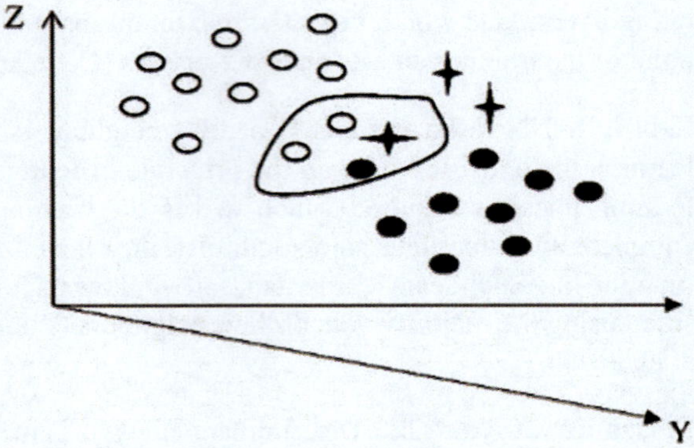

Random Forest. Random forest is a supervised machine learning approach used for both classification and regression tasks. It constructs forests and makes them random. The forest it constructs is a decision trees ensemble usually trained with bagging approach. Rather than seeking the significant or dominant features during the splitting of a node, it seeks the most significant feature from a random subset of features (Chan and Paelinckx 2008).

5 Performance of Evaluation Measures

Evaluation methods are the key elements in assessing the performance of classification models. The designed classification model might give satisfactory results when evaluated using any one of the metrics but provides unsatisfactory results when evaluated using other metrics. Thus, assessment of the classification model using a single evaluation measure is not sufficient. Table 4 describes several assessment methods for evaluating the effectiveness of the designed model (Pathan et al. 2018; Tharwat 2018).

Table 4 Summarized analysis of classification approaches employed in the computerized analysis

References	Yr	Type of Image	Segmentation	Feature Selection	Classification Models	Classification	Evaluated Results
Yu et al. 2016)	2017	1250 dermoscopic	Fully Convolutional Residual Network	Low level feature algorithm	Deep Residual Network	M/B	Accuracy: 85.6% Sensitivity: 54.8%, Specificity: 93.2%
Dalila et al. 2017)	2017	172 dermoscopic	Ant Colony Based Segmentation	Relief Algorithm	KNN, ANN	M/B	Accuracy: 85.23% for KNN, Accuracy: 93.63% for ANN
Kasmi and Mokrani 2016)	2016	200 dermoscopic	Geodesic active contour	(NO)	TDS	M/B	Accuracy: 94.1, Sensitivity: 91.3%, Specificity: 95.9%
Oliveira et al. 2016b)	2016	408 dermoscopic	Chan-Vese	(NO)	Support Vector Machine	M/B	Accuracy: 74.37%
Abbas et al. 2016)	2016	350 dermoscopic	Circular center of each PSL	(NO)	MV-SVM	M/B	Accuracy: 93.1%, Sensitivity: 94.1%, Specificity: 84.1%
Amelard et al. 2015)	2015	206 dermoscopic	(NO)	(NO)	LIBSVM	M/B	Accuracy: 83.58%, 81.37%, 81.18%
Abuzaghleh et al. 2015)	2015	200 dermoscopic	Otsu thresholding + Active Contour using Sparse- Field level - set method	(NO)	LIBSVM	M/B/Atypical	Benign = 96.4%, Atypical = 95.8%, Melanoma = 97.6%

(continued)

Table 4 (continued)

References	Yr	Type of Image	Segmentation	Feature Selection	Classification Models	Classification	Evaluated Results
Codella 2015)	2015	200 dermoscopic	(NO)	(NO)	Support Vector Machine	M/B	Accuracy: 91.1%, Sensitivity: 97.1%, Specificity: 65.1%, Precision: 92.1%., F-measure: 94.2%, Auc: 95.2%
Giotis et al. 2015)	2015	170 dermoscopic	K-MEANS clustering	(NO)	Ensemble method (CLAM, CIA-LVQ,Naïve Bayes	M/N	Accuracy: 81.0%, Precision: 0.741%, Negative Predicted Value: 85.8%
Shimizu et al. 2015)	2015	968 dermoscopic	Threshold	Incremental Stepwise (828/25)	Linear Classifier	M/N	Melanoma: Detection Rate: 90.47%; Nevus: Detection Rate: 82.52%, bcc: 82.60%-Detection Rate sk: 80.62%-Detection Rate
Rastgoo et al. 2015)	2015	180 dermoscopic	Threshold	PCA (NO/NO)	Random Forest	Melanoma/ dysplastic nevus	Sensitivity: 98.0%, Specificity: 70.1%

(continued)

Table 4 (continued)

References	Yr	Type of Image	Segmentation	Feature Selection	Classification Models	Classification	Evaluated Results
Barata et al. 2015)	2015	Database1: 200 Database2: 482 dermoscopic	(NO)	Fusion Strategy (NO/NP)	Random Forest	M/N	Sensitivity: DB1: 98.0%; Specificity: 90.1%; Sensitivity: DB2: 83.3%, Specificity: 76.2%
Møllersen et al. (2017)	2015	210 dermoscopic	(NO)	Wrapper & Filter (59/19)	DA	Classification	Correct Rate: 81.2%, Sensitivity: 83.0%, Specificity: 80.4%
Amelard et al. 2015)	2015	206 macroscopic	(NO)	– (62/–)	Support Vector Machine	M/B	Accuracy: 83.58%, Sensitivity: 91.02%, Specificity: 73.46%
Zortea et al. 2014)	2014	206 dermoscopic	(NO)	Sequential Forward Select (53/7.6)	DA	M/B	Sensitivity: 86.1%, Specificity: 52.3%, Correct Rate: 63.2%
Schaefer et al. 2014)	2014	564 dermoscopic	Threshold region = growing & merging	Fast Correlation Based Filter (437/74)	Support Vector Machine	M/B	Accuracy: 93.84%, Sensitivity: 93.75%, Specificity: 93.83%
Abbas et al. 2013a)	2013	120 dermoscopic	Threshold + Improved Dynamic Programming	Sequential Floating Forward Selection	Support Vector Machine	M/N	Sensitivity: 88.3%, Specificity: 91.4%

(continued)

Table 4 (continued)

References	Yr	Type of Image	Segmentation	Feature Selection	Classification Models	Classification	Evaluated Results
Abbas et al. 2013b)	2013	350 dermoscopic	450 * 450 ROI is selected	PCA	ML-SVM, ML-KNN, AdaBoost MC	Reticular, Globular, Cobblestone, Homogeneous, Parallel ridge, Starburst, Multicomponent	Sensitivity: 89.3%, Specificity: 93.76%
Ma and Staunton 2013)	2013	134 macroscopic	(NO)	Correlation Analysis (25/13)	ANN	M/B	Sensitivity: 83.0%, Specificity: 90.1%, Auc: 89.2%
Cavalcanti et al. 2013)	2013	152 macroscopic	Threshold	(stage 1: 52; stage 2: 12)	stage one: KNN; stage two: maximum likelihood	M/B	Accuracy: 99.35%, Sensitivity:100%, Specificity:97.79%
Barata et al. 2014)	2013	176 dermoscopic	Threshold	Individual and combined feature analysis	AdaBoost	M/N	Sensitivity:96.2%, Specificity:80.1%
Abbas et al. 2013a)	2013	120 dermoscopic	Dynamic programming	SFFS (NM/NM)	Support Vector Machine	M/N	Melanoma: Sensitivity:88.2%, Specificity:91.20%, Auc: 88.1%; Nevus: Sensitivity:86.6%, Specificity:88.3%, Auc:82.5%
Garnavi et al. 2012)	2012	289 dermoscopic	Global threshold + adaptive histogram threshold	Gain Ratio method	SVM + RF + LMT + HNB classifiers	M/B	Accuracy: 91.27%, Auc: 93.8%

(continued)

Table 4 (continued)

References	Yr	Type of Image	Segmentation	Feature Selection	Classification Models	Classification	Evaluated Results
Garnavi et al. 2012)	2012	289 dermoscopic	Threshold	GRFS (35,455/23	Random Forest	M/B	Accuracy: 91.27%, Auc: 93.8%

where M/N = Melanoma/Nevus, DA = Discriminant analysis, acc = Accuracy, se = Sensitivity, sp = Specificity, ppv = Positive Predicted Value (Precision), npv = Negative Predicted Value, fm = F = Measure, cr = Correct Rate, dr = Detection Rate, Auc = Area Under the ROC Curve, bcc = Basal Cell Carcinoma, sk = Seborrheic Keratosis, M/B = Melanoma/Benign

Table 5 Classifier evaluation parameters (Tharwat 2018)

Measures	Formula	Description
ACC	$\frac{TN+TP}{FN+FP+TN+TP}$	Proportion of correctly classified instances out of total instances
SE (TPR)	$TPR + TNR - 1$	Ratio of correctly classified positive instances
SP (TNR)	$\frac{TN}{FP+TN}$	Ratio of correctly classified negative instances
PPV	$\frac{TP}{TP+FP}$	Ratio of correctly classified positive samples out of total positive predictive instances
NPV	$\frac{TN}{TN+FN}$	Ratio of correctly classified negative instances out of total negative predictive instances
FPR	1 - TNR	Complement of specificity
FNR	1-TPR	Complement of sensitivity
YI	$TPR + TNR - 1$	It combines the TPR and TNR into a single measure
FM	$\frac{2TP}{2TP+FP+FN}$	Harmonic mean of PPV and TPR
GM	$\sqrt{TPR * TNR}$	Evaluation of sensitivity and specificity

where ACC = Accuracy, SE = Sensitivity, SP = Specificity, PPV = Positive Predicted Value (Precision), NPV = Negative Predicted Value, FPR = False Positive Rate, FNR = False Negative Rate, YI = Youden's Index, FM = F-Measure, and GM = Geometric-mean

6 Conclusion and Future Trends

This study represents an overview of the research conducted in the computerized analysis of dermatological images. The aim of this paper is to describe all the stages of a CAD system for the detection of skin cancer using several classification approaches. This review gives key insights into the recent developments in computerized analysis of Pigmented Skin Lesions by employing dermoscopic images. A thorough evaluation of cancer statistics constitutes of the domain facets is discussed. Recent evolving computational approaches for segmentation and feature extraction are elaborated in brief. Additionally, classifiers for the identification of skin lesions are discussed along with performance evaluation measures. This review concludes various methods focusing on the classification of skin lesions employed in automated computer-Aided Diagnosis system. This analysis proves to be helpful for dermatologists in diagnosing skin cancers.

Though numerous researches have been conducted addressing detection and prevention methods regarding skin cancer still, novel methods might need to be addressed and designed to fill the research gaps for performance improvement of the CAD systems (Table 5).

References

Abbas Q, Celebi ME, Serrano C, et al (2013b) Pattern classification of dermoscopy images: a perceptually uniform model. Pattern Recognit 46:86–97. https://doi.org/10.1016/j.patcog.2012.07.027

Abbas Q, Emre Celebi M, Garcia IF, Ahmad W (2013a) Melanoma recognition framework based on expert definition of ABCD for dermoscopic images. Ski Res Technol 19:1–10. https://doi.org/10.1111/j.1600-0846.2012.00614.x

Abbas Q, Sadaf M, Akram A (2016) Prediction of dermoscopy patterns for recognition of both melanocytic and non-melanocytic skin lesions. https://doi.org/10.3390/computers5030013

Abuzaghleh O, Member S, Barkana BD (2015) Noninvasive real-time automated skin lesion analysis system for melanoma early detection and prevention. 3

Ali A, Couceiro MS, Hassenian AE (2014) Melanoma detection using fuzzy c-means clustering coupled with mathematical morphology. 73–78

Amelard R, Glaister J, Wong A, Clausi DA (2015) High-Level Intuitive Features (HLIFs) for intuitive skin lesion description. IEEE Trans Biomed Eng 62:820–831. https://doi.org/10.1109/TBME.2014.2365518

Argenziano G, Soyer HP, De Giorgi V, et al (2000) Interactive atlas of dermoscopy (Book and CD-ROM)

Dermoscopy Atlas. https://www.dermoscopyatlas.com/. Accessed 22 Sep 2018

Atlas of Dermatology. https://www.dermnet.com/. Accessed 22 Sep 2018

Barata C, Ruela M, Francisco M et al (2014) Two systems for the detection of melanomas in dermoscopy images using texture and color features. IEEE Syst J 8:965–979. https://doi.org/10.1109/JSYST.2013.2271540

Barata C, Celebi ME, Marques JS (2015) Melanoma detection algorithm based on feature fusion. 2653–2656

Barzegari M, Ghaninezhad H, Mansoori P et al (2005) Computer-aided dermoscopy for diagnosis of melanoma. BMC Dermatol 5:1–4. https://doi.org/10.1186/1471-5945-5-8

Berseth M (2017) ISIC 2017: skin lesion analysis towards melanoma detection. 1–4

Borja-Cacho D, Matthews J (2008) Scratching the surface of skin development. NANO 6:2166–2171. https://doi.org/10.1021/nl061786n.Core-Shell

Cadenas JM, Garrido MC, Martínez R (2013) Feature subset selection filter–wrapper based on low quality data. 40:6241–6252. https://doi.org/10.1016/j.eswa.2013.05.051

Cavalcanti PG, Scharcanski J, Baranoski GVG (2013) Expert systems with applications a two-stage approach for discriminating melanocytic skin lesions using standard cameras. Expert Syst Appl 40:4054–4064. https://doi.org/10.1016/j.eswa.2013.01.002

Chandrashekar G, Sahin F (2014) A survey on feature selection methods. Comput Electr Eng 40:16–28. https://doi.org/10.1016/j.compeleceng.2013.11.024

Chang Y, Stanley RJ, Moss RH, Van Stoecker W (2005) A systematic heuristic approach for feature selection for melanoma discrimination using clinical images. Ski Res Technol 11:165–178. https://doi.org/10.1111/j.1600-0846.2005.00116.x

Chan JCW, Paelinckx D (2008) Evaluation of random forest and adaboost tree-based ensemble classification and spectral band selection for ecotope mapping using airborne hyperspectral imagery. Remote Sens Environ 112:2999–3011. https://doi.org/10.1016/j.rse.2008.02.011

Codella N (2015) Accurate and scalable system for automatic detection of malignant melanoma. https://doi.org/10.1201/b19107-11

Codella NCF, Gutman D, Celebi ME, et al (2018) Skin lesion analysis toward melanoma detection: a challenge at the international symposium on biomedical imaging (ISBI) 2016, hosted by the international skin imaging collaboration (ISIC). In: Proceedings of the international symposium on biomedical imaging 2018, pp 168–172. https://doi.org/10.1109/ISBI.2018.8363547

Criminisi A (2011) Decision forests: a unified framework for classification, regression, density estimation, manifold learning and semi-supervised learning. Found Trends® Comput Graph Vis 7:81–227. https://doi.org/10.1561/0600000035

D'Orazio J, Jarrett S, Amaro-Ortiz A, Scott T (2013) UV radiation and the skin. Int J Mol Sci 14:12222–12248. https://doi.org/10.3390/ijms140612222

Dalila F, Zohra A, Reda K, Hocine C (2017) Segmentation and classification of melanoma and benign skin lesions. Opt—Int J Light Electron Opt 140:749–761. https://doi.org/10.1016/j.ijleo.2017.04.084

DermIS. https://www.dermis.net/dermisroot/en/home/index.htm. Accessed 22 Sep 2018

Di Leo G, Fabbrocini G, Paolillo A, et al (2009) Towards an automatic diagnosis system for skin lesions: estimation of blue-whitish veil and regression structures. In: 2009 6th international multi-conference on systems signals devices, SSD 2009, vol 1, pp 1–6. https://doi.org/10.1109/SSD.2009.4956770

Ercal F, Chawla A, Stoecker WV, et al (1994) Neural network diagnosis of malignant melanoma from color images. IEEE Trans Biomed Eng 41:837–845. https://doi.org/10.1109/10.312091

Erickson BJ, Korfiatis P, Akkus Z, Kline TL (2017) Machine learning for medical imaging. RadioGraphics 37:505–515. https://doi.org/10.1148/rg.2017160130

Garnavi R, Aldeen M, Bailey J (2012) Computer-aided diagnosis of melanoma using border- and wavelet-texture analysis. IEEE Trans Inf Technol Biomed 16:1239–1252. https://doi.org/10.1109/TITB.2012.2212282

Gehler PV, Schölkopf B (2009) An introduction to kernel learning algorithms. Max-Planck-Gesellschaft, Biol Kybern 25–48. https://doi.org/10.1109/72.914517

Giotis I, Molders N, Land S, et al (2015) MED-NODE: a computer-assisted melanoma diagnosis system using non-dermoscopic images. Expert Syst Appl 42:6578–6585. https://doi.org/10.1016/j.eswa.2015.04.034

Haralick RM, Shanmugam K, Dinstein I (1973) Textural features for image classification. IEEE Trans Syst Man Cybern SMC 3:610–621

Huberty CJ (2018) Discriminant analysis. 45:543–598

Dermofit Image Library—Edinburgh Innovations. https://licensing.eri.ed.ac.uk/i/software/dermofit-image-library.html. Accessed 22 Sep 2018

Jaglan P, Dass R, Duhan M (2019) A comparative analysis of various image segmentation techniques. In: Proceedings of 2nd International Conference on Communication, Computing and Networking. Springer, Singapore, pp 359–374

Jaleel JA, Salim S (2013) Computer aided detection of skin cancer. 1137–1142

Jerant AF, Johnson JT, Demastes Sheridan C, Caffrey TJ (2000) Early detection and treatment of skin cancer. Am Fam Phys 62

Kasmi R, Mokrani K (2016) Classification of malignant melanoma and benign skin lesions: implementation of automatic ABCD rule. 10:448–455. https://doi.org/10.1049/iet-ipr.2015.0385

Kaufman HL (2005) The melanoma book: a complete guide to prevention and treatment. Gotham

Korotkov K, Garcia R (2012) Computerized analysis of pigmented skin lesions: a review. Artif Intell Med 56:69–90. https://doi.org/10.1016/j.artmed.2012.08.002

Lee CH, Gutierrez F, Dou D (2011) Calculating feature weights in naive Bayes with Kullback-Leibler measure. In: Proceedings of the IEEE international conference on data mining, ICDM, pp 1146–1151. https://doi.org/10.1109/ICDM.2011.29

Lew MS (2013) Principles of visual information retrieval. p 152

Lowe NJ (2006) An overview of ultraviolet radiation, sunscreens, and photo-induced dermatoses. Dermatol Clin 24:9–17. https://doi.org/10.1016/j.det.2005.08.001

Mark Elwood J, Jopson J (1997) Melanoma and sun exposure: an overview of published studies. Int J Cancer 73:198–203. https://doi.org/10.1002/(SICI)1097-0215(19971009)73:2%3c198::AID-IJC6%3e3.0.CO;2-R

Ma L, Staunton RC (2013) Analysis of the contour structural irregularity of skin lesions using wavelet decomposition. Pattern Recognit 46:98–106. https://doi.org/10.1016/j.patcog.2012.07.001

Mehta P, Shah B (2016) Review on techniques and steps of computer aided skin cancer diagnosis. Procedia Comput Sci 85:309–316. https://doi.org/10.1016/j.procs.2016.05.238

Melanoma M, The FOR, The ROF (2007) Malignant melanoma in the 21st century, part 1: epidemiology, risk factors, screening, prevention, and diagnosis. 82:364–380
Mendonca T, Ferreira PM, Marques JS, et al (2013) PH^2—a dermoscopic image database for research and benchmarking. In: 2013 annual international conference of the IEEE engineering in medicine and biology society, pp 5437–5440. https://doi.org/10.1109/EMBC.2013.6610779
Møllersen K, Hindberg K, Schopf TR et al, Skin Lesion 9 improved diagnostics for general practice by computer-aided diagnostics
Okur E, Turkan M (2018) A survey on automated melanoma detection. Eng Appl Artif Intell 73:50–67. https://doi.org/10.1016/j.engappai.2018.04.028
Oliveira RB, Filho ME, Ma Z et al (2016a) Computational methods for the image segmentation of pigmented skin lesions: a review. Comput Methods Programs Biomed 131:127–141. https://doi.org/10.1016/j.cmpb.2016.03.032
Oliveira RB, Marranghello N, Pereira AS, Tavares JMR (2016b) A computational approach for detecting pigmented skin lesions in macroscopic images. 1–30
Pathan S, Prabhu KG, Siddalingaswamy PC (2018) Techniques and algorithms for computer aided diagnosis of pigmented skin lesions—a review. Biomed Signal Process Control 39:237–262. https://doi.org/10.1016/j.bspc.2017.07.010
Pereira AS, Manuel J, Tavares RS (2016) Computational methods for the image segmentation of pigmented skin lesions: a review. 131:127–141. https://doi.org/10.1016/j.cmpb.2016.03.032
Rastgoo M, Garcia R, Morel O, Marzani F (2015) Automatic differentiation of melanoma from dysplastic nevi. Comput Med Imaging Graph 43:44–52. https://doi.org/10.1016/j.compmedimag.2015.02.011
Rodríguez R, Sossa JH (2017) Mathematical techniques for biomedical image segmentation. Elsevier
Rourrur ASG, Rvise SNS (1982) Alternative k-nearest neighbour rules in supervised pattern recognition. 136:15–27
Schaefer G, Krawczyk B, Celebi ME, Iyatomi H (2014) An ensemble classification approach for melanoma diagnosis. Memetic Comput 6:233–240. https://doi.org/10.1007/s12293-014-0144-8
Search results. https://www.dermquest.com/results/?q=Malignantmelanoma. Accessed 22 Sep 2018
Shimizu K, Iyatomi H, Celebi ME et al (2015) Four-class classification of skin lesions with task decomposition strategy. 62:274–283
Siegel R, Miller KD, Ahmedin J (2017) Cancer statistics . Ca Cancer J 67:7–30. https://doi.org/10.3322/caac.21387
Singh D, Gautam D, Ahmed M (2014) Detection techniques for melanoma diagnosis: a performance evaluation. In: 2014 international conference on signal propagation and computer technology ICSPCT 2014, pp 567–572. https://doi.org/10.1109/ICSPCT.2014.6884948
Singh BK, Verma K, Panigrahi L, Thoke AS (2017) Integrating radiologist feedback with computer aided diagnostic systems for breast cancer risk prediction in ultrasonic images: An experimental investigation in machine learning paradigm. Expert Syst Appl 90:209–223. https://doi.org/10.1016/j.eswa.2017.08.020
Slominski A, Wortsman J (2013) Neuroendocrinology of the Skin1. Endocr Rev 21:457–487. https://doi.org/10.1210/edrv.21.5.0410
Slominski AT, Zmijewski MA, Skobowiat C, et al (2012) Sensing the environment: regulation of local and global homeostasis by the skin's neuroendocrine system
Society AC (2011) Colorectal cancer facts & figures 2011–2013
Tharwat A (2018) Classification assessment methods. Appl Comput Informatics. https://doi.org/10.1016/j.aci.2018.08.003
Vanaja S (2014) Analysis of feature selection algorithms on classification: a survey. 96:28–35
(2017) What is melanoma skin cancer? Am Cancer Soc 1–98
Yu L, Member S, Chen H, et al (2016) Automated melanoma recognition in dermoscopy images via very deep residual networks. 0062:1–11. https://doi.org/10.1109/TMI.2016.2642839

Zitová B, Flusser J (2003) Image registration methods: a survey. Image Vis Comput 21:977–1000. https://doi.org/10.1016/S0262-8856(03)00137-9

Zortea M, Schopf TR, Thon K et al (2014) Artificial intelligence in medicine performance of a dermoscopy-based computer vision system for the diagnosis of pigmented skin lesions compared with visual evaluation by experienced dermatologists. Artif Intell Med 60:13–26. https://doi.org/10.1016/j.artmed.2013.11.006

Medical Diagnosis of Coronary Artery Disease Using Fuzzy Rule-Based Classification Approach

Namrata Singh and Pradeep Singh

Abstract Coronary Artery Disease (CAD) is one of the leading causes of morbidity and mortality worldwide including India. Although recent advances in modern medical science have led to better diagnosis and treatment of CAD, yet its early detection is still a challenge. Fuzzy classification approaches are used to deal with uncertainty inherent in medical field. These fuzzy rule-based systems are extremely effective tools in disease diagnosis as they are capable to develop potential linguistic models. The aim of this paper is to initially develop a fuzzy rule-based classification system (FRBCS) based on clinical and epidemiological variables of patients and then to determine its accuracy in the diagnosis of CAD. The membership functions for medical attributes were chosen after extensive review of related literature. The rules were formulated as per the opinion of expert physicians. The present work describes the risk factors accountable for CAD, fuzzy modeling of clinical variables, rule evaluation and defuzzification of the fuzzified outputs to crisp values. The accuracy of the proposed fuzzy if–then rule classification system is 89%. Further, the present approach can assist medical practitioners in diagnosing CAD more precisely based on the fuzzy rules.

Keywords Fuzzy rule-based classification systems · Medical decision support · Coronary artery disease diagnosis · Classification · Cardiovascular diseases

1 Introduction

The development of Coronary artery disease (CAD) is described as the slow growth of atherosclerotic plaques in coronary arteries, causing luminal stenosis. This further leads to occlusion, thus resulting in myocardial infarction (MI) (Tsipouras et al.

N. Singh (✉) · P. Singh
Department of Computer Science and Engineering, National Institute of Technology, Raipur 492001, Chhattisgarh, India
e-mail: nsingh.phd2016.cs@nitrr.ac.in

P. Singh
e-mail: psingh.cs@nitrr.ac.in

A. A. Rizvanov et al. (eds.), *Advances in Biomedical Engineering and Technology*, Lecture Notes in Bioengineering,
https://doi.org/10.1007/978-981-15-6329-4_27

2008). Probably the most common cause of sudden cardiac death is attributed to cardiovascular diseases worldwide. Considering the epidemiology and pathophysiology of CAD, the early identification and effective modification of risk factors of coronary heart disease, the prevention of its growth and further complications, its diagnosis and timely treatment are of utmost importance to combat the disease (Alizadehsani et al. 2018). One of the widely used gold standard technique for the diagnosis of CAD is coronary angiography (CA). Since CA is an invasive and expensive method, therefore for the diagnosis of CAD other noninvasive techniques are being utilized in the clinical practice.

In the past years, numerous computer-aided diagnosis methodologies related to CAD have been proposed in the research works (Alizadehsani et al. 2016; Anooj 2012). Thus, noninvasive methods capable of predicting the presence of CAD using easily acquired attributes and providing an understanding of the decisions made would be of immense clinical significance (Verma et al. 2016, 2018). Fuzzy modeling can be used to handle the fuzziness inherent in biomedical problems and its integration with data mining gives the desired interpretation for the acquired decisions.

Many of the state-of-the-art fuzzy rule-based classification approaches and computational methods have been developed for detecting CAD. Sabahi (2018) proposed a novel bimodal fuzzy analytic hierarchy process (BFAHP) for dealing with uncertainty in multiple criteria decision-making (MCDM). This approach calculates the fuzzy validity by amalgamating the validities of suitable risk factors based on collective intelligence and expert knowledge. The Bayesian formulation is used to compute the fuzzy probability of risk factors. BFAHP is applied to a real dataset for CHD risk assessment which identifies diastolic blood pressure and high-density lipoprotein as the significant risk factors. Sanz et al. (2014) developed a system which predicts the risk of cardiovascular disease within the next 10 years. The approach provides both an interpretable model and a diagnosis for describing the decision, helpful for the physicians. The proposed methodology aggregates FRBCS with interval-valued fuzzy sets that consist of modeling of linguistic labels of the learner and applying genetic tuning for optimization of FRBCS.

The recent Clinical Decision Support System (CDSS) proposed by Nazari et al. (2018) utilizes Fuzzy Inference System (FIS) and Fuzzy Analytic Hierarchy Process (FAHP) for assessment of heart disease. Pal et al. (2012) developed a system for screening and early detection of CAD by utilizing easily accessible clinical parameters and laboratory tests. Muthukaruppan and Er (2012) proposed a hybrid particle swarm optimization-(PSO) based fuzzy expert system for the CAD diagnosis. Singh and Singh (2019) utilized various machine learning techniques for cardiac arrhythmia classification. Further, rule-based classification proves to be immensely helpful for classification of other diseases too (Singh and Singh 2017). In this contribution, the proposed FRBS was constructed with formulation of 30 fuzzy rules and Mamdani fuzzy inference approach was applied as a decision-making technique for CAD classification.

In this paper, we propose a fuzzy rule-based classification system (FRBCS) for the diagnosis of CAD. Section 2 describes the fuzzy rule-based methodology, Sect. 3 illustrates about results and discussion, and finally the conclusion is provided in Sect. 4.

2 Fuzzy Rule-Based Methodology

The methodology proposed for the creation of fuzzy decision support system utilizes the Mamdani approach. In order to generate a diagnosis, an unidentified case is given to the obtained decision support system. The fuzzy expert system is described in the following steps.

2.1 Variables Selection

The variables selected as input in the fuzzy expert system were 6 namely, age, fasting blood sugar (FBS), triglyceride level (TGL), low-density lipoprotein (LDL), systolic blood pressure (SBP), and high-density lipoprotein (HDL). Further, these input parameters were fuzzified ranging from 0 to 1.

2.2 Fuzzification

The trapezoidal membership function is used for the given interval range because of its capacity and generality to incorporate more fuzzy information.

2.3 Knowledge Base (IF–THEN Rules Formulation)

A set of rules was used for relating the fuzzy variables to the outcome classification. For modeling of linguistic variables, the Mamdani type fuzzy rule base is utilized, formulated in the following way: IF … Clinical variable 1 AND Clinical variable 2… THEN… Normal or CAD. Some of clinical fuzzy rules are illustrated below:

(a) *If (Age is Young) and (FBS is Good) and (TGL is Normal) and (LDL is Normal) and (SBP is Normal) and (HDL is High) then (Output is No)*
(b) *If (Age is Young) and (FBS is Bad) and (TGL is Moderate) and (LDL is Moderate) and (SBP is Middle) and (HDL is Moderate) then (Output is Yes)*

3 Results and Discussion

3.1 Patients

The coronary artery disease data (Z-Alizadeh Sani Dataset) used in this paper is obtained from the UCI machine learning (Dheeru and Karra Taniskidou 2017) database containing 303 patient samples, where each instance had 54 features. The features were arranged in four groups, namely symptom and examination, demographic, laboratory and echo, and ECG features. Each sample was classified into two classes, viz. class 1 (normal) and class 2 (CAD) having 87 and 216 subjects, respectively. If a patient's coronary artery diameter narrowing is greater than or equal to 50%, then he/she is categorized as having CAD, or otherwise as normal. Out of 54 features, 6 suitable risk factors accountable for CAD are recognized and used as input to the FIS. These factors are age, fasting blood sugar (FBS), triglyceride level (TGL), low-density lipoprotein (LDL), systolic blood pressure (SBP), and high-density lipoprotein (HDL). We used membership functions for medical variables according to literature review (Ross 2010; Mohammadpour et al. 2015; D'Acierno et al. 2013; Peña-Reyes and Sipper 2002). For each of these 6 variables, we took 3 fuzzy sets namely for age (young, middle, and old), FBS (good, acceptable, and bad), TGL (normal, moderate, and high), LDL (normal, middle, and high), SBP (low, middle, and high), and LDL (low, middle, and high).

The trapezoidal membership function is applied for each fuzzy set as shown in Fig. 1, and the ranges of linguistic variables are given in Table 1. In Table 1, SI denotes the support interval and CI denotes the core interval of the range of different levels of linguistic variables for various risk factors. Further, Fig. 2 demonstrates the output value for the given input values.

3.2 Comparison with the Prior Work

Comparative performance analysis of the proposed fuzzy rule-based classification system in classifying CAD to existing works is shown in Table 2.

In Fig. 3, the confusion matrix demonstrates the tabular fuzzy rule-based system classification results summary of the true or actual labels versus the predicted ones. The classifier evaluation has been performed on test data using three evaluation metrics viz. accuracy, sensitivity, and specificity. The accuracy, sensitivity, and specificity of the classification system are 89%, 98.04%, and 86.49%, respectively.

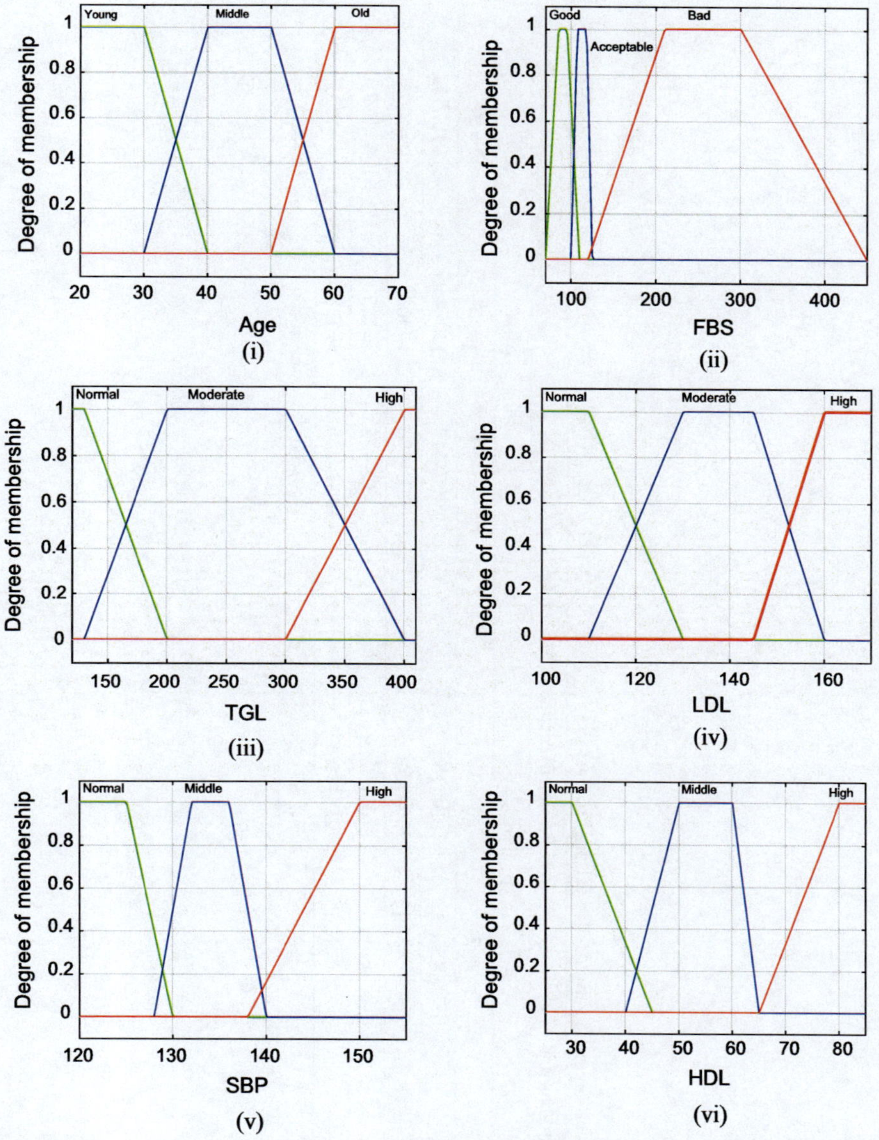

Fig. 1 Graph of membership function **a** age, **b** fasting blood sugar, **c** triglyceride level, **d** LDL, **e** systolic blood pressure, **f** HDL

4 Conclusion

We have presented a fuzzy rule-based classification system for CAD diagnosis. One of the most significant results of this work is the classification of CAD while balancing the tradeoff between interpretability and accuracy. The major findings of

Table 1 Range of different levels of linguistic variables for various risk factors

Risk factor	Young/Good/Normal		Middle/Acceptable/Moderate		Old/Bad/High	
	SI	CI	SI	CI	SI	CI
Age (years)	<=40	<=30	30–60	40–50	>=50	>=60
FBS (mg/dl)	70–110	85–95	100–125	107–118	120–450	210–300
TGL (mg/dl)	<=200	<=130	130–400	200–300	>=300	>=400
LDL (mg/dl)	<=130	<=110	110–160	130–145	>=145	>=160
SBP (mm-Hg)	<=130	<=125	128–140	132–136	>=138	>=150
HDL (mg/dl)	<=45	<=30	40–65	50–60	>=65	>=80

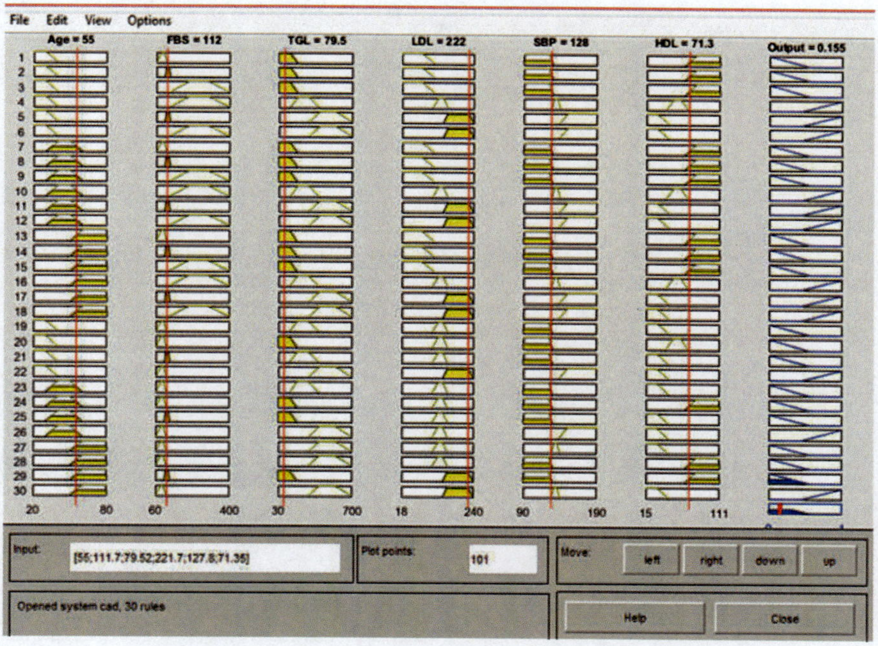

Fig. 2 Delivered output based on different inputs

this approach are the selection of suitable shape and number of membership functions and fuzzy rules which improve the classification accuracy of the overall system, thus leading to higher quality of precision. The classification accuracy depends upon various factors such as number of input features, type of membership functions, appropriate formulation and combination of fuzzy rules. These rules describe the relationship between the symptoms and the diagnosis outcome in a more accurate and understandable way. Thus, the proposed fuzzy expert system achieves an accuracy of 89%, sensitivity 98.04%, and specificity 86.49%. Further, rule pruning and rule minimization that affect the system's decision-making capabilities can be considered as future work.

Table 2 Comparison of performances of FRBCS in CAD diagnosis

Author (year)	Approach	Dataset	Performance Parameters	
Sabahi (2018)	BFAHP	Real	ACC	85.91
		Cleveland		87.31
		Hungarian		86.57
		Switzerland		85.22
		Long Beach VA		85.62
Sanz et al. (2014)	IVFS_Amp+ K_α (Chi)	Real	ACC	73.82
	IVFS_Amp+ K_α (FH-GBML)		ACC	73.71
Pal et al. (2012)	Fuzzy expert system	Real	ACC	84.20
			SPEC	83.33
			SENS	95.85
Muthukaruppan and Er (2012)	PSO-based fuzzy expert system	Hungarian and Cleveland	ACC	93.27
			SPEC	93.3
			SENS	93.2
Paul et al. (2018)	Adaptive weighted FRBS	Cleveland	ACC	92.31
		Hungarian		95.56
		Switzerland		89.47
		Long Beach VA		91.80
		Heart		92.68
Reddy and Khare (2017)	OFBAT + RBFL	Cleveland	ACC	68.55
		Hungarian		66.67
		Switzerland		78
Priyatharshini and Chitrakala (2019)	SL-FRBS	Cleveland	ACC	90.7
			SENS	91.5
Proposed work	FRBCS	Z-Alizadeh Sani	ACC	89
			SPEC	86.49
			SENS	98.04

Fig. 3 Confusion matrix of FRBCS

		Predicted Labels	
		Normal	CAD
True Label	Normal	50	1
	CAD	20	128

References

Alizadehsani R, Zangooei MH, Hosseini MJ, Habibi J, Khosravi A, Roshanzamir M, Khozeimeh F, Sarrafzadegan N, Nahavandi S (2016) Coronary artery disease detection using computational intelligence methods. Knowl-Based Syst 109:187–197

Alizadehsani R, Hosseini MJ, Khosravi A, Khozeimeh F, Roshanzamir M, Sarrafzadegan N, Nahavandi S (2018) Non-invasive detection of coronary artery disease in high-risk patients based on the stenosis prediction of separate coronary arteries. Comput Methods Programs Biomed 162:119–127

Anooj PK (2012) Clinical decision support system: risk level prediction of heart disease using weighted fuzzy rules. J King Saud Univ—Comput Inf Sci 24:27–40

D'Acierno A, Esposito M, De Pietro G (2013) An extensible six-step methodology to automatically generate fuzzy DSSs for diagnostic applications. BMC Bioinformatics 14:S4

Dheeru D, Karra Taniskidou E (2017) UCI machine learning repository. https://archive.ics.uci.edu/ml

Mohammadpour RA, Abedi SM, Bagheri S, Ghaemian A (2015) Fuzzy rule-based classification system for assessing coronary artery disease. Comput Math Methods Med 2015:1–8

Muthukaruppan S, Er MJ (2012) A hybrid particle swarm optimization based fuzzy expert system for the diagnosis of coronary artery disease. Expert Syst Appl 39:11657–11665

Nazari S, Fallah M, Kazemipoor H, Salehipour A (2018) A fuzzy inference- fuzzy analytic hierarchy process-based clinical decision support system for diagnosis of heart diseases. Expert Syst Appl 95:261–271

Pal D, Mandana KM, Pal S, Sarkar D, Chakraborty C (2012) Fuzzy expert system approach for coronary artery disease screening using clinical parameters. Knowl-Based Syst 36:162–174

Paul AK, Shill PC, Rabin MRI, Murase K (2018) Adaptive weighted fuzzy rule-based system for the risk level assessment of heart disease. Appl Intell 48:1739–1756. https://doi.org/10.1007/s10489-017-1037-6

Peña-Reyes CA, Sipper M (2002) Combining evolutionary and fuzzy techniques in medical diagnosis. In: Schmitt M, Teodorescu H, Jain A, Jain A, Jain S, Jain LC (eds) Computational intelligence processing in medical diagnosis. Studies in fuzziness and soft computing. Physica, Heidelberg, pp 391–426

Priyatharshini R, Chitrakala S (2019) A self-learning fuzzy rule-based system for risk-level assessment of coronary heart disease. IETE J Res 65:288–297. https://doi.org/10.1080/03772063.2018.1431062

Reddy GT, Khare N (2017) An efficient system for heart disease prediction using hybrid OFBAT with rule-based fuzzy logic model. J Circuits, Syst Comput 26:1750061. https://doi.org/10.1142/S021812661750061X

Ross TJ (2010) Fuzzy logic with engineering applications. Wiley, Chichester, UK

Sabahi F (2018) Bimodal fuzzy analytic hierarchy process (BFAHP) for coronary heart disease risk assessment. J Biomed Inform 83:204–216

Sanz JA, Galar M, Jurio A, Brugos A, Pagola M, Bustince H (2014) Medical diagnosis of cardiovascular diseases using an interval-valued fuzzy rule-based classification system. Appl Soft Comput 20:103–111

Singh N, Singh P (2017) Rule based approach for prediction of chronic kidney disease: a comparative study. Biomed Pharmacol J 10:867–874

Singh N, Singh P (2019) Cardiac arrhythmia classification using machine learning techniques. In: Engineering vibration, communication and information processing. Springer, Singapore. pp 469–480

Tsipouras MG, Exarchos TP, Fotiadis DI, Kotsia AP, Vakalis KV, Naka KK, Michalis LK (2008) Automated diagnosis of coronary artery disease based on data mining and fuzzy modeling. IEEE Trans Inf Technol Biomed 12:447–458

Verma L, Srivastava S, Negi PC (2016) A hybrid data mining model to predict coronary artery disease cases using non-invasive clinical data. J Med Syst 40:178
Verma L, Srivastava S, Negi PC (2018) An intelligent noninvasive model for coronary artery disease detection. Complex Intell Syst 4:11–18

Segmentation of Lungs in Thoracic CTs Using K-means Clustering and Morphological Operations

Satya Prakash Sahu, Rahul Kumar, Narendra D. Londhe, and Shrish Verma

Abstract Lung segmentation is the most essential and significant step for computer-aided diagnosis (CAD) systems for the detection of cancer in the lung region at the prior phase. The efficacy of CAD system is highly dependent on how the lungs are properly and accurately segmented. Effective lung segmentation reduces the various challenges for the CAD system to detect the juxtapleural nodules. This paper emphasizes the segmentation of lungs in CT images using the K-means approach with automatic threshold and morphological operations. Total ten subjects have been taken from the public dataset LIDC-IDRI including two subjects of juxtapleural nodules. The proposed method achieved a performance of 97.522% overlap ratio and 0.9685 Jaccard's similarity index values.

Keywords CAD system · Lung segmentation · Computed tomography · K-means · Nodule detection · Morphological operations

1 Introduction

Among the different types of cancer, lung cancer remains the most common cancer worldwide. There were 1.8 million new lung cancer cases estimated to occur in 2012 (Ferlay et al. 2013). In the US, from the American Cancer Society, Cancer Facts and Figures 2017 (American Cancer Society 2017), there were approximately 222,500 new cases of cancer and 155,870 related deaths that accounts for one out of four cancer-related deaths.

Lung cancer is leading to the highest mortality rate; this has urged researchers to perform a diagnosis at the early stages. The survival rate may be greatly enhanced, i.e., by 70–80% if lung cancer is detected in the first stage (Swensen et al. 2002). The various challenges in lung segmentation are due to inhomogeneities in the lung region and similar densities in other pulmonary structures like veins, arteries, bronchi, and bronchioles. However, the most challenging task for researchers is to detect

S. P. Sahu (✉) · R. Kumar · N. D. Londhe · S. Verma
National Institute of Technology, Raipur, India
e-mail: spsahu.it@nitrr.ac.in

A. A. Rizvanov et al. (eds.), *Advances in Biomedical Engineering and Technology*, Lecture Notes in Bioengineering,
https://doi.org/10.1007/978-981-15-6329-4_28

the nodules attached to pleural surface (i.e., juxtapleural). If the lung segmentation process does not define the lung boundaries correctly, the juxtapleural nodule may be missed. Armato et al. showed that approximately 5–17% true lung nodules are not detected due to poor and inaccurate lung segmentation (Armato and Sensakovic 2004). Thus, there is a need for effective lung segmentation technique for the enhancement in the efficiency of CAD system, i.e., truly capable of identifying the true lung nodules.

In this paper, a hybrid method using k-means clustering and automatic thresholding has been proposed for lung segmentation to delineate the boundary of lungs and to extract the region of interest (ROI) area. Then, the broken borders are corrected by applying morphological closing operations to include the juxtaplueral nodules to minimize the chance of over-segmentation. Figure 1 shows the process flow diagram for the adopted method.

2 Related Work

Threshold-based approach is the most popular method for image segmentation, which utilizes the intensity (gray level) value of the image. Since the lung regions are having lower gray level (−500 HU approximately) with respect to other anatomical structures in the thorax region, hence, with the application of optimum threshold, the segmentation of lungs has been adopted by a number of authors in their research articles. Hu et al. (2001) used iterative thresholding and some morphological operations for lung segmentation. A similar threshold-based approach was given by Gao et al. (2007), i.e., region growing and morphological smoothing. Wei et al. (2009) proposed a threshold using histogram analysis to segment the lung region. Adaptive fuzzy thresholding was given by Ye et al. (2009) in lung segmentation from CT data. Helen et al. (2011) enhanced the performance of 2D Otsu-based thresholding algorithm for the segmentation of pulmonary parenchyma in CT lung images using particle swarm optimization (PSO). Sahu et al. (2017, 2019) have given the hybrid method based on fuzzy clustering, thresholding, and morphological operations. Active contour-based method (ACM) is the next significant method that has been greatly utilized and used in various research studies (Xu and Prince 1998; Wang et al. 2010; Chan and Vese 2001; Cui et al. 2013; Athertya and Kumar 2014). This method relies on the minimization of energy function through dynamic contour iteratively. However, the problem of initialization and convergence are the challenges with this method which had been addressed by a number of authors in their articles. Xu and Prince (1998) worked for better convergence and introduced an external force, i.e., gradient vector flow (GVF) to guide the contour. Wang et al. (2010) has given the improved version for these issues through normally biased GVF. Chan and Vese (2001) has given ACM without edges and used curve evolution techniques for the initialization problem. Cui et al. (2013) given the ACM based on statistics of a local region for the segmentation of medical images. The fully automatic method using ACM is exploited by Athertya and Kumar (2014) that reduces the interaction of the user.

The other useful technique that uses the information of earlier shape and parameters of lungs for obtaining the ROI from the thorax region is shape-based segmentation. The shape parameters such as contour or edges and points of previous data are characterized for the derivation of variational energy framework for the segmentation of lungs. Annangi et al. (2010) proposed a region-based active contour method with the use of prior shape and low-level features to handle the issue of local minima. Kockelkorn et al. (2010) proposed interactive techniques for lung segmentation in CT data with severe abnormalities where prior shape term are trained using k-NN classifier thereby correction in classification results. Sofka et al. (2011) adopted a pattern recognition technique with the combination of statistical shape model and anatomical information for the robust lung segmentation. A novel approach is given by Sun et al. (2011) based on the robust active shape model, and further, a constrained optimal surface finding method has been used to adapt the initial segmentation result. For the initialization of active shape models, Gill et al. (2014) gave the feature-based atlas approach for segmentation in CT images.

3 Material and Methods

3.1 Data

The public dataset of the Lung Imaging Database consortium-Image Database Resource Initiative (LIDC-IDRI) has been accessed for the experimental database (Armato et al. 2015). This dataset contains 1018 cases of Lung CT (Computer Tomography) scans (Armato et al. 2011). The LIDC dataset also contains the annotated information in the attached XML file provided by four radiologists (experts) for each CT scan (Clark et al. 2013). For the proposed method, ten patients have been acquired through the LIDC dataset with the following specifications: DICOM Image of size 512×512, the image intensity value from -600 to 1600 Hounsfield Unit (HU), tube current of 265–570 mA and tube voltage of 120 kVp.

3.2 Data Preprocessing

The experimental database contains CT scans of ten patients with some cases of CTs having juxtapleural lung nodules. The contoured outline or boundary tracing for lung region for each patient of all the slices in different cases is done by an expert as the representation of ground truth. For improving the accuracy in the further operations, the pixel values of the image are converted to double format followed by image enhancement for better contrast. Pixel value of the image needs to be represented in

HU value for setting appropriate window level and window width are set through the. As per the DICOM file, the Rescale Slope and Rescale Intercept are done by a linear transformation. Linear transformation leads to an effective range of HU values for the images; afterward, the HU value greater than the upper range will be shown in white, and HU value that is lesser than the lower range will be shown in black. Then the operations of image processing for the conversion of 16-bits image into an 8-bits image will be applied.

3.3 *Segmentation of Lungs*

The process of lung segmentation is proposed as follows:

A. K-means clustering is applied to the input CT image thereby generating the grayscale-masked image. (Algorithm and derivations explained in Sect. 3.3.1)
B. The automatic thresholding method applied to the output of K-means masks for the conversion into binary, and then the background subtraction for ROI is performed. (explained in Sect. 3.3.2)
C. The large airways and other vessels within the lung region are removed by the hole-filling algorithm. Let S be the set whose boundary pixel is labeled by 1 s (i.e., 8-connected boundaries) and each boundary enclosing a hole (background region). With the given point p in each hole, the objective is to fill the holes by 1 s. This algorithm fills the holes in an image through the process of dilation, intersection, and complementation.
   The algorithm is followed by morphological reconstruction applied to the lung portion and the lighter border is also removed by reducing the overall intensity of the border structures.
D. For handling the juxtapleural cases, the morphological closing operations are applied in all lung CT images.
E. The final segmented lung regions are obtained by masking the output image with an input grayscale image.

3.3.1 K-means Algorithm

K-means is an unsupervised and simple clustering approach given by Hartigan and Wong (1979). The procedure follows an easy way of partitioning a given dataset (*n*-object or data points) into a certain number of *k*-clusters. This approach is utilized here for the segmentation process since it classifies the objects into some desired number of groups or clusters based on some selected features. The process of clustering is minimizing the sum of squares of distance iteratively from the considered data points to the corresponding centroid of the cluster.

Given a set of data points $(x_1, x_2, \ldots, x_n)$, where each data point is having *d*-dimension real vector, *k*-means clustering objective is to partition the *n*-data points

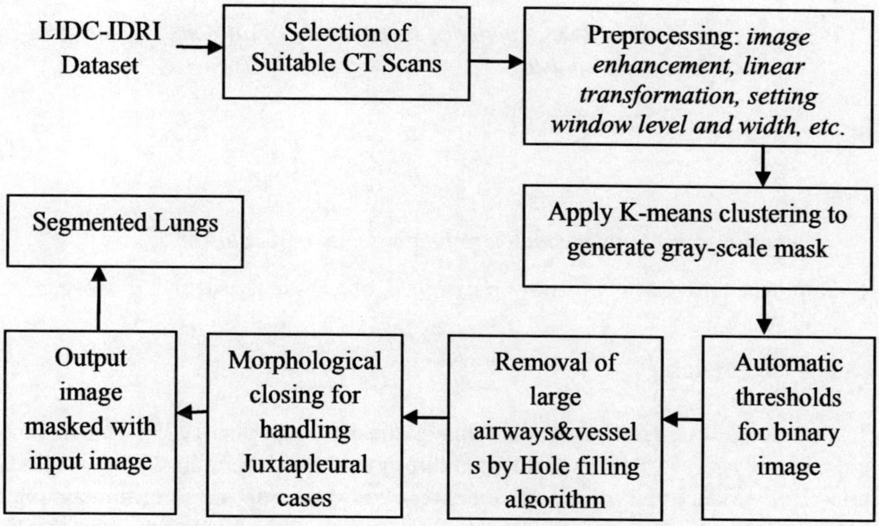

Fig. 1 Process flow diagram

into $k(\leq n)$ clusters or groups $G = \{G_1, G_2, \ldots, G_k\}$ to minimize the sum of squares of distance (squared *Euclidean distance*) within a cluster. The data points are clustered by minimizing the following objective function (F) around the centroids (c_i, for $i = 1, 2, \ldots, k$)

$$F = \sum_{i=1}^{k} \sum_{x \in G_i} (x - c_i)^2 \tag{1}$$

where c_i is the mean point or centroid of all points $x \in G_i$. The algorithm is as follows:

1. *Intensity distribution of the image is computed for various intensities.*
   a. *for each data point* x_i:
      i. *nearest centroid* c_z *is calculated, i.e., Euclidean distance between data point* x_i *(intensity of an instance point) and all the cluster centers* c_j *(centroid intensities)*

$$c_z = \arg\min ||x_i - c_j||^2 \tag{2}$$

      ii. *data point* x_i *is assigned to cluster j, whose Euclidean distance is minimum from the* jth *cluster center among all the cluster centers.*
   b. *for each cluster* $j = 1,2, \ldots, k$
2. *Initialize cluster centers (centroids)* $c_1, c_2, \ldots, c_k$ *with k random intensities.*
3. *Repeat until convergence or shifting of cluster centers are observed*

i. *new centroid intensity* c_j*, computed as the mean of all points* x_i *assigned to cluster* j *in above step using*

$$c_j = \frac{1}{n_j} \sum_{i=1}^{n_j} x_i \tag{3}$$

where n_j *represents the number of data points in* jth *cluster.*

4. *Stop, when no data points were reassigned, otherwise repeat step 3 above.*

3.3.2 Thresholding

Otsu's method is utilized here for automatic thresholding (Otsu 1975). This method finds the gray-level value, t^* (estimated threshold), which minimizes the weighted intra-class variance and maximizes the inter-class variance. The algorithm assumes that the histogram of images is bimodal. Let the pixels of the image be represented in as 1, 2, …, L gray levels. The normalized histogram can be viewed as a probability distribution (p_j):

$$p_j = n_j / N,\ p_j \geq 0,\ \sum_{j=1}^{L} p_j = 1 \tag{4}$$

where n_j denotes the number of a pixel having the gray level$j$ and the total number of pixel $N = n_1 + n_2 + \ldots + n_L$.

Assuming the threshold at 't', the pixels of an image will be classified in class C_o (object as white) and C_b (background as black) then, $C_b = [0,1,2,\ldots,t]$ and $C_o = [t + 1, t + 2,\ldots,L - 1]$. The class occurrence probabilities for background and object will be:

$$q_b = \sum_{j=1}^{t} p_j = q(t) \tag{5}$$

$$q_o = \sum_{j=t+1}^{L-1} p_j = 1 - q(t) \tag{6}$$

and, gray-level mean value for background and object will be given by:

$$\mu_b = \sum_{j=1}^{t} j p_j / q_b = \mu(t) / q(t) \tag{7}$$

$$\mu_0 = \sum_{j=t+1}^{L-1} jp_j / q_0 = \frac{\mu_T - \mu(t)}{1 - q(t)} \tag{8}$$

where $q(t)$ as per Eq. (5) is 0th order and

$\mu(t) = \sum_{j=1}^{t} jp_j$ is first-order cumulative moments histogram up to the tth level,

and $\mu_T = \mu(L) = \sum_{j=1}^{L} jp_j$.

The optimum threshold will be obtained by maximizing the inter-class variance $\sigma_B^2(t)$. The generalized optimal threshold $t*$ is obtained by histogram analysis and testing each gray level for which the possibility of the threshold t that maximizes $\sigma_B^2(t)$ is given by:

$$\sigma_B^2(t*) = \max_{1 \le t < L} \sigma_B^2 \tag{9}$$

4 Experimental Results and Analysis

A total of ten subjects including two subjects having juxtapleural nodules from the dataset of LIDC have been taken for the experiment. The total numbers of 1800 CT slices approximately have been examined from the subjects taken. The lung contours (parenchyma or boundaries) were manually segmented by experts under the supervision of a thoracic radiologist for all the slices of ten subjects. For the support of manual segmentation tasks, an open-source software system "The Medical Imaging Interaction Toolkit" (MITK 2016.11 Release Workbench and Toolkit 2016.11) has been used (MITK 2016). This application framework also supports for correction of borders by some adjustments in contour if any inaccuracies are observed. These manually segmented contours were taken as ground truth or the references for individual images. Finally, the overall segmentation accuracies of the proposed method are analyzed and compared to the reference standard using different evaluation measures.

4.1 Evaluation Measures

The following metrics were used for accuracy of segmentation results: over-segmentation rate (OS) and under-segmentation rate (US), overlap ratio (OR), and Jaccard's similarity index (JI).

The *OS* is given as the region or volume obtained from the automated method but is absent in reference standard image and vice versa in case of *under-segmentation* rate (Pu et al. 2008; Hoover et al. 1996). Let V_a and V_m be the volume generated in the binary mask by our automated method and volume based on manual segmentation or reference standard, respectively. The *OR* is given as the relative overlap between the volumes of automated method (V_a) and the volumes of reference standard (V_m) and calculated by the following equation (vanRikxoort et al. 2009).

$$OR(V_a, V_m) = \frac{|V_a \cap V_m|}{|V_a \cup V_m|} \tag{10}$$

Another evaluation measure used for the segmentation accuracy is Jaccard's similarity index (*JI*) (Sampat et al. 2009). This method is based on the comparison of similarity between ROIs (number of pixels or voxels) segmented from the automated method (S_a) and the ground truth (S_m) region. It is defined as the overlap or coincidence degree between S_a and S_m and can be given as follows:

$$JI(S_a, S_m) = \frac{|S_a \cap S_m|}{|S_a| + |S_m| - |S_a \cap S_m|} \tag{11}$$

The values of $JI(S_a, S_m)$ lie in [0,1] and higher values indicate better segmentation accuracies.

4.2 Result Analysis

The performance has been tested on acquired ten CT subjects (including two subjects of juxtapleural nodules) having approximately 1800 CT slice images from the dataset of LIDC-IDRI. The sample results with the flow process of the proposed method have been shown in Fig. 2. Using the evaluation measures discussed in Sect. 4.1, Fig. 3a indicates the bar plots of over-segmentation and under-segmentation rate with overlap ratio difference and Fig. 3b indicates the accuracy in terms of Jaccard's similarity index of ten patients, where patients 1 and 2 are the pleural nodules cases. The results show the proposed method achieved mean values of over-segmentation rate, under-segmentation rate, and overlap ratio difference of 0.527, 1.951, and 2.478, respectively. And, the accuracy achieved in terms of Jaccard's similarity index of 0.9685. Also for the patient having juxtapleural nodules, the proposed method shows the effectiveness and includes all juxtaplural nodules inside the ROIs with the accuracy of overall overlap ratio of 95.496% and Jaccard's similarity index value of 0.9497.

The literature reveals various methods and approaches for lung segmentation given by researchers, but only a few are effective to deal with the case of juxtapleural nodules. In Table 1, the performance of the proposed method has been compared with the other method given in research articles for the lung segmentation. The fair comparison of methods would be possible if the same datasets and other standards of images were used. However, as compared to other methods, in general, our automated method is achieving good results and giving higher segmentation accuracies.

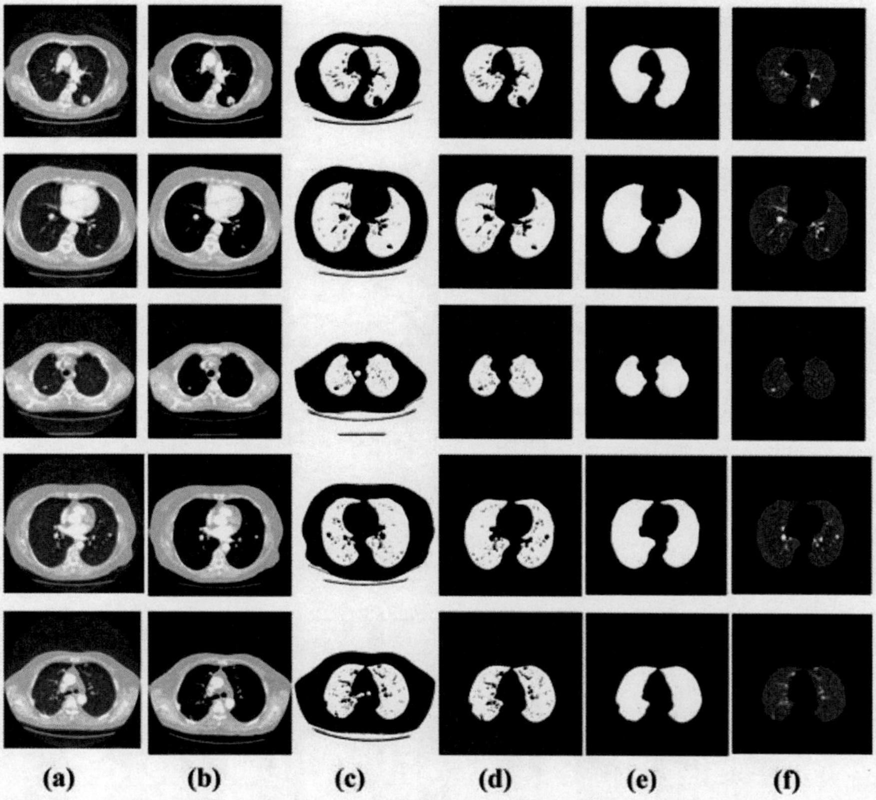

Fig. 2 The sample lung segmentation results of the proposed method are as follows: Column **a** are five original lung CT images and ROI images from the top to the bottom, where row 1 and row 5 are the images containing juxtapleural nodules; **b** grayscale-masked images after applying K-means; **c** binary images of (**b**) using automatic threshold; **d** background-subtracted images; **e** the final lung parenchyma mask after removing the large airways and vessels; and **f** final result of the proposed method—segmented lungs

5 Conclusion

Lung segmentation is a very critical and significant step not only in the CAD system for lung cancer but also for the other pulmonary diseases. The efficacy of the proposed method was demonstrated in ten cases (approximately 1800 CT images) of lung thoracic CTs from the LIDC dataset including two cases of CTs having juxtapleural nodules.

The experimental results show that our method can correctly re-include all juxtapleural nodules into the ROI of lungs with the minimum rate of over and under-segmentation. The proposed method is achieving the overall performance in an overlap ratio of 97.522% overlap ratio and segmentation accuracy of 0.9685 in terms of Jaccard's similarity index.

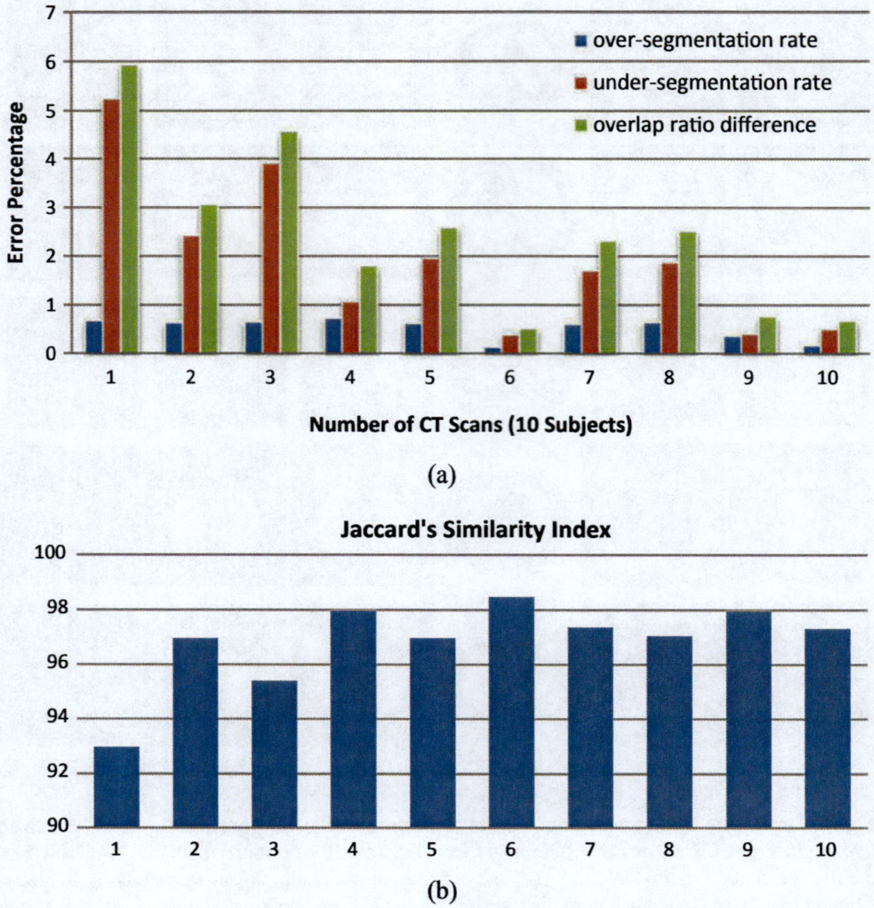

Fig. 3 Proposed method's bar plot of 10 subjects: **a** Over-segmentation rate, Under-segmentation rate, and Overlap ratio difference; **b** Jaccard's similarity index

The proposed method takes an average processing time of 0.65 s per CT slice using MATLAB software in a system with CPU Intel Core i5 3.1 GHz and 8 GB RAM. Accurate lung segmentation is still a challenge in the presence of severe pathologies so the effectiveness of this approach for other pulmonary diseases and abnormalities to be tested in the future.

Table 1 Comparative performance of the proposed method

Authors	Database	OR average (%)	OSaverage (%)	US average (%)	JI
Wei et al. (2013)	97 CTs with 25 juxtapleuralcases	95.24	–	–	–
Zhou et al. (2014)	20 CTswith 65 juxtapleuralcases	95.81	–	–	–
Liao et al. (2016)	80 CTs of 4 categories non, benign, malignant and pleural nodules with 20 cases in each category	92.22	–	–	non-nodules—0.941 benign—0.9342 malignant—0.9257 pleural nodules—0.8821
Proposed method	10 CTs with 2 juxtapleural cases (a total of 1800 CT slice images approx.)	97.52	0.527	1.951	overall—0.9685 pleural nodules—0.9497

References

American Cancer Society, Cancer facts and figures, 2017

Annangi P, Thiruvenkadam S, Raja A, Xu H, Sun X, Mao L (2010) A region based active contour method for x-ray lung segmentation using prior shape and low level features. In: 2010 IEEE international symposium on biomedical imaging: from nano to macro. IEEE, pp 892–895

Armato SG, Sensakovic WF (2004) Automated lung segmentation for thoracic CT: impact on computer-aided diagnosis1. Acad Radiol 11(9):1011–1021

Armato SG, McLennan G, Bidaut L, McNitt-Gray MF, Meyer CR, Reeves AP, Zhao B, Aberle DR, Henschke CI, Hoffman EA, Kazerooni EA (2011) The lung image database consortium (LIDC) and image database resource initiative (IDRI): a completed reference database of lung nodules on CT scans. Med Phys 38(2):915–931

Armato III, Samuel G, McLennan G, Bidaut L, McNitt-Gray MF, Meyer CR, Reeves AP, Clarke LP (2015) Data from LIDC-IDRI. The Cancer Imaging Archive. https://doi.org/101.7937/K9/TCIA.2015.LO9QL9SX

Athertya JS, Kumar GS (2014) Automatic initialization for segmentation of medical images based on active contour. In: 2014 IEEE conference on biomedical engineering and sciences (IECBES). IEEE, pp 446–451

Chan TF, Vese LA (2001) Active contours without edges. IEEE Trans Image Process 10(2):266–277

Clark K, Vendt B, Smith K, Freymann J, Kirby J, Koppel P, Moore S, Phillips S, Maffitt D, Pringle M, Tarbox L (2013) The Cancer Imaging Archive (TCIA): maintaining and operating a public information repository. J Digit Imaging 26(6):1045–1057

Cui W, Wang Y, Lei T, Fan Y, Feng Y (2013) Local region statistics-based active contour model for medical image segmentation. In: 2013 seventh international conference on image and graphics (ICIG). IEEE, pp 205–210

Ferlay J, Soerjomataram I, Ervik M, Dikshit R, Eser S, Mathers C, et al (2013) Lyon, France: International Agency for Research on Cancer; 2013. GLOBOCAN 2012 v1.0, Cancer Incidence and Mortality Worldwide: IARC CancerBase No. 11. https://globocan.iarc.fr. Accessed 04 Oct 2016

Gao Q, Wang S, Zhao D, Liu J (2007) Accurate lung segmentation for X-ray CT images. In: Third international conference on natural computation, 2007. ICNC 2007, vol 2. IEEE, pp 275–279

Gill G, Toews M, Beichel R (2014). Robust initialization of active shape models for lung segmentation in CT scans: a feature-based atlas approach. J Biomed Imaging 13

Hartigan J, Wong M (1979) Algorithm AS 136: A K-Means Clustering Algorithm. J Roy Stat Soc. Series C (Appl Stat) 28(1):100–108. https://doi.org/10.2307/2346830

Helen R, Kamaraj N, Selvi K, Raman VR (2011) Segmentation of pulmonary parenchyma in CT lung images based on 2D Otsu optimized by PSO. In: International conference on emerging trends in electrical and computer technology (ICETECT). IEEE, pp 536–541

Hoover A, Jean-Baptiste G, Jiang X, Flynn PJ, Bunke H, Goldgof DB, Bowyer K, Eggert DW, Fitzgibbon A, Fisher RB (1996) An experimental comparison of range image segmentation algorithms. IEEE Trans Pattern Anal Mach Intell 18(7):673–689

Hu S, Hoffman EA, Reinhardt JM (2001) Automatic lung segmentation for accurate quantitation of volumetric X-ray CT images. IEEE Trans Med Imaging 20(6):490–498

Kockelkorn TT, van Rikxoort EM, Grutters JC, van Ginneken B (2010) Interactive lung segmentation in CT scans with severe abnormalities. In: 2010 IEEE international symposium on biomedical imaging: from nano to macro. IEEE, pp 564–567

Liao X, Zhao J, Jiao C, Lei L, Qiang Y, Cui Q (2016) A segmentation method for lung parenchyma image sequences based on superpixels and a self-generating neural forest. PLoS ONE 11(8):e0160556

MITK (2016) The medical imaging interaction toolkit, German cancer research center, division of medical and biological informatics. https://mitk.org/wiki/MITK

Otsu N (1975) A threshold selection method from gray-level histograms. Automatica. 11(285–296):23–27

Pu J, Roos J, Chin AY, Napel S, Rubin GD, Paik DS (2008) Adaptive border marching algorithm: automatic lung segmentation on chest CT images. Comput Med Imaging Graph 32(6):452–462

Sahu SP, Agrawal P, Londhe ND (2017) A new hybrid approach using fuzzy clustering and morphological operations for lung segmentation in thoracic CT images. Biomed Pharmacol J 10(4):1949–1961

Sahu SP, Agrawal P, Londhe ND, Verma S (2019) Lung segmentation of CT images using fuzzy C-means for the detection of cancer in early stages. In: Advances in data and information sciences 2019. Springer, Singapore, pp 167–176

Sampat MP, Wang Z, Gupta S, Bovik AC, Markey MK (2009) Complex wavelet structural similarity: a new image similarity index. IEEE Trans Image Process 18(11):2385–2401

Sofka M, Wetzl J, Birkbeck N, Zhang J, Kohlberger T, Kaftan J, Declerck J, Zhou SK (2011) Multi-stage learning for robust lung segmentation in challenging CT volumes. In: International conference on medical image computing and computer-assisted intervention on September 2011. Springer Berlin Heidelberg, pp 667–674

Sun S, Bauer C, Beichel R (2011) Automated 3-D segmentation of lungs with lung cancer in CT data using a novel robust active shape model approach. IEEE Trans Med Imaging 31(2):449–460

Swensen SJ, Jett JR, Sloan JA, Midthun DE, Hartman TE, Sykes AM, Aughenbaugh GL, Zink FE, Hillman SL, Noetzel GR, Marks RS (2002) Screening for lung cancer with low-dose spiral computed tomography. Am J Respir Crit Care Med 165(4):508–513

vanRikxoort EM, de Hoop B, Viergever MA, Prokop M, van Ginneken B (2009) Automatic lung segmentation from thoracic computed tomography scans using a hybrid approach with error detection. Med Phys 36(7):2934–2947

Wang Y, Liu L, Zhang H, Cao Z, Lu S (2010) Image segmentation using active contours with normally biased GVF external force. IEEE Signal Process Lett 17(10):875–878

Wei Q, Hu Y, Gelfand G, MacGregor JH (2009) Segmentation of lung lobes in high-resolution isotropic CT images. IEEE Trans Biomed Eng 56(5):1383–1393

Wei Y, Shen G, Li JJ (2013) A fully automatic method for lung parenchyma segmentation and repairing. J Digit Imaging 26(3):483–495

Xu C, Prince JL (1998) Snakes, shapes, and gradient vector flow. IEEE Trans Image Process 7(3):359–369

Ye X, Lin X, Dehmeshki J, Slabaugh G, Beddoe G (2009) Shape-based computer-aided detection of lung nodules in thoracic CT images. IEEE Trans Biomed Eng 56(7):1810–1820

Zhou S, Cheng Y, Tamura S (2014) Automated lung segmentation and smoothing techniques for inclusion of juxtapleural nodules and pulmonary vessels on chest CT images. Biomed Signal Process Control 30(13):62–70

Univariate Feature Selection Techniques for Classification of Epileptic EEG Signals

Moushmi Kar and Laxmikant Dewangan

Abstract Feature selection methods can be applied to enhance the EEG classification more accurately along with its applicability for the large volume of data. Univariate filter feature selection techniques are used to find accurate classifier models for epileptic and non-epileptic classifications. In this paper, the sample feature subsets are obtained from four different feature selection techniques, namely ReilefF, GR, one attribute and symmetrical uncertainty. We choose the top 20 features among every feature selection technique for the reduction of dimensionality and complexity. In the classification phase, the problem is classified with three different classification algorithms such as naive Bayes, random forest, J48 to classify the EEG signals. The performance evolution of classifiers is carried out by applying K-fold cross-validation and holdout methods for determining accuracy, recall and false positive rate along with MCC. Finally, total 178 features and a reduced set of top 20 best features are applied to the different classifiers for binary classification. Random forest classifier achieves the highest accuracy of 97.122%, recall of 96.112%, false positive rate of 97.352%, AUC 96.732% and MCC 0.909% with selected best 20 features. Random forest classifier with a combination of symmetrical uncertainty feature selection and holdout–cross-validation protocol obtained the best performance. So, results obtained by this experiment show that this methodology has the potential to contribute to the epileptic seizure classification with minimal complication.

Keywords EEG · Epilepsy · Feature selection · Cross-validation · Univariate

M. Kar (✉) · L. Dewangan
Department of Electronics and Telecommunication, Chouksey Engineering College, Bilaspur, C.G., India
e-mail: moushmikar@gmail.com

L. Dewangan
e-mail: laxmikant.dewangan@gmail.com

A. A. Rizvanov et al. (eds.), *Advances in Biomedical Engineering and Technology*, Lecture Notes in Bioengineering,
https://doi.org/10.1007/978-981-15-6329-4_29

1 Introduction

Epilepsy is occurred due to the formation of abnormal electric pulse in the brain. Besides, there are some imbalances of chemicals called neurotransmitters, which are responsible for brain signalling. These two factors individually or in a combination of both cause epilepsy (Blume et al. 2001). So the normal nature of neuronal activity is being disturbed in epilepsy that creates various behavioural changes (Quyen et al. 2000). It is determined by recorded EEG signals, having characteristic information related to it. Epilepsy creates clear disturbance and leaves its signature on standard EEG signals. Brain activity is monitored for the long term using EEG as it is the only convenient way. Recent advancement in the neuro-stimulating system for epilepsy is determined based on the EEG. EEG is used to detect abnormalities of brain functioning (Yang and Rothman 2001; Yang et al. 2002; Litt and Echauz 2002; Anderson et al. 2008; Skarpaas and Morrell 2009).

There are two discrete classifications of seizures, namely (i) clinical and (ii) sub-clinical. Any behavioural changes related to the seizure are assigned to clinical seizures. If the physical changes happened during any epileptic seizures are negligible or even no change, then this is called electrographic or sub-clinical seizures.

The majority of seizures are being captured by continuous monitoring over a long period of time of an epileptic patient with simultaneous observation of the behavioural patterns in video using EEG signals. This is one of the convenient ways. However, this technique is extremely difficult, expensive and not a handy arrangement. Automatic detection of seizure and spike techniques can help the clinical staff. So there is a requirement for a computerized technique for the identification and classification of seizure.

For this reason, it is essential to have an automated recognition and classification of epilepsy using machine learning techniques (Niedermeyer and Lopes da Silva 2005). However, execution of the present machine learning frameworks is lagging behind the necessities of ongoing purpose due to the limitation forced by data size and lack of faith of clinical staff on the systems. So the improvement of the execution of the machine learning framework has opened up a basic research area. A specialist interface has become more or less obligatory for a computerized finding framework that can be authentic and reliable for clinical experts. This article develops a feature selection technique for the determination of most relevant and optimal features having high discriminatory power to differentiate epileptic and non-epileptic seizures and compare its performance with existing ones.

Now, the selected relevant features are applied to three different classifier techniques to achieve higher classification accuracy. Next, we compare the performance of classifier models to various feature selection methods and obtained the best-suited classifier as well as the best feature selection techniques with low complexity and less time-consuming.

1.1 Related Work

The research on epileptic seizures is briefly divided into three steps, which are (i) emphasize pre-processing methods, (ii) focus on feature selection and (iii) classification framework for evaluation and prediction. In the beginning step, many groups of researchers introduced an efficient way to deal with raw EEG signals for the extraction of features (Bhattacharyya et al. 2017). Feature extraction is a part of pre-processing methods where time and frequency domain techniques are one of the categories of it. In this technique, EEG signals are decomposed into sub-band by downsampling.

For the feature extraction, Mi. et al. used a discrete wavelet transform (Mi et al. 2017). Other methods by Acharya et al. such as statistical features (Acharya et al. 2013) from non-linear dynamics are used for feature extraction. Tewfik applied fractional dimension (Acharya et al. 2013) and Mi. et al. applied entropy (Tawfik et al. 2016), which were also used for feature extraction. The above-mentioned references experimentally proved that the features that used DWT can accurately separate epileptic seizure signals.

The performance of classifier can be improved by relevant feature selection techniques reported by many researchers. For example, Harpale and Bairagi (2018) reported the classification of epileptic seizures by an adaptive method. Alzami et al. (2018) proposed a feature selection techniques with customized differential evolution to obtain the best possible feature subset to improve the accuracy. The authors applied an adaptive hybrid feature selection for classifying the epilepsy samples. They used DWT as pre-processing and mRMR, Fisher, chi-square and ReliefF as feature selection techniques. They reported 100% of accuracy with fivefold cross-validation method. ReliefF methods have been reported for epileptic seizure classification and evaluate the significance of each feature by Pippa et al. (2016). For accurate classification, Guo et al. (2011) applied a genetic algorithm to select the relevant feature subsets and classification using ANN. The accuracy of classifier is reported 95.2%, sensitivity of 98.17% and specificity of 92.12% using wavelet energy-based features. The same group performed classification using wavelet transform and ApEn features and got accuracy of 95.2%, sensitivity of 98.17% and specificity of 92.12% using relative accuracy of 99.85%, sensitivity of 100% and specificity of 99.2%. Li et al. (2017) have reported a neural network function to get better accuracy based on feature selection.

The last but important steps are classification for detection and prediction of epilepsy after accurately classified the EEG signals. Rabbi et al. used machine learning methods such as artificial neural network (ANN) and support vector machine (SVM) for the classification of EEG signals (Rabbi and Fazel-Rezai 2012). Chua et al. performed classification using GMM and SVM classifiers with HOS-based features and obtained an accuracy of 93.11% and 92.67%, respectively (Chua et al. 2009). Hekim proposed a classification of EEG signals by using two ANN models and average powers of Power Spectral Densities (PSD) (Hekim 2012). Pachori and Patidar reported a new method for the classification of seizure and non-seizure

EEG signals (Pachori and Patidar 2014). Fergus et al. applied a machine learning approach with k-class nearest-neighbor classifier that differentiates epileptic seizure from non-epileptic seizures (Pachori and Patidar 2014).

Some researchers used multiple classifiers, which provided a more vigorous, steady and precise outcome than a single classifier. In this area, Hassan et al. detected epileptic seizures by applying wavelet transform with a tuneable Q factor (Hassan et al. 2016). Li et al. used neural network ensembles that impede the detection algorithm from falling into a local minimum (Li et al. 2017). Recent works and its methodology are summarized in Table 1. Though several works have been reported in the recent past as highlighted above, still there are some restrictions and challenges as recognized from past approaches. Firstly, more number of features are used in the classifier system. The computational complexity is increased with more number of features. This research explores the appropriate feature selection technique for the selected relevant features. The conventional approaches for feature selection have shown a lack of potential in terms of information and decision support. Hence, top N criteria should be used to select the best possible set of features. Secondly, the results reported in most of the published works are for large training data size and small

Table 1 Performance comparison of classifiers as reported in previous studies

Author	Methodology	Statistical parameters
Guler and Ubeyli (2005)	Wavelet transform and ANFIS features extraction and ANN (Artificial neural network)classifier	Classification accuracy: 98.68%
Kannathal et al. (2005)	Adaptive Neuro-Fuzzy Inference System (ANFIS)	Classification accuracy: 90%
Subasi (2006)	WT and ANFIS Mixture expert model (a modular neural network)	Accuracy: 95%, False positive rate: 93.7% Recall: 94.3%
Guo et al. (2010)	ANN classifier and wavelet energy based features	accuracy: 95.2%, Recall: 98.17%, false positive rate: 92.12%
Acharya et al. (2012)	Features CD, FD, H and ApEn in Fuzzy classifiers	Accuracy: 99.7%, Recall: 100%, false positive rate: 100%
Gopan et al. (2013)	Neuro-fuzzy classifier(adaptive)	Accuracy: 100%
Gajic et al. (2014)	Wavelet Transform and quadratic classifier	Accuracy: 95%
Al-Ghayab et al. (2016)	least square support vector machine	Accuracy: 99.9%, Recall: 99.80% false positive rate: 100%
Vidyaratne and Iftekharuddin (2017)	FD and HWPT energy features and vector machine classifier	Accuracy: 99.8%, Recall: 96%

testing data size. However, in order to utilize the reported machine learning framework in clinical practice, their performance should be tested on large test data size. So, improvements of classification algorithms are required for accurately differentiating neurological disorders.

The present work is divided into four sections. The second section represents the methodology used in this study. The third part contains the results followed by conclusions and future works of this study in the fourth part.

2 Material and Methods

The classification of epileptic seizures from non-epileptic seizures is still a challenging task, as every patient shows slightly different behaviours in terms of their neurological response in EEG. In this paper, we try to find an appropriate feature selection approach for epileptic classification. Comparison of the different classifier approaches with suitable feature selection is shown in Fig. 1.

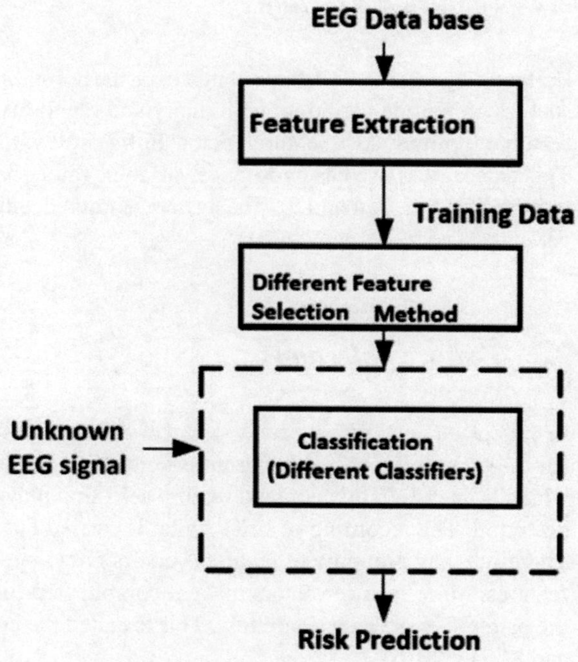

Fig. 1 Block diagram of the proposed methodology

2.1 Dataset

In this study, we have used a freely available EEG dataset from the University of Bonn (Andrzejak et al. 2001). Five sets of data (denoted as A–E) having 100 single-channel EEG segments and the duration of each segment are 23.6 s. The time signals are sampled at 4097 data points and these data points give the values of EEG recording at singular data points. Total 178 input vectors are shown in Y (where $Y = 1, 2, 3, 4$ and 5).

The EEG dataset consists of five sets (denoted as A–E). For healthy datasets, five healthy volunteers have participated in an awaken state with open eyes (A) and closed eyes (B). All the subjects falling in classes 2, 3, 4 and 5 are participants with no epileptic seizures. Subject 1 has a recording of seizure activity. The data are divided into five classes, but most of the time researchers have done binary classification (Andrzejak et al. 2001). https://www.meb.unibonn.de/epileptologie/science/physik/eegdata.html.

2.2 Feature Extractions

Epileptic signals can be differentiated from the normal signal by time-domain features such as amplitude variation, periodicity and regularity. Here, we used time-domain feature amplitude as a feature vector. In this study, total 178 feature sets are used. This dataset was used as a response variable. Total 178 input vectors are shown in Y (where $Y = 1, 2, 3, 4$ and 5). The system is trained and tested with a database of five patients (Andrzejak et al. 2001).

2.3 Feature Selection

In epilepsy detection, we need to know which features are sufficient and appropriate for the tasks, although it is not an easy task to get the prior knowledge of feature which is useful. In this context, feature selection plays a significant role in epilepsy detection. The recording of EEG signal is carried out on 32–96 electrode channels. Therefore, any grouping of features vectors grows speedily with possibly redundant features. More feature vectors make it complicated further for classifiers to identify the patterns in the data accurately. This is called the curse of dimensionality (Keogh and Mueen 2016).

The feature selection algorithms are generally categorized as supervised and unsupervised models. Supervised feature selections are further divided into filter models, wrapper models and embedded models. Filter models are independent of the classifier algorithm; it relies on the characteristics of data. It calculates the entire feature subsets or rank individual features. While there are plenty of filter algorithms described in

Table 2 Common filter univariate methods for feature selection

Name	Filter class	Applicable to task	study
One attribute	Univariate	Classification	Holte (1993)
Gain ratio	Univariate	Classification	Witten and Frank (2011)
Symmetrical uncertainty	Univariate	Classification	Yu and Liu (2003)
ReliefF	Univariate	Regression	Yu and Liu (2003)
Chi square	Univariate	Classification	Witten and Frank (2011)

text, a list of common methods which is used in the classification process is given in Table 2. Filter methods have two types of features, which are univariate filter and multivariate filters. The univariate filters calculate (and usually rank) a solo feature, while multivariate filters evaluate a complete feature datasets. Multivariate filters depend on search strategy. Filter methods are comparatively faster and independent of given algorithm, easy to implement and avoid overfitting (Jovic et al. 2015).

Wrapper methods evaluate the value of given subset by inductive algorithm. It wraps a classifier up in feature selection algorithm (Kohavi and John 1997). This model calculates datasets based on classification output. For every subset, the evaluation is repeated and the production of datasets is dependent on search process. Wrappers are depending on modelling algorithms, and that is why they are much slower then filter method to find sufficiently good subsets. EEG signals used in data analysis are time-consuming and may contain a large number of irrelevant data. In the present work, the relevant feature subsets are used for accurate classification. Furthermore, a new feature selection model is proposed to overcome the drawbacks of conventional feature selection techniques, which uses only a single evaluation criterion.

For the classification difficulty, filter feature selection has been used to select the features that maximizes the classifier performance and capable of differentiating samples of different classes.

In this study, univariate feature selection methods (ReliefF, GR, One attribute, Symmetrical Uncertainty) are utilized to estimate the significance of particular features using top N-choice rule.

Different Feature Selections. These techniques are ensembled of feature selection techniques wherein four filter univariate methods are used, namely ReilefF, GR, One attribute and symmetrical uncertainty. The feature selection models applied in this study is shown in Fig. 2. Here, all 178 features from EEG signal are offered to four distinct filter techniques for classification purpose. For the given feature space, each module generates a list of features that are found most relevant using specific choice. A majority choice is then used to combine the choices of individual feature selection modules due to its simplicity and robust performance. The feature that receives the

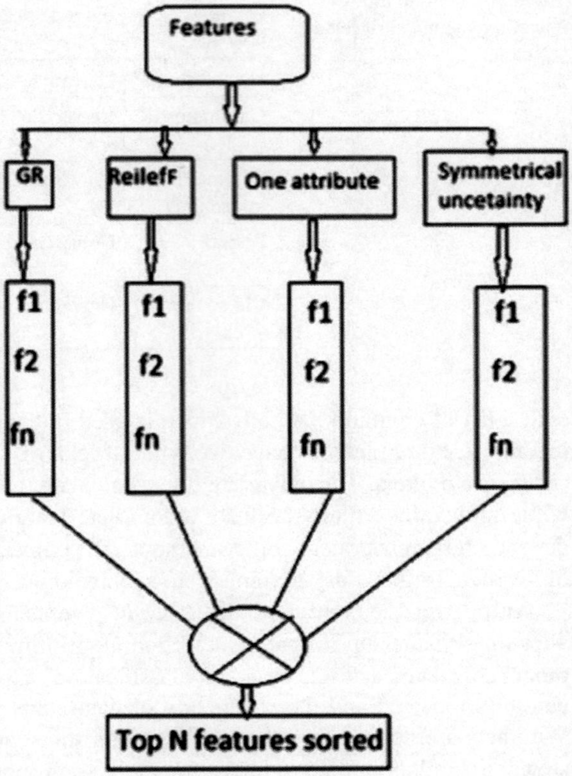

Fig. 2 Block diagram of significant features selection procedure

maximum number of choices is considered to be the most relevant one. Based on the choice received to individual features, top N features are finally selected. The top 20 most relevant features are generated by each method using the individual voting selection criteria.

The methods used in this paper are described below:

(a) ReliefF feature selection method: ReleifF used random selection for weight evaluation. Monte Carlo algorithm is the base of ReliefF feature selection method. According to the relevance of class, this algorithm weighs each features.
(b) Gain Ratio feature selection method (GR): This method is used to compensate the biasness of information gain attributes. It is a non-symmetrical measurement, but information gain is a symmetrical measure. Information gain is biased toward feature with more values rather than features that are more informative. GR is given by

$$GR = \frac{IG}{H(X)} \tag{1}$$

where *IG* is the information gain divided by the entropy of *X*. *GR* values always fall into [0 1] due this normalization.

(c) One Attribute feature selection method: This method was proposed by Holte (Holte 1993) in which he established a single rule for each feature in the training data and then found out the rule with the smallest error.

(d) Symmetrical Uncertainty feature selection method: It evaluates the fitness of features for attribute selection by evaluating feature and the target class. More importance is given to the highest value of SU. It is symmetrical in nature, so it reduces the number of comparison needs.

2.4 Statistical Measures and Classification of EEG Signals

In this paper, EEG signals are classified by three popular classification techniques, namely naive Bayes, random forest and J48 classifier, which are explained below. Next, the performance of the classifiers is compared to the proposed feature selection technique.

Naive Bayes: This naive Bayes model of classifiers is based on the Bayes theorem. This method is a collection of algorithms. It makes use of all the attributes contained in the dataset that are mutually independent, and that is why it is called naive (Rish 2001).

Random forest classifier: Random forest classifier is a randomly selected subset of training set to create a set of decision trees. To decide the final classes of the test object, random forest aggregates the votes from different decision trees. The improvement in the classification performance is enhanced by raising a group of trees and using the most popular tree class (Breiman 2001).

J48 classifier: C4.5 decision tree classifier is defined as J48 classifier. In this classifier, internal nodes indicate the different features, the branches linking the nodes tell us the probable values that these features can have in the experiential samples and the terminal nodes indicate the ultimate results of classification of the dependent variables (Quinlan 1993).

To evaluate the unbiased estimation of generalization error, we measure the prediction of the system on testing sets. The output evaluated from the test data sets can be used to compare with the other classifier performances. For the classification purpose, the training and testing datasets are partitioned by a common methodology. There are different partition methods reported so far, but in this study, we applied two method, namely holdout and cross-validation methods.

Holdout Method: Training and testing data sets are separated by two sets which are disjointed with each other. Generally, two-third is used as training and one-third as a testing (Kohavi 1995).

K-fold Cross-Validation: This method is advancement over holdout method. In this method, every time one of the k datasets is applied as a testing dataset and k-1 dataset as a training dataset. The advantage of this method is that each subset is getting in whole process training and testing once in lifetime. In this research paper, two different cross-validating setups such tenfold cross-validation (k10) and fivefold cross-validations (k5) are used for performance measurements.

In this work, three different classifiers are applied to compare and calculate the performance of feature selection techniques using some parameters such as accuracy (ACC), Recall(SEN), false positive rate (SPE), area under curve (AUC) and Mathew's correlation coefficient. These parameters are defined below:

A. Accuracy (ACC): Classification accuracy is the number of correct predictions, i.e. epileptic datasets are predicted as epileptic, and non-epileptic samples are predicted as non-epileptic. It is evaluated using the following equation:

$$\begin{aligned} \text{ACC} &= \frac{\text{No. of True positves } + \text{ No. of True Negatives}}{\text{Total no. of samples}} \times 100 \text{ Accuracy} \\ &= \frac{\text{No. of correct prediction of samples}}{\text{Total no. of prediction samples}} \times 100 \end{aligned} \tag{2}$$

B. Recall (SEN): Recall is also knows as true positive rate that is defined as the proportion of positive epileptic data correctly predicted as epileptic seizure. Evaluation of Recall is given below

$$\text{SEN} = \frac{\text{No.oftruepositives}}{\text{No.oftruepositives} + \text{No.offalsenegatives}} \times 100 \tag{3}$$

C. False positive rate (SPE): It is also known as true negative rate or false positive rate. The EEG samples correctly predicted as epileptic samples are defined as false positive rate. Evaluation of false positive rate is:

$$\text{SPE} = \frac{\text{No.oftruenegatives}}{\text{No.oftruenegatives} + \text{No.offalsepositives}} \times 100 \tag{4}$$

D. Area under receiver operating characteristic curve (AUC): This is mostly applied in binary classification problems. AUC is defined as the classifier probability that is arbitrarily selected as positive sample instead of negative sample.

$$\text{AUC} = \frac{1}{2}[\text{Recall} + \text{SPE}] \times 100 \tag{5}$$

Matthew's correlation coefficient (MCC): It is utilized as quality measurement criteria for binary classification of machine learning techniques. It is defined as the correlation coefficients between the predicted and experiential values.

$$\text{MCC} = \frac{\text{No. of true positives} \times \text{No. of true negatives} - \text{No. of false negatives} \times \text{No. of false positives}}{\sqrt{\begin{array}{c}(\text{No. of true positives} + \text{No. of false positives})(\text{No. of true positives} + \text{No. of false negatives}) \\ (\text{No. of true negatives} + \text{No. of false positives})(\text{No. of true negatives} + \text{No. of false negatives})\end{array}}} \times 100 \tag{6}$$

The prediction results of classifiers are presented by an easy and definite way called confusion matrix. In the binary classification, confusion matrix represents the false positives, false negatives, true positives and true negatives. A true positive represent the amount of epileptic samples accurately predicted as epileptic, while a false negative represents the amount of epileptic samples wrongly classified as non-epileptic. Similarly, a true negative represents the number of non-epileptic samples correctly classified as epileptic, while a false positive represents the number of non-epileptic samples wrongly classified as epileptic. In this study, the experiments have been carried out using WEKA software. The experimental work carried out on WEKA software is shown in Fig. 3.

3 Result and Discussion

In our experiment, we applied ReliefF, GR, One attribute and symmetrical uncertainty feature selection techniques. The results of feature selection models are mentioned in Table 2. In this study, top 20 features were applied in filter models, which were used for the classification and are shown in Table 3. It is observed that the selected features and ranks of feature subsets specified by different feature selection models are slightly differing. For example, feature X41 is chosen as one of the relevant features by most of the feature selection techniques, whereas it is not present in the best 20 features selected by ReliefF selection method. However, many features are significant to only one feature selection model such as X73, X74, X72, etc. Thus, selection of related features by applying a single condition may omit some important features, which may affect the prediction of classification adversely. In this study, we used top 20 features from each feature selection method to classify different classifiers.

Tables 4, 5 and 6 show different classifier models' classification performance without any feature selection approach and using tenfold, fivefold and holdout–cross-validation techniques. In this part, without applying any feature selection model, whole features have been used for the classification purpose. It is observed that

Table 3 Top 20 features subsets selected by univariate feature selection models

Filter features selection models	Features selected
Gain ratio	X175, X138, X3, X64, X142, X156, X32, X119, X173, X123, X16, X111, X41, X30, X38, X97, X171, X163, X127, X128
One attribute	X37, X127, X159, X107, X23, X160, X169, X2, X65, X21, X158, X141, X170, X106, X112, X108, X173, X110, X166, X85,
ReliefF	X73, X74, X72, X33, X75, X32, X124, X131, X34, X69, X71, X172, X173, X125, X178, X130, X76, X36, X70, X37
Symmetrical uncertainty	X175, X3, X138, X64, X142, X156, X173, X32, X111, X16, X123, X38, X41, X128, X127, X119, X30, X163, X97, X19

Table 4 Statistical measures of different classifier models for 178 features using tenfold approach

Classifier types	Statistical measures (%)				
	ACC	Recall	SPE	AUC	MCC
Naive Bayes	95.739	89.520	97.286	93.403	0.867
Random forest	97.122	96.112	97.352	96.732	0.909
J48	94.818	87.968	96.478	92.223	0.837

Table 5 Statistical measures of different classifier models for 178 features using fivefold approaches

Classifier types	Statistical measure (%)				
	ACC	Recall	SPE	AUC	MCC
Naive Bayes	95.739	89.520	97.286	93.403	0.867
Random forest	97.122	96.112	97.352	96.732	0.909
J48	94.818	87.968	96.478	92.223	0.837

Table 6 Statistical measures of different classifier models for 178 features using holdout

Classifier types	Statistical measure(%)				
	ACC	Recall	SPE	AUC	MCC
Naive Bayes	96.189	90.318	97.664	93.991	0.881
Random forest	97.443	96.712	97.610	97.161	0.919
J48	94.271	86.807	96.066	91.437	0.819

the random forest classifiers produce a better classification accuracy of 97.122% for tenfold technique. Other performance measures, i.e. Recall, false positive rate, AUC and MCC for the proposed classification approach are found to be 96.112%, 97.352%, 96.732% and 0.909%, respectively, for tenfold which are higher than that of other classifiers. The accuracy achieved by random forest classifier is 97.122% for fivefold similar to that of tenfold partition and 97.443% for holdout partition. However, the other statistical measures such as Recall, false positive rate, AUC and MCC are 96.712%, 97.610%, 97.161% and 0.919%, respectively, of 178 features for holdout partition performed a better classification.

Tables 7, 8 and 9 show the statistical measures of different classifier types when only top 20 relevant features are using one attribute feature selection technique. In this feature selection technique, again random forest outperformed other classifiers with an accuracy of 96.235%, 96.408% and 96.522% for tenfold, fivefold and holdout data partition systems, respectively. The other statistical measures such as Recall, false positive rate, AUC and MCC are 92.999%, 96.988%, 94.993% and 0.880%, respectively, for tenfold data partition. Similarly, Recall, false positive rate, AUC and MCC for random forest classifier with one attribute feature selection technique

Table 7 Statistical measures of different classifier models for top 20 features for one attribute feature selection technique using tenfold data partition rule

Classifier types	Statistical measures (%)				
	ACC	Recall	SPE	AUC	MCC
Naive Bayes	95.409	89.483	96.845	93.164	0.855
Random forest	96.235	92.999	96.988	94.993	0.880
J48	94.157	87.581	95.682	91.632	0.814

Table 8 Statistical measures of different classifier models for top 20 features for one attribute feature selection technique using fivefold data partition rule

Classifier types	Statistical measures (%)				
	ACC	Recall	SPE	AUC	MCC
Naive Bayes	95.391	89.439	96.834	93.136	0.855
Random forest	96.408	92.945	97.227	95.086	0.886
J48	94.235	87.357	95.853	91.605	0.817

Table 9 Statistical measures of different classifier models for top 20 features for one attribute feature selection technique using holdout data partition rule

Classifier types	Statistical measures (%)				
	ACC	Recall	SPE	AUC	MCC
Naive Bayes	95.806	89.615	97.348	93.482	0.869
Random forest	96.522	92.838	97.402	95.120	0.890
J48	94.065	85.438	96.203	90.821	0.814

are 92.945%, 97.227%, 95.086% and 0.886% for fivefold and 92.838%, 97.402%, 95.120% and 0.890% for holdout, respectively.

Tables 10, 11 and 12 show the statistical measures of different classifier types when only top 20 relevant features are using ReliefF feature selection technique. In this feature selection technique, again random forest outperformed other classifiers with an accuracy of 95.304, 95.330 and 95.755% for tenfold, fivefold and holdout data partition systems. We can observe a better accuracy with holdout method from the

Table 10 Statistical measures of different classifier models for top 20 features for ReliefF Ranking feature selection technique using tenfold data partition rule

Classifier types	Statistical measures (%)				
	ACC	Recall	SPE	AUC	MCC
Naive Bayes	94.861	88.370	96.420	92.395	0.838
Random forest	95.304	90.854	96.330	93.592	0.850
J48	93.365	85.366	95.229	90.297	0.789

Table 11 Statistical measures of different classifier models for top 20 features for ReliefF Ranking feature selection technique using fivefold data partition rule

Classifier types	Statistical measures (%)				
	ACC	Recall	SPE	AUC	MCC
Naive Bayes	94.852	88.399	96.399	92.399	0.837
Random forest	95.330	90.829	96.371	93.600	0.851
J48	92.826	84.414	94.753	89.583	0.771

Table 12 Statistical measures of different classifier models for top 20 features for ReliefF Ranking feature selection technique using holdout data partition rule

Classifier types	Statistical measures (%)				
	ACC	Recall	SPE	AUC	MCC
Naive Bayes	95.524	89.987	96.858	93.422	0.837
Random forest	95.755	91.734	96.690	94.212	0.851
J48	93.734	86.431	95.430	90.931	0.771

Table 13 Statistical measures of different classifier models for top 20 features for Symmetrical uncertainty feature selection technique using tenfold data partition rule

Classifier types	Statistical measures (%)				
	ACC	Recall	SPE	AUC	MCC
Naive Bayes	95.470	89.657	96.878	93.267	0.857
Random forest	96.844	94.366	97.424	95.895	0.900
J48	94.174	86.514	96.019	91.266	0.816

Table 14 Statistical measures of different classifier models for top 20 features for Symmetrical uncertainty feature selection technique using fivefold data partition rule

Classifier types	Statistical measures (%)				
	ACC	Recall	SPE	AUC	MCC
Naive Bayes	95.470	89.621	96.888	93.255	0.857
Random forest	96.791	94.147	97.412	95.780	0.869
J48	94.174	86.514	96.019	91.266	0.816

above-mentioned result. The other statistical measures such as Recall, false positive rate, AUC and MCC are 90.854%, 96.330%, 93.592% and 0.850%, respectively, for tenfold data partition. Similarly, Recall, false positive rate, AUC and MCC for random forest classifier with one attribute feature selection technique are 90.829%, 96.371%, 93.600% and 0.851% for fivefold and 91.734%, 96.690%, 94.212% and 0.851% for holdout, respectively.

Table 15 Statistical measures of different classifier models for top 20 features for Symmetrical uncertainty feature selection technique using holdout data partition rule

Classifier types	Statistical measures (%)				
	ACC	Recall	SPE	AUC	MCC
Naive Bayes	96.010	90.755	97.295	94.025	0.875
Random forest	96.905	94.244	97.534	95.889	0.902
J48	94.400	87.671	95.990	91.830	0.822

Tables 13, 14 and 15 show the statistical measures of different Classifiers types when only top 20 relevant features are using symmetrical uncertainty as a feature selection technique. In this feature selection technique, again random forest classifier gave the best accuracy of 96.844%, 96.791% and 96.905% for tenfold, fivefold and holdout data partition system, respectively. It is found that split data partition approach produce the highest classification accuracy, almost similar to all 178 features. The other statistical measures such as Recall, false positive rate, AUC and MCC are 94.366%, 97.424%, 95.895% and 0.900%, respectively, for tenfold data partition. Similarly, Recall, false positive rate, AUC and MCC for random forest classifier with symmetrical uncertainty feature selection technique are 94.147%, 97.412%, 95.780% and 0.869% for fivefold and 94.244%, 97.534%, 95.889% and 0.902% for holdout, respectively.

Tables 16, 17 and 18 show the statistical measures of different classifier types using only top 20 relevant features for gain ratio feature selection technique. It is observed that random forest classifier produced the highest classification accuracy of 95.965%, 95.939% and 96.113% for tenfold, fivefold and holdout approaches,

Table 16 Statistical measures of different classifier models for top 20 features for Gain Ratio feature selection technique using tenfold data partition rule

Classifier types	Statistical measure(%)				
	ACC	Recall	SPE	AUC	MCC
Naive Bayes	95.180	89.078	96.654	92.866	0.848
Random forest	95.965	92.110	96.867	94.489	0.872
J48	94.096	86.460	95.923	91.191	0.813

Table 17 Statistical measures of different classifier models for top 20 features for Gain Ratio feature selection technique using fivefold data partition rule

Classifier model	Statistical measure (%)				
	ACC	Recall	SPE	AUC	MCC
Naive Bayes	95.150	88.958	96.653	92.806	0.847
Random forest	95.939	91.716	96.936	94.326	0.871
J48	93.896	86.154	95.737	90.946	0.807

Table 18 Statistical measures of different classifier models for top 20 features for Gain Ratio feature selection technique using holdout data partition rule

Classifier model	Statistical measure				
	ACC	Recall	SPE	AUC	MCC
Naive Bayes	95.729	89.883	97.165	93.524	0.866
Random forest	96.113	91.777	97.148	94.463	0.877
J48	93.760	84.845	95.953	90.399	0.804

respectively, amongst all classifiers. The other statistical measures such as Recall, false positive rate, AUC and MCC are 92.110%, 96.867%, 94.489% and 0.872%, respectively, for tenfold data partition. Similarly, Recall, false positive rate, AUC and MCC for random forest classifier with symmetrical uncertainty feature selection technique are 91.716%, 96.936%, 94.326% and 0.871% for fivefold and 91.777%, 97.148%, 94.463% and 0.877% for holdout, respectively.

It is found that random forest classifier performed best among all the classifiers. But, if we compare better feature selection method, then symmetrical uncertainty feature selection approaches the highest classification as compared to all feature selection techniques.

From the results obtained from the different feature selection techniques (Tables 19, 20 and 21), it is observed that symmetrical uncertainty feature selection gave a better performance than other filter-based feature selection techniques. In

Table 19 Statistical measures of Random forest classifier models for top 20 features for different feature selection technique using tenfold data partition rule

Feature selection method	Statistical measure (%)				
	ACC	Recall	SPE	AUC	MCC
Gain ratio	95.965	92.110	96.867	94.489	0.872
One attribute	96.235	92.999	96.988	94.993	0.880
ReliefF	95.304	90.854	96.330	93.592	0.850
Symmetrical uncertainty	96.844	94.366	97.424	95.895	0.900

Table 20 Statistical measures of Random forest classifier models for top 20 features for different feature selection technique using fivefold data partition rule

Feature selection method	Statistical measure (%)				
	ACC	Recall	SPE	AUC	MCC
Gain ratio	95.939	91.716	96.936	94.326	0.871
One attribute	96.408	92.945	97.227	95.086	0.886
ReliefF	95.330	90.829	96.371	93.600	0.851
Symmetrical uncertainty	96.791	94.147	97.412	95.780	0.869

Table 21 Statistical measures of random forest classifier models for top 20 features for different feature selection techniques using holdout data partition rule

Feature selection method	Statistical measure (%)				
	ACC	Recall	SPE	AUC	MCC
Gain ratio	96.113	91.777	97.148	94.463	0.877
One attribute	96.522	92.838	97.402	95.120	0.890
ReliefF	95.755	90.734	96.690	93.212	0.851
Symmetrical uncertainty	96.905	94.244	97.534	95.889	0.902

this paper, random forest classifier achieves the highest classification measurements when symmetrical uncertainty feature selection technique is used.

Figure 3 shows the performance of random forest classifier with different feature selection techniques. The graph shows that the accuracy is improved with symmetrical uncertainty feature selection technique along with holdout method.

The performance accuracy of previous studies and in this studies are compared to same datasets are shown in Table 22.

It can be concluded from above results that the highest classification accuracy and false positive rate are obtained with top N features with random forest classifiers and symmetrical uncertainty feature selection techniques. This study can be useful for the classification of epileptic and nonepileptic datasets (Fig. 4).

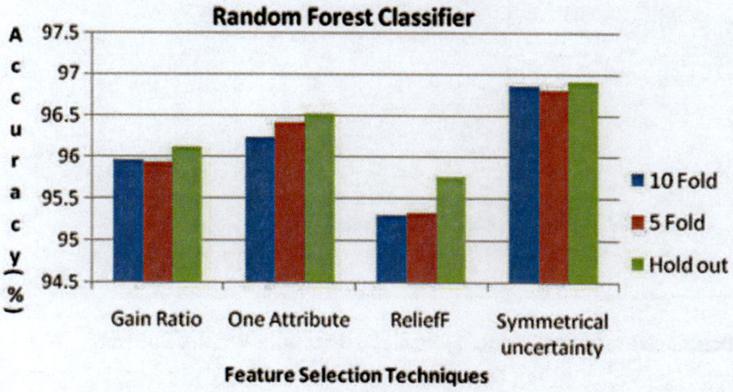

Fig. 3 The performance of Random forest classifier with different feature selection techniques and cross-validation method

Table 22 Comparison chart of classification accuracies found by previous researchers for epileptic datasets

Author	Methodology	Statistical parameters
Kannathal et al. (2005)	Adaptive Neuro-Fuzzy Inference system (ANFIS)	Classification accuracy: 90%
Subasi (2006)	WT and ANFISMixture expert model (a modular neural network)	Accuracy: 95%, False positive rate: 93.7% Recall: 94.3%
Guo et al. (2010)	ANN classifier and wavelet energy based features	Accuracy: 95.2%, Recall: 98.17%, false positive rate: 92.12%
Gajic et al. (2104)	Wavelet Transform and quadratic classifier	Accuracy: 95%,
Vidyaratne and Iftekharuddin (2017)	FD and HWPT energy features and vector machine classifier	Accuracy: 99.8%, Recall: 96%,
Our methods	Random forest classifiers with symmetrical uncertainty feature selection techniques	Accuracy: 96.905%, SPE: 97.534

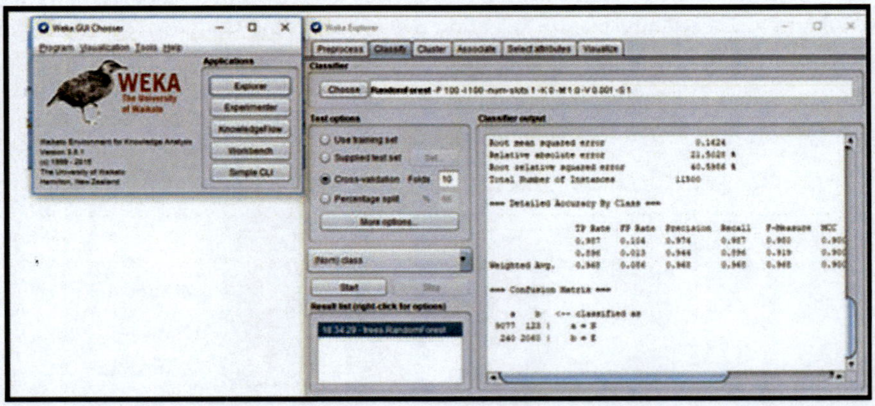

Fig. 4 Experimental result of Random forest classifier with WEKA Software

4 Conclusion and Future Perspectives

4.1 Conclusion

In this study, we have investigated the problems related to the detection of epilepsy by three different classifiers using top N feature selection criteria. Compared to the traditional single feature selection criterion, our approaches performed better to remove the redundancy of the datasets. We employed filter method as feature selection technique because it is easy in implementation and has low complexity as compared to wrapper method. In this study, the algorithm employs four different filter feature selection methods and opted top 20 features among them to evaluate the significance of various features. The result shows that symmetrical uncertainty feature selection performs very well over other feature selection models, achieving the highest classification accuracy of 97.122% for random forest classifier.

The random forest classifier achieved the highest accuracy of 97.122%, Recall—96.112%, false positive rate—97.352%, AUC—96.732% and MCC—0.909% for a subset of 20 best features. The most relevant features (top 20) improved the performance of random forest classifier with symmetrical uncertainty feature selection technique.

4.2 Future Perspectives

Seizure characteristic training has many variations. So, it is a challenging task to make a generalized system. Furthermore, improvement in the performance of classifiers may be available in future work using multivariate and other non-linear filter features selection techniques. For the advance seizure prediction in the future, the combination of filter and wrapper feature selection techniques may have produced advantages with robust datasets and advanced classifier algorithms. The prediction of seizure before its occurrence is still a difficult task and needs further active rigorous research.

References

Acharya UR, Sree SV, Alvin APC, Suri JS (2012) Application of non-linear and wavelet based features for the automated identification of epileptic EEG signals. Int J Neural Syst 22(2):1250002-1-14

Acharya UR, Sree SV, Swapna G, Martis RJ, Suri JS (2013) Automated EEG analysis of epilepsy: a review. Knowl Based Syst 45:147–165

Al-Ghayab HR, Li Y, Abdulla S, Diykh M, Wan X-K (2016) Classification of epileptic EEG signals based on simple random sampling and sequential feature selection. Brain Inform 3:85–91

Alzami F et al (2018) Adaptive hybrid feature selection-based classifier ensemble for epileptic seizure classification. IEEE Access 6:29132–29145

Anderson WS, Kossoff EH, Bergey GK, Jallo GI (2008) Implantation of a responsive neurostimulator device in patients with refractory epilepsy. Neurosurgical Focus 25(3):E12

Andrzejak RG, Lehnertz K, Mormann F, Rieke C, David P, Elger CE (2001) Indications of nonlinear deterministic and finite-dimensional structures in time series of brain electrical activity: dependence on recording region and brain state. Phys Rev E 64:061907

Bhattacharyya A, Pachori RB, Upadhyay A, Acharya U (2017) R: Tunable-Q wavelet transform based multiscale entropy measure for automatedclassification of epileptic EEG signals. Appl Sci 7(4):385

Blume W Lüders H, Mizrahi E, Tassinari C, van Emde Boas W, Engel J (2001) Glossary of descriptive terminology for ictal semiology: report of the ILAE task force on classification and terminology. Epilepsia 42(9):1212–1218

Breiman L (2001) Random forests. Mach Learn 45(1):5–32

Chua KC, Chandran V, Acharya UR, Lim CM (2009) Automatic identification of epileptic EEG signals using higher order spectra. J Eng Med 223(4):485–495

Gajic D, Djurovic Z, Di Gennaro S, Gustafsson F (2014) Classification of EEG signals for detection of epileptic seizures based on wavelets and statistical pattern recognition, biomedical engineering: applications. Basis Commun 26(2)

Gopan KG, Harsha A, Joseph LA, Kollialil ES (2013) Adaptive neuro-fuzzy classifier for 'Petit Mal'epilepsy detection using mean teager energy. In: 2013 international conference on advances in computing, communications and informatics (ICACCI). Mysore, pp 752–757

Guler I, Ubeyli ED (2005) Adaptiveneuro-fuzzy inference system for classification of EEG signals using wavelet coefficients. J Neurosci Methods 113–121

Guo L, Rivero D, Dorado J, Rabunal JR, Pazos A (2010) Automatic epileptic seizure detection in EEGs based on line length feature and artificial neural networks. J Neurosci Methods 191(1):101–109

Guo L, Rivero D, Munteanu CR, Pazos A (2011) Automatic feature extraction using genetic programming: an application to epileptic EEG classification. Expert Syst Appl 38(8):10425–10436

Harpale V, Bairagi V (2018) An adaptive method for feature selection and extraction for classification of epileptic EEG signal in significant states. J King Saud Univ Comput Inf Sci 1319–1578

Hassan AR, Siuly S, Zhang Y (2016) Epileptic seizure detection in EEG signals using tunable-Q factor wavelet transform and bootstrap aggregating. Comput Methods Programs Biomed 137:247–259

Hekim M (2012) ANN-based classification of EEG signals using the average power based on rectangle approximation window, PRZEGLĄD ELEKTROTECHNICZNY (Electrical Review). 210–215. ISSN 0033-2097

Holte RC (1993) Very simple classification rules perform well on most commonly used datasets. Mach Learn 11:63–91

Jovic A, Brkic K, Bogunovic N (2015) A review of feature selection methods with applications. In: 2015 38th international convention on information and communication technology, electronics and microelectronics (MIPRO). IEEE, Opatija, Croatia, pp 1200–1205

Kannathal N, Choo ML, Acharya UR, Sadasivan PK (2005) Entropies for detection of epilepsy in EEG. Comput Methods Programs Biomed 80(3):187–194

Keogh E, Mueen A (2016) Curse of dimensionality. Encyclopedia of machine learning. Springer, pp 257–258

Kohavi R (1995) A study of cross-validation and bootstrap for accuracy estimation and model selection. In: Proceedings of the 14th international joint conference on Artificial intelligence, vol 2. Morgan Kaufmann Publishers, Montreal, Quebec, Canada, pp 1137–1143

Kohavi R, John GH (1997) Wrappers for feature subset selection. Artif Intell 97(1–2):273–324

Litt B, Echauz J (2002) Prediction of epileptic seizures. Lancet Neurology 1(1):22–30

Li M, Chen W, Zhang T (2017) Automatic epileptic EEG detection using DT-CWT based non-linear features. Biomed Signal Process Control 34:114–125

Mi L, Chen W, Zhang T (2017) Classification of epilepsy EEG signals using DWT-based envelope analysis and neural network ensemble. Biomed Signal Process Control 31:357–365

Niedermeyer E, Lopes da Silva F (2005) Electroencephalography: basic principles, clinical applications, and related fields. 5th edn. Lippincott Williams & Wilkins. ISBN 0781751268

Pachori RB, Patidar S (2014) Epileptic seizure classification in EEG signals using second-order difference plot of intrinsic mode functions. Comput Methods Programs Biomed 113(2):494–502

Pippa E et al (2016) Improving classification of epileptic and non-epileptic EEG events by feature selection. Neurocomputing 171:576–585

Quinlan R (1993) C4.5: programs for machine learning. Mach Learn 16(3):235–240

Quyen MLV, Navarro V, Baulac M, Renault B, Martinerie J (2000) Anticipation of epileptic seizures from standard EEG recordings. The Lancet 361(9361):970–971

Rabbi AF, Fazel-Rezai R (2012) A fuzzy logic system for seizure onset detection in intracranial EEG. Comput Intell Neurosci 2012:1–12

Rish I (2001) An empirical study of the Naive Bayes classifier. In: IJCAI 2001 workshop on empirical methods in artificial intelligence, vol 3, pp 41–46

Skarpaas TL, Morrell MJ (2009) Intracranial stimulation therapy for epilepsy. Neurotherapeutics 6(2):238–243

Subasi A (2006) Automatic detection of epileptic seizure using dynamic fuzzyneural networks. Expert Syst Appl 31(2):320–328

Tawfik NS, Youssef SM, Kholief M (2016) A hybrid automated detection of epileptic seizures in EEG records. Comput Electr Eng 53:177–190

Vidyaratne LS, Iftekharuddin KM (2017) Real-time epileptic seizure detection using EEG. IEEE Trans Neural Syst Rehabil Eng 25(11):2146–2156

Witten IH, Frank E (2011) Data mining: Practical machine learning tools and techniques, 2nd edn. Morgan Kaufmann Elsevier, San Francisco CA, USA

Yang XF, Rothman SM (2001) Focal cooling rapidly terminates experimental neocortical seizures. Ann Neurol 49(6):721–726

Yang XF, Duffy DW, Morley RE, Rothman SM (2002) Neocortical seizure termination by focal cooling: temperature dependence and automated seizure detection. Epilepsia 43(3):240–245

Yu L, Liu H (2003) Feature selection for high-dimensional data: a fast correlation-based filter solution. In: 20th international proceeding of international conference on machine learning (ICML-2003). AAAI Press, Washington DC, USA, pp 856–863

Feature Selection for Classification of Breast Cancer in Histopathology Images: A Comparative Investigation Using Wavelet-Based Color Features

Kushangi Atrey, Bikesh Kumar Singh, and Narendra K. Bodhey

Abstract Breast cancer is taken into account as one of the foremost common kinds of cancer occurring in women worldwide. This work is intended to develop an automated classification system in order to classify the breast tumor as benign or malignant. An online database with 645 benign and 1370 malignant histopathological images with 40X magnification is used in this work. A total of 174 color features based on wavelet transform are extracted. To work out the foremost relevant attributes for classifying breast tumors, feature selection is carried out. The best subset is used for feature choice (wrapper), whereas prime 18 features are used for filter-based methods. A back-propagation artificial neural network (BPANN) is used as the classifier. Various performance measures like accuracy, sensitivity, specificity, etc., are used for evaluating the performance of BPANN classifier after applying different feature selection techniques. Classification accuracy of 98.56% with tenfold data division protocol is obtained when all 174 color wavelet transform-based features were used for classification. However, by using different feature selection techniques, the highest accuracy is obtained as 95.83% for the wrapper approach when 18 relevant features were used.

Keywords Breast cancer · Color features · Histopathological images · Feature selection

K. Atrey (✉) · B. K. Singh
Department of Biomedical Engineering, National Institute of Technology, Raipur, CG, India
e-mail: katrey.phd2017.bme@nitrr.ac.in

N. K. Bodhey
Department of Radiodiagnosis, All India Institute of Medical Sciences, Raipur, CG, India

A. A. Rizvanov et al. (eds.), *Advances in Biomedical Engineering and Technology*, Lecture Notes in Bioengineering,
https://doi.org/10.1007/978-981-15-6329-4_30

1 Introduction

1.1 Background

Breast cancer is one of the foremost types of cancer resulting in death among women worldwide (Wei et al. 2017). Around 5,08,000 women died worldwide in 2011 due to breast cancer. In the US, over 2,66,000 new cases are expected for the year 2018 (Breast cancer—Statistics and facts statista 2018). The tumors in breast are broadly classified as benign and malignant. Various types of methods used for the diagnosis of breast cancer are mammograms (X-rays), magnetic resonance imaging, ultrasound, etc. (Spanhol et al. 2016). However, a biopsy is carried out by the pathologist for determining whether the tumor is cancerous or not. The elaborated analysis of a diagnostic test (biopsy) sample that is completed by a specialist (pathologist) is histopathology (Aswathy and Jagannath 2017). It requires microscopic examination of tissue to study the disease.

The manual approach requires an experienced pathologist. Also, due to the inter- and intra-variability among radiologists, automatic analysis is needed for accurate analysis. Hence, the researchers started working with CAD (Computer-aided diagnosis) systems.

We have considered a database with 645 benign and 1370 malignant histopathology images of breast cancer in this paper. A total of 174 color features based on wavelet transform are extracted. Also, several attribute selection techniques are applied to decide the most relevant features. The feature evaluation is carried out using a machine learning model based on the back-propagation neural network (BPANN).

1.2 Related Work

Recent studies in the related area have been reported in this section.

Belsare et al. (2015) conducted experiments on 70 images of both benign and malignant categories using a linear discriminant classifier to classify the breast histopathology images (Belsare et al. 2015). They reported that a linear discriminant analyzer (LDA) performed better as compared to other classifiers (support vector machine and k-nearest neighbor).

Spanhol et al. (2016) divided the dataset as 70% for train and 30% for test data (Spanhol et al. 2016). They showed that Parameter-Free Threshold Adjacency Statistics (PFTAS) performs best for 200X magnification using SVM with a recognition rate of 85.1%. For 40X magnification, a rate of 100% was achieved for oracle using Quadratic linear analysis (QDA).

Convolutional neural network (CNN) was utilized by Bayramoglu et al. (2016) for malignancy determination and magnification level of image (Bayramoglu et al. 2016). By performing experiments on a dataset of 7909 histopathology images of breast cancer, they achieved an average recognition rate of 83.25% using fivefold data division protocol.

Kahya et al. (2017) proposed a classifier based on an adaptive sparse support vector machine (ASSVM) and a feature selection technique to classify images of different magnifications to classify histopathology images of breast cancer. Classification accuracies of 95.37 and 94.97% were achieved for training and testing respectively for 40X magnification. The results showed ASSVM performed well as compared to other classifiers like SSVM-lasso and SVM, etc. (Kahya et al. 2017). Also, ASSVM also showed better performance for other magnifications as per the features selected by each method.

Swathy et al. (2017) performed classification on MITOS-ATYPIA-14 dataset with an artificial neural network (ANN) as classifier. Mean square error (MSE) was used as a performance function (Rajan and Swathy 2017) and fivefold cross-validation was implemented. They trained ANN with the Levenberg–Marquardt back-propagation algorithm using gray-level co-occurrence matrix (GLCM) and chip histogram-based texture features and obtained an accuracy of 96.5%.

By using images of 400X magnification of 588 benign and 1232 malignant categories, Samah et al. (2017) proposed a method to classify cancer cells as benign or malignant. Using k-nearest neighbor classifier, they compared various feature extractors (Samah et al. 2017). The highest accuracy was achieved as 85.62% with pyramid-structure wavelet transform as compared to GLCM, Local Binary Patterns (LBP), and tree-structure wavelet transform.

Using deep convolutional neural networks (BiCNN model) Wei et al. (2017) obtained a classification accuracy of about 97% for magnification of 40X to classify breast cancer (Wei et al. 2017).

In another approach by Araujo et al. (2017), classification is performed using Convolutional Neural Networks (CNNs). First, in the pre-processing step, the images were normalized then the features extracted by CNN were used to train SVM (Araujo et al. 2017). Threefold cross-validation was used with radial basis function kernel function. Majority voting was applied to obtain results with accuracies of 83.3% for carcinoma/non-carcinoma (image-wise classification) and 77.8% for four classes. The sensitivity was achieved as 95.6% using CNN with SVM.

Motlagh et al. (2018) collected data from two different types of sources. They used RGB color-map to preserve different features and tissue structures of histopathological images (Motlagh et al. 2018). They used inceptions and ResNet (Residual network) architectures by implementing a deep learning approach with 10% test and 90% train data set. The correctly classified malignant and benign cancer type was obtained as 98.4%.

Another approach was proposed for the classification and detection of breast in histopathological images using deep CNNs by Al Rahhal (2018). They divided the dataset into 30% test and 70% train data (Al Rahhal 2018). High patient-level classification accuracy of 86.80% was achieved with VGGm (Visual geometry group) model.

CNNs were proposed by Bardou et al. (2018) for the classification of breast cancer on 7909 microscopic images consisting of two types of breast tumors (Bardou et al.

2018). Dense Scale Invariant Feature Transform (DSIFT) features and Speeded-Up Robust Features (SURF) were used as local descriptors. For multi-class classification, the accuracy between 83.31 and 88.23% was achieved, while for binary classification, the accuracy achieved was between 96.15 and 98.33%.

Using the structural and statistical information from the 7909 histopathology images available online, Nahid et al. (2018) proposed the use of CNN, a Long Short-Term Memory (LSTM), and a combination of both to classify breast cancer (Nahid et al. 2018). SVM and Softmax layers were used after feature extraction. The best value of F-measure was obtained for both 40X and 100X, the best precision value was obtained for the 40x dataset as 96.00% and the value of best accuracy was achieved on the 200X dataset as 91.00%.

Breast Cancer Histology Challenge (BACH) 2018 dataset consisting of 400 histology microscopy images was used by Golatkar et al. (2018) that utilized a deep learning-based method for classification. They used a fine-tuning Inception-v3 CNN (Golatkar et al. 2018). For classifying benign versus malignant, 93% accuracy was achieved, while an average accuracy over the four different classes was achieved as 85%.

After reviewing the related studies, the most commonly used classifiers are found as SVM, CNN, ANN k-NN, etc. Also, the studies reported mostly used grayscale features for classification. Some features are also obtained using the co-occurrence matrix. Several color features based on wavelet transform, extracting the mean and standard deviation coefficients are also used. We intend to develop a model to classify breast tumors as benign or malignant using BPANN algorithm with different data division protocol and achieving classification accuracy with and without applying feature selection techniques and compare the results.

1.3 Contributions

This work focuses on the classification of breast tumors as benign or malignant using two different approaches: First, using all 174 color wavelet transform features and second, using the various feature selection techniques to obtain the relevant features. The classification accuracy obtained after classification with BPANN is compared. The classification accuracy with all 174 color wavelet transform features using ANN is achieved as 98.56%, while the highest accuracy obtained using different feature selection techniques is 95.83% for wrapper approach with 18 relevant features.

1.4 Organization of Paper

In the present work, materials and methods used are reported in Sect. 2. Results and discussions are explained in Sect. 3. In Sect. 4, conclusions and future scope have been discussed.

Table 1 Details of all 174 features

S.No	Name of features	Details of features
1	Mean	R/G/B Plane: mean1 to mean 87 (Ghazali et al. 2007)
2	Standard deviation	R/G/B Plane: std1 to std 87 (Ghazali et al. 2007)

2 Materials and Methods

2.1 Data

We have used an online publicly available dataset (BreakHis dataset) for this study. It consists of 7909 images of benign and malignant tumors. This dataset contains the histopathological images broadly of two types: benign and malignant breast tumors (Spanhol et al. 2016) with magnification 40X, 100X, 200X, and 400X of each type. The size of each image is 700 × 460. For this study, we have used 645 benign and 1370 malignant images of 40X magnification. During feature extraction, the images are resized to 400 × 400.

2.2 Feature Extraction

In order to find out the most important data by reducing the dimension of input to represent the image in an efficient way and to differentiate each class, feature extraction is carried out. In this study, we have extracted a total of 174 wavelet transform-based color features. BPANN is employed to perform classification on these features.

Color features. As color features adds more information to an image and it also shows better stability, it is considered as one of the most widely used features (Kumar and Chauhan 2013; Anusha et al. 2014). A mathematical way of representing data is wavelet analysis. To calculate the coefficient distribution over selected mother of wavelet, feature extraction from wavelet transformation is carried out. Discrete Wavelet Transform (DWT)-based image processing technique is implemented for this (Ghazali et al. 2007). The feature vector is built using the mean and standard deviation values of wavelet transform coefficients of an image. An RGB color plane is used to represent each pixel as a three-element vector (Table 1).

2.3 Feature Selection

Several data sets are multidimensional and larger in size. Hence, reducing the attributes is one of the important tasks, as larger dataset if used for classification requires more computation time and may sometimes lead to an interpretation of wrong results.

After all 174 features are extracted, it is important to find out the most relevant attributes to classify breast tumors. For this purpose, feature selection is carried out using WEKA. Broadly, the feature choice strategies are classified as filter- and wrapper-based techniques. Filter-based techniques use rank to select relevant features based upon their ordering, while wrapper method generates the subset of features.

For this study, filter methods like information gain, gain ratio, and wrapper-based method J48 are used. They work as follows:

Information Gain: It is a filter-based feature selection algorithm for ranking the features. It measures the dependency between the feature and class label (Phyu and Oo 2016; Porkodi 2014).

For a feature X with class label Y, IG is given by

$$IG = H(X) - H(X|Y) \tag{1}$$

where $H(X)$ is the entropy of X, while the entropy of X after observing Y is defined by $H(X|Y)$.

Information gain is evaluated and features with a high value of information gain are selected as relevant features.

Gain ratio: It enhances information gain by offering a normalized score of a feature's contribution to an optimal information gain-based classification decision. It is represented by

$$GR = \frac{IG}{H(X)} \tag{2}$$

For predicting variable Y, normalization of information gain is to be done by dividing by the entropy of X, and vice versa (Porkodi 2014). A high value of the gain ratio for selected features shows the actual feature is more helpful for classification.

J48: In WEKA data mining tool, J48 classifier is a decision tree that is used for classification. It uses the wrapper approach in which the induction algorithm is regarded as a black box (Phyu and Oo 2016; Kohavi and John 1997). The dataset is divided into training sets and the algorithm uses separate sets of features removed from the data. The attribute subset, which has the highest value of assessment, is considered as the final set. Here, tenfold cross-validation is considered as the evaluation function. The search method is used to find the state which is having a maximum evaluation. In this study, best-first search method is used.

The best subset is obtained using wrapper approach and the top 18 features are selected for filter-based methods (Table 2).

Table 2 Eighteen most relevant features subset

Category	Feature selection technique	Selected features
Filter approach	Information gain	mean1, mean2, mean11, mean5, mean10, mean4, mean9, mean3, mean13, mean6, mean7, std30, mean8, mean15, std31, mean12, mean14, std32
Filter approach	Gain ratio	std66, mean49, mean12, mean19, mean16, mean34, mean22, std63, mean23, mean10, mean25, mean14, mean46, mean15, mean24, mean20, mean83, mean85
Wrapper approach	J48	mean15, mean47, mean56, mean73, mean74, mean78, std9, std19, std23, std32, std35, std37, std42, std44, std66, std67, std76, std78

2.4 *Classification*

BPANN is used for the classification. The classification is done using tenfold data division protocol.

Back-Propagation Artificial Neural Networks (BPANN). For this study, we have considered BPANNs. It is considered as one of the common types of neural networks. An input layer is considered in which the input is 174 wavelet transform features and the 18 features selected by applying various feature selection techniques. One hidden layer is considered with 20 hidden neurons and an output layer is considered as shown in Fig. 1. The function of hidden neurons is to get involved between the input and network output (Haykin 1994). The transfer function used at the output of hidden layer is hyperbolic tangent sigmoid transfer function, while log sigmoid at the output layer. The number of maximum iterations is set to 1000. For training the neural network, trainlm is used. This training function is Levenberg–Marquardt method of updating the weights and bias.

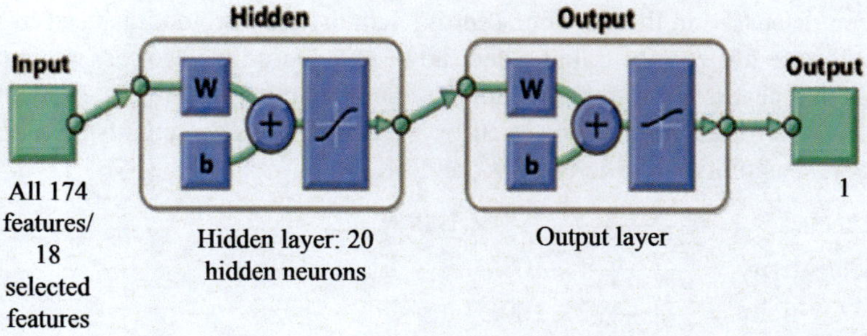

Fig. 1 Architecture of neural network

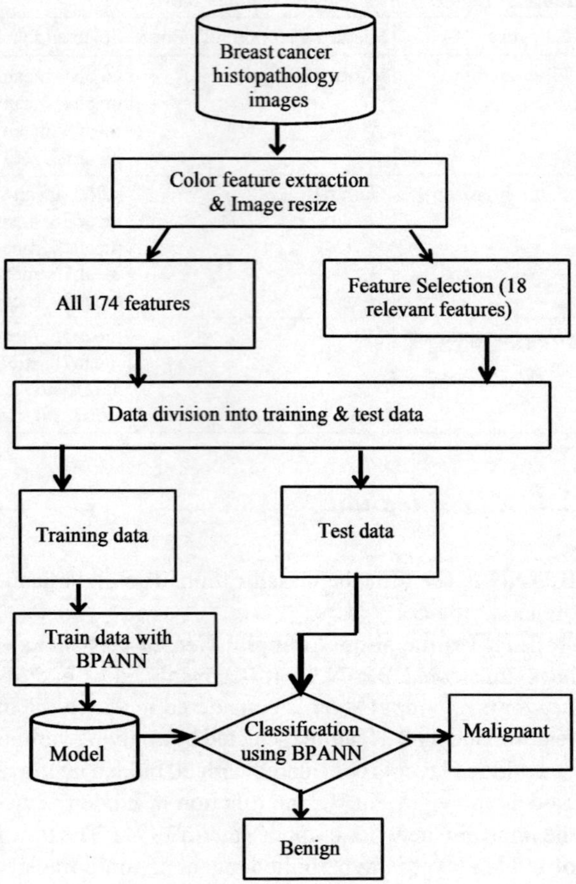

Fig. 2 Representation of the proposed approach

The representation of the proposed approach is shown in Fig. 2. To classify the tumor as benign, 0 is used, while to classify the malignant tumors, 1 is used.

Performance Evaluation. The result of performance measures using BPANN has been discussed in this section. Tenfold data division protocol is used in which nine folds are utilized for training and the last fold is utilized for assessment. The process is repeated ten times, abandoning one distinctive fold for assessment each time. The performance measures can be evaluated from the confusion matrix as accuracy, sensitivity, specificity, AUC, and MCC.

(i) Accuracy $= \frac{(TP+TN)}{(TP+FN+TN+FP)} \times 100$(3)
(ii) Sensitivity $= \frac{TP}{(FN+TP)} \times 100$(4)
(iii) Specificity $= \frac{TN}{(TN+FP)} \times 100$(5)
(iv) AUC (Area under the receiver operating characteristic curve) $= \frac{1}{2}(\frac{TP}{(FN+TP)} + \frac{TN}{(TN+FP)}) \times 100$(6)
(v) MCC (Matthew's correlation coefficient) =

$$\frac{(TP\times TN)-(FP\times FN)}{\sqrt{(TP+FP)(TP+FN)(TN+FP)(TN+FN)}}\times 100 \quad (7)$$

where *TP*, *FP*, *FN*, and *TN* denote true positive, false positive, false negative, and true negative values, respectively.

3 Results and Discussions

BreakHis dataset is considered and histopathology images have been utilized in this work. A total of 2015 images of both benign and malignant categories have been considered with 40X magnification after resizing to 400 × 400. Feature extraction is performed using wavelet transform and means, standard deviation coefficients were acquired for R/G/B color plane. Several feature selection techniques were also applied to find out the most relevant attributes. Eighteen such features were obtained. All 174 wavelet transform features and the 18 features obtained using feature selection were the input to ANN classifier.

Tables 3 and 4 show the various performance measures (%) using mean squared error as a performance function for tenfold data division protocol. The results are obtained using trainlm as the training function and 20 hidden neurons. Using all 174 color features, the classification accuracy of 98.56% is obtained while using several feature selection techniques the highest accuracy of 95.83% is achieved with the wrapper approach. With other filter-based methods accuracies of 91.75% (*IG*) and 95.38% (*GR*) is obtained.

The performance of the proposed method is compared with some recent studies based on the literature review. Some authors utilized CNNs, deep convolutional neural networks, ANNs (Al Rahhal et al. 2018; Wei et al. 2017; Swathy et al. 2017) as classifiers and achieved classification accuracies of 86.80%, 97%, and 96.5%, respectively. Samah et al. 2017, and Kahya et al. 2017, used the concept of wavelet

Table 3 Performance of BPANN classifier with all 174 features

Features	Performance measures (%)				
	Accuracy	Sensitivity	Specificity	AUC	MCC
All 174 color wavelet transform features	98.56	99.27	97.04	98.15	96.61

Table 4 Performance of BPANN classifier with different feature selection techniques

Feature selection technique	Performance measures (%)				
	Accuracy	Sensitivity	Specificity	AUC	MCC
Information gain	91.75	95.40	82.28	88.84	79.46
Gain ratio	95.38	97.37	90.21	93.79	88.72
J48	95.83	97.66	91.31	94.48	89.88

transform with SVM and k-nearest neighbor as classifiers and obtained classification accuracies of 85.62% for pyramid structure wavelet transforms and between 93.62 and 96.28% (for test and train data with different magnifications), respectively. Not many authors have used feature selection techniques. Also, mostly the work is carried out on grayscale images. So in this paper, we have used wavelet transform-based color features and achieved a classification accuracy of 98.56% for all 174 color features with ANN while using several feature selection techniques the highest accuracy of 95.83% is obtained with wrapper approach.

4 Conclusion and Future Scope

A classification model has been presented in this work using the BPANN algorithm. The performance of the classifier was evaluated based on accuracy, sensitivity, specificity, AUC, and MCC. The proposed classifier presented a classification accuracy of 98.56% when all 174 color wavelet transform features were used. The accuracy was reduced to 95.83% by selecting 18 relevant features.

In future, some more feature selection techniques can be implemented to enhance the performance of classifiers by increasing the accuracy.

References

Al Rahhal MM (2018) Breast cancer classification in histopathological images using convolutional neural network. Breast Cancer 9(3)

Anusha MV, Reddy MVU, Ramashri DT (2014) Content based image retrieval using color moments and texture, vol 3, no 2, pp 2812–2815

Araujo T, Aresta G, Castro E, Rouco J, Aguiar P, Eloy C, Polonia A, Campilho A (2017) Classification of breast cancer histology images using convolutional neural networks. PLoS ONE 12(6):e0177544

Aswathy MA, Jagannath M (2017) Detection of breast cancer on digital histopathology images: present status and future possibilities. Inform Med Unlocked 8:74–79

Bardou D, Zhang K, Ahmad SM (2018) Classification of breast cancer based on histology images using convolutional neural networks. IEEE Access 6:24680–24693

Bayramoglu N, Kannala J, Heikkila J (2016) Deep learning for magnification independent breast cancer histopathology image classification. In: 2016 23rd international conference on pattern recognition (ICPR). IEEE, pp 2440–2445

Belsare AD, Mushrif MM, Pangarkar MA, Meshram N (2015) Classification of breast cancer histopathology images using texture feature analysis. In: TENCON 2015–2015 IEEE Region 10 Conference. IEEE, pp 1–5

Breast cancer—Statistics and facts statista. https://www.statista.com/topics/4337/breast-cancer-in-the-us/. Accessed 09 Oct 2018.

Ghazali KH, Mansor MF, Mustafa MM, Hussain A (2007) Feature extraction technique using discrete wavelet transform for image classification. In: SCOReD 2007. 5th student conference on research and development, 2007. IEEE, pp 1–4

Golatkar A, Anand D, Sethi A (2018) Classification of breast cancer histology using deep learning. In: International conference image analysis and recognition. Springer, Cham, pp 837–844

Haykin S (1994) Neural networks: a comprehensive foundation, 2nd ed. Prentice Hall PTR

Kahya MA, Al-Hayani W, Algamal ZY (2017) Classification of breast cancer histopathology images based on adaptive sparse support vector machine. J Appl Math Bioinform 7(1):49–69

Kohavi R, John GH (1997) Wrappers for feature subset selection. Artif Intell 97:273–324

Kumar S, Chauhan A (2013) Feature extraction techniques based on color images. In: Special conference issue: National conference on cloud computing & big data, vol 214, pp 208

Motlagh NH, Jannesary M, Aboulkheyr H, Khosravi P, Elemento O, Totonchi M, Hajirasouliha I (2018) Breast cancer histopathological image classification: a deep learning approach. BioRxiv 242818

Nahid AA, Mehrabi MA, KongY (2018) Histopathological breast cancer image classification by deep neural network techniques guided by local clustering. BioMed Res Int

Phyu TZ, Oo NN (2016) Performance comparison of feature selection methods. In: MATEC web of conferences, vol 42, p 06002. EDP Sciences, pp 1–4

Porkodi R (2014) Comparison of filter based feature selection algorithms: an overview. Int J Innov Res Technol Sci (IJIRTS) 2(2):108–113

Rajan RK, Swathy C.K (2017) Breast cancer detection using histopathology and thermography images. Int J Innov Res Sci, Eng Technol 6(4):6868–6875

Samah AA, Fauzi MFA, Mansor S (2017) Classification of benign and malignant tumors in histopathology images. In: 2017 IEEE international conference on signal and image processing applications (ICSIPA). IEEE, pp 102–106

Spanhol FA, Oliveira LS, Caroline Petitjean C, Heutte L (2016) A dataset for breast cancer histopathological image classification. IEEE Trans Biomed Eng 63(7):1455–1462

Wei B, Han Z He X, Yin Y (2017) Deep learning model based breast cancer histopathological image classification. In: 2017 IEEE 2nd international conference on cloud computing and big data analysis (ICCCBDA). IEEE, pp 348–353

Numerical Study on Particle Deposition in Healthy Human Airways and Airways with Glomus Tumor

Digamber Singh, Anuj Jain, and Akshoy Ranjan Paul

Abstract The numerical simulations were carried out to identify the particle deposition and flow characteristics in the healthy airways and airways with a glomus tumor present in the trachea with an asymmetric unsteady state normal breathing of 10 liter per minute. The presence of tumor was observed to affect the airflow characteristics in the respiratory tract as well as the drug delivery at the diseased spot. A three-dimensional geometrical model of a human respiratory tract was reconstructed from the computed tomography (CT) scan data of a 48-year-old Indian male and an artificial glomus tumor was also constructed above 2 cm of a carina on the posterior side of the tracheal wall. The probable area of inflammation due to toxic particle deposition is identified effectively assisting the medical practitioners to make a prognosis of a person suffering from lung tumor.

Keywords Air quality index (AQI) · Computer tomography (CT) · Pulmonary function test (PFT) · Transient · Computational fluid dynamics (CFD) · (LRN) k–ω turbulence model

1 Introduction

The identification of the particle transport and deposition ratio in diseased airways, which are the emerging areas of biomedical engineering, assists the medical practitioners. However, the poor air quality index (AQI 201–300) is associated with inflammation/erosion in airways due to deposition of fine toxic particles ($PM_{2.5}$), such issue indicated the worldwide burden of 3.17 million mortality, as per the current report

D. Singh (✉) · A. Jain · A. R. Paul
Educational Institute, Motilal Nehru National Institute of Technology Allahabad, Prayagraj 211004, Uttar Pradesh, India
e-mail: digamber35@gmail.com

A. A. Rizvanov et al. (eds.), *Advances in Biomedical Engineering and Technology*, Lecture Notes in Bioengineering,
https://doi.org/10.1007/978-981-15-6329-4_31

of world health organization (WHO) in the year of 2015. The mortality loss due to pulmonary diseases will be the fourth largest issue in the world until the 2030 year. This issue has been investigated in this study. The presence of glomus tumors (cancer) in the airways is intensively affected the airflow and particle deposition. Although, the pollutant source was one of the major causes of cardiovascular diseases, and also it has decayed the quality of life and increased the burden of mortality loss (Mathers et al. 2006). Therefore, the drastic changes in the particle density of $PM_{2.5}$ lead to the severe lung diseases such as COPD, Cancer, and Asthma, etc. In this article, such issues are targeted. The tumor in airways, which eventually forms an unwanted mass (growth of cells) in the tracheobronchial, is known as lung tumor. The obstruction in the larynx, trachea, and main bronchi are influenced by the airflow path and reduce the lungs forced vital capacity (FVC).

In the previous studies, the effect of lung tumor at various locations in the human airways was discussed by Segal et al. (1999). The airflow pattern in airways was simulated with a tumor on carina using the Weibel model (1963). It was identified that the tumor has a significant effect on bifurcation junction's flow; moreover, the effect of tracheal tumor size and flow rate on local airflow field of airways, with the unique features of asymmetric breathing, was emphasized in the past research but they were not considered. The asymmetric nature of the flow and the natural breathing flow were studied by Riazuddin et al. (2010). In this article, the asymmetric nature of the flow is considered as a flow boundary condition at the inlet, which shows the novelty of research. In general, multiple small-size tumors are present in the trachea. It is having primarily three main components: (1) round tumor cells, (2) blood supply cells, and (3) infected cells. The different types of tumor (size) listed in Table 1.

In the literature, few researchers have discussed the effect on particle deposition and internal flow due to tracheal obstruction (tumor), based on a simplified airway model. In a recent study, Srivastav et al. (2014), used a realistic (CT-scan) tracheal model for the study of surface wall shear and particle deposition, but it is not considering the complete respiratory model, it is lacking the real findings. Hence, there is no literature available on such an issue. The study discussed the effect of a glomus tumor on the particle deposition and internal flow of airways using computational fluid dynamics (CFD). The computational results showed the particle deposition ratio, flow velocity vortices, and flow streamlines in this study. It is very useful for the diagnosis of cardiovascular diseases and also for the effective target drug delivery method.

Table 1 Different sizes of the tumor

Authors	Tumor dimension (cm^3)	The shape of tracheal glomus tumor
Watanabe et al. (1998)	2.0 × 1.6 × 1.4	Polypoid
Colaunt et al. (2008)	2.0 × 1.0 × 1.0	Polypoid

1.1 Human Upper Respiratory Tract

The reconstruction of human upper airways is done using medical imagining (CT-scan) technique; however, the two-dimensional images are pixel size and thickness can vary between 256 × 256 to 512 × 512 and 1.0 × 1.0 mm to 0.5 × 0.5 mm. In order to have high-pixel images with a better visualization, CT technique is widely used to understand the clinical or pathological aspects of respiration. The computational three-dimensional model is shown in Fig. 1. The quality of the images depends on threshold values of tissues based on the density and threshold potential of a green mask. Therefore, it is associated with depth or slice thickness. This may allow good detection of a tissue structure, which is the best advantage of using CT.

The glomus tumor in the trachea is artificially constructed using three-dimensional modeling software as shown in Fig. 1b. The authors are using the similar dimensions $2.0 \times 1.6 \times 1.4\ cm^3$ reported by Watanabe et al. (1998), for tumor surgery, the computationally similar tumor was created in the trachea (2 cm above the carina on the posterior side of the tracheal wall).

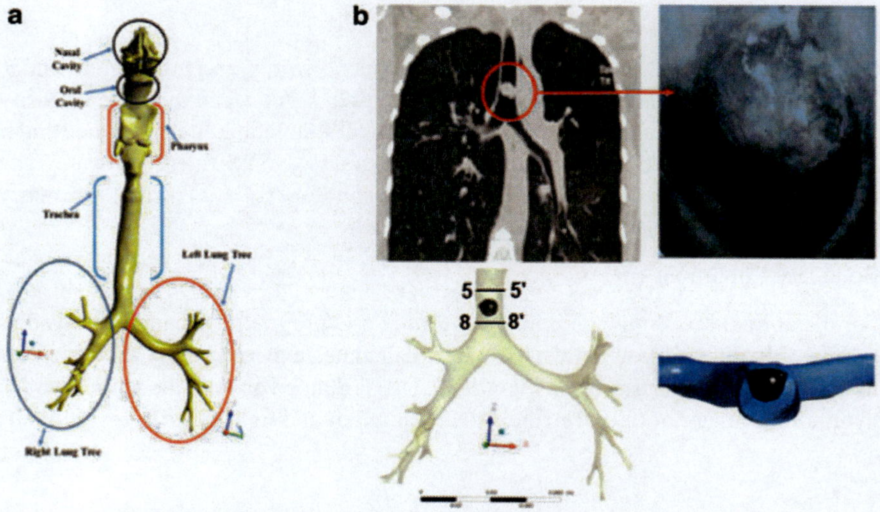

Fig. 1 **a** Human airway model, **b** glomus tumor in the trachea

2 Numerical Methodology

2.1 Governing Equations

For the studies of asymmetric cyclic transient airflow in airways, the following assumptions are considered: (I) incompressible airflow and (II) the wall of the modal considered as rigid hydraulically.

Continuity Equation:

$$\frac{\partial u_i}{\partial x_j} = 0 \tag{1}$$

Momentum Equation:

$$\frac{\partial (u_i u_j)}{\partial x_i} = -\frac{1}{\rho}\frac{\partial p}{\partial x_i} + \frac{\mu}{\rho}\frac{\partial^2 u_i}{\partial x_i \partial x_j} \tag{2}$$

where u_i, u_j $(i, j = 1, 2, 3)$ are the velocity components in x, y, and z directions, and p, ρ, μ stand for the pressure, air mass density ($= 1.225$ kg/m^3), and dynamic viscosity ($= 1.789 \times 10^{-5}$ kg/ms), respectively. The numerical simulations are performed under light breathing conditions by using the commercial ANSYS-fluent software. The numerical results are compared for maximum inspiratory ($t_1 = 2.0$ s) and expiratory flow ($t_5 = 5.0$ s).

Boundary Conditions

For the normal breathing condition (10 L/min), a C-program has been employed as an inlet velocity profile as shown in Fig. 2 is calculated experimentally. However, the individual velocity user-defined function (UDF) is defined on both the nostrils with a hydraulic diameter of 0.0141 m and 0.0126 m in left and right nostrils, respectively.

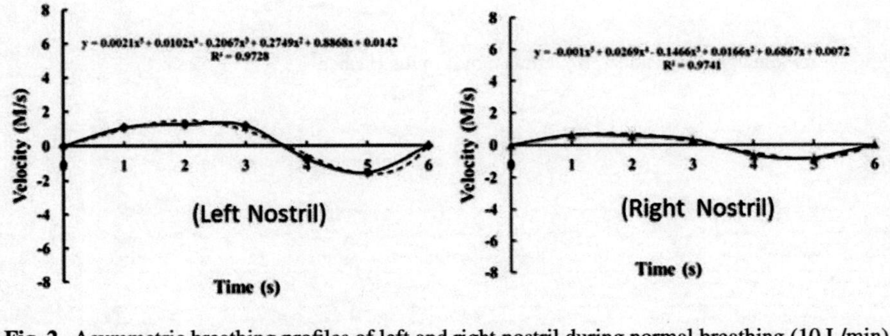

Fig. 2 Asymmetric breathing profiles of left and right nostril during normal breathing (10 L/min)

The value of turbulent intensity (I) is chosen between 5 and 10%. There is no significant difference observed in the results, the no-slip boundary condition is applied to flow walls of the HURT model ($u_i = 0$). A zero-gauge pressure is established at all outlets. The transient cyclic numerical airflow simulation was performed with turbulent parameters.

2.2 Numerical Grid Generation and Solution Schemes

The numerical grid is constructed with the 'Ansys ICEM' module. The human airways grid is generated with unstructured tetrahedral elements, which is having a large volume of mesh. Therefore, to explore the numerical grid generation technique and reduce the grid skewness of tetrahedral grid elements cells, it is converted into a polyhedral grid. As a result, the grid size was reduced to half, it is saving computational efforts and time (ANSYS fluent guide-18). Meanwhile, four grid sizes of the polyhedral element, 4.5, 4, 3.5, and 3 mm, were tested for inspiratory airflow 30 L/min. The numerical grid is independent with 4 mm of element size, and the final numerical grid is having 68, 90,000 elements as shown in Fig. 3.

The airway asymmetric fluid flow dynamics was simulated with the finite-volume method. The internal fluid dynamics of human airways was solved using a low Reynolds number (LRN) k–ω turbulence model. The governing equations of k–ω turbulence model are written as follows:

For (k) turbulent kinetic energy;

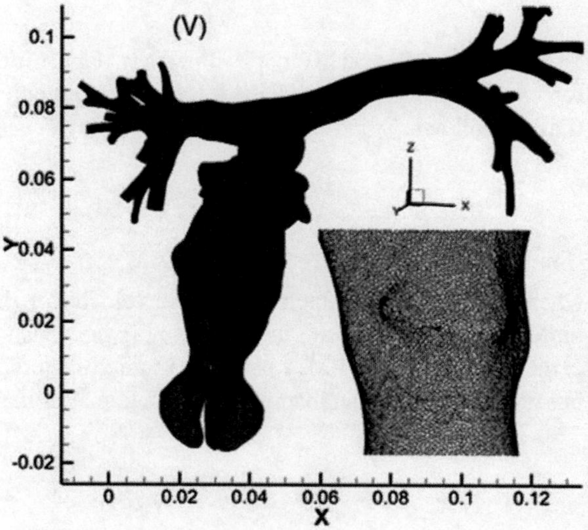

Fig. 3 (V) Numerical mesh of HURT model at the glottis region

$$\frac{\partial}{\partial t}(\rho k) + \frac{\partial}{\partial x_i}(\rho k u_i) = \frac{\partial}{\partial x_j}\left(\Gamma_k \frac{\partial k}{\partial x_j}\right) \tag{3}$$

For (ω) specific dissipation rate;

$$\frac{\partial}{\partial t}(\rho \omega) + \frac{\partial}{\partial x_i}(\rho \omega u_i) = \frac{\partial}{\partial x_j}\left(\Gamma_\omega \frac{\partial \omega}{\partial x_j}\right) \tag{4}$$

where the airflow diffusivities in turbulence model are written as follows:

$$\Gamma_k = \mu + \frac{\mu_t}{\sigma_k}, \Gamma_\omega = \mu + \frac{\mu_t}{\sigma_\omega} \tag{5}$$

Also, σ_k and σ_ω are the turbulent Prandtl numbers for the used turbulence model and μ_t turbulent viscosity is written as follows:

$$\mu_t = \alpha * \frac{\rho k}{\omega} \tag{6}$$

where the value of $\alpha* = 0.555$ for low Reynolds number flow.

The pressure–velocity coupling SIMPLE-C was set to confirm, and the fluent solution residual has converged up to 10^{-5} with an adaptive time method.

2.3 *Discrete-Phase Model*

To solve the trajectory of inspired particle deposition in HURT, the effect of drag forces and gravity integrating the force balance equation is written on the Lagrangian frameas follows:

$$\frac{du_p}{dt} = F_D(u - u_p) + \left[\frac{g(\rho_p - \rho)}{\rho_p}\right] \tag{7}$$

where u is the fluid velocity, u_p is the velocity of the inspired particles, μ is the molecular viscosity of the fluid (air), ρ_p is the density of inspired particles, and d_p is the diameter of the $PM_{2.5}$ particles in micrometers. FD ($u - u_p$) is the drag force per unit injected on the inspired particulate matter mass written as follows:

$$F_D = \frac{18\mu}{\rho_p d_p^p}\frac{C_D\mathrm{Re}}{24} \tag{8}$$

Re is the relative airflow Reynolds number written as follows:

$$\mathrm{Re} \equiv \frac{\rho d_p |u_p - u|}{\mu} \tag{9}$$

The drag coefficient C_D is written as follows:

$$C_D = a_1 + \frac{a_2}{\mathrm{Re}} + \frac{a_3}{\mathrm{Re}^2} \tag{10}$$

where a_1, a_2, and a_3 are the model constants.

The total mass of the $PM_{2.5}$ particle, 6.275×10^{-08} kg/s, was considered under poor AQI (201–300), and the density of particles was considered as equivalent to water (1000 kg/m^3).

2.4 Particle Deposition Model Validation

To quantify the inspired particle deposition efficiency in the nasal cavity, the deposition efficiency of suspended particles is defined as the ratio of the mass of particles deposited in the trap region/mass of particles injected in the airway. Different micro-sized particles were injected in the human airways, which have a very small diameter (1–10 μm). The numerical results were compared to the experimental data, which was reported by Kelly et al. (2004), as shown in Fig. 4.

In the realistic airways, the sudden anatomical and morphological changes occurred in the external nostril of the nasal cavity. As the inspired airflow velocity increased, the outcome of particle deposition also increased due to the inertial

Fig. 4 Different micro-sized particle deposition efficiencies in the nasal cavity

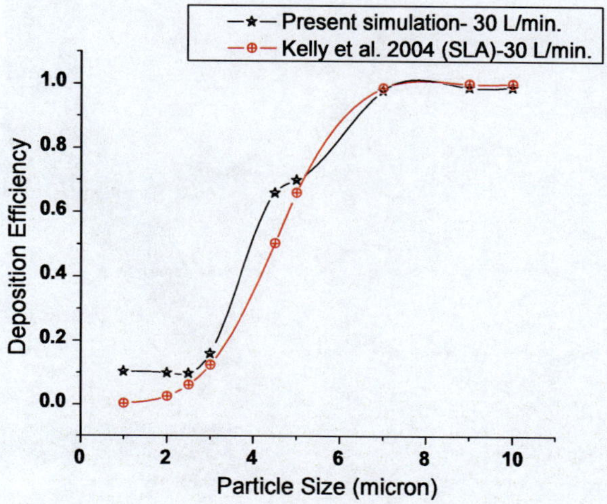

effect. The overall numerical findings were showing a reasonable agreement with experimental results available in the literature.

3 Results and Discussion

The airflow characteristics and particle deposition analysis were studied in the human airway model.

In Fig. 5, the airflow distribution was seen in both a healthy and a diseased human airway at maximum inspiratory ($t_1 = 2.0$ s) and expiratory ($t_5 = 5.0$ s) flows during normal breathing (10 L/min). To understand the effect of obstruction due to tumor on suspended micro-sized particle deposition in the human airways, we need to first understand the flow vortices, which are generated due to the sudden contraction of cross-sectional area. The massive velocity magnitude has been observed and it

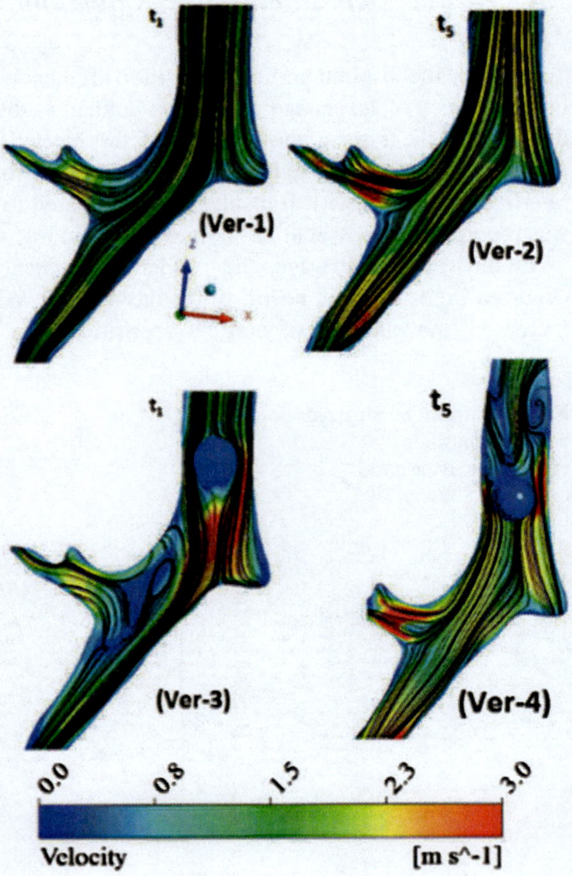

Fig. 5 Velocity contours (m/s) and streamlines on the vertical (Ver) plane of human airways from the front side during light breath (10 L/min)

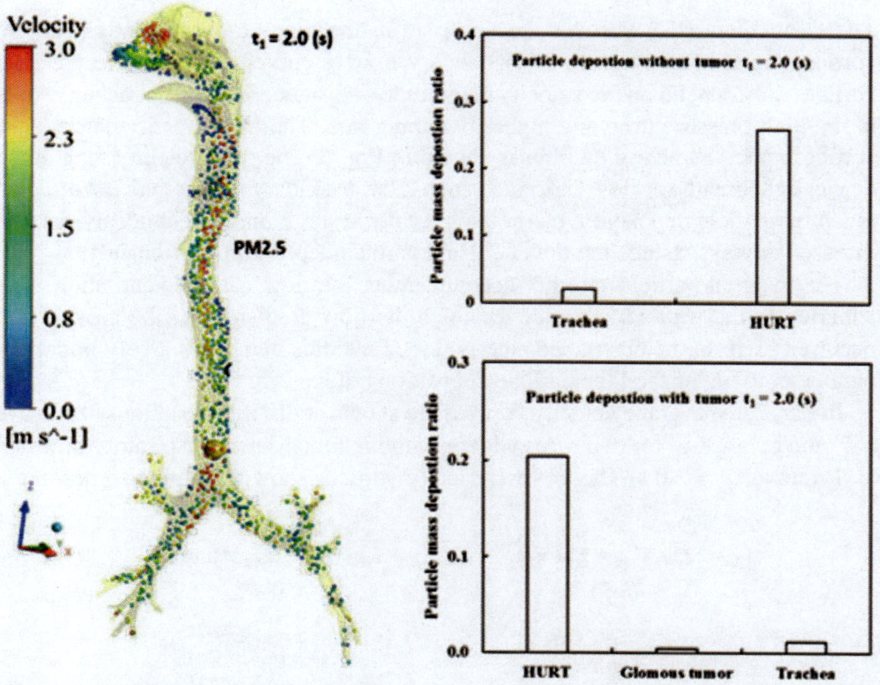

Fig. 6 $PM_{2.5}$ particle transport pattern in a diseased airway and deposition ratio with or without tumor in human airways at the $t_1 = 2.0$ s

is constructed inside the secondary vortices of the airways. Also, an asymmetric distribution of flow occurred in the primary bronchus, which is influenced by the particle deposition and transport pattern in the airways shown in Fig. 6. The particle deposition ratio is compared in healthy and diseased airways considering inspiratory flow during a normal breath. In diseased airways, particle deposition is increased up to 10–15%; hence, the repetitive breathing may also increase the deposition efficiency.

The constricted flow path due to the presence of a tumor causes the higher WSS and increases the breathing efforts. Therefore, during light breathing, the velocity streamlines are different before and after near the obstruction (tumor) because of a significant reduction in energy. It may also reduce the flow velocity and centrifugal forces, which may propel the particles in the airways. It is seen in the plane (Ver-3). On the upper side of the plane, the recirculatory vortices are much stronger during inhalation.

- **Ver-1** = Stream flow lines without tumor at maximum inspired airflow at $t_1 = 2.0$ s.
- **Ver-2** = Stream flow lines without tumor at maximum exhaled airflow at $t_5 = 5.0$ s.
- **Ver-3** = Stream flow lines with tumor at maximum inspired airflow at $t_1 = 2.0$ s.
- **Ver-4** = Stream flow lines with tumor at maximum exhaled airflow at $t_5 = 5.0$ s.

The Dean Flow (De) vortices are observed in the plane 6-6 constantly. The De vortices are used to evaluate the secondary flow in curved pipes, where the flow vortices only depend on the velocity magnitude or distance. Tumor in trachea results in the high-pressure drop and higher breathing rate. The higher recirculatory flow vortices appeared near the tumor is shown in Fig. 6. The flow velocity magnitude during light breath was low (1.25 m/s) so that the secondary vortices are not affected much by particle deposition, but in a higher flow rate, it may affect deeply. So, the diseased airways change the flow field and particle deposition significantly.

The inspired particles are not accompanying the flow path of ventilation. It is adhering to the wall surface, due to which the flow gradient near the airway wall occurred high. In the obstructed airway flow, Reynolds number is a very important parameter to understand the particle deposition efficiency.

In Fig. 7, the average velocity vortices are shown at different locations just above 5-5' and below 8-8' for two cases with or without tumor during inspiratory flow rate 10 L/min at $t_1 = 2.0$ s. The airflow velocity vortices show the flow disorder in the

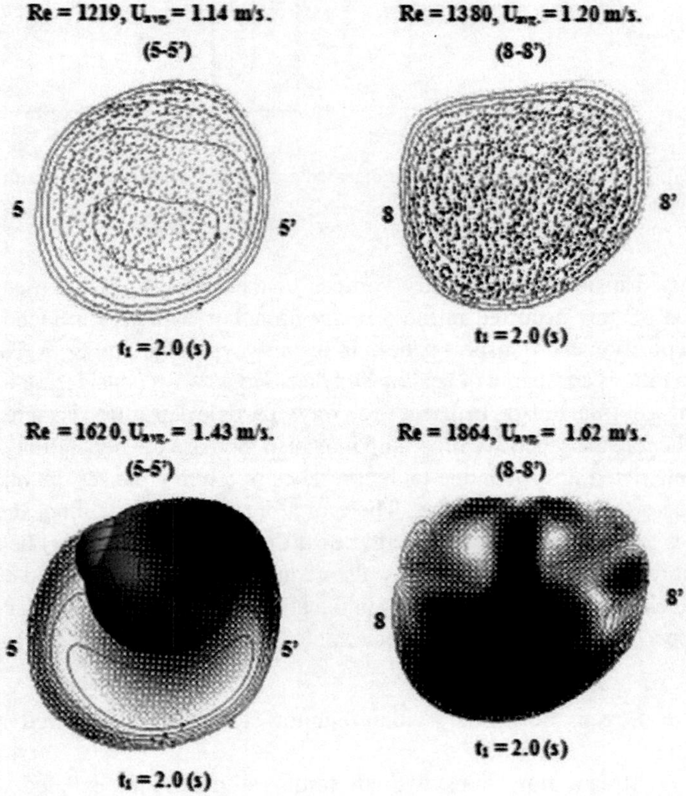

Fig. 7 Particle deposition and secondary vortices in the human airways during inspiratory flow under normal asymmetric breath (10 L/min)

downstream of flow as compared to the upstream in airways. The flow recirculation zones are formed in the downstream. However, due to large pressure drop in the posterior side of the trachea, a pair of counter-rotating vortices has been generated at the cross section of 8-8'. At this section, a higher Reynolds number flow leads to the particle deposition in the downstream of airways at $t_1 = 2.0$ s.

4 Conclusions

1. A sudden contraction of the cross-sectional area is due to tumor in the trachea. The massive velocity magnitude is observed in the contraction region, as a comparison of the particle deposition ratios in both the healthy and the diseased airways. During inspiratory flow, the particle deposition efficiency increased by 10–15%. Hence, the repetitive breathing may increase the deposition as well.
2. The obstruction in the main flow field accompanied the higher friction in the airways path, and it is increasing the breathing effort. During light breathing, the velocity streamlines are different before and after because of a significant loss of energy.
3. The airflow velocity vortices show a high airflow disturbance in the downstream as compared to the upstream flow. The recirculation zones occurred in the downstream of the airways. Due to a sudden drop in pressure, a pair of counter-rotating vortices are generated along with the posterior side of the trachea and created the asymmetricity inflow distribution in both bronchi.

Acknowledgements This work was supported by the project on "Effects of Respirable Suspended Particulate Matter (RSPM) on Chronic Obstructive Pulmonary Diseases (COPD) Using Computational Fluid Dynamics (CFD)" sanctioned by the Council of Science & Technology, (UP). Letter no. 505/21/05/15/AMD//F-60, PF/2015.

Conflict of Interest The author declared no conflicts of interest publication of this article.

References

Ansys-Fluent Version 18 (2018) User guide. Ansys Inc., USA

Colaut F, Toniolo L, Scapinello A, Pozzobon M (2008) Tracheal glomus tumor successfully resected with rigid bronchoscopy: a case report. J Thorac Oncol 3:1065–1067

Kelly James T, Asgharian B, Kimbell JS, Wong BA (2004) Particle deposition in human nasal airway replicas manufactured by different methods. Part 1: inertial regime particles. Aerosol Sci Technol 38:1063–1071

Mathers CD, Loncar D (2006) Projection of global mortality and burden of diseases from 2002 to 2030. PLoS Med 3:e442

Riazuddin VN, Zubair M, Abdullah MZ, Ismail R, Shuaib IL, Hamid SA, Ahmad KA (2010) Numerical study of inspiratory and expiratory flow in a human nasal cavity. J Med Biol Eng 31:201–206

Srivastav VK, Kumar A, Shukla SK, Paul AR, Bhatt AD, Jain A (2014) Airflow and aerosol drug delivery in a CT-scan based human respiratory tract with tumor using CFD. J Appl Fluid Mech 7(2):345–356

Segal RA, Guan X, Shearer M, Martonen TB (1999) Mathematical model of airflow in the lungs of children I: effect of tumor size and locations. J Theor Med 2:199–213

Weibel ER (1963) Morphometry of the human lungs. Academics, New York

Watanabe M, Keigo T, Katsuaki O, Teruhiro A, Susumu T, Hideyyki S, Shinsuke A (1998) Successful resection of a glomus tumor arising from the lower trachea: report of a case. Jpn J Surg 28:332–334

Virtual Reality Therapy in Prolonging Attention Spans for ADHD

S. Sushmitha, B. Tanushree Devi, V. Mahesh, B. Geethanjali, K. Arun Kumar, and P. G. Pavithran

Abstract Attention-Deficit/Hyperactivity Disorder is a major brain disorder that brings with it, the defects of inattention or hyperactivity or sometimes a concoction. This makes it difficult to concentrate on any task. In order to increase the attention span VR-Based task was designed and compared with Conventional task. The EEG data from 10 healthy and 10 ADHD participants were recruited for this study and they were asked to perform both the tasks. The increase in theta/alpha ratio was noted for the ADHD group while performing paper-based and VR task but not seen during rest. This may be due to the usage of the same virtual environment for all the participants; the virtual screen based on the user's choice would yield an improvement in task performance. The results of this study can thus be effectively exploited in the creation of an Attention Enhancement System that is in line with the child's interest profile.

Keywords ADHD · EEG · Theta/alpha ratio · Virtual reality

1 Introduction

Attention-deficit/hyperactivity disorder (ADHD) is a brain disorder marked by an ongoing pattern of inattention and/or hyperactivity impulsivity that interferes with functioning or development (https://www.nimh.nih.gov/health/topics/attention-deficit-hyperactivity-disorder-adhd/index.shtml). An inattentive person wanders off task, lacks persistence, has difficulty for sustained attention, and is disorganized; and these problems are not due to defiance or lack of comprehension. A hyperactive person moves constantly even in situations that aren't appropriate. In adults, it is marked with extreme restlessness or wearing others out with constant activity. Impulsivity causes hasty actions that occur at the moment without first thinking about them and that may have a high potential for harm. Although there is no global

S. Sushmitha · B. Tanushree Devi · V. Mahesh (✉) · B. Geethanjali · K. Arun Kumar · P. G. Pavithran
Department of Biomedical Engineering, SSN College of Engineering, Chennai, India
e-mail: maheshv@ssn.edu.in

A. A. Rizvanov et al. (eds.), *Advances in Biomedical Engineering and Technology*, Lecture Notes in Bioengineering,
https://doi.org/10.1007/978-981-15-6329-4_32

consensus on the prevalence of attention-deficit/hyperactivity disorder (ADHD) in children, adolescents, and/or adults, meta-regression analyses have estimated the worldwide prevalence at between 5.29 and 7.1% in children and adolescents, and at 3.4% (range 1.2–7.3%) in adults (https://www.chadd.org/understanding-adhd/about-adhd/data-and-statistics/general-prevalence.aspx).

Therapeutic tools for ADHD, used at present include medication, which produces undesirable physiological and psychological side effects (Yeh et al. 2012). This study focuses on providing therapy using virtual reality, thus eliminating the need for conventional methods and its accompanying side effects. The motivation for this study is to enhance the attention span of ADHD children to enjoy a happy adulthood. Also, it is derived from a strong societal need to make therapy more accessible and reliable for these children. Creating an easy use Virtual Reality based rehabilitation environment along with the reward and motivation would help the ADHD to increase the attention span. Along with that, recording the brain function while performing the various attention-based tasks would provide insight into ADHD participants' brain.

Various techniques such as EEG and functional Magnetic Resonance imaging (Eswaran et al. 2018) were used; the Electroencephalogram (EEG) is the depiction of the electrical activity occurring at the surface of the brain. This is in the range of microvolts and is characterized by varying frequencies and amplitudes (https://emedicine.medscape.com/article/1139332-overview). The range of frequencies is used to identify spectral features such as alpha, beta, theta, delta, etc. As a rule of the thumb, higher levels of theta activity are noticed in ADHD Children. theta by alpha ratio, used to characterize task performance, is found to be elevated in children with ADHD (Heinrich et al. 2014). EEG is the best means to record the various EEG bands' changes over time and thereby preserving the temporal resolution (Geethanjali et al. 2018). Most of the research studies used Conventional Therapy (paper-based task) (Ritu et al. 2017; Aguiar et al. 2010) to increase the attention span. In the present, VR-Based Therapy was designed and compared with Conventional Therapy.

The remainder of the paper is organized in the following manner. Section 2 discusses the Materials and Methods used in the study. Section 3 renders the results and discussions of the study. Section 4 provides conclusions of the Study.

2 Materials and Methods

2.1 Participant Selection

The control and the experimental groups were selected from the pool of participants, with the age ranging from 4 to 13 years, children suffering from ADHD are considered as experimental (10 participants) and other groups as a normal healthy child (10 participants). Further, a preliminary diagnostic test was assigned to validate the fact that they had/didn't have clinical ADHD. The standard test used was Vanderbilt's Criteria (Sanders et al. 2009). The participants suffering from ADHD were provided by Swabhimaan Trust, Chennai. This study has been approved by the Institutional Human Ethics Committee, SSN College of Engineering (IHEC/SSN CE/Pr.No.18/26.10.2018) and informed consent was obtained from the parents of the children. The participants were categorized based on the results of this test (Sanders et al. 2009).

2.2 Selection of Tasks

Paper-Based Task/Therapy. The Porteus Maze Test (Fig. 1) is a low-verbal test that involves the executive functioning of the brain, which engages planning and inhibition. It is a measure of reflection/impulsivity. The main purpose of the test is to reveal any haphazard, impulsive, or overconfident habits of action, or a tendency to become absorbed in the task of finding the way through the maze as to neglect other considerations. In the Porteus Maze task, subjects are presented with mazes of increasing difficulty. During the task, the participants are presented with 6 mazes in order of increasing difficulty. Planning their way out of the mazes will reflect the improved problem-solving ability and prolonged attention (Ritu et al. 2017; https://vlp.mpiwg-berlin.mpg.de/; Aguiar et al. 2010).

Virtual Reality Task/Therapy. The Virtual Environment was developed using Unity-3D (www.nichq.org/~/media/files/resources/nichqvanderbiltassessment-full). The game was created using Unity-3D, a popular game engine. It is a 'Desktop-VR'

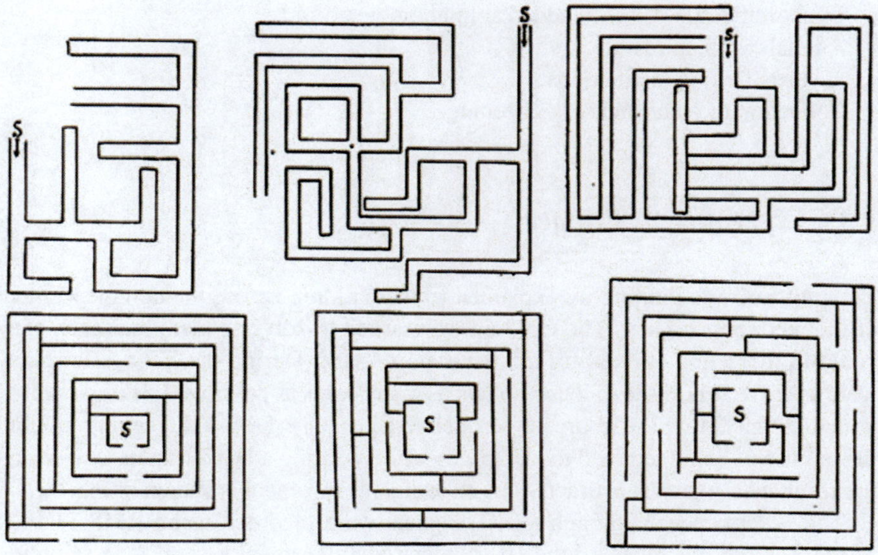

Fig. 1 Porteus Maze—paper-based task (Ritu et al. 2017)

Fig. 2 Created virtual reality platform (coin collection as reward)

and can also be realized on a Head Mount Device. It is equipped with shifting landscapes (Fig. 2), animation, and plenty of colors which, in turn, not only make the environment interactive and pleasing but also prevent stagnation and eventual loss of focus. The game features collection of coins and score-keeping (Fig. 2) which is a kind of Motivation and Reward-based biofeedback system. The motivation for the child comes from wanting to collect all the coins and a sense of competition and the reward is a high score. Keeping in mind the task at hand, the child manages to remain focused through the process. A game is created that involved the following:

- Creation of a terrain and a sky.
- Creation of hills.
- Creation of fire (particle system), log, brick houses, etc.
- Creation of multiple water bodies and a bridge.
- Teleportation device with added animation.
- Animals with animation.
- 3D texts issuing instructions.
- Collection of coins and score-keeping.

2.3 Experimental Protocol

The protocol was thoroughly explained to the children before the commencement of the recording session. The children were asked to solve as many mazes as they could in the 2 min allotted for the paper-based task. During the VR session, they were asked to follow the trail and collect as many coins as possible. It is important to note that the Virtual Environment was delivered as a Desktop-VR, keeping in mind the hassles of being tethered to electrodes while wearing a Head Mount Device and the resultant movement artifacts. The experimental protocol is shown in Fig. 3.

The Electroencephalograph (EEG) signals were recorded using RMS SUPER SPEC 32 EEG machine and the 10–20 electrode placement system. The sampling rate was kept at 256 Hz/channel and 19 channels were used for both the tasks. The

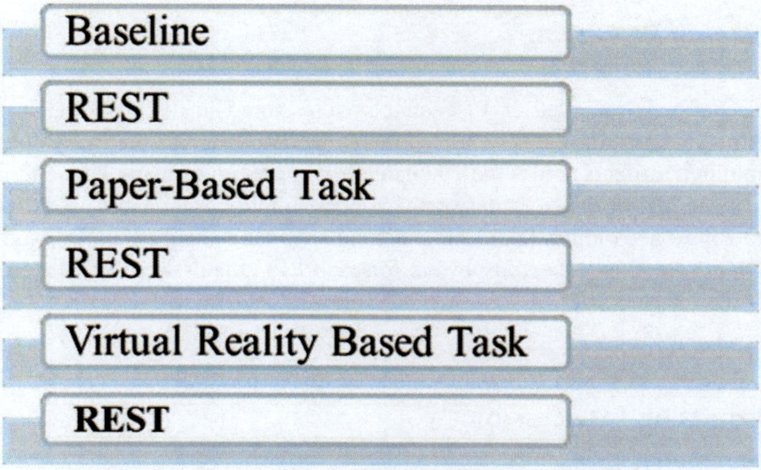

Fig. 3 Experimental protocol

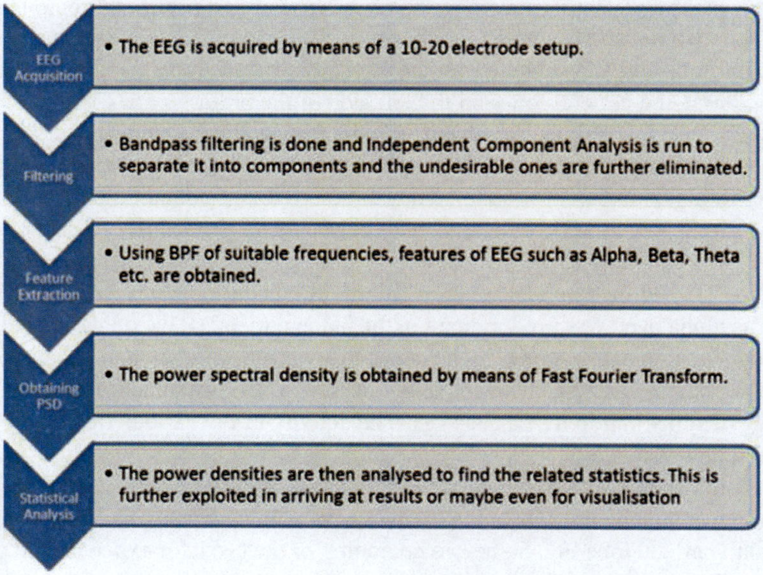

Fig. 4 Process flow

electrode locations were, namely, Fp1, Fp2, F7, F3, Fz, F4, F8, T3, C3, Cz, C4, T4, T5, P3, Pz, P4, T6, O1, O2, and A1 and A2 were used as reference electrodes. Ag/Agcl was used as an adhesive for electrode placement and the impedance was kept less than 5 kohms.

2.4 Signal Processing

The EEG signals recorded at 19 channels were stored in the EEG machine and the offline analysis was carried out. The recorded EEG signals that were corrupted with common artifacts across the 19 channels were removed using ICA and implemented by EEGLAB. Then Butterworth bandpass filter of order 3 was used to extract various EEG bands alpha (8–13 Hz), beta (13–40 Hz), theta (4–8 Hz), and delta (0.1–4 Hz). The mean spectral powers for each EEG bands were calculated using MATLAB.

3 Results and Discussion

3.1 Visualization of ADHD Brain

High levels of theta activity were found in the frontal and prefrontal regions of the brain and also noted at central and parietal lobe. High alpha activity is seen in the prefrontal lobe and frontal lobe (middle of the brain). High beta activity is noticed in the central and parietal lobes on the left side of the brain.

The variation in various EEG bands (alpha, beta, and theta) before the beginning of the tasks for the control group is shown in Fig. 5a. The first two images depict the activity in the theta band (4, 6 Hz), The next three images depict the activity in the alpha band (8, 10, 12 Hz). The last three images depict the activity in the beta band (15, 22, 29 Hz). The theta band was dominant in the left central electrode locations (C3 and Cz) and this was not seen at frontal electrode locations (F3, F4, F7, and F8). The alpha band was predominant in the in the parietal (P3 and P4) and occipital electrode locations (O1 and O2). The beta band was prevailing at parietal and central electrode locations regions. The above figure gives us the visualization of the brain of average, healthy children of ages 4–13. The activity in the various frequency bands can be seen above.

The high levels of alpha, beta, and theta bands were observed in the prefrontal lobe and frontal lobe (middle of the brain) and it is almost evenly spreading from the central to parietal lobe Fig. 5b before beginning of the tasks for experimental group. And no changes were noted for any of the tasks for both control and experimental groups. Based on the results, for control group the alpha, beta, and theta values are average establishing a sense of balance, orderliness, and restraint. Whereas for ADHD, both the theta and alpha components were high for all the tasks and this indicates high levels of disorderliness, chaos, and inattention. Beta value is very low indicating a lack of focus and awareness.

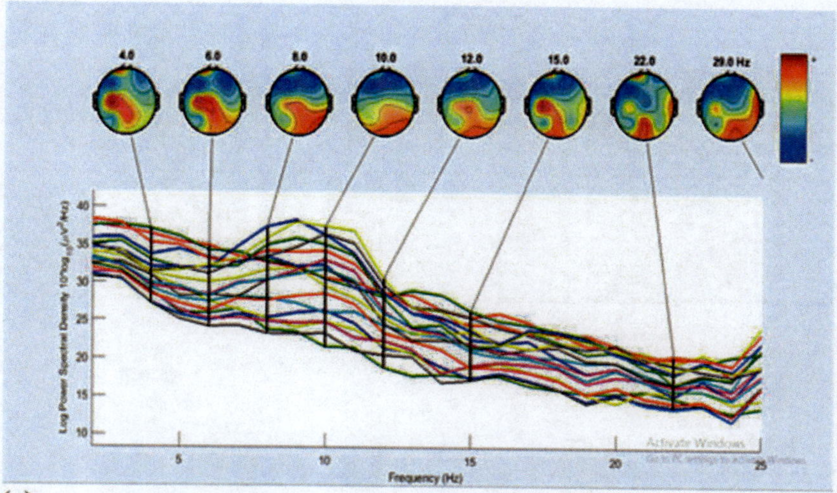

(a)

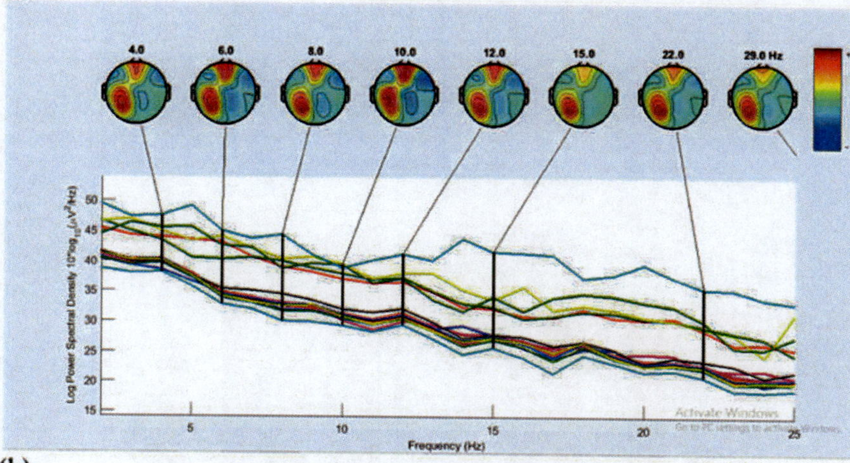

(b)

Fig. 5 **a** Shows the variation in various EEG bands (alpha, beta, and theta) before the beginning of the task for the control group **b** experimental group

3.2 Theta/Alpha Ratio

The theta by alpha ratio was calculated during paper-based and VR tasks and it was compared with rest. For VR-based task as in Fig. 6a, at frontal (F7, Fz, and F8), temporal (T3, T4, T5, T6), parietal (P3, Pz), and occipital (O1 and O2) electrodes location were statistically ($p < 0.05$) high when compared to rest for the experimental group, but no significant changes were seen in any of electrodes location for the control group. Similarly, for the paper-based task as in Fig. 6b, at frontal (F7, Fz, and F8), temporal (T3, T4, T5, T6), parietal (P3, Pz) and occipital (O1 and O2) electrodes

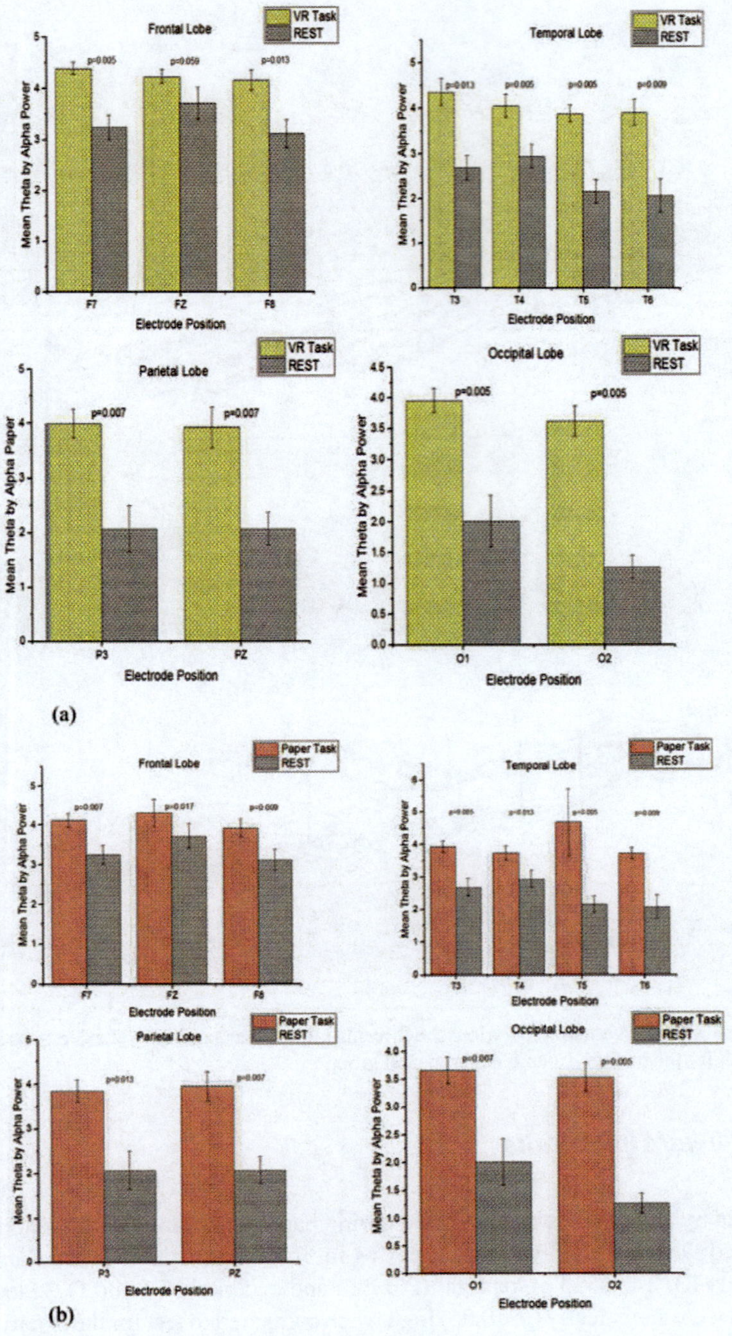

Fig. 6 **a** VR task versus REST **b** paper task versus REST—Theta/Alpha Power

location were statistically ($p < 0.05$) high when compared to rest shown for the experimental group, but no significant changes were seen in any of the electrodes' location for the control group.

The neuropsychological functions of the ADHD were assessed by Porteus maze (Aguiar et al. 2010) so the same task was used as a standard paper-based task. The attention level of the children affected with ADHD differs with different types of tasks. When presented with a paper task and computer-based task, a significant difference in their attention levels is observed with increased attention levels for computer-based tasks (Geethanjali et al. 2018). Virtual Reality can be used as rehabilitation therapy in treating ADHD children (Yeh et al. 2012), an easy usable virtual environment will advance the attention span (Eswaran et al. 2018). Based on this, a virtual environment was created using Unity 3D software. Consistent elevation in theta activity and alpha activity is observed in ADHD children. The corresponding increase of theta/alpha ratio serves as a marker for ADHD (Heinrich et al. 2014) This was noted for the experimental group while performing paper-based and VR task and not seen during rest. This was the reason for the ADHD children to pay attention to the task at hand and respond only when necessary. The study concludes there were no significant changes in brain activity while performing paper- and VR-based tasks. This may be due to the usage of the same virtual environment for all the participants; the virtual screen based on the user's choice would improve the performance.

4 Conclusion

Virtual reality has been a proven tool for the rehabilitation process for many disorders. In this work, an attempt was made to achieve VR based neuro-rehabilitation support for ADHD children. During the paper and virtual reality based tasks, there was no significant decrease observed in theta/alpha ratio in EEG for ADHD participants. So the user-specific virtual environment along with rein enforcement techniques would further enhance the rehabilitation process for ADHD children.

References

Aguiar A, Eubig PA, Schantz SL (2010) Attention deficit/hyperactivity disorder: a focused overview for children's environmental health researchers. NCBI

Cho BH, Lee JM, Ku JH et al (2002) Attention enhancement system using virtual reality and EEG biofeedback. In: Proceedings of the IEEE virtual reality (VR02). USA

Eswaran VSB, Veezhinathan M, Balasubramanian G, Taneja A (2018) Virtual reality therapy for mental stress reduction. J Clin Diagn Res 12(10):JC11–JC16

Geethanjali B, Adalarasu K, Jagannath M (2018) Music induced emotion and music processing in the brain—a review. J Clin Diagn Res 12(1):VE01–VE03

Heinrich H, Busch K, Studer P, Moll GH, Kratz O, Erbe K (2014) EEG spectral analysis of attention in ADHD. Front Hum NeuroSci

Information on ADHD: https://www.nimh.nih.gov/health/topics/attention-deficit-hyperactivity-disorder-adhd/index.shtml

Information on EEG: https://emedicine.medscape.com/article/1139332-overview

Information on Porteus Maze: https://vlp.mpiwg-berlin.mpg.de/

Parsons TD, Bowerly T, Buckwalter JG et al (2007) A controlled clinical comparison of attentionperformance in children with ADHD in a virtual reality classroom compared to standard neuropsychological methods. Child Neuropsychol 13(4):363–381

Ritu V, Keerthana M, Geethanjali B, Mahesh V (2017) A functional connectivity based approach to visualize the event related changes in depression through cognitive information processing during working memory tasks. In: IEEE 16th international conference on cognitive informatics & cognitive computing (ICCI*CC). Oxford, UK

Sanders A, Holtzer R, Lipton R, Hall C, Verghese J (2009) Egocentric and exocentric navigation skills in older adults, the journals of gerontology. Series A, Biol Sci Med Sci 63:1356–1363. https://doi.org/10.1093/gerona/63.12.1356

Statistics on ADHD: https://www.chadd.org/understanding-adhd/about-adhd/data-and-statistics/general-prevalence.aspx

Vanderbilt Assessment Scale: www.nichq.org/~/media/files/resources/nichqvanderbiltassessment-full

Yeh S-C, Fan Y-C, Liu P-C, Tsai C-F, Rizzo A (2012) An innovative ADHD assessment system using virtual reality. In: IEEE EMBS conference on biomedical engineering and sciences. Langkawi, pp 78–83

Multispectral NIRS System Design to Analyze Hemoglobin Concentration on Plantar Foot Surface

Resham Raj Shivwanshi, N. P. Guhan Seshadri, and R. Periyasamy

Abstract Diabetic complications in the lower extremity are common and diverse. They are associated with significant morbidity and mortality and impact heavily upon the public health system. 15% of diabetes patient will suffer from the foot sole or ankle ulcers throughout the disease time and about 14% of ulcers lead to amputation. Loss of sensation (neuropathy) at the foot areas, peripheral arterial disease, and poor microcirculation at foot are the major risk factors that suppress the oxygen availability to the tissues and lead to ulcerations. The present study focused on designing a multispectral near-infrared system using three laser diodes of wavelength 780, 830 and 880 nm to measure the changes in hemoglobin concentrations. Ten healthy volunteers were selected for this study and three standard risk foot areas were considered. Standard induced ischemic stimulus was given to assess the response of the designed system. Further, this study analyzed the correlation between hemoglobin concentration with tissue hardness and temperature at measuring foot site. Results show that the designed multispectral NIR system was well able to assess the changes in hemoglobin concentration at foot sole areas. There was no significant correlation between tissue hardness and hemoglobin concentrations in healthy volunteers, simultaneously temperature at the measuring site shows a negative relationship with hemoglobin concentrations.

Keywords Near-infrared spectroscopy · Multispectral system · Plantar foot complications · Early detection of foot ulcers · Tissue oxygenation · Tissue temperature

R. R. Shivwanshi · N. P. Guhan Seshadri
National Institute of Technology Raipur, Raipur, Chhattisgarh, India
e-mail: guhan131192@gmail.com

R. Periyasamy (✉)
National Institute of Technology Trichy, Trichy, Tamil Nadu, India
e-mail: periyasamyr@nitt.edu

A. A. Rizvanov et al. (eds.), *Advances in Biomedical Engineering and Technology*, Lecture Notes in Bioengineering,
https://doi.org/10.1007/978-981-15-6329-4_33

1 Introduction

With the advent of modern technology, people are getting involved in sundry life-changing activities. Diabetic complications in the lower extremity are common and diverse. They are associated with significant morbidity and mortality and impact heavily upon the public health system (Naidoo et al. 2015). The people having diabetes are increasing all over the world due to population growth, improper food, fatness, and scarcity of physical activities. Diabetic foot infections are a well-recognized risk factor for hospitalization and amputation. According to a recent meta-analysis, one in every 30 hospitalized patients at any given time is affected by a diabetic foot infection. Study shows that 15% of diabetes patient will suffer from the foot sole or ankle ulcers throughout the disease time and about 14% of ulcers lead to amputation (Reiber et al. 1998). Loss of sensation (neuropathy) at the foot areas, peripheral arterial disease, and poor microcirculation at foot are the major risk factors that suppress the oxygen availability to the tissues and lead to ulcerations. An early and accurate diagnosis is required to ensure an effective treatment and reduce the risk of minor and major amputation (Giurato et al. 2017).

At present, there is no device or method available which can monitor and assess the hemoglobin concentrations at foot sole areas to identify the risky regions (regions with a diminished supply of oxygen) before it gets affected. The prevalence of complications can be intercepted by keep analyzing hemoglobin and tissue oxygen concentration. Near-Infrared spectroscopy (NIRS) is becoming a more propitious noninvasive technology to understand the foot sole risk area. It renders biological and systemic perusal of cells and tissues of the particular area. Simultaneously, temperature measurement of the skin is useful to analyze impaired areas. Studies have indicated that temperature assessment and hardness assessment of foot soles can be one of the important parameters for assessing the risk of foot ulceration (Liu et al. 2013). To analyze ischemic diseases, promising techniques are introduced with the potential to routinely assess tissue oxygenation. Several methods and techniques are presented in the past two decades, which are related to the assessment of impaired plantar foot and ulceration. Most of the studies are reported that the risk of LEA (lower extremity amputation) can be reduced by early detection of foot ulcers. The clinically most widely accepted technique is NIRS (Scheeren et al. 2012).

Recently, many researchers (Li and Zhang 2014; Khalil et al. 2013; Zhang et al. 2013) are focused in developing a multi-wavelength near-infrared system to assess the tissue oxygenation. But, at present, there is no device or method available which can monitor and assess the hemoglobin concentrations at foot sole areas to identify the risky regions (regions with a diminished supply of oxygen) before it gets affected. Therefore, the present study aimed at designing a multispectral NIR probe using laser diodes of wavelength 780, 830, and 880 nm with fixed source-detector distance to assess the changes in hemoglobin concentrations in the plantar foot surface. We further analyzed the correlation between hemoglobin concentrations with the tissue hardness and temperature at the foot sole areas in order to assess the areas with reduced oxygen supply to the tissues.

2 Instrumentation and Experiments

The system uses three laser diodes in the near-infrared region of wavelength 780, 830, and 880 nm as an optical source of power less than 20 mW to illuminate the plantar foot surface of the foot. These wavelengths were selectively chosen in the way it covers the absorption bands of oxyhemoglobin (HbO_2) and deoxyhemoglobin (dHb) and it mainly resulting from dHb, not from myoglobin. The detector is a PIN photodiode (QSD2030F) with peak sensitivity at 880 nm. The source and detector placed side by side to carry out measurement in reflectance mode and separated by a fixed distance of 1.5 cm. A transimpedance amplifier (see Fig. 1) is integrated with a photodiode to convert the diode current into the voltage signal. The laser diode and detector have been calibrated using a power meter and wavelength meter before the start of recording. Figure 2 shows the designed multispectral NIR system with optode assembly.

The reflected signal from the foot was sampled at a rate of 1KHz and taken to LabVIEW using a data acquisition card for further analysis. In LabVIEW, the signal is again downsampled to 100 Hz and then filtered with a low-pass filter with a cutoff frequency of 40 Hz (Yong et al. 2005). In order to remove further noises, the signal is again filtered using a moving average filter with an order of 5. The changes in the concentration of HbO_2 and dHb were calculated by implementing the modified Beer–Lambert law in LabVIEW (Cope 1991; Zhang et al. 2013).

$$O_{\text{dens}} = \text{Log}(I_l/T_l) = \varepsilon_{cD} \tag{1}$$

where O_{dens} is the optical density, I_l is the light intensity of incident light, T_l is the light intensity of transmitted light, and D is the distance travelled by light.

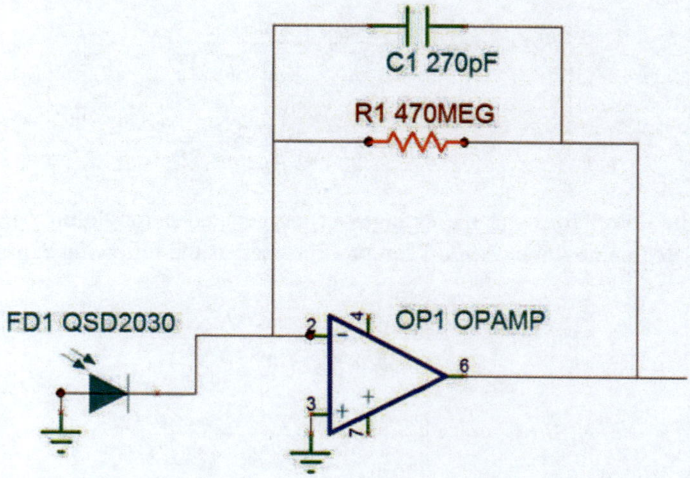

Fig. 1 Transimpedance amplifier design

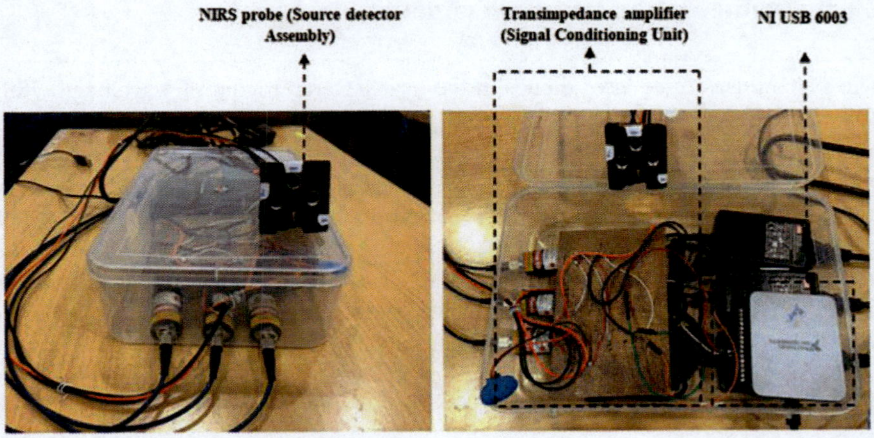

Fig. 2 Multispectral NIR system

With the help of this formula, the concentration of oxygenated and deoxygenated blood can be obtained (Zhang et al. 2013). The further calculation is as follows:

$$O_{\text{dens}} = \left\{\varepsilon^{\lambda}_{hb}[dhb] + \varepsilon^{\lambda}_{hbO_2}[hbO]\right\}D \tag{2}$$

where O_{dens} is absorbance at wavelength λ, $\varepsilon^{\lambda}_{hb}$ and $\varepsilon^{\lambda}_{hbO_2}$ are the extinction coefficient at λ wavelength for molar concentrations of deoxygenated hemoglobin, [*dhb*], and oxygenated hemoglobin, [*hbO*], respectively (Kim et al. 2005).

The concentration of oxygenated and deoxygenated hemoglobin can be evaluated using the following expression:

$$hbO = \frac{O_{dens}\lambda_1 \cdot \varepsilon^{\lambda_2}_{dhb} - O_{dens}\lambda_2 \cdot \varepsilon^{\lambda_1}_{dhb}}{(\varepsilon^{\lambda_2}_{dhb} \cdot \varepsilon^{\lambda_1}_{hbO} - \varepsilon^{\lambda_1}_{dhb} \cdot \varepsilon^{\lambda_2}_{hbO})D} \tag{3}$$

$$dhb = \frac{O_{dens}\lambda_1 \cdot \varepsilon^{\lambda_2}_{hbO} - O_{dens}\lambda_2 \cdot \varepsilon^{\lambda_1}_{hbO}}{(\varepsilon^{\lambda_1}_{dhb} \cdot \varepsilon^{\lambda_2}_{hbO} - \varepsilon^{\lambda_2}_{dhb} \cdot \varepsilon^{\lambda_1}_{hbO})D} \tag{4}$$

With the above formula, the changes in oxygenated hemoglobin $\Delta[hbO]$ and deoxygenated hemoglobin $\Delta[dhb]$ can be expressed as the following expression.

$$\Delta[hbO] = \frac{\Delta O_{dens}\lambda_1 \cdot \varepsilon^{\lambda_2}_{dhb} - \Delta O_{dens}\lambda_2 \cdot \varepsilon^{\lambda_1}_{dhb}}{(\varepsilon^{\lambda_2}_{dhb} \cdot \varepsilon^{\lambda_1}_{hbO} - \varepsilon^{\lambda_1}_{dhb} \cdot \varepsilon^{\lambda_2}_{hbO})D} \tag{5}$$

$$\Delta[dhb] = \frac{\Delta O_{dens}\lambda_1 \cdot \varepsilon^{\lambda_2}_{hbO} - \Delta O_{dens}\lambda_2 \cdot \varepsilon^{\lambda_1}_{hbo}}{(\varepsilon^{\lambda_2}_{dhb} \cdot \varepsilon^{\lambda_1}_{hbO} - \varepsilon^{\lambda_1}_{dhb} \cdot \varepsilon^{\lambda_2}_{hbO})D} \tag{6}$$

3 Experiment Protocol

Ten healthy volunteers were selected for the experiment and their mean age was 27.5 (±2). The human foot is anatomically divided into 10 significant areas (Cavanagh et al. 1987) as shown in Fig. 3. Three standard areas (areas 2, 5, and 8) were considered for the current study (shown in Table 1) and changes in the concentration of HbO and dHb were measured in the foot sole of healthy volunteers. Vascular occlusion technique was used to assess the response for the induced ischemic stimulus by NIRS system (Li et al. 2013). The pressure cuff was wrapped around the calf muscles of the volunteer and inflated above arterial occlusion level (>180 mmHg) to halt the blood flow and the responses were recorded using NIRS system (see Fig. 4).

Before the beginning of the experiment, volunteers were asked to wash their feet so that the reflected light should not be affected by any dirt present in the foot sole.

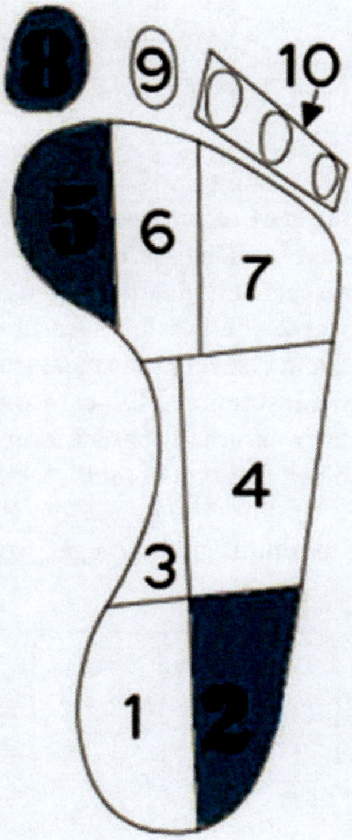

Fig. 3 Anatomical division of foot

Table 1 Foot sole area used in this study

Foot area	Anatomical name
2	Calcaneum
5	Ball of great joint
8	Great toe

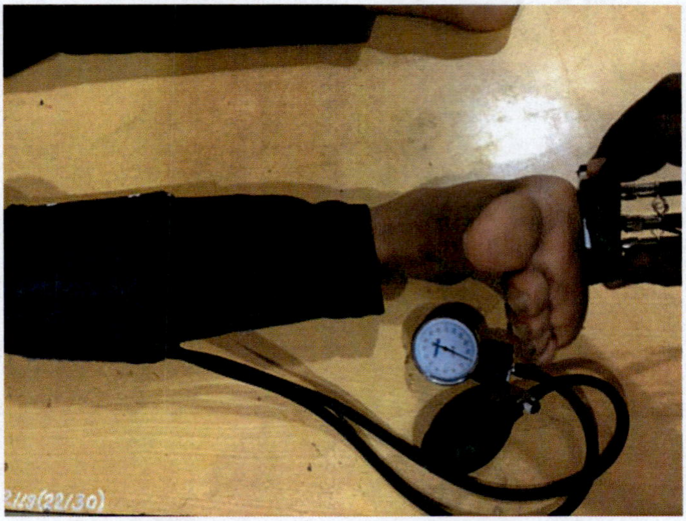

Fig. 4 Probe placed at the foot sole of a healthy volunteer

15 min of relaxation period in the supine position was given to volunteer before the start of measurement. Tissue hardness and temperature at the measuring sites were also measured using the shoremeter and noncontact infrared thermometer, respectively, before the start of measurement. The experiment protocol followed was 2 min baseline measurement, 3 min cuff occlusion measurement followed by 2 min recovery time measurement. This research protocol followed the guidelines of the Institutional Ethical Committee NIT Raipur and all the participants have given their informed consent form at the beginning of the experiment. Figure 5 shows the block diagram of multispectral near-infrared system and its components.

In Fig. 5, the laser diode is in the first stage. The diode is used as an infrared transmitter and it ranges from 780 to 880 nm wavelength. There are three diodes

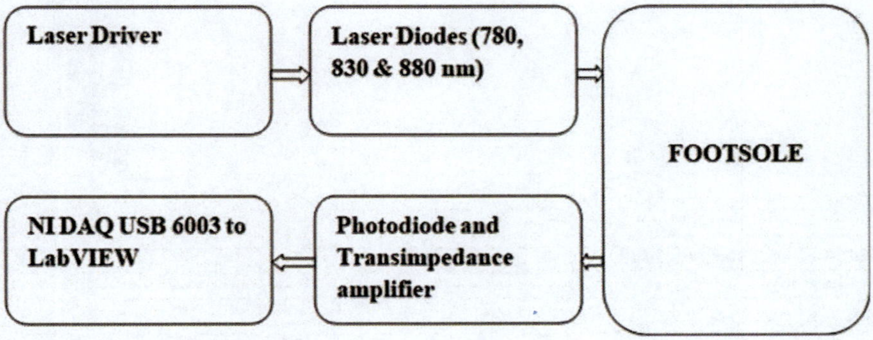

Fig. 5 Block diagram of multispectral NIRS system

placed in the device which are able to transmit infrared radiation of 780, 830, and 880 nm consequently. The transmitted wave enters into the foot sole area, selected for the study. This radiated energy will reflect back toward the sensor diode, while some parts of energy get absorbed. The photodiode sensor having the ability to convert the photonic energy into electrical signal. These electrical signals can be further amplified int the amplifier section. We have implied the current to voltage converting circuit, which will convert the current signal into voltage range. This variable voltage signal is fed into the NI DAQ system to control and interface the circuit with the computer. USB6003 is used to connect the entire system through personal computer, where it will process in LabVIEW to acquire the desired output.

The absorption of the infrared red light by different tissues can be monitored with the help of this spectroscope and it provides hemoglobin concentration around the area under measurement. The overall system can be connected to a computer and a dedicated LabVIEW program provides appropriate results in the form of graphs and adequate numeric value.

4 Result

The result of the advanced NIRS system is given in the following graph which is acquired from the healthy subject. It shows the ability of designed hardware and software systems to render the appropriate response for a clinical study.

Figure 6 shows typical responses of changes in the concentration of [HbO_2] and [dHb] during arterial cuff occlusion measured by the NIRS system at the measuring sites (area 2, 8, and 5) of one subject. From the response, it was observed that the concentration of HbO_2 was decreased during the occlusion period and it shows a lack of oxygen perfusion to the tissue. In contrast, the concentration of dHb rising during the occlusion period and remain high during the entire period of occlusion and return to its baseline after the release of cuff pressure. Through the designed NIRS system, the present study mainly investigates the correlation between changes in concentration of hemoglobin with the tissue hardness and temperature at the measuring sites.

Figures 7, 8 and 9 show the regression analysis plot between tissue hardness and hemoglobin concentrations at different measuring areas (2, 5, and 8) at the plantar foot surface. It is observed that tissue hardness and hemoglobin concentration were not significantly correlated in healthy volunteers. Similarly, regression analysis has been done between temperature and hemoglobin concentrations. It is observed that the concentrations of dHb and HbO_2 were showing a negative relationship as shown in Fig. 10 with a significant level of less than 0.05 in area 5.

Fig. 6 Foot sole responses during 3 min arterial occlusion in one typical subject at **a** area 2, **b** area 5, **c** area 8

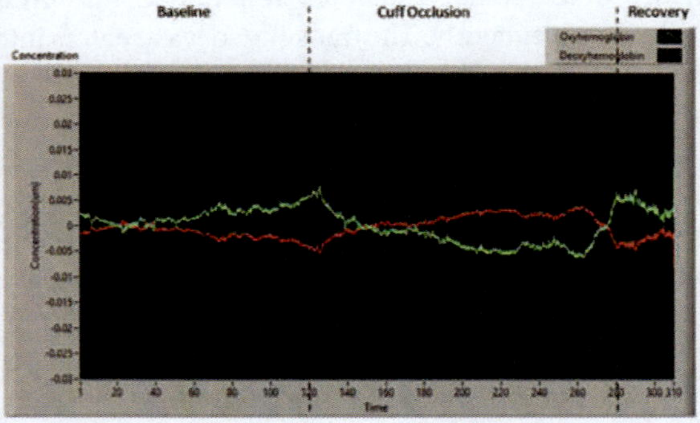

(a)

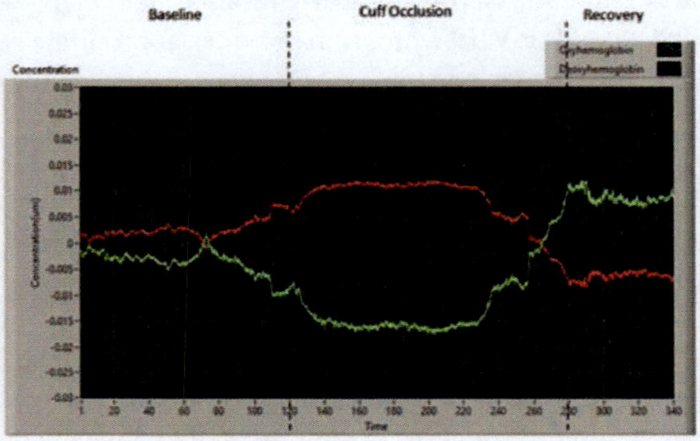

(b)

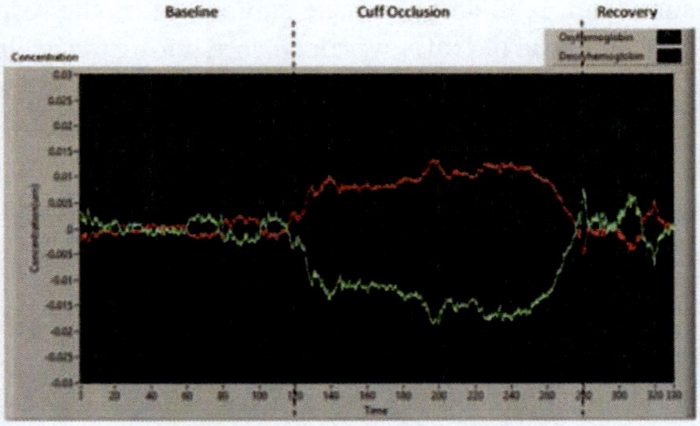

(c)

5 Discussion

Reduced blood circulation in the lower extremities increases the probability of peripheral nerves exposed to pathological changes in the lower extremities, which leads to peripheral arteriosclerosis, and consequently foot ulcerations (Forbes et al. 2013; Flemming and Cullum 2001). The present study analyzed this condition by using a

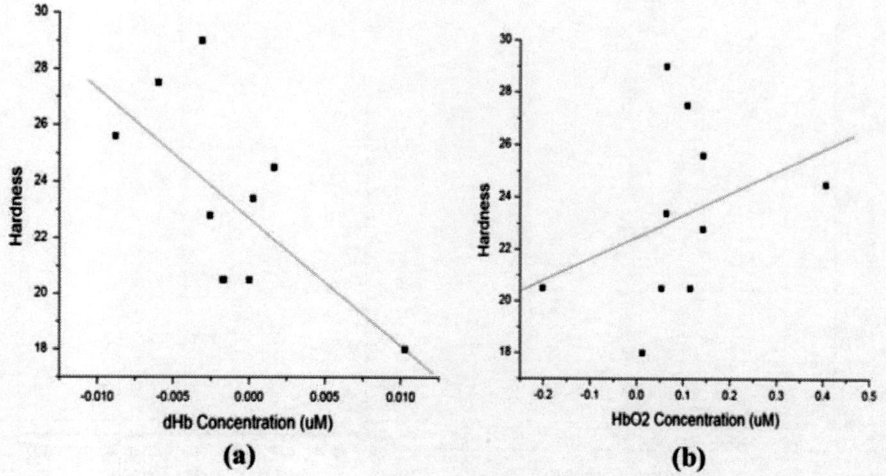

Fig. 7 Regression plot between tissue hardness and changes in **a** dHb concentration ($r = -0.45$), **b** HbO_2 concentration ($r = 0.36$) at area 2

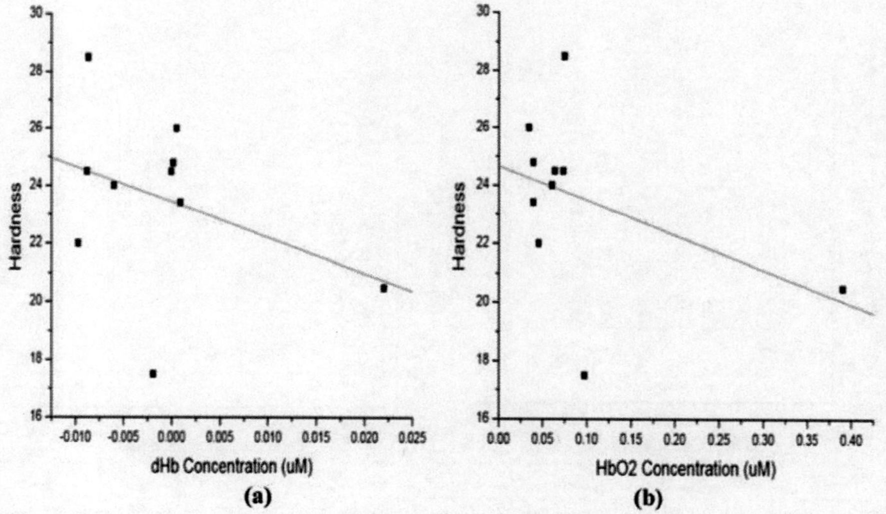

Fig. 8 Regression plot between tissue hardness and changes in **a** dHb concentration ($r = -0.37$), **b** HbO_2 concentration ($r = -0.46$) at area 5

vascular occlusion technique in healthy volunteers to evaluate the response of the designed NIRS system. The responses were recorded and further analyzed to quantify the changes in hemoglobin concentrations. In diabetes patients, vascular disease and neuropathy in the lower limbs lead to the formation of chronic ulcers from what is often initially a minor trauma. Lin et al. (2013) stated that the poor supply of oxygen to the tissues leads to thickening of the basal layer of blood wall and ends in

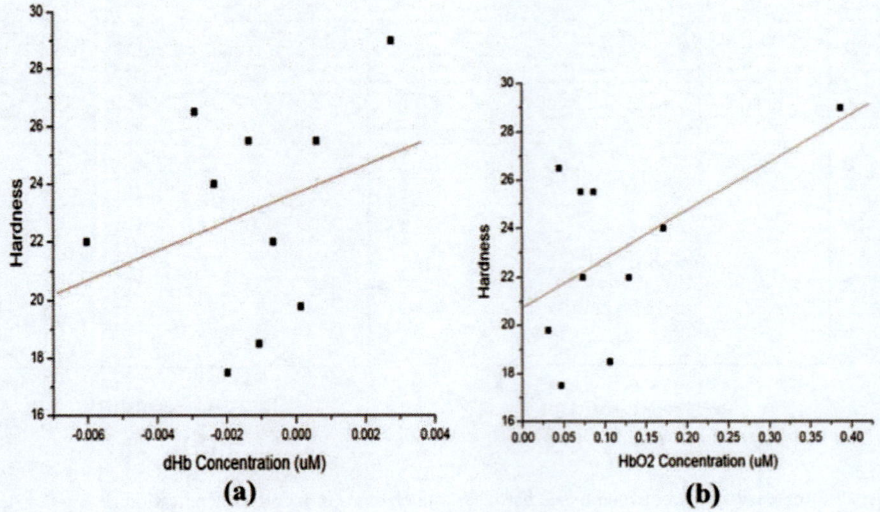

Fig. 9 Regression plot between tissue hardness and changes in **a** dHb concentration ($r = 0.35$), **b** HbO_2 concentration ($r = 0.48$) at area 8

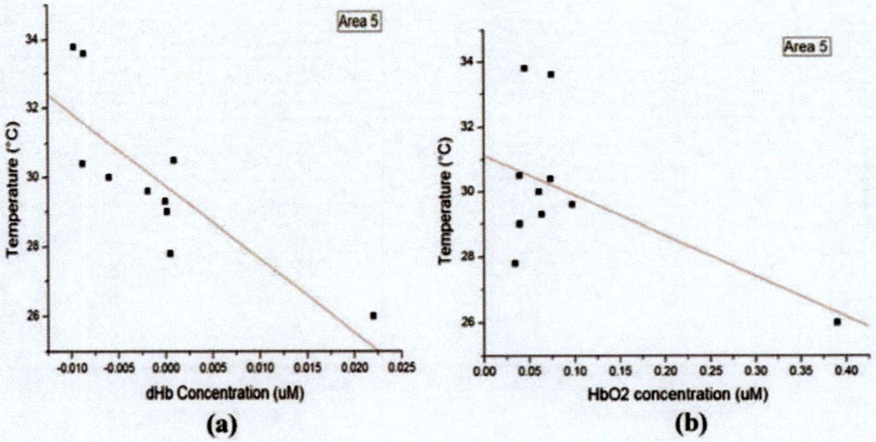

Fig. 10 Regression plot between temperature and changes in **a** dHb concentration ($r = -0.81, p = 0.03$), **b** HbO_2 concentration ($r = -0.56, p = 0.05$) at area 5

cell necrosis form the underlying tissues excessively hard. It is vascular abnormalities which obstruct the vascular system either on the macro-circulatory or micro-circulatory level, which further reduces the oxygen availability to the tissues (Cade 2008; Strandness and Eidt 2000). Considering this, the present study further analyzed the correlation between the foot tissue hardness and temperature with the derived hemoglobin concentrations at selected standard foot risk areas. The result shows that there is no significant correlation between hemoglobin concentrations and tissue hardness in any of the three areas. This shows the possibility of the tissues to adapt and regenerate oxygen diffusion into the tissues of healthy volunteers. Similarly, there was no significant correlation between temperature and hemoglobin concentration in area 2 and area 8, whereas area 5 shows the negative correlation of temperature and hemoglobin concentrations. This shows that increased temperature decreases the affinity of hemoglobin for oxygen and it has to unload oxygen before it enters cell, thereby it enhances metabolic activity. The present study designed a multispectral NIRS system which is solely to quantify the hemoglobin concentration changes at foot. The response of the device was assessed using vascular occlusion technique and further analyzed the correlation of tissue hardness, temperature with hemoglobin concentrations. This shows promising results that this system can be used to analyze the changes in hemoglobin concentration which will be helpful in predicting vascular complications at the foot. In the future, the system will be further analyzed in patients with lower extremity vascular complications and diabetes. The sample size considered in this study is less and to validate our design further, more samples will be considered.

6 Conclusion

NIRS is a noninvasive, easy applicable tool to measure the tissue oxygenation at the regional level and helps in assessing oxygen supply to the tissues. At present, there is no device or method available which can monitor and assess the hemoglobin concentrations at foot sole areas to identify the risky regions before it gets affected. So, in the present study, we designed a NIRS probe that solely measures the hemoglobin concentrations at foot sole areas and can be used at diabetic foot clinics to monitor and identify risk areas (improper supply of oxygen) at foot sole. The NIRS probe has been tested on only ten healthy volunteers to validate the outcomes. In order to enhance the study in the future, we will include more normal subjects and diabetic patients to conclude that the designed NIRS probe can be employed to assess hemoglobin concentrations and to ascertain the risk areas with a diminished supply of oxygen on foot sole significantly. Further, the probe can be validated with the available clinical gold standard techniques.

Acknowledgements The authors are grateful to the Science and Engineering Research Board (SERB), Ministry of Science and Technology, Govt. of India, New Delhi for providing financial support (ECR/2015/000161).

References

Cade WT (2008) Diabetes-related microvascular and macrovascular diseases in the physical therapy setting. Phys Ther 88(11):1322–1335

Cavanagh PR, Rodgers MM, liboshi A (1987) Pressure distribution under symptom-free feet during barefoot standing. Foot Ankle 7(5):262–278

Cope M (1991) The application of near infrared spectroscopy to non invasive monitoring of cerebral oxygenation in the newborn infant. Doctoral dissertation, PhD dissertation in University of London

Flemming K, Cullum N (2001) Electromagnetic therapy for treating venous leg ulcers. Cochrane Database Syst Rev (1)

Forbes JM, Cooper ME (2013) Mechanisms of diabetic complications. Physiol Rev 93(1):137–188

Giurato L, Meloni M, Izzo V, Uccioli L (2017) Osteomyelitis in diabetic foot: a comprehensive overview. World J Diabetes 8(4):135

Khalil MA, Hoi J, Kim HK, Hielscher AH (2013) Dynamic contact-free continuous-wave diffuse optical tomography system for the detection of vascular dynamics within the foot. In: SPIE BiOS, international society for optics and photonics, pp 85781H–85781H

Kim JG, Xia M, Liu H (2005) Extinction coefficients of hemoglobin for near-infrared spectroscopy of tissue. IEEE Eng Med Biol Mag 24(2):118–121

Li J, Zhang X (2014) A low-cost CCD-based imager for mapping venous oxygenation. Session 2A0:624.

Li T, Lin Y, Shang Y, He L, Huang C, Szabunio M, Yu G (2013) Simultaneous measurement of deep tissue blood flow and oxygenation using noncontact diffuse correlation spectroscopy flow-oximeter. Sci Rep 3:1358

Lin OH, Lai JY, Tsai HY (2013) Preventing diabetes extremity vascular disease with blood oxygen saturation images. Int J Instrum Sci 2(A):1–7

Liu C, van der Heijden F, Klein ME, van Baal JG, Bus SA, van Netten JJ (2013) Infrared dermal thermography on diabetic feet soles to predict ulcerations: a case study

Naidoo P, Liu VJ, Mautone M, Bergin S (2015) Lower limb complications of diabetes mellitus: a comprehensive review with clinicopathological insights from a dedicated high-risk diabetic foot multidisciplinary team. Br J Radiol 88(1053):20150135

Reiber G, Lipsky B, Gibbons G (1998) The burden of diabetic foot ulcers. Am J Surg 176(2):5S–10S

Scheeren TWL, Schober P, Schwarte LA (2012) Monitoring tissue oxygenation by near infrared spectroscopy (NIRS): background and current applications. J Clin Monit Comput 26(4):279–287

Strandness DE, Eidt JF (2000) Peripheral vascular disease. Circulation 102:IV46–51

Yong K, Cobb JE, Claremont DJ (2005) Near infrared spectroscopy of the diabetic foot

Zhang Y, Guo C, Zhao J, Yang C, Wang K, Sun J (2013) Non-invasive measurement of haemodynamic changes with a multi-wavelength NIRS system. J Comput Inform Syst 9(13):5265–5272

Comparative Investigation of Different Classification Techniques for Epilepsy Detection Using EEG Signals

Sunandan Mandal, Manvendra Thakur, Kavita Thakur, and Bikesh Kumar Singh

Abstract Among the major brain abnormalities that have been identified, various remedial strategies are proposed to tackle most of such conditions. One of the serious abnormalities of the nervous system is epilepsy, which causes electrical distraction and strains the neural system. Usually, epilepsy is determined by the neurologist by analyzing the EEG signals grabbed from the brain. The task is very challenging as it requires continuous examination and connotation of the EEG signal of an epileptic patient. Hence, the development of efficient automatic systems is currently a bottom neck issue, which can recognize the epileptic seizure attack and distinguish between normal condition and epileptic seizure condition. Over the years, efforts have been made to establish an automatic system for precise detection and classification of an epileptic seizure. In the present work, a range of statistical, entropy, and fractal-based attributes are calculated from the coefficients of selected wavelet. The classifiers used in the paper are trained on the selected features as well as full features. The objective of the study is the selection of appropriate features and applying Tree, SVM, KNN, BPANN algorithm for the classification. Features are ranked with the help of Information Gain, Relief F, and Correlation attributes ranking techniques. Using Information Gain and Relief F feature ranking technique with 10-fold cross-validation, the classification accuracy of 100% was achieved by Backpropagation Artificial Neural Network (BPANN) for classifying epileptic and normal EEG signals.

Keywords EEG · Epileptic seizure · K-nearest neighbor · Wavelet transform · Backpropagation artificial neural network · Support vector machine

---

S. Mandal (✉) · K. Thakur
School of Studies in Electronics & Photonics, Pt. Ravishankar Shukla University, Raipur, India
e-mail: sunandan.mandal12@gmail.com

M. Thakur
Computer Science and Engineering, Kalinga Institute of Industrial Technology, Bhubaneswar, India

B. K. Singh
Department of Biomedical Engineering, National Institute of Technology, Raipur, India

A. A. Rizvanov et al. (eds.), *Advances in Biomedical Engineering and Technology*, Lecture Notes in Bioengineering,
https://doi.org/10.1007/978-981-15-6329-4_34

1 Introduction

The human neural system gets active within a few months of prenatal growth and generates electrical signals that describe cognitive status as well as the status of all the parts of the human body also. If the neural system is not working properly then it affects adversely the human daily life. The disturbance in the neural system may be the cause of brain-related disease or inborn or any severe accident that causes brain injury. An epileptic seizure is a neuronal disease that also affects and dominates over the crucial decision of life (Epilepsy Foundation 2018). World's 1% of the whole population gets affect by this brain-related noncontagious disease (World Health Organization 2018). During the treatment of severe epilepsy disease, the patient has to be in continuous observation of neurologists and medical assistant. EEG (Electroencephalogram) is a simple and common way for continuous observation of brain functioning. Continuous observation of dynamic patterns of EEG is very serious and difficult as well as tedious for any professional. Therefore, in the past few decades, huge research works are taking place to model an automatic system for this time-consuming job. Some literature reviews of recent and major research in this field are reported in related work.

Ghosh-Dastidar et al. (2008) proposed Principal Component Analysis (PCA) and Enhanced Cosine RBFNN (Radial Basis Function Neural Network) based epilepsy and seizure identification model. In this work, EEGs were band-limited to 0–60 Hz and decomposed into physiological EEG sub-bands using fourth-level decomposition with Daubechies wavelet transform of fourth order. Features like CD (Correlation Dimension), LLE (Largest Lyapunov Exponent), and STD (Standard Deviation) were computed form band-limited EEGs and EEG subbands. Feature space was reduced with the help of PCA and applied in the input of enhanced cosine RBFNN classifier. The classification accuracy rate of 99.3% was achieved for normal and interictal EEGs. Gandhi et al. (2010) reported DWT (Discrete Wavelet Transform) based automatic epileptic detection model using PNN (Probabilistic Neural Network). The EEG segments of both normal as well as epileptic were decomposed up to the sixth level using db4 (Daubechies) mother wavelet. Different features like Standard Deviation (STD), Energy, and Entropy were extracted from wavelet coefficients. This expert model achieved a detection accuracy of 99.33%. Übeyli (2010) proposed Lyapunov exponents and PNN-based epilepsy detection model. Some statistical values like mean, maximum (max), minimum (min), and STD were calculated from the extracted Lyapunov exponents. The overall accuracy rate was reached to be 98.05% using PNN. In Guo et al. (2010), a multiwavelet transform-based epileptic seizure identification model was reported. ApEn (Approximate Entropy) was measured from the wavelet coefficient and ANN (Artificial Neural Network) was used for classification. Classification accuracy (CA) was found to be 99.85% for normal and epileptic EEGs. An automatic system of epileptic seizure detection using Higher Order Spectra (HOS) features was reported by Acharya et al. (2011). Here the HOS features were extracted from approximate and detailed coefficients of WPD (Wavelet packet Decomposition). ANOVA (Analysis of Variance) was applied to select the significant features. The selected features were further used for the classification of Normal, Interictal, and Ictal EEGs. The performance of several classifiers like SVM (Support Vector Machine) with linear, polynomial, and RBF kernels, fuzzy, KNN (K-Nearest Neighbor), PNN, DT (Decision Tree), GMM (Gaussian Mixture Model), and NBC (Naïve Bayes Classifier) were studied with threefold Cross-Validation (CV) technique. Fuzzy classifier was shown to have the highest detection accuracy of 98.5%.

Further, a Simple Random Sampling (SRS) technique based EEG signal discrimination approach using LS-SVM (Least Square Support Vector Machine) classifier was reported in Siuly et al. (2011). In this work, nine statistical parameters, namely, mean, mode, min, max, median, first quartile, third quartile, standard deviation, and IQR were also extracted from the sub-sample. An average classification rate of 95.58% was achieved using LS-SVM. Acharya et al. (2012) reported epileptic seizure detection using nonlinear and wavelet-based features. In this work, nonlinear features like HOS, ApEn, SampEn (Sample Entropy), FD (Fractal Dimension), and Hurst Exponent were extracted. The significant features were selected using ANOVA. Performance of six classifiers namely, DT, GMM, KNN, SVM, RBFPNN, and Fuzzy Sugeno was evaluated using 10-fold CV technique. 99.7% CA rate was achieved using a fuzzy classifier. Empirical Mode Decomposition (EMD) based epileptic seizure detection has been reported in Alam and Bhuiyan (2013). In this work, some statistical attributes, namely, variance, kurtosis, and skewness were calculated from decomposed EEG. CA rate of 100% was achieved using ANN classifier with a hold out CV technique. Chen (2014) proposed automatic EEG seizure identification using DTCWT (Dual-Tree Complex Wavelet) based Fourier features and KNN classifier. CA rate of 100% was achieved for normal and epileptic EEGs. A Genetic Programming based epileptic identification has been reported in Bhardwaj et al. (2016). In this work, EMD and Constructive Genetic Programming (CGP), Standard GP, Semantic search based GP were applied for classification. Al Ghayab et al. (2016) proposed SRS with sequential attribute selection for the epileptic EEG identification system. CA rate of 99.90% was achieved for normal and epileptic binary classification using attributes, namely, max, min, mode, mean, median, first quartile, second quartile, range, and standard deviation. Diykh et al. (2017) reported community structure detection in epileptic and normal EEGs using Weighted Complex Networks. In this work, the performance of KNN, LS-SVM, k-means, and naïve Bayes classifier was evaluated. The highest mean accuracy of 98% was achieved in LS-SVM. ATFFWT (Analytic Time-Frequency Flexible Wavelet Transform) with fractal features based approaches for epileptic identification has been proposed in Sharma et al. (2017). Significant features were elected using a t-test and applied to RBF kernel based LS-SVM classifier. Recently Alickovic et al. (2018) reported an automatic epileptic seizure detection system. In this work, EEG segments were preprocessed using MSPCA (Multiscale Principal Component Analysis). In this work, WPD and DWT wavelet decomposition were also applied. Some statistical attributes, namely, mean, skewness, kurtosis, average power, standard deviation, and absolute mean were also measured from extracted wavelet coefficients. CA rate of 100% was obtained for SVM, MLP (Multilayer Perceptron), Random Forest, and KNN classifier.

This paper contains four sections. Section two explains about materials and methods applied in this work. The results and discussion part are explained in section three and the last section presents conclusions.

2 Materials and Methods

This section comprises various subsections involved in epileptic seizure identification from EEG signals.

2.1 *Data Description*

The EEG database used for this research work is accessed from the website of the University of Bonn, Germany (EEG 2017). This database contains 500 EEG segments with 4097 data points. Five healthy subjects and five epileptic patients participated in the database creation. These EEG segments are divided into five sets, namely, Set A, B, C, D, and E. EEG segments of Set A and B belong to normal EEG signals during eye open and close conditions, respectively. Similarly, EEG segments of Set C and D belong to interictal EEG signals recorded from an epileptic patient during no epileptic seizure. The segments of Set E belong to ictal EEG signals recorded during an epileptic seizure. These EEG signals were recorded with a 128-channel system using a 173.61 Hz sampling rate. EEG segments are preprocessed and noise removed during the segmentation of the 23.6 s long EEG signal. More details about the database are described in Andrzejak et al. (2001).

2.2 *Feature Extraction*

In the present work, time as well as frequency-domain features are utilized concurrently for epilepsy detection. Some statistical features, namely, mean, kurtosis, skewness, entropy, variance, STD, min, max, range, crest factor, and form factor are measured from both normal as well as epileptic EEG segments. Fractal Dimensional (FD) attributes using Katz and Higuchi algorithm are calculated. Permutation and ApEn features are also measured from the EEG segments. These 15 features are also pulled out from DWT coefficients. Nine mother wavelets, namely, Daubechies (db4), Biorthogonal (bior 3.1 and bior 3.2), Coiflets (Coif 1–Coif 5), and Haar are used for generating third-level coefficients. Table 1 shows the overview of features utilized for the present work.

Table 1 Overview of features extracted in present work

Feature category	Number of features	Feature name
Without wavelet	15	Mean, kurtosis, skewness, entropy, variance, standard deviation, min, max, range, crest factor, form factor, Katz FD, ApEn, Higuchi FD, permutation entropy (P1–P15)
Third-level approximate coefficient db4 wavelet	15	Mean, kurtosis, skewness, entropy, variance, standard deviation, min, max, range, crest factor, form factor, Katz FD, ApEn, Higuchi FD, permutation entropy (P16–P30)
Third-level approximate coefficient bior 3.1 wavelet	15	Mean, kurtosis, skewness, entropy, variance, standard deviation, min, max, range, crest factor, form factor, Katz FD, ApEn, Higuchi FD, permutation entropy (P31–P45)
Third-level approximate coefficient bior 3.2 wavelet	15	Mean, kurtosis, skewness, entropy, variance, standard deviation, min, max, range, crest factor, form factor, Katz FD, ApEn, Higuchi FD, permutation entropy (P46–P60)
Third-level approximate coefficient Coif 1 wavelet	15	Mean, kurtosis, skewness, entropy, variance, standard deviation, min, max, range, crest factor, form factor, Katz FD, ApEn, Higuchi FD, permutation entropy (P61–P75)
Third-level approximate coefficient Coif 2 wavelet	15	Mean, kurtosis, skewness, entropy, variance, standard deviation, min, max, range, crest factor, form factor, Katz FD, ApEn, Higuchi FD, Permutation Entropy (P76–P90)
Third-level approximate coefficient Coif 3 wavelet	15	Mean, kurtosis, skewness, entropy, variance, standard deviation, min, max, range, crest factor, form factor, Katz FD, ApEn, Higuchi FD, permutation entropy (P91–P105)
Third-level approximate coefficient Coif 4 wavelet	15	Mean, kurtosis, skewness, entropy, variance, standard deviation, min, max, range, crest factor, form factor, Katz FD, ApEn, Higuchi FD, permutation entropy (P106–P120)
Third-level approximate coefficient Coif 5 wavelet	15	Mean, kurtosis, skewness, entropy, variance, standard deviation, min, max, range, crest factor, form factor, Katz FD, ApEn, Higuchi FD, permutation entropy (P121–P135)

(continued)

Table 1 (continued)

Feature category	Number of features	Feature name
Third-level approximate coefficient Haar wavelet	15	Mean, kurtosis, skewness, entropy, variance, standard deviation, min, max, range, crest factor, form factor, Katz FD, ApEn, Higuchi FD, permutation entropy (P136–P150)

2.3 Feature Selection and Classification Technique

Reduction of the features as well as computation complexity of any classification model needed a feature selection algorithm. This process also helps to progress the performance of the classifier. In the present work, three feature ranking algorithms are applied as feature selection techniques, namely, Information Gain (Info Gain), Correlation, and Relief F. The performance of these feature selection techniques are discussed in next section (see Sect. 3). BPANN, Decision Tree, SVM, and KNN are used for classification purposes (Singh et al. 2017). In the present work, the BPANN is used with MSE (Mean Squared Error) as a performance function and SCG (Scaled Conjugate Gradient) as a training function. The classifiers like Decision Tree, SVM, and KNN are used with default settings made available by MATLAB® R2016a software.

2.4 Proposed Model

The architecture of the proposed model for epilepsy detection is depicted in Fig. 1. This model is partitioned into two major parts, namely, online and offline system as shown in the left and right side of the figure, respectively. Both major parts have a similar task that is preprocessing. The next task in the offline system is feature extraction. 150 attributes are pulled out from normal EEG as well as epileptic EEG segments. Most relevant features are selected in step, i.e., feature selection. From these N relevant features, the classifier is supervised and generates learning parameters. In the online part, after preprocessing the next task is the extraction of relevant N features only. These N features are put as input to the classifier and decision has been taken according to the learning of classifier.

2.5 Performance Evaluation

In the present work, the k-fold CV technique is used for performance evaluation. K-fold CV technique divides the complete data segments into k subset. From this k subset, $(k - 1)$ set is used as training dataset and 1 subset is for testing test. $K = 10$

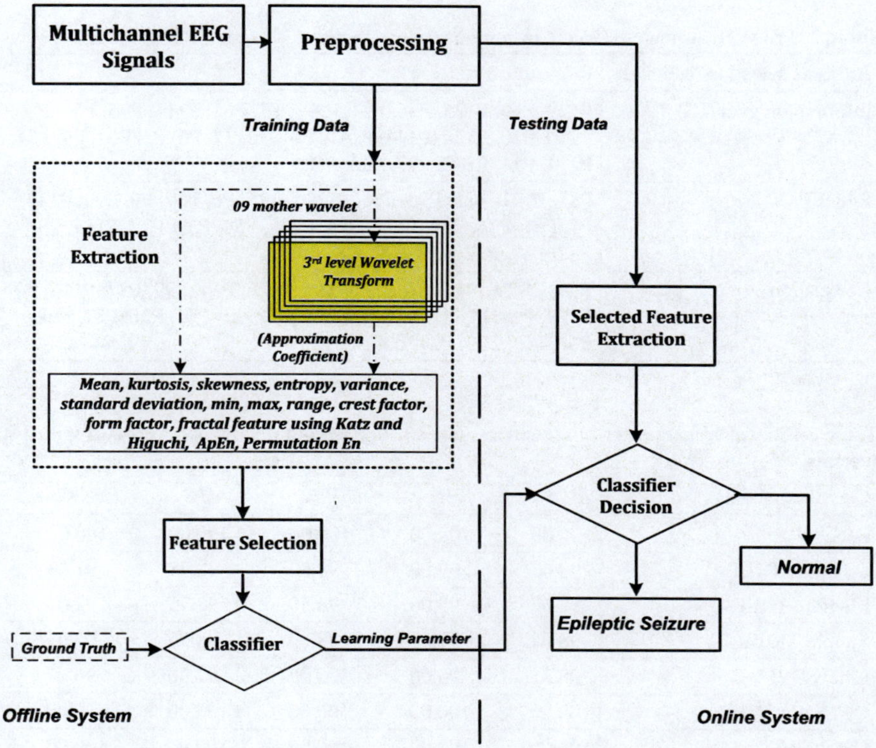

Fig. 1 Architecture of proposed methods

CV is the most commonly used k-fold technique, which is also applied in present work. The performance of feature selection techniques and classifiers are compared using statistical parameters, namely, accuracy (ACC), sensitivity (SEN), specificity (SPEC), area under receiver operating characteristic curve (AUC), and Mathew's Correlation Coefficient (MCC). The definition and mathematical expressions are referred from Singh et al. (2017).

3 Results and Discussions

In the present work, the Classification Problem (CP) is studied over three different feature selections and 16 classification techniques. Here the classification problem is normal versus epileptic EEGs, i.e., AB versus E only. Using feature selection, we have selected only the top 30 features for CP for all three attribute selection algorithms. The top 30 attributes selected using three attribute selection techniques as shown in Table 2.

Table 3 shows the performances of 16 classification techniques for AB versus E

Table 2 Prime 30 attributes selected by attribute selection techniques for CP (AB vs. E)

Attribute selection technique	Prime attributes
Information gain (IG)	P129, P69, P125, P96, P21, P99, P81, P65, P111, P95, P66, P80, P110, P20, P126, P144, P141, P140, P127, P6, P5, P67, P9, P84, P114, P24, P142, P97, P82, P68
Relief F (RLF)	P81, P141, P21, P96, P66, P111, P126, P6, P69, P9, P99, P144, P7, P51, P36, P84, P114, P24, P129, P54, P39, P22, P112, P8, P82, P67, P52, P38, P97, P53
Correlation	P96, P66, P141, P81, P21, P111, P126, P69, P99, P84, P114, P6, P129, P24, P144, P9, P51, P22, P8, P112, P54, P36, P82, P97, P68, P67, P23, P38, P39, P98

Table 3 Statistical parameter of classifier models using 10-fold CV for AB versus E with 150 features

Classifier model	ACC	SEN	SPEC	AUC	MCC
BPANN	100.00	100.00	100.00	100.00	100.00
Complex TREE	98.67	99.00	98.50	98.75	97.03
Medium TREE	98.67	99.00	98.50	98.75	97.03
Simple TREE	98.67	99.00	98.50	98.75	97.03
Linear SVM	98.33	95.00	100.00	97.50	96.27
Quadratic SVM	100.00	100.00	100.00	100.00	100.00
Cubic SVM	100.00	100.00	100.00	100.00	100.00
Fine Gaussian SVM	96.33	100.00	94.50	97.25	92.27
Medium Gaussian SVM	100.00	100.00	100.00	100.00	100.00
Coarse Gaussian SVM	94.33	83.00	100.00	91.50	87.46
Fine KNN	98.33	95.00	100.00	97.50	96.27
Medium KNN	95.33	86.00	100.00	93.00	89.65
Coarse KNN	83.33	50.00	100.00	75.00	63.25
Cosine KNN	98.00	94.00	100.00	97.00	95.53
Cubic KNN	94.33	83.00	100.00	91.50	87.46
Weighted KNN	96.33	89.00	100.00	94.50	91.85

using all the features with 10-fold CV protocol. BPANN, Quadratic SVM, Cubic SVM, and Medium Gaussian SVM classification models showed a 100% classification rate. The other performance parameters, i.e., sensitivity, specificity, AUC, and MCC also reached 100%. It is observed that except for three classifiers models, namely, Coarse Gaussian SVM, Coarse KNN, and Cubic KNN all the other classifier models achieved greater than or equal to 95% classification rate. Unsatisfactory performance, i.e., below 90% classification accuracy is shown by only one classifier model, namely, coarse KNN. It is also illustrated from Table 3 that the ten classifier model shows accuracy performance greater than equal to 98%.

Table 4 Statistical parameter of classifier models using information gain feature selection and 10-fold CV for top 30 attributes of AB versus E

Classifier model	ACC	SEN	SPEC	AUC	MCC
BPANN	100.00	100.00	100.00	100.00	100.00
Complex TREE	99.00	99.00	99.00	99.00	97.76
Medium TREE	99.00	99.00	99.00	99.00	97.76
Simple TREE	99.00	99.00	99.00	99.00	97.76
Linear SVM	98.67	96.00	100.00	98.00	97.01
Quadratic SVM	98.67	96.00	100.00	98.00	97.01
Cubic SVM	98.67	96.00	100.00	98.00	97.01
Fine Gaussian SVM	99.33	98.00	100.00	99.00	98.50
Medium Gaussian SVM	98.67	96.00	100.00	98.00	97.01
Coarse Gaussian SVM	96.33	89.00	100.00	94.50	91.85
Fine KNN	99.33	98.00	100.00	99.00	98.50
Medium KNN	98.67	96.00	100.00	98.00	97.01
Coarse KNN	91.33	74.00	100.00	87.00	80.92
Cosine KNN	97.67	93.00	100.00	96.50	94.79
Cubic KNN	98.67	96.00	100.00	98.00	97.01
Weighted KNN	99.33	98.00	100.00	99.00	98.50

Tables 4, 5 and 6 shows the performances of 16 classification models for AB versus E with 10-fold CV technique and prime 30 attributes achieved from three different attribute selection algorithms, namely, Information Gain, Relief F, and Correlation, respectively. From the results of all attribute algorithms, BPANN shows the highest classification rate of greater than 99% with only 30 attributes. Tables 4 and 5 shows 100% Classifier performance, i.e., for Information Gain and Relief F algorithm using BPANN which is similar to full features result (see Table 3). We also perceive from Tables 4, 5 and 6 that 13 classifier model shows an accuracy rate greater than or equal to 98% in all three attribute selection algorithms. The three classifier models, namely, Coarse Gaussian SVM, Coarse KNN, and Cosine KNN show below 98% accuracy in all three attribute selection algorithms. No classifier shows unsatisfactory performance, i.e., below 90% in any attribute selection algorithms.

Table 7 shows the comparison of present work with other works. From the table, it is clear that our model outperforms other models. It is also observed that mostly SVM and Neural Network based classifier are utilized in this particular classification problem.

Table 5 Statistical parameter of classifier models using Relief F feature selection and 10-fold CV for top 30 attributes of AB versus E

Classifier model	ACC	SEN	SPEC	AUC	MCC
BPANN	100.00	100.00	100.00	100.00	100.00
Complex TREE	98.67	98.00	99.00	98.50	97.00
Medium TREE	98.67	98.00	99.00	98.50	97.00
Simple TREE	98.67	98.00	99.00	98.50	97.00
Linear SVM	98.00	94.00	100.00	97.00	95.53
Quadratic SVM	98.33	95.00	100.00	97.50	96.27
Cubic SVM	98.67	96.00	100.00	98.00	97.01
Fine Gaussian SVM	99.33	98.00	100.00	99.00	98.50
Medium Gaussian SVM	98.67	96.00	100.00	98.00	97.01
Coarse Gaussian SVM	96.33	89.00	100.00	94.50	91.85
Fine KNN	99.33	98.00	100.00	99.00	98.50
Medium KNN	98.67	96.00	100.00	98.00	97.01
Coarse KNN	92.00	76.00	100.00	88.00	82.38
Cosine KNN	97.33	92.00	100.00	96.00	94.05
Cubic KNN	98.00	94.00	100.00	97.00	95.53
Weighted KNN	99.00	97.00	100.00	98.50	97.76

Table 6 Statistical parameter of classifier models using correlation feature selection and 10-fold CV for top 30 attributes of AB versus E

Classifier model	ACC	SEN	SPEC	AUC	MCC
BPANN	99.67	99.00	100.00	99.50	99.25
Complex TREE	99.00	99.00	99.00	99.00	97.76
Medium TREE	99.00	99.00	99.00	99.00	97.76
Simple TREE	99.00	99.00	99.00	99.00	97.76
Linear SVM	98.67	96.00	100.00	98.00	97.01
Quadratic SVM	98.67	96.00	100.00	98.00	97.01
Cubic SVM	99.00	97.00	100.00	98.50	97.76
Fine Gaussian SVM	99.33	98.00	100.00	99.00	98.50
Medium Gaussian SVM	98.67	96.00	100.00	98.00	97.01
Coarse Gaussian SVM	96.33	89.00	100.00	94.50	91.85
Fine KNN	99.33	98.00	100.00	99.00	98.50
Medium KNN	98.67	96.00	100.00	98.00	97.01
Coarse KNN	91.67	75.00	100.00	87.50	81.65
Cosine KNN	97.00	91.00	100.00	95.50	93.32
Cubic KNN	98.67	96.00	100.00	98.00	97.01
Weighted KNN	99.00	97.00	100.00	98.50	97.76

Table 7 A comparison of classification accuracy with other approaches for binary classification

Author and year	Method	Classification problem	Classification accuracy (%)
Guo et al. (2010)	Multiwavelet transform, ApEn, and MLPNN	ABCD versus E	98.27
Diykh et al. (2017)	Weighted complex network based feature, and LS-SVM	ABCD versus E	97.90
Nigam and Graupe (2004)	Nonlinear preprocessing and diagnostic neural network	A versus E	97.20
Tzallas et al. (2007)	Time-frequency analysis and ANN	ABCD versus E	97.73
Nicolaou and Georgiou (2012)	Permutation entropy and linear SVM	A versus E	93.55
Kabir et al. (2016)	OAT (optimum allocation technique) based feature and logistic model tree	AB versus CD versus E	95.33
Present work	DWT-based feature and BPANN	AB versus E	100.00
Present work	DWT-based feature, Information gain and BPANN	AB versus E	100.00
Present work	DWT-based feature, Relief F and BPANN	AB versus E	100.00

4 Conclusions

In this work, three different classification problems are studied with 3 attribute selection algorithms and 16 classification techniques. Using Information Gain and Relief F feature ranking technique with 10-fold CV protocol, the CA rate of 100% was achieved with BPANN to classify epileptic and normal EEG signals. The proposed model shows good performance. This work can be useful for designing an automated epilepsy detection system to assist the neurologist and medical professionals.

References

Acharya UR, Sree SV, Suri JS (2011) Automatic detection of epileptic EEG signals using higher order cumulant features. Int J Neural Syst 21(05):403–414

Acharya UR, Sree SV, Alvin APC, Suri JS (2012) Application of non-linear and wavelet based features for the automated identification of epileptic EEG signals. Int J Neural Syst 22(2):1250002-1–14

Al Ghayab HR, Li Y, Abdulla S, Diykh M, Wan X (2016) Classification of epileptic EEG signals based on simple random sampling and sequential feature selection. Brain Inform 3(2):85–91

Alam SMS, Bhuiyan MIH (2013) Detection of seizure and epilepsy using higher order statistics in the EMD domain. IEEE J Biomed Health Inform 17:312–318

Alickovic E, Kevric J, Subasi A (2018) Performance evaluation of empirical mode decomposition, discrete wavelet transform, and wavelet packed decomposition for automated epileptic seizure detection and prediction. Biomed Signal Process Control 39:94–102

Andrzejak RG, Lehnertz K, Mormann F, Rieke C, David P, Elger CE (2001) Indications of nonlinear deterministic and finite-dimensional structures in time series of brain electrical activity: dependence on recording region and brain state. Phys Rev E 64:061907-1–061907-8

Bhardwaj A, Tiwari A, Krishna R, Varma V (2016) A novel genetic programming approach for epileptic seizure detection. Comput Methods Programs Biomed 124:2–18

Chen G (2014) Automatic EEG seizure detection using dual-tree complex wavelet-Fourier features. Expert Syst Appl 41(5):2391–2394

Diykh M, Li Y, Wen P (2017) Classify epileptic EEG signals using weighted complex networks based community structure detection. Expert Syst Appl 90:87–100

EEG (2017) EEG time series download page. http://epileptologie-bonn.de/cms/front_content.php?idcat=193&lang=3&changelang=3. Accessed 20 July 2017

Epilepsy Foundation (2018). What is epilepsy? https://www.epilepsy.com/learn/about-epilepsy-basics/what-epilepsy. Accessed 08 June 2018

Gandhi T, Panigrahi BK, Bhatia M, Anand S (2010) Expert model for detection of epileptic activity in EEG signature. Expert Syst Appl 37(4):3513–3520

Ghosh-Dastidar S, Adeli H, Dadmehr N (2008) Principal component analysis-enhanced cosine radial basis function neural network for robust epilepsy and seizure detection. IEEE Trans Biomed Eng 55(2):512–518

Guo L, Rivero D, Pazos A (2010) Epileptic seizure detection using multiwavelet transform based approximate entropy and artificial neural networks. J Neurosci Methods 193(1):156–163

Kabir E, Siuly S, Zhang Y (2016) Epileptic seizure detection from EEG signals using logistic model trees. Brain Inform 3(2):93–100

Nicolaou N, Georgiou J (2012) Detection of epileptic electroencephalogram based on permutation entropy and support vector machines. Expert Syst Appl 39(1):202–209

Nigam V, Graupe D (2004) A neural-network-based detection of epilepsy. Neurol Res 26(1):55–60

Sharma M, Pachori RB, Acharya UR (2017) A new approach to characterize epileptic seizures using analytic time-frequency flexible wavelet transform and fractal dimension. Pattern Recogn Lett 94:172–179

Singh BK, Verma K, Panigrahi L, Thoke AS (2017) Integrating radiologist feedback with computer aided diagnostic systems for breast cancer risk prediction in ultrasonic images: An experimental investigation in machine learning paradigm. Expert Syst Appl 90:209–223

Siuly S, Li Y, Wen P (2011) EEG signal classification based on simple random sampling technique with least square support vector machine. Int J Biomed Eng Technol 7(4):390–409

Tzallas A, Tsipouras M, Fotiadis D (2007) Automatic seizure detection based on time–frequency analysis and artificial neural networks. Comput Intell Neurosci 13. Article ID 80510

Übeyli ED (2010) Lyapunov exponents/probabilistic neural networks for analysis of EEG signals. Expert Syst Appl 37(2):985–992

World Health Organization (2018). Fact sheet on epilepsy. http://www.who.int/en/news-room/fact-sheets/detail/epilepsy. Accessed 15 June 2018

Designing a Low-Cost Spin-Drying Desiccation Technique Using 3D Printed Whirligig Model for Preservation of Human Umbilical Cord Blood-Derived Mesenchymal Stem Cells

Sharda Gupta, Akalabya Bissoyi, Pradeep Kumar Patra, and Arindam Bit

Abstract In modern stem cell therapy, stem cells preservation is one of the major challenges. The present study aims to develop an advanced desiccation technique for the preservation of mesenchymal stem cells. Cryopreservation of human mesenchymal stem cells (MSCs) is one of the options for its long-term storage. The established techniques of preserving MSCs using cryopreservation aims to preserve cells at ultra-low temperatures using either liquid nitrogen or a deep freezer. However, limitations include cross-contamination, regular maintenance, the requirement of continuous liquid nitrogen supply, and electricity cost. A new technique of storing dried samples at ambient temperature has been developed utilizing desiccation. In this study, we desiccated the mesenchymal stem cells using a low-cost, power-free 3D printed whirligig model. It is an improved and effective method for preserving many different types of cells. We concluded that trehalose was introduced efficiently into the cells using fluid-phase endocytosis. We observed the effect of drying with trehalose solution at spinning speeds ranging from 2000 to 4000 rpm. To characterize the desiccation actuated cell death, we performed proliferation and viability analysis of post-desiccated MSCs by MTT assay. 60% of the cells were shown to survive this whirligig model based spin-drying in the presence of trehalose following immediate rehydration. Additionally, we explored apoptosis using the Annexin V-FITC detection kit and the presence of reactive oxygen species (ROS) by flow cytometry using DCFH-DA dye. In brief, this work provides a plausible methodology for preserving MSCs by desiccation.

Keywords Desiccation · MSCs · Low cost · 3D printing · Trehalose

---

S. Gupta · A. Bissoyi · A. Bit (✉)
Department of Biomedical Engineering, National Institute of Technology, Raipur, India
e-mail: arinbit.bme@nitrr.ac.in

P. K. Patra
Department of Biochemistry, Pt. JNM Medical College, Raipur, India

A. A. Rizvanov et al. (eds.), *Advances in Biomedical Engineering and Technology*, Lecture Notes in Bioengineering,
https://doi.org/10.1007/978-981-15-6329-4_35

1 Introduction

Currently, Mesenchymal Stem Cells (MSCs) are used for the treatment of many types of diseases. MSCs are able to differentiate into different cell types including osteoblasts, adipocytes, and chondrocytes (Bissoyi et al. 2016). In order to utilize MSCs for the field of tissue engineering and regenerative medicine, MSCs are preserved so that they can be utilized whenever required. For the long-term preservation of cells or tissues, cryopreservation has been accomplished as a traditional technique by using liquid nitrogen or deep freezer. Here, all chemical or enzymatic reactions cease, thus preserving the cells for clinical applications. The cells many times cannot tolerate such a subzero environment and may experience lethal damage. However, with the addition of cryoprotectants, cells can survive such a low temperature. Advanced methods are being invented to lessen the use of cryoprotectants as many of these have inherent toxicity. As the overexposure of MSCs to DMSO or other synthetic cryoprotectants can cause toxicity and ultimately lead to cell death (Morris et al. 2016). Recently, a new technique of ambient temperature dried cell storage has been developed. It is an improved and effective alternative as desiccated cells cannot grow and divide but remain in a viable state and can be preserved for a long time (Bissoyi et al. 2016). The use of liquid nitrogen, although very useful for transportation of cells is dangerous to handle. Moreover, the development of an appropriate technique for the long-term storage of MSCs is of paramount importance.

Cells can be stored and transported in the desiccated form Gordon et al. (2001). This method rectifies the inherent problems of preserving MSCs by simplifying the method of sample storage, maintenance, and transport. After water removal from cells (i.e., desiccation), mammalian cells can be stored more conveniently. The intracellular water removal (drying of cells) can damage cellular structure (cell membrane and proteins) as molecular interaction may change. The water molecule that maintains spacing between polar phospholipid head groups gets removed with the dehydration of cell membrane causing lipid fatty acid chain and polar head groups to come close and membrane may collapse. This results in a change of membrane from the fluid phase to gel phase eventually leading to an increase of membrane phase transition temperature (Tm). In addition, during rehydration, soluble cell content may leak through the cell membrane as membrane experiences another phase transition (Meyers 2006).

An improved desiccation tolerance is seen in cells containing trehalose. Trehalose replaces water molecules (which are an essential structural element of cell structures like cell membrane and proteins) and helps prevents protein denaturation and aggregation while desiccation. In dehydrated samples, trehalose, a non-reducing disaccharide of glucose has the potential to replace hydrogen-bonded water molecules resulting in the reduction of phase transition (Puhlev et al. 2001). Due to glycosidic bond stability, it has a high glass transition temperature allowing living cells to be in glassy state at ambient temperature (Meyers 2006). Research studies have evaluated the efficiency of trehalose for desiccation (Hill-Bator et al. 2014). During drying, trehalose replaces the water shell around macromolecules and prevents damaging effects (Guo et al. 2000). Trehalose stabilizes membrane and proteins in the dry state and is preferred over other sugars as it causes lower viscosity by the formation of glassy skin at the interface, which leads to entrapment of a large volume of water. Recently the introduction of trehalose into intracellular space and different techniques such as thermal poration (He et al. 2006; Beattie et al. 1997), electroporation (Zhou et al. 2010; Shirakashi et al. 2002), P2X7 channels (Elliott et al. 2006; Virginio et al. 1999), fluid-phase endocytosis (Oliver et al. 2004) have been reported.

Cells can be desiccated either using spin-drying technique or passive drying technique. Passive drying techniques lead to problems of cracking, skin formation, and variable water loss kinetics. During passive drying, a glassy skin is formed after a certain extent of dryness, so preventing further desiccation and resulting in non-uniformity in water distribution (Chakraborty et al. 2011). Since the uniformity of drying is essential, the spin-drying method overcomes this disadvantage by using the centrifugal force to force the evaporation of water. This method leads to rapid and uniform drying (Chakraborty et al. 2011; Abazari et al. 2014). Bhamla and co-authors have used the paper-based human-powered centrifuge, which efficiently the biological samples (Bhamla et al. 2017). Based on this model, we have designed a 3D printed whirligig model to dry the stem cell samples after the introduction of trehalose.

2 Materials and Methods

Reagents and Consumables. Polydimethylsiloxane(PDMS), PLA, 2′,7′-dichlorofluorescein diacetate (DCFH-DA) dye, and 3,4,4, HEPES [5-dimethylthiazol-2-y1]-2,5-diphenyltetrazolium bromide (MTT), Anthrone reagent were purchased from Sigma Aldrich(USA); Ficol-plaque solution, trehalose, and formaldehyde were obtained from Himedia labs (India); Dulbecco's Modified Eagle Medium (DMEM), antibiotic (Penicillin-streptomycin), 0.25% Trypsin/EDTA, Phosphate Buffer Saline (PBS), and Fetal bovine serum (Hi-FBS) were from Gibco (BRL, USA); Phalloidin-Alexa Fluor 488, and Annexin V-FITC, were from Invitrogen; CD44, CD73, CD90, CD105, CD34, CD45, and HLA-DR surface markers, from BD Pharminogen (Becton Dickenson, San Jose, CA); Reagents like ethanol, acetone, methanol, acetic acid, and glutaraldehyde were procured from Merck (India); Liquid nitrogen and silica gel was from local vendor; and all plastic wares, were obtained from BD Falcon (USA).

Spin-Drying Whirligig Disc. The spin-drying whirligig disc developed here was made up of Polylactic Acid (PLA) polymer with fittings to hold and cover the cover-slips during rotation as shown in Fig. 1. At each end of the string, PVC pipes were attached as a handle. The spin-drying disc was rotated to remove the water content in the cells at 2000–4000 RPM for 7 min. A 3D digital model of the disc was created in Computer-Aided Design (CAD) software and saved as in STL file format. This file was then sliced into 2D cross-sectional layers using slicer software as G-code.

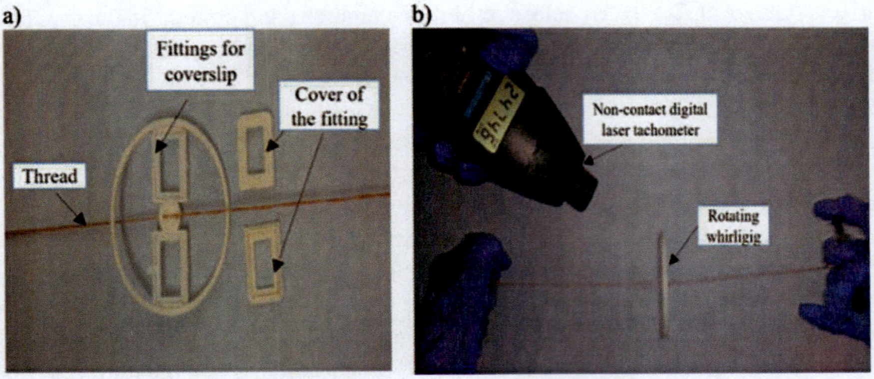

Fig. 1 Spin-drying disc dimensions with design outlook. Schematic diagram showing the three-dimensional construction of the spin-drying disc using a 3D printer **a** schematic showing the complete spin-drying disc, **b** measurement of rotation per minute (rpm) of rotating whirligig using non-contact digital laser tachometer

Finally, the G-code obtained allows the printing of a 3D model (Gupta et al. 2018; Gupta and Bit 2018).

Desiccator Device. In this study, a desiccator box was built. In the spin-drying process, centrifugal force was applied to the sample which causes the removal of water and rapid desiccation of the sample for that space is provided as shown in Fig. 3 for the spinning of the disc in the sheet. The thermal-insulated box was attached to the acrylic sheet into which liquid nitrogen was filled. The desiccation box having a spin-drying disc was provided with a continuous supply of liquid nitrogen to create an inert environment. To enhance the ability of cells to resist desiccation, cells were desiccated under vacuum. This helps to avoid damage that can occur due to oxygen-free radicals (Puhlev et al. 2001).

Cell Isolation, Expansion of Human Umbilical Cord (hUCB)-Derived Mesenchymal Stem Cell and Incorporation of Trehalose as Desiccant. Cell isolation process was approved by the Institutional Ethical Committee of NIT Raipur (NITRR/IEC/2/2015). The isolation of mononuclear cells from the human umbilical blood cord (hUCB) was carried out by the simple procedure of Ficoll-Hypaque density gradient technique. Then, the cells were maintained in Dulbecco's modified Eagle medium supplemented with 10% fetal bovine serum (FBS), and 2% antibiotic-penicillin and seeded in a culture flask. Cultures were maintained at 37 °C in 5% humidified CO_2 environment with a change of culture media twice in a week. The sub-culturing was done after the 80% cell confluence. For trehalose loading, the fluid-phase endocytosis technique was used as per Oliver et al. (Oliver et al. 2004). Briefly, cells were cultured in culture medium along with 0, 25, 50, 100, or 125 mM trehalose at 37 °C for 24 h. After incubation, cells were washed with DPBS and trypsinized. After trypsinization, cells were washed with DPBS and centrifuged at 167 g for 10 min. Pellet was obtained and 3 ml of DPBS was added to it. Anthrone test was carried out to quantify the amount of trehalose taken up by the cell. 6 mL of anthrone reagent was added to cultured cells and heated at 100 °C for 3 min. The mixture was allowed to cool at room temperature. The absorbance was measured at 620 nm and the graph was plotted against control (cells unloaded with trehalose). Anthrone test is very effective and precise for measuring sugars present in the sample. A boundary of PDMS (polydimethylsiloxane) was prepared on the edges of the gelatin-coated coverslip. The coverslip was sterilized with ethanol (70% v/v), dried under UV light and washed with culture media to remove the remaining ethanol and the cells were seeded (5×10^6 cells/cm^2) onto coverslip as shown in Fig. 2.

Immunophenotypic Characterization. The cultured cells were detached using trypsin, washed with PBS, and permeabilized using paraformaldehyde. To evaluate the expression pattern of stem-cell-specific surface antigens, cells were immuno-labeled with human monoclonal antibodies against positive (CD44, CD73, CD90, CD105) and negative (CD34, CD45, and CD14) surface markers of MSCs. The markers were then examined using a flow cytometer.

Analysis of MSCs Adherence and Filamentous (F) Actin Distribution in Post-Thaw MSCs. Cytoskeleton has a vital part in cell morphology, adhesion, signaling, and growth. Cells were allowed to migrate, divide, and maintain its shape due to any change in cytoskeletal (33). Evaluation of the cell adherence on the surface of cover-slip is measured by the F-actin distribution in desiccated MSCs using Phalloidin-Alexa Fluor 488 staining. Washing of the post-desiccated cells is done with PBS (pH 7.4), cultured over coverslip, and are fixed with a 4% formaldehyde solution. To stain the F-actin in MSCs, a freshly prepared phalloidin solution was added onto the coverslip. The cells were examined under confocal microscopy after washing with mounting medium DABCO (pH 8.7).

Proliferation and Viability Analysis of Desiccated MSCs by MTT Assay. Cell proliferation was assessed using MTT [3-(4, 5-dimethythiazol-2-yl)-2, 5-diphenyl tetrazolium bromide] assay. Briefly, post-thaw MSCs were seeded and grown in

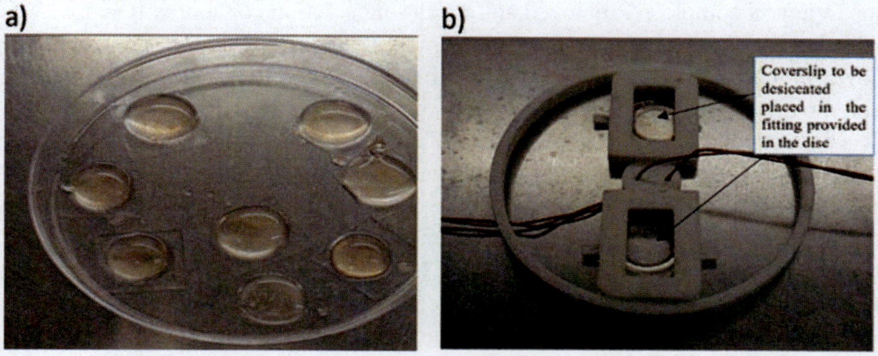

Fig. 2 Schematic image of **a** cell seeding over coverslip, **b** image of the complete spin-drying disc into which cells containing coverslip were placed

96-well culture plate. Then, 10 μl MTT solution (5 mg/ml) was added to the well-containing cells. After incubation for 4 h at 37 °C in a 5% CO_2 atmosphere, acidic isopropanol was added to each well to dissolve the dark blue formazan crystals. The supernatant was collected to evaluate absorbance at 570 nm using Multi-Well Microplate Reader (Bio-Tek ELx800). The freshly isolated MSCs were used as the control. The cell viability was expressed as a percentage of control.

Statistical Methods. All these experiments were performed in triplets. The values were given as mean ± standard deviation (S.D.). The statistical differentiation between the experimental groups is analyzed by one-way analysis of variance (ANOVA) succeeded by posthoc Tukey's Multiple Comparison Test using GraphPad Prism 5 (GraphPad). $p < 0.05$ is considered as significant. To analyze time course related deviation, two-way ANOVA is generally used.

3 Results and Discussion

As dry preservation of cells at ambient temperature increases desiccation tolerance of cells. Hence, a new method to preserve cells using the spin-drying disc is developed. In this study, we have demonstrated that desiccated stem cells can be stored and preserved effectively. We have done experimental observations of the effect of drying with trehalose solution at a spinning speed ranging from 2000 to 4000 rpm. The three-dimensional spin-drying whirling disc prepared can be used to place the slide containing cells. The cells when reach 80% confluence were trypsinized and seeded into glass coverslip. The coverslip was subjected to desiccation after the introduction of trehalose through fluid-phase endocytosis technique. Vacuum was created by placing silica gels at the bottom surface of the desiccator box. If glass coverslip has sufficient concentration of cells, next the cells were spin dried by rotating the whirling disc inside the desiccation box (see Fig. 3). For spin-drying, the culture medium was completely removed and the drying solution containing trehalose was added. A sufficient amount of trehalose must be incorporated into the cells. The coverslip containing cell to be dried was placed on the fitting provided in the disc and the disc was placed in the space provided for its rotation in the desiccator box. The desiccation box having a spin-drying disc was provided with a continuous supply of liquid nitrogen to create an inert environment. The disc was rotated at a spinning speed of 2000–4000 rpm for 4 min whose rpm was measured using a non-contact digital tachometer. For the removal of moisture content, silicon gels were kept at the bottom space of the desiccator box. The dried sample obtained after spinning was observed under a confocal microscope. An improved desiccation tolerance was seen in cells containing trehalose. In spin-drying technique, drying occurs rapidly and uniformly. The control sample taken as reference had not been dried. The chief findings of this experiment are that to preserve cells at ambient temperature, dry preservation of cells can be done which does not affect the cell viability. Trehalose is introduced to cells using fluid-phase endocytosis technique. For the ingestion of trehalose into cells, trehalose was added to the sample and incubated. Cell samples

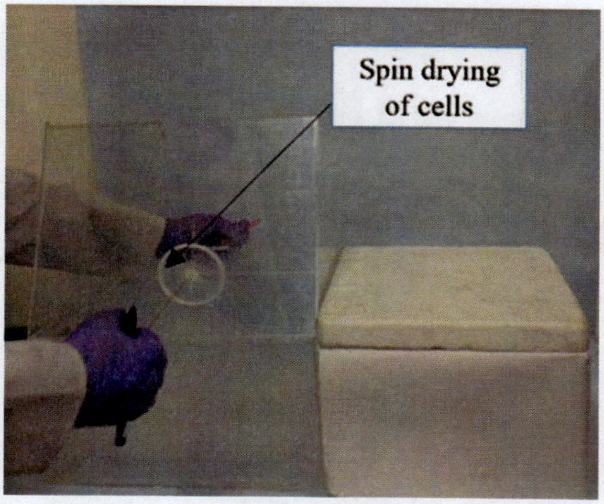

Fig. 3 Experimental setup showing transparent desiccator box made up of acrylic sheet into which a spin-drying of cells seeded coverslip was done by rotating spin-drying whirling disc in the space provided and is attached to a thermo-insulated box supplied with liquid nitrogen. The bed of the desiccator box carries silica gel used as a desiccant for proper drying of the sample. The whole procedure was carried out under the biosafety cabinet

were spin-dried in air having relatively low humidity. A sufficient amount of trehalose must be incorporated into the cells. A decrease in viability of cells over time was seen in incompletely dried cells. The spin-dried cells show approximately the same growth characterization as of control cells.

Immunophenotypic Characterization. The hUCB-derived MSCs were cultured for isolation of MSCs utilizing their adhering capacity. The surface markers are well known for describing MSCs presence as positive and negative. The cells were then characterized under flow cytometry. The result revealed that the fourth passage cultured hMSCs showed less than 1% expression of hematopoietic CD34 (0.52%), pan leukocyte CD45 (0.43%) and HLA-DR (0%) markers whereas the expression of the surface markers CD73 (95%), CD90 (93%), and CD105 (92%). The x-axis indicates fluorescence intensity, and the y-axis, cell counts (see Fig. 4).

Effect of Spin-Dried Desiccation on the Post-Thaw Culture of MSCs. To characterize the desiccation actuated cell death in MSCs, MTT assay was done and the result is obtained. The cell viability results of MSCs are calculated by cryopreservation without desiccation, desiccation with spin-drying, and desiccation without spin-drying. The result indicates significant differences in the viability of cells in MSCs desiccated with spin-drying and without spin-drying with respect to control and cryopreserved cells. The data state that the post-thawed MSCs showed a decrease in cell viability when simply cryopreserved or after cryopreservation with desiccation without spin-drying with respect to non-cryopreserved control MSCs. To assess the effect of spin-drying-based desiccation on cells, the MTT assay was performed. The viability of cells on first day is more in control sample but slowly cells cryopreserved

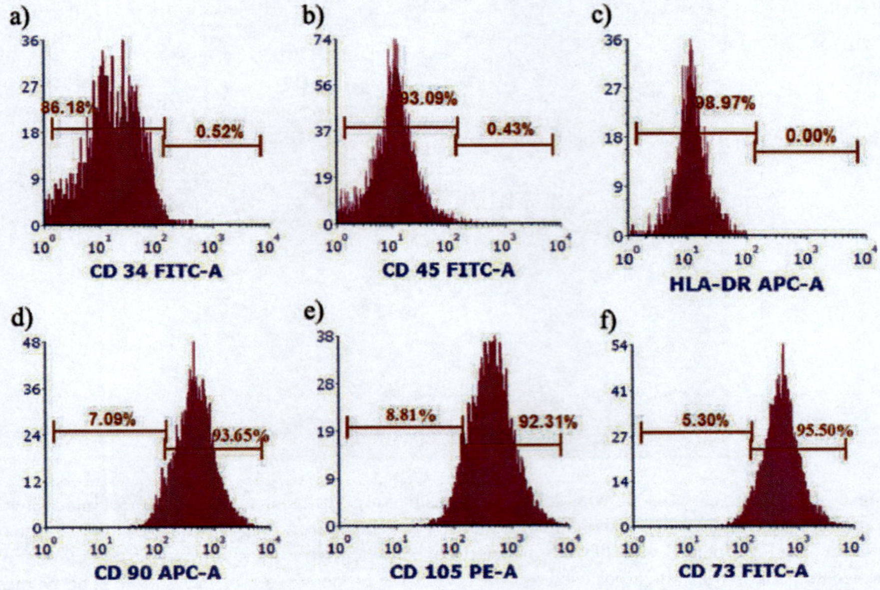

Fig. 4 MSCs re-stained with **a** FITC-A labeled (CD34), **b** FITC-A labeled (CD45), **c** APC-A labeled HLA-DR, **d** APC-A labeled (CD90), **e** FITC-A labeled (CD73) and **f** PE-A labeled (CD105)

along with desiccation with spin-drying showed the presence of more viable cells. The viability of cells cryopreserved without desiccation was over 12.1%, 15.23%, and 22.45% on day 1, 2, and 3, respectively. In addition, the viability of cells desiccated with spin-drying was 20.32%, 22.71%, and 34.7% on day 1, 2, and 3 respectively. Viability of cells desiccated without spin-drying on day 1, 2, and 3 was 12.14%, 10.36%, and 10.23%, respectively. Even the viability of normal cells without cryopreservation taken as control was still 29%, 33.1%, and 30.63% on day 1, 2, and 3 respectively. There is a significant difference between cell viability of cell cryopreserved with spin-drying and without spin-drying. Thus the data stated that cryopreservation after desiccation with spin-drying is an ideal method for preservation and storage of human MSCs (see Fig. 5).

Analysis of MSCs Adherence and Filamentous (F) Actin Distribution in Post-Thaw MSCs. Fluorescence image analysis is done to characterize the changes in the cytoskeleton structure of MSCs cryopreserved with different desiccation processes either with spin-drying or without spin-drying. Non-cryopreserved MSCs serve as a control. It was seen that MSCs cryopreserved along with spin-drying desiccation do not undergo any deformation after 48 h of culture and maintained their original shape. However, MSCs cryopreserved without spin-drying were not able to recover their original shape even after 48 h of culture. During the progress of culture, MSCs cryopreserved after desiccation with spin-drying, proliferated, and became long shuttle-shaped. These cells had a more rounded shape compared to non-cryopreserved cells. MSCs cryopreserved without desiccation are not able to recover their shape. Various

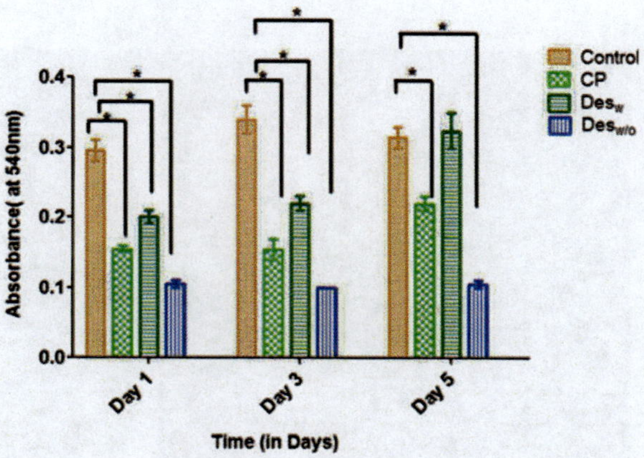

Fig. 5 The bar diagram showing MTT assay of post desiccation MSCs after 1, 3, and 5 days indicating metabolic activities. Cell viability analysis by MTT assay kit in MSCs simply cryopreserved, cryopreserved along with spin-drying desiccation, and cryopreserved without spin-drying desiccation. Non-cryopreserved cells were taken as control

functions are carried out by cytoskeleton, which includes giving shape to the cell, contraction of muscle, cell signaling, regulation of cell dynamics, and cell motility. Cell viability, cell proliferation, and differentiation are some of its major properties. Any loss in function of the cell membrane interferes with the functioning of the transport system such as pH regulatory system in the cell membrane. Mitochondrial transport system is essential for the oxidative phosphorylation, which is the major energy-generating pathway. Any disruption in the cell organelles can disturb the transport system. Hence, cytoskeleton integrity was analyzed by confocal microscopy. It was demonstrated that the cell morphology of non-cryopreserved MSCs is retained with an intact cell membrane. Immunofluorescence images showed more cell attachment in non-cryopreserved MSCs as compared to cryopreserved MSCs irrespective of the method used in desiccation. It is observed that cryopreservation with spin-dried desiccation showed maximum retention of cytoskeleton and improved the cell viability of post-thawed MSCs (see Fig. 6).

4 Conclusion

The viability and functionality of cells get greatly affected by the technique used for preservation. Spin-drying desiccation with the help of 3D printer prepared whirligig is an excellent technique for the preservation of human umbilical cord-derived MSCs. It helps in rapid and uniform drying of cells to very low moisture content. Spin-drying desiccation preserves cell functionality and viability more efficiently as compared to desiccation without spin-drying. Under ideal environmental conditions, trehalose works ordinarily like any other oligosaccharide in preserving biomaterials. However, under optimized conditions trehalose works very efficiently and thus is a preferred

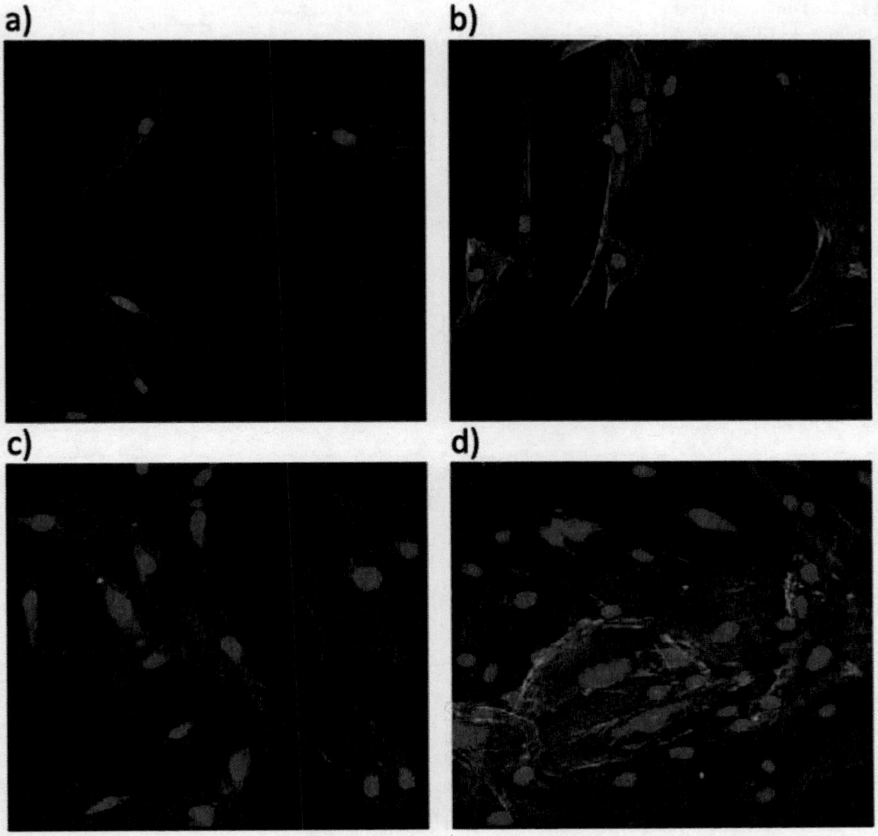

Fig. 6 Images showing the cytoskeletal organization. Laser confocal images (**a**–**d**). The cell nucleus was stained with DAPI and cytoskeleton with Phalloidin **a** MSC seeded on glass coverslip after desiccation, **b** static control, **c** MSCs seeded on coverslip after 3 days, **d** static control after 3 days. Cytoskeleton and nucleus are pseudocolors in green and blue, respectively

desiccating sugar. The performance of the cell viability of the dried sample was comparable to that of conventional means. In the future, we will make a study on the effect of trehalose induced into intracellular space of MSCs through transporter Vector (TRET1).

Acknowledgements This study supported by the grant from the Department of Science and Technology (ECR/2017/001115) New Delhi, India.

Declaration The authors declare that they have no conflict of interest.

References

Abazari A, Chakraborty N, Hand S et al (2014) A Raman microspectroscopy study of water and trehalose in spin-dried cells. Biophys J 107:2253–2262. https://doi.org/10.1016/j.bpj.2014.09.032

Beattie GM, Crowe JH, Lopez AD et al (1997) Trehalose: a cryoprotectant that enhances recovery and preserves function of human pancreatic islets after long-term storage. Diabetes 46:519–523

Bhamla MS, Benson B, Chai C et al (2017) Hand-powered ultralow-cost paper centrifuge. Nat Biomed Eng 1:0009. https://doi.org/10.1038/s41551-016-0009

Bissoyi A, Kumar A, Rizvanov AA et al (2016) Recent advances and future direction in lyophilisation and desiccation of mesenchymal stem cells. Stem Cells Int 2016:3604203. https://doi.org/10.1155/2016/3604203

Chakraborty N, Chang A, Elmoazzen H et al (2011) A spin-drying technique for lyopreservation of mammalian cells. Ann Biomed Eng 39:1582–1591. https://doi.org/10.1007/s10439-011-0253-1

Elliott GD, Liu X-H, Cusick JL et al (2006) Trehalose uptake through P2X7 purinergic channels provides dehydration protection. Cryobiology 52:114–127. https://doi.org/10.1016/j.cryobiol.2005.10.009

Gordon SL, Oppenheimer SR, Mackay AM et al (2001) Recovery of human mesenchymal stem cells following dehydration and rehydration. Cryobiology 43:182–187. https://doi.org/10.1006/cryo.2001.2361

Guo N, Puhlev I, Brown DR et al (2000) Trehalose expression confers desiccation tolerance on human cells. Nat Biotechnol 18:168–171. https://doi.org/10.1038/72616

Gupta S, Bit A (2018) Rapid prototyping for polymeric gels. Polymeric Gels. Elsevier, pp 397–439

Gupta S, Bissoyi A, Bit A (2018) A review on 3D printable techniques for tissue engineering. Bionanoscience 1–16. https://doi.org/10.1007/s12668-018-0525-4

He X, Amin AA, Fowler A, Toner M (2006) Thermally induced introduction of trehalose into primary rat hepatocytes. Cell Preserv Technol 4:178–187. https://doi.org/10.1089/cpt.2006.4.178

Hill-Bator A, Misiuk-Hojło M, Marycz K, Grzesiak J (2014) Trehalose-based eye drops preserve viability and functionality of cultured human corneal epithelial cells during desiccation. Biomed Res Int 2014:292139. https://doi.org/10.1155/2014/292139

Meyers SA (2006) Dry storage of sperm: applications in primates and domestic animals. Reprod Fertil Dev 18:1. https://doi.org/10.1071/RD05116

Morris TJ, Picken A, Sharp DMC et al (2016) The effect of Me2SO overexposure during cryopreservation on HOS TE85 and hMSC viability, growth and quality. Cryobiology 73:367–375. https://doi.org/10.1016/j.cryobiol.2016.09.004

Oliver AE, Jamil K, Crowe JH, Tablin F (2004) Loading human mesenchymal stem cells with trehalose by fluid-phase endocytosis. Cell Preserv Technol 2:35–49. https://doi.org/10.1089/153834404322708745

Puhlev I, Guo N, Brown DR, Levine F (2001) Desiccation tolerance in human cells. Cryobiology 42:207–217. https://doi.org/10.1006/cryo.2001.2324

Shirakashi R, Köstner CM, Müller KJ et al (2002) Intracellular delivery of trehalose into mammalian cells by electropermeabilization. J Membr Biol 189:45–54. https://doi.org/10.1007/s00232-002-1003-y

Virginio C, MacKenzie A, North RA, Surprenant A (1999) Kinetics of cell lysis, dye uptake and permeability changes in cells expressing the rat $P2X_7$ receptor. J Physiol (Lond) 519:335–346. https://doi.org/10.1111/j.1469-7793.1999.0335m.x

Zhou X, Yuan J, Liu J, Liu B (2010) Loading trehalose into red blood cells by electroporation and its application in freeze-drying. Cryo Lett 31:147–156

Fabrication of Hydroxyapatite-Chitosan-Silk Fibroin Based Composite Film as Bone Tissue Regeneration Material

Sharda Gupta, Rupsha Mukherjee, Rajendra Kumar Jangle, Deependra Singh, Manju Singh, and Arindam Bit

Abstract The composite film prepared from a combination of biomaterials such as natural, ceramic, and synthetic polymers can support the growth of different types of tissues. Silk fibroin (SF) used in this study is extracted from ecorace of Tasar cocoon (Railey, Daba, and Baraf), which is locally available in our State (Chhattisgarh). The silk fibroin from these three different ecoraces was allowed to blend with hydroxyapatite (HAP) and chitosan (CS). According to the literature, SF and CS are nontoxic, non-immunogenic, biodegradable, and biocompatible. Whereas hydroxyapatite (HAP) possesses good biocompatibility with bone tissue and have the property of osteoconductivity. Hence, this tri-composite film unites both organic and inorganic components to promote osteoinduction and osteoconduction property in an ideal configuration. The structural characterization of the film was analyzed using Fourier transform infrared spectroscopy (FTIR), SEM techniques, and the hemocompatibility of the composite film was calculated to test the compatibility of the constructed film. It is expected that the resulting film when seeded with mesenchymal stem cells would become a highly functional bone tissue substitute and further optimization work will be done to achieve this goal.

Keywords Hydroxyapatite · Non-mulberry silk · Bone tissue engineering · Tri-composite film

1 Introduction

Bone can be regenerated by itself but still depending upon the type of defect it is required to depend upon bone substitutes in many circumstances. An autogenous bone graft is considered as gold standard as this approach allows bone formation

S. Gupta · R. Mukherjee · A. Bit (✉)
Department of Biomedical Engineering, National Institute of Technology, Raipur, India
e-mail: arinbit.bme@nitrr.ac.in

R. K. Jangle · D. Singh · M. Singh
University Institute of Pharmacy, Pt. Ravishankar Shukla University, Raipur, Chhattisgarh, India

A. A. Rizvanov et al. (eds.), *Advances in Biomedical Engineering and Technology*, Lecture Notes in Bioengineering,
https://doi.org/10.1007/978-981-15-6329-4_36

by osteoconduction and osteoinduction (Wiltfang et al. 2004; Misch 2010). But to avoid the unfavorable condition of donor morbidity, it is better to use bone substitutes. Many types of bone substitutes for orthopedics such as polymer based or ceramic based are available. The ceramic bone substitute has attracted interest in bone tissue engineering due to its good chemical resistivity, create inert bone like environment, significant strength, and impact resistance is nowadays widely used as bone-substitute (Smith 1963; Campana et al. 2014).

For the development of a new bone substitute for bone tissue replacement, its biological property and structure must be considered. At the ultra-structural level, bone is a composite material made up of apatite and collagen (Rajkhowa et al. 2010). Also, the composite material must have improved cellular interaction property, good mechanical property, and bioactivity.

The silk fibroin has osteogenic potential due to its similarity to collagen-I (structural protein present in bone) (Gupta et al. 2016). The silk fibroin obtained from the "Antheraea Mylitta", a tropical tussar silkworm due to its biocompatibility, biodegradability, elasticity, mechanical stability RGD (arginine-glycine-aspartic acid) motif is widely used for tissue engineering. This RGD motif allows the osteogenic differentiation of mesenchymal stem cell (MSCs) (Bhattacharjee et al. 2017). The silk fibroin protein can be blended with other polymers to overcome its disadvantages. For blending silk material with other biomaterials, a proper solvent is required. The silk protein being insoluble in formic acid but the regenerated silk powder can be dissolved in formic acid (Melke et al. 2016; Silva et al. 2013). Hence, in this study formic acid is used as a solvent for silk blend. The use of silk alone is not suitable as film material. Therefore, a combination of materials is used to form a polyelectrolyte complex. This is done to enhance the properties of materials to make it more suitable for reconstruction purposes.

Hydroxyapatite is a naturally found mineral with formula Ca10(PO4)6(OH)2 that is widely used as a ceramic bone substitute for bone tissue engineering as it is a good carrier of osteoinductive growth factor (Um et al. 2001, 2003). Also it is osteoconductive, inert, and biocompatible. However, it is extremely rigid. This property of brittleness put a limit in its use for treating load-bearing bone defects. The incorporation of HAP into silk film will overcome the shortcomings of the brittleness of HAP by enhancing the mechanical properties of HAP.

Chitosan has brilliant bioactive properties, and the addition of silk fibroin helps to control the degradation of chitosan. Chitosan incorporation has antibacterial effects (Heise et al. 1990), whereas silk fibroin shows increased growth and adhesion of osteoblast-like cell (MG63) on the film surface (Dutta et al. 2015).

The aim of this study is to prepare a tri-composite scaffold (made up of silk fibroin solution, chitosan, and HAP) that would provide promising effects for bone tissue regeneration area. The prepared composite was further analyzed and characterized for studying the structural, mechanical, and hemocompatible properties.

2 Materials and Methods

Non-mulberry silk cocoons (Baraf, Railey, Daba ecorace of Antheraea Mylitta species) were purchased from Chhattisgarh Khadi Gram Udyog Board, Raipur, Chhattisgarh (India); Lithium bromide (LiBr) were procured from Srlchem (India); Sodium carbonate (Na_2CO_3), acetic acid were purchased by Loba Chemie, Maharashtra, India; Dialysis membrane, Chitosan was from Himedia.

Extraction of Silk Fibroin from Cocoon. Cocoons were cut into small pieces and were boiled in 0.5 wt% Na_2CO_3 for 60 min, washed thoroughly and dried overnight. Preheat the dry fiber at 90 °C for few minutes in distilled water. The Baraf silk fibers (BSF), Daba silk fibers (DSF), and Railey silk fibers (RSF) extracted were then dissolved in the prepared 9.3 M Lithium Bromide (LiBr) solution at 90 °C for 1 h 30 min, 4 h, and 2 h, respectively. Colored and viscous solutions were obtained; it is observed that the dark brown colored solution is found for Railey cocoon, pale yellow solution for Baraf cocoon, and amber-colored solution for Daba ecorace of tussar cocoon. The dissolved viscous silk fibroin solution of three different ecoraces of non-mulberry was allowed to dialyze for 72 h against ultrapure water using a dialyzing membrane. The resulting fibroin solution was centrifuged at 4500 rpm to remove the debris. The supernatant is poured in a round bottom flask and kept in a deep freezer at −20 °C for 24 h. The silk protein was lyophilized at −110 °C with 480 torr pressure for 72 h. to obtain freeze-dried silk powder (Fig. 1).

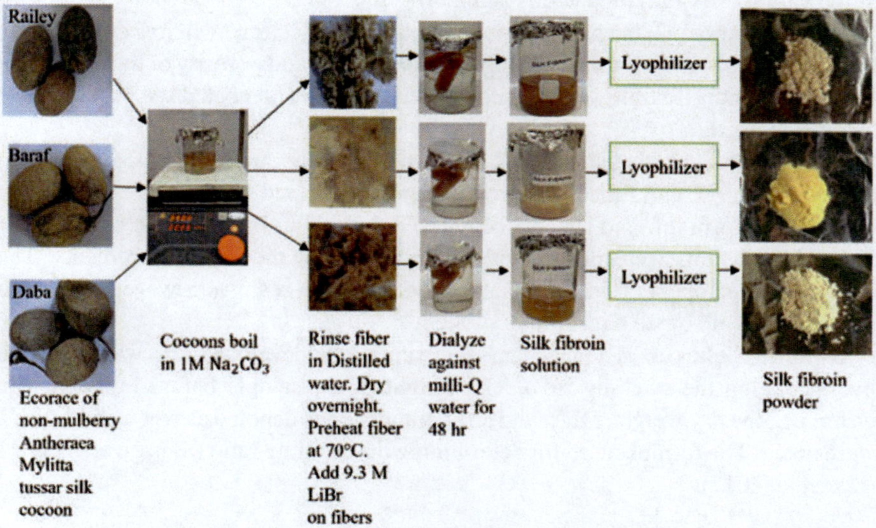

Fig. 1 Schematic representation of stepwise extraction of silk fibroin from different ecoraces of non-mulberry Antheraea Mylitta tussar cocoon

Table 1 The composition of various silk films

Film type	Baraf silk fibroin solution	Daba silk fibroin solution	Railey silk fibroin solution	Hydroxyapatite	Chitosan
B/CS/HAP	+	–	–	+	+
D/CS/HAP	–	+	–	+	+
R/CS/HAP	–	–	+	+	+

Preparation of Chitosan Solution. Chitosan solution prepared as discussed earlier (Xiao et al. 2008). Briefly, a concentration of 4% was prepared by dissolving chitosan (0.2 g) into 2% acetic acid (30 ml) with stirring for 5 h to get a perfectly transparent solution.

Preparation of Hydroxyapatite. Hydroxyapatite synthesis was described elsewhere (Begam et al. 2017). Briefly, the solution of calcium hydroxide and orthophosphoric acid were prepared using distilled water and both were stirred separately for 30 min. Then orthophosphoric acid was added dropwise in stirring condition in calcium hydroxide solution. The complete synthesis was carried out at 80 °C at 11–12 pH, dried overnight, filtered, washed with distilled water, and dried at 80 °C for 24 h. The HAP powder was obtained after calcination of the crushed cake at 800 °C.

Preparation of Regenerated Silk/HAP/Chitosan Composite Film. The lyophilized silk powder was dissolved in formic acid to prepare 4% w/v silk solution. The blended solution was prepared by blending the chitosan solution with regenerated silk fibroin solution and 1% HAP slurry for the next 4 h. The prepared solution was poured into Petri plate and dried at a hot air oven at 70 °C for 8 h. The composition of the various silk film prepared is summarized in Table 1.

Morphological Characterization. Scanning Electron Microscopy (SEM) [ZEISS-EVO18] is used for analyzing the morphology and porosity of the composite film. The composite film was mounted on metal stubs and coated with gold before being examined.

Structural Analysis using FTIR Spectroscopy. The structural property, molecular composition, and functional groups of the developed films were evaluated by Fourier transform infrared spectroscopy (FTIR) using an Infrared Microscope. The SFCSHAPM films were placed within the holder for the IR measurements. The equipment was operated in absorbance mode with 32 scans at a wavelength from 400 to 4000 cm^{-1} with a resolution of 8 cm^{-1}.

Swelling Behavior. The mechanical testing of the developed film was evaluated by calculating the swelling ratio. The weight of the sample before immersion in water, i.e., the dry weight, DRY, and after immersion is denoted as wet weight, WET were noted. The formula used for determining the swelling ratio (S) is given by Eq. 1 (Lee et al. 2017).

$$S = \frac{W_w - W_d \times 100}{Wd} \tag{1}$$

where

W_w: mass of the swollen wet hydrogel sample

W_d: mass of the dried sample.

Hemocompatibility Study of the Composite Film. Blood sample was taken from a healthy volunteer at Institutional Dispensary (NIT Raipur) with approval from the Institutional Ethical Committee of NIT Raipur (NITRR/IEC/6/2015). The composite films were evaluated for hemocompatibility by hemolysis studies in which human blood is collected in a Sodium Citrate solution (3.8 gm %) in 10:1 ratio and is diluted with normal saline (4:5) (Krishna Samanta and Chanda 2018). The film was cut into small pieces and put inside a standard test tube containing 10 ml of saline and incubated at 37 °C for 30 min. To the test tube, 0.2 ml of diluted blood were added, mixed gently, and incubated for 60 min. The positive control sample is prepared by diluting 0.2 ml of human blood in 10 ml of 0.1% Na_2CO_3 solution and incubates for 60 min at 37 °C. The Negative control sample was prepared by diluting blood in normal saline solution at the above same condition. After incubation, all test tubes were centrifuged for 5 min at 500 rpm and the supernatant is carefully transferred to cuvette or quartz glass cell. The UV-V is spectrophotometer is put on and the optical density (O.D.) is measured at 545 nm. The hemocompatibility was then calculated using Eq. 2:

$$\text{Hemocompatibility} = \frac{\text{O.D. (Test)} - \text{O.D. (Neg)}}{\text{O.D. (Pos)} - \text{O.D. (Neg)}} \quad (2)$$

where O.D. (Neg) is the optical density of negative control and O.D. (Pos) is the optical density of positive control. The optical density of the test sample was measured using a spectrophotometer.

3 Results and Discussion

3.1 Mechanical Testing

The surface morphology of blended tri-composite non-mulberry silk fibroin film was characterized using SEM images. The SEM image of blended silk film is shown in Fig. 2. The SEM images reveal that surface of all types of films is porous and has rough morphology. The surface of Daba ecorace was rather rough. The rough surface allows more surface area for cell adhesion increasing the rate of cell differentiation and proliferation.

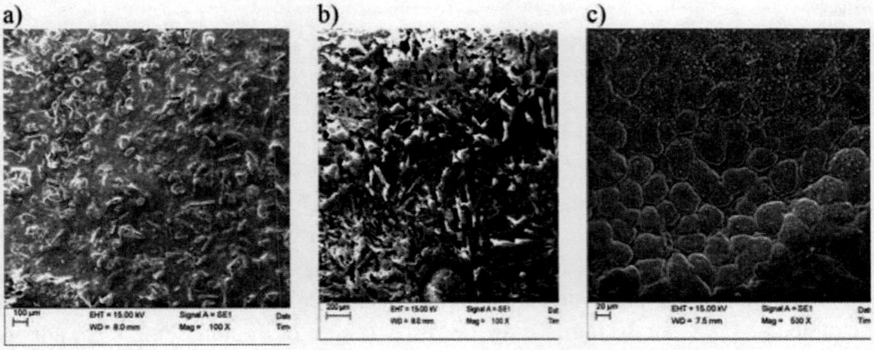

Fig. 2 SEM image of silk blended chitosan/HAP/Silk fibroin solution **a** B/CS/HAP, **b** D/CS/HAP, **c** R/CS/HAP

Table 2 Comparative study of swelling ratio

Sample		Dry weight (g)	Wet weight (g)	Swelling ratio (%)	Average swelling ratio (%)
B.1	D/CS/HAP	0.018	0.035	94.44	114.01
B.2		0.024	0.056	133.33	
B.3		0.028	0.060	114.285	
D.1	B/CS/HAP	0.024	0.107	77.57	61.63
D.2		0.037	0.115	67.82	
D.3		0.020	0.099	39.5	
R.1	R/CS/HAP	0.025	0.037	48	38.33
R.2		0.014	0.020	42	
R.3		0.020	0.025	25	

3.2 *Swelling Ratio*

The swelling ratio of silk fibroin/chitosan thin films was calculated according to the given formula as discussed elsewhere. The silk fibroin thin film samples were tested to find its water-absorbing capacity. The results are shown in Table 2.

The results obtained from mechanical testing evidently describe the excellent mechanical properties exhibited by D/CS/HAP.

3.3 *Fourier Transform Infrared Spectroscopy*

FTIR was used to assess the secondary structure of the silk fibroin biopolymer. The different sample absorbs different infrared radiation wavelength depending upon the protein chain conformation. The absorption bands corresponding to Amide I

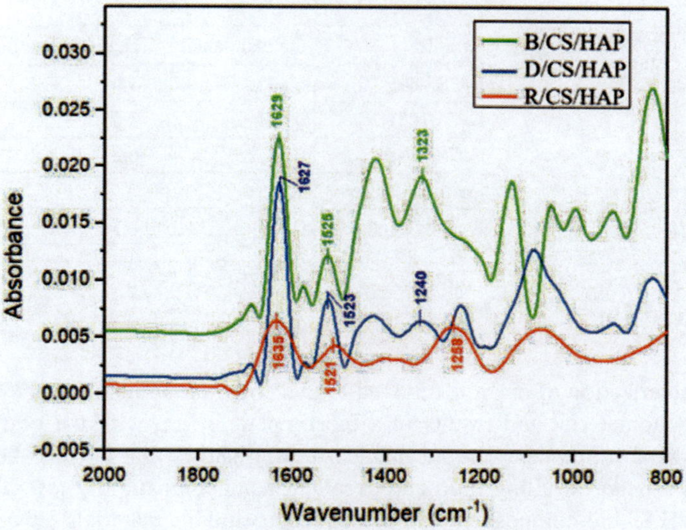

Fig. 3 Comparison of FTIR spectra of the tussar silk ecorace blended CS/HAP composite films

represents C=O stretching of peptide groups and lies in the range between 1700 and 1600 cm^{-1}, Amide II band arises due to C–N stretching in combination with N–H bending falls in the range between 1540 and 1520 cm^{-1} and Amide III band occurs in 1220–1330 cm^{-1} range due to C–N stretching and C=O bending vibration. The blended silk films showed a vibration band for Amide I at 1629 cm^{-1}, 1627 cm^{-1}, and 1635 cm^{-1} for B/CS/HAP, D/CS/HAP, and R/CS/HAP, respectively. The peak at 1525 cm^{-1}, 1523 cm^{-1}, and 1521 cm^{-1} for B/CS/HAP, D/CS/HAP, respectively, representing Amide II band. And amide III band at 1323 cm^{-1}, 1240 cm^{-1}, and 1258 cm^{-1} for B/CS/HAP, D/CS/HAP, respectively (Fig. 3).

3.4 Hemocompatibility Test

The silk fibroin thin films were tested for hemocompatibility. The Baraf, Daba, and Railey samples were tested against standard blood samples. The positive (+) control was made from human blood and saline solution. The negative (–) control was obtained using only a saline solution. The optical densities of control solutions and silk solutions were measured using spectrophotometer at 545 nm. The calculated results are shown in Table 3.

From the above result, it was seen that Daba shows excellent hemocompatibility properties. Hence, it can be preferred over other silk samples for use as a bone substitute.

Table 3 Hemocompatibility of thin film samples

Sample	Optical density	Hemocompatibility (%)
B/CS/HAP	0.641	4.9
D/CS/HAP	0.601	3.6
R/CS/HAP	0.347	6.2
Positive control	1.221	–
Negative control	0.611	–

4 Conclusion

The characterization of bio-films proved to overcome the limitations of silk fibroin and HAP biomaterial and results in a biocompatible film with the best material and biological properties. The mechanical testing shows that D/CS/HAP composition has a better swelling ratio and excellent hemocompatibility test. Therefore, the D/CS/HAP bio-composite could be applied as suitable materials for bone tissue substitutes in the future.

Acknowledgements This study supported by the grant from the Department of Science and Technology (ECR/2017/001115) New Delhi, India.

Declaration The authors declare that they have no conflict of interest.

References

Begam H, Nandi SK, Chanda A, Kundu B (2017) Effect of bone morphogenetic protein on Zn-HAp and Zn-HAp/collagen composite: a systematic in vivo study. Res Vet Sci 115:1–9. https://doi.org/10.1016/j.rvsc.2017.01.012

Bhattacharjee P, Kundu B, Naskar D et al (2017) Silk scaffolds in bone tissue engineering: An overview. Acta Biomater 63:1–17. https://doi.org/10.1016/j.actbio.2017.09.027

Campana V, Milano G, Pagano E et al (2014) Bone substitutes in orthopaedic surgery: from basic science to clinical practice. J Mater Sci Mater Med 25:2445–2461. https://doi.org/10.1007/s10856-014-5240-2

Dutta SR, Passi D, Singh P, Bhuibhar A (2015) Ceramic and non-ceramic hydroxyapatite as a bone graft material: a brief review. Ir J Med Sci 184:101–106. https://doi.org/10.1007/s11845-014-1199-8

Gupta P, Adhikary M, Joseph Christakiran M et al (2016) Biomimetic, osteoconductive non-mulberry silk fiber reinforced tricomposite scaffolds for bone tissue engineering. ACS Appl Mater Interfaces 8:30797–30810. https://doi.org/10.1021/acsami.6b11366

Heise U, Osborn JF, Duwe F (1990) Hydroxyapatite ceramic as a bone substitute. Int Orthop. https://doi.org/10.1007/BF00178768

Krishna Samanta S, Chanda A (2018) Study on the structure and properties of crystalline pure and doped β-Tri calcium phosphate ceramics. Mater Today Proc 5:2330–2338. https://doi.org/10.1016/j.matpr.2017.09.237

Lee JM, Sultan MT, Kim SH et al (2017) Artificial auricular cartilage using silk fibroin and polyvinyl alcohol hydrogel. Int J Mol Sci. https://doi.org/10.3390/ijms18081707

Melke J, Midha S, Ghosh S et al (2016) Silk fibroin as biomaterial for bone tissue engineering. Acta Biomater 31:1–16. https://doi.org/10.1016/j.actbio.2015.09.005

Misch CM (2010) Autogenous bone: is it still the gold standard? Implant Dent 19:361. https://doi.org/10.1097/ID.0b013e3181f8115b

Rajkhowa R, Gil ES, Kluge J et al (2010) Reinforcing silk scaffolds with silk particles. Macromol Biosci 10:599–611. https://doi.org/10.1002/mabi.200900358

Silva SS, Popa EG, Gomes ME et al (2013) Silk hydrogels from non-mulberry and mulberry silkworm cocoons processed with ionic liquids. Acta Biomater 9:8972–8982. https://doi.org/10.1016/j.actbio.2013.06.044

Smith L (1963) Ceramic-plastic material as a bone substitute. Arch Surg 87:653. https://doi.org/10.1001/archsurg.1963.01310160115023

Um IC, Kweon HY, Park YH, Hudson S (2001) Structural characteristics and properties of the regenerated silk fibroin prepared from formic acid. Int J Biol Macromol 29:91–97. https://doi.org/10.1016/S0141-8130(01)00159-3

Um IC, Kweon HY, Lee KG, Park YH (2003) The role of formic acid in solution stability and crystallization of silk protein polymer. Int J Biol Macromol 33:203–213. https://doi.org/10.1016/j.ijbiomac.2003.08.004

Wiltfang J, Kloss FR, Kessler P et al (2004) Effects of platelet-rich plasma on bone healing in combination with autogenous bone and bone substitutes in critical-size defects. Clin Oral Implants Res 15:187–193. https://doi.org/10.1111/j.1600-0501.2004.00980.x

Xiao X, Liu R, Huang Q (2008) Preparation and characterization of nano-hydroxyapatite/polymer composite scaffolds. J Mater Sci Mater Med 19:3429–3435. https://doi.org/10.1007/s10856-008-3499-x

Prediction of Hydroxyurea Effect on Sickle Cell Anemia Patients Using Machine Learning Method

Bikesh Kumar Singh, Apoorva Ojha, Kshirodra Kumar Bhoi, Akalabya Bissoyi, and Pradeep Kumar Patra

Abstract The objective of this paper is to predict the effect of hydroxyurea of individuals suffering from Sickle Cell Disease (SCD) using predictive machine learning techniques. Nowadays worldwide sickle cell disease becomes the most common inherited disease, which remains associated with long life morbidity and makes life expectancy low or we can say it reduces the lifespan of the concerned person. To overcome such diseases, a chemotherapeutic drug Hydroxyurea is taken by the concerned patient to re-mediate clinical problems of sickle cell disease. This paper discusses the use of a machine learning approach to predict the effects of the drug Hydroxyurea (HU) treatment in different sickle cell anemic patients. Fetal hemoglobin before and after treatment of HU establishes the basis for classification. Support vector machine has been used for classification and the dataset is rigorously analyzed for different data division methods and we got the maximum accuracy level of 94%.

Keywords Sickle cell anemia · Hydroxyurea · Support vector machine

1 Introduction

Sickle Cell Disease (SCD) is one of the widespread genetic diseases having a significant impact on the worth and expectancy of a patient's life. In countries like India, the sickle cell gene is mainly dispersed in Chhattisgarh, Madhya Pradesh, Orissa, Maharashtra, Jharkhand, some parts of Andhra Pradesh, Tamil Nadu, Karnataka, and Kerala (Patra et al. 2011). The contingency of the sickle cell attribute is as high as 30% of the entire population in some of the communities (SCIC 2017). Sickle

B. K. Singh · A. Ojha · K. K. Bhoi (✉) · A. Bissoyi
Department of Biomedical Engineering, National Institute of Technology Raipur, Raipur, India
e-mail: kkbhoi.1993@gmail.com

P. K. Patra (✉)
Chhattisgarh Institute of Medical Sciences (CIMS), Bilaspur, Chhattisgarh, India
e-mail: pkp1964@yahoo.co.in

A. A. Rizvanov et al. (eds.), *Advances in Biomedical Engineering and Technology*, Lecture Notes in Bioengineering,
https://doi.org/10.1007/978-981-15-6329-4_37

gene is most common among several groups' tribal citizens in our country with the occurrence of heterozygotes varying from 1 to 40% (Colah et al. 2015). Patients suffering from SCD often complain of body pain, anemia, jaundice, repeated blood transfusion, pneumonia, repeated infection, etc. Thus, patients with SCD need timely treatment and care which otherwise may reduce their lifespan. A recent report of the American Society of Hematology on SCD suggests that the existing procedures for care and treatment of SCD are not adequate and need improvements by developing new tools for patient care, early diagnosis, and planned treatment of this disease (Report on Sickle Cell Disease 2016). The current standard methods for treatment of SCD include: (i) Hydroxyurea (HU) medication has demonstrated desirable effects in reducing the symptoms associated with SCD by invigorating the production of fetal hemoglobin that helps to avert the formation of sickle cells (Khalaf et al. 2016).

Sickle Cell Chlorosis is a blood disorder that is genetically inherited by the progeny from his parents. This disease occurs due to the inheritance of two abnormal copies of the hemoglobin gene, present in chromosome 11, one from each parent. If one of the originators has sickle cell anemia and the other has sickle cell trait, then there are equal chances for both the cases in the child. Due to the presence of anomalies in oxygen-carrying protein hemoglobin, RBCs become rigid and sickle-shaped. Its indications are described by the agony in limbs, legs, back, stomach area, chest, or head that kept going at least around two hours, prompted a demand for therapeutic consideration, and was owing to sickle cell illness (Sickle Cell Anemia-Treatment 2017). As indicated by WHO, about 5% of the total populace carries genes responsible for hemoglobinopathies.

Most of the hospitals and healthcare sectors in India employ a manual approach for analyzing patient's input for SCD. Manual exploration is time consuming, subject to observational errors, inter- and intraobserver variability and stressful for patients. Further, the efficacy of such an approach is entirely dependent on the clinician's experience and expertise. This imposes a serious hitch especially in rural areas where expert clinicians may not be available. Hence, the development of an intelligent diagnosis system is needed to assist clinicians in predicting the appropriate dosage of HU, monitoring its effectiveness/treatment response in individuals suffering from SCD, and suggest a specific treatment plan for the same. The advance in medical information systems is playing a key role in medicine and biology. Artificial intelligence and machine learning approaches have found numerous applications in medicine including huge medical data analysis, assisting medical professionals in disease diagnosis, utilizing patient symptoms and characteristics for effective treatment planning, reducing inter- and intraobservational errors, etc.

The United States Foods and Drug Administration (FDA) has approved Hydroxyurea as an effective drug which reduces the constancy of painful episodes. It hikes the position of HbF and the hemoglobin level. It usually diminished the rate of painful episodes by 50% (Agrawal et al. 2014). Traditionally, when the treatment of a patient with hydroxyurea is commenced, we can't say whether the patient will respond positively to it or not. The patients have to undergo therapy without knowing it will be effective or not.

Thus to eliminate this uncertainty and to deal with this variability in the effectiveness of Hydroxyurea, we needed a procedure to predict the response of the patients, so that for non-responders another approach of treatment could be adopted. Artificial Neural Network and Machine Learning applications have come in play to serve this purpose.

Machine learning method is a field of artificial intelligence which provides the computers the capacity to gain from experiences, and then efficiently predict the output. Initially, the computer is facilitated with some data or observations so that a pattern could be identified and learned automatically, then it makes informed decisions and predictions. Many works have been done using ANN and different approaches of machine learning classification methods to accurately predict the responses of Sickle Cell Anemia patients. However, there remains the scope of achieving better accuracy.

Valafar et al. (2000) presents the study of the effectiveness of Hydroxyurea using 23 features using Artificial Neural Networks (ANN) pattern recognition methods, supported by an effective Student T-Test. The resulting accuracy was about 85%. The effective proof that Hydroxyurea elevates the level of Fetal Hemoglobin (HbF) and detailed study of the relation between the HbF and response of sickle cell anemia patient to HU is given in Steinberg et al. (1997). Khalaf et al. (2016) employs various neural network architectures, significantly backpropagation trained feed-forward networks to classify the sickle cell disease dataset using the ROC curve for analysis.

Thus in this paper, we proposed different classification methods using feature selection and majority voting to achieve accuracy of 94%. We have employed supervised learning algorithms, i.e., using a training dataset with well-defined input and corresponding outputs, to generalize the pattern observed and use it for further predictions.

2 Materials and Methods

Support Vector Machine (SVM) is a regulated learning strategy or supervised learning method utilized for both classification and regression, though in this study we have used it for classification. Each of the examples of the training set is categorized into different classes, the SVM model then determines the output class for the new data. The SVM model represents the training dataset as the scatter plot in which the different class is clearly separated from each other and when the new data is provided, it is also plotted and then determine in which classes region it falls under by making use of a support vector. Support Vector Machines constructs a set of hyperplanes that most efficiently divide the training dataset. SVM algorithm uses a set of mathematical functions called kernel generating the desired output. Different kernels that can be used are Linear, Polynomial, Nonlinear kernels, Sigmoids, etc. For this paper, the dataset has been trained with several classifier out of which the Quadratic SVM gave the best result.

Quadratic SVM kernel is a type of polynomial kernel defined as

$$K(x, y) = x^T * y + c^d \tag{1}$$

where d is the degree of the polynomial, x and y are vectors in the information space. For Quadratic SVM, $d = 2$. For the dataset under study in this paper, we have two output classes, class 1 (Responsive) denotes that the HbF level in the patient is increased by 15 and class 0 (non-responsive) for the increase in HbF level below 20%. So here class 1 is predicted as responders and class 0 is predicted as non-responders. Thus, class 1 implies that Hydroxyurea has a significant effect on the patient.

2.1 Data Acquisition

The data of 304 patients is used in this study. There were 18 parameters under which each patient was characterized, namely, Age, Sex, HbF, HbA2, HbS, Hematocrit, Hb, MCV, MCH, RDW, TLC, DLC (L, M, G), Plc, EnzSGPT, EnzSGOT. The sex is encoded as Female = 1 and Male = 2. Hydroxyurea affects the level of these parameters in the patient's body which is recorded after each visit. The HbF value is used for standardizing the output class. The study is approved by the Institutional Ethical Committee (IEC) (Approval No. 102/IEC/CIMS/2019; 735/IEC-AIIMSRPR/2019; No./NITRR/IEC/2019/06). All the ethical guidelines like informed consent and consent for participation were followed as per the approval for IECs.

2.2 Simulation Methodology

Software's Used. WEKA—Waikato Environment for Knowledge Analysis (Weka) is a series of machine learning programmes written in Java, created at the University of Waikato, New Zealand. It is free programming, which we have utilized to highlight determination in this exploration.

MATLAB—Matrix Laboratory is a multi-worldview numerical processing condition created by MathWorks. MATLAB allows lattice controls, plotting of capacities and information, the procedure of calculations, production of UIs, and interfacing with projects written in different dialects, including C, C++, C, Java, Fortran, and Python. It has been utilized for classification.

3 Result and Discussion

3.1 Features Selection

The first step is Feature Selection done through the use of software WEKA through both methods Filter Based and Wrapper Based. The process of feature selection is crucial as it performs statistical tests on attributes and determines which attributes are more related to the output. The Filter-Based Feature Selection method provides various algorithms to arrange the feature and rank them according to their predictive power. Various methods employed in filter-based selection are correlation ranking filter, gain ratio feature evaluator, information gain ranking filter, OneR feature evaluator, principal components, relief ranking filter, and symmetrical uncertainty ranking filter (Tables 1 and 2).

Table 1 Table describing the parameters of sickle cell anemia patients

Features number	Features	Description
1	Age	Age of the patient at the time of analysis
2	Sex	Male of female
3	HbA2	Haemoglobin A2
4	HbS	sickle haemoglobin
5	Hematocrit	Ratio of the volume of red blood cells to the total volume of blood
6	Hb	Total haemoglobin concentration
7	MCV	Mean corpuscular volume
8	MCH	Mean corpuscular haemoglobin
9	MCHC	Mean corpuscular haemoglobin concentration
10	RDW	Red blood cell distribution width
11	TLC	Total leukocytes counts
12	DLC (L/M/G)	Differential leukocytes counts
13	Plc	Platelet count
14	Enz SGPT	Enzyme SGPT
15	Enz SGOT	Enzyme SGOT
16	HbF	Fetal haemoglobin

Table 2 Results of filter-based method

Serial number	Name of methods	Rank of features
1	Correlation ranking filter	10, 5, 7, 6, 3, 11, 8, 14, 2, 15, 1, 7, 4, 1, 13, 16, 12, 9
2	Gain ratio feature evaluator	4, 10, 5, 6, 17, 8, 3, 2, 7, 9, 16, 14, 15, 13, 11, 12, 1
3	Information gain ranking filter	4, 10, 5, 6, 17, 8, 3, 2, 7, 9, 16, 14, 15, 13, 11, 12, 1
4	OneR feature evaluator	3, 4, 6, 2, 10, 1, 11, 16, 15, 6, 8, 12, 17, 7, 13, 9, 14
5	Principal components	1, 2, 3, 4, 5, 6, 7, 8, 9, 10, 11, 12, 13
6	Relief ranking filter	4, 1, 3, 6, 5, 14, 8, 7, 15, 11, 10, 9, 12, 1, 16, 17, 13
7	Symmetrical uncertainty ranking filter	4, 10, 5, 6, 17, 8, 3, 2, 7, 9, 16, 14, 15, 13, 11, 12, 1

Table 3 Results of filter-based method

Serial number	Name of method	Selected subset
1	CFS subset evaluator	4, 10

The Wrapper-Based Feature Selection showcases the subset of input features that are most relevant in the prediction of output. We used CFS Subset Evaluator for our study (Table 3).

3.2 *Classification*

Support Vector Machine algorithms were carried out for each of the feature selection results and compared to evaluate the best accuracy achieved. The patients whose Hbf value was increased by 15% after the HU treatment were referred to as class 1 or else as class 0. The algorithms like Cross-Validation, Hold Out, Leave One Out, Repeated Hold Out, and Resubstitution were used for prediction and the maximum accuracy was delivered by Information Gain and OneR feature selection method using Hold Out classification method (Figs. 1 and 2).

3.3 *Accuracy, Sensitivity, Specificity, and ROC Curve*

This paper gives the conception of accuracy, specificity, and sensitivity for the circumstances of disease diagnosis. We made this test to know how good the test for the prediction of a positive disease is. Accuracy is the term used to measure how the

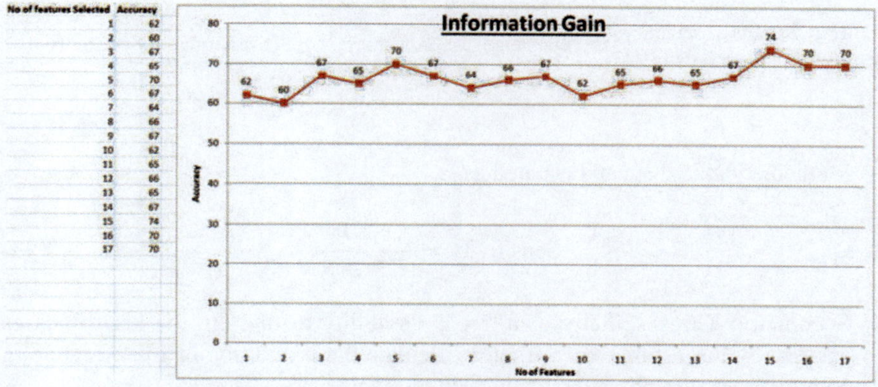

Fig. 1 Plot for the number of features selected in order of rank of information gain against accuracy

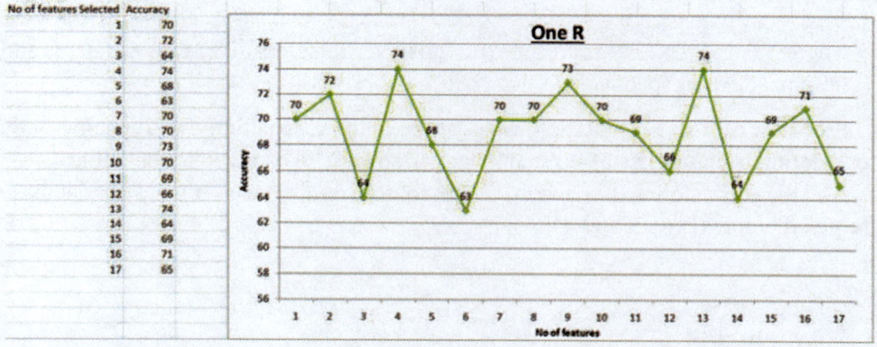

Fig. 2 Plot for the number of features selected in order of rank of OneR against accuracy

correctness of a diagnostic test identifiers, specificity is used for estimation of how similar patients without the disease can be correctly detected and sensitivity evaluates how accurate the test is for detecting a positive disease. ROC curve is a graphical representation between sensitivity and specificity and it helps to conclude the optimal model by determining the best threshold for the diagnostic test.

- Accuracy: The accuracy of a test is its ability to make difference between the patient and healthy cases accurately. For the estimation of the test, we should calculate the proper ratio of true positive, and true negative in evaluated cases. Mathematically, it can be stated as

$$\text{Accuracy} = \frac{TP + TN}{TP + TN + FP} \tag{2}$$

- Sensitivity: The sensitivity of a test is the ability to find patient cases accurately. To calculate it, we should calculate the ratio of true positive in-patient cases.

Table 4 Results of accuracy, sensitivity, and specificity

Percentage	Accuracy	Sensitivity	Specificity
80/20	0.94	0.95	0.91

Mathematically, this can be stated as

$$\text{Sensitivity} = \frac{TP}{TP + FN} \tag{3}$$

- Specificity: The specificity of a test is its ability to find out the healthy cases accurately. For estimation, we must calculate the proportion of true negative in healthy cases. Mathematically, this can be stated as

$$\text{Specificity} = \frac{TN}{TN + FP} \tag{4}$$

where TP, FP, TN, FN represent true positive, false positive, true negative, and false negative, respectively.

From our database calculation, we got the accuracy, sensitivity, and specificity by considering training data as 80% and validation as 20% as given in the Table 4.

And the confusion matrix and ROC plot for the training dataset from our database are given below (Figs. 3 and 4).

4 Conclusion

From the experiment, we analyzed that hydroxyurea effects on each patient to a different extent. In our experimental data, we used a 15% rise in the HbF level that is considered as a responder and classifies the sickle cell anemia patient accordingly either as class 1 or class 0. After examining the results of the classification technique for calculating accuracy of prediction whether sickle cell anemia patients are responding to the drugs or not, we attained the best accuracy of 94% for our collected experimental dataset.

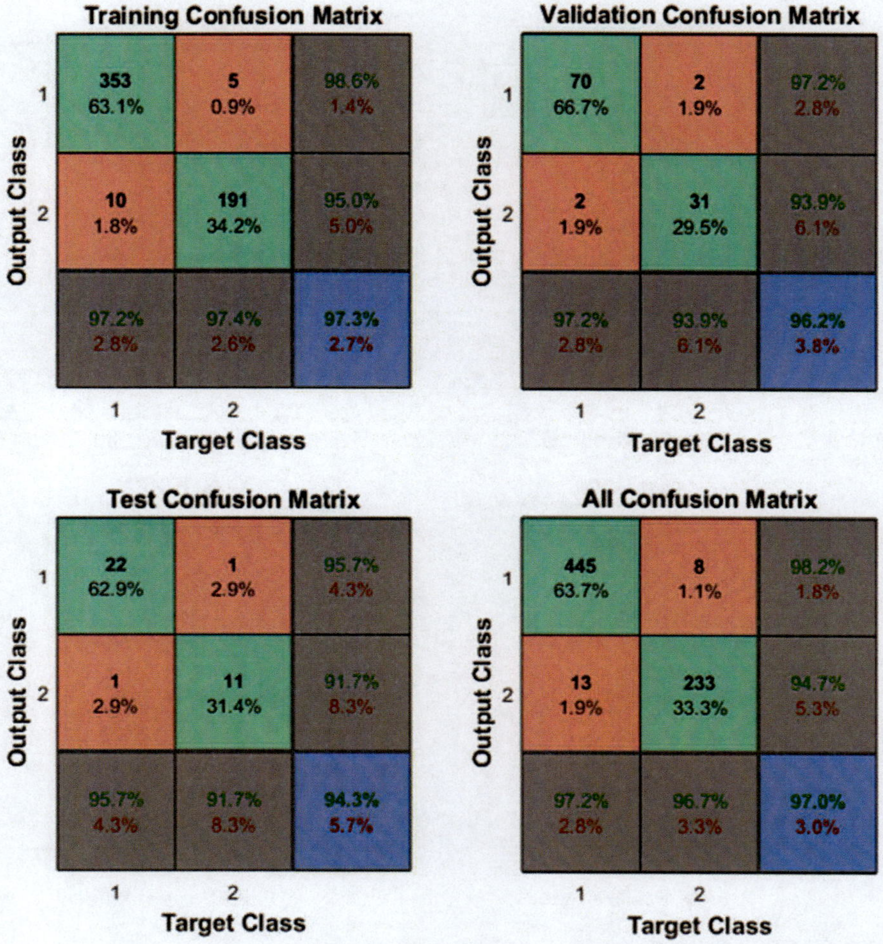

Fig. 3 The plot of confusion matrix for training, testing, and validation data

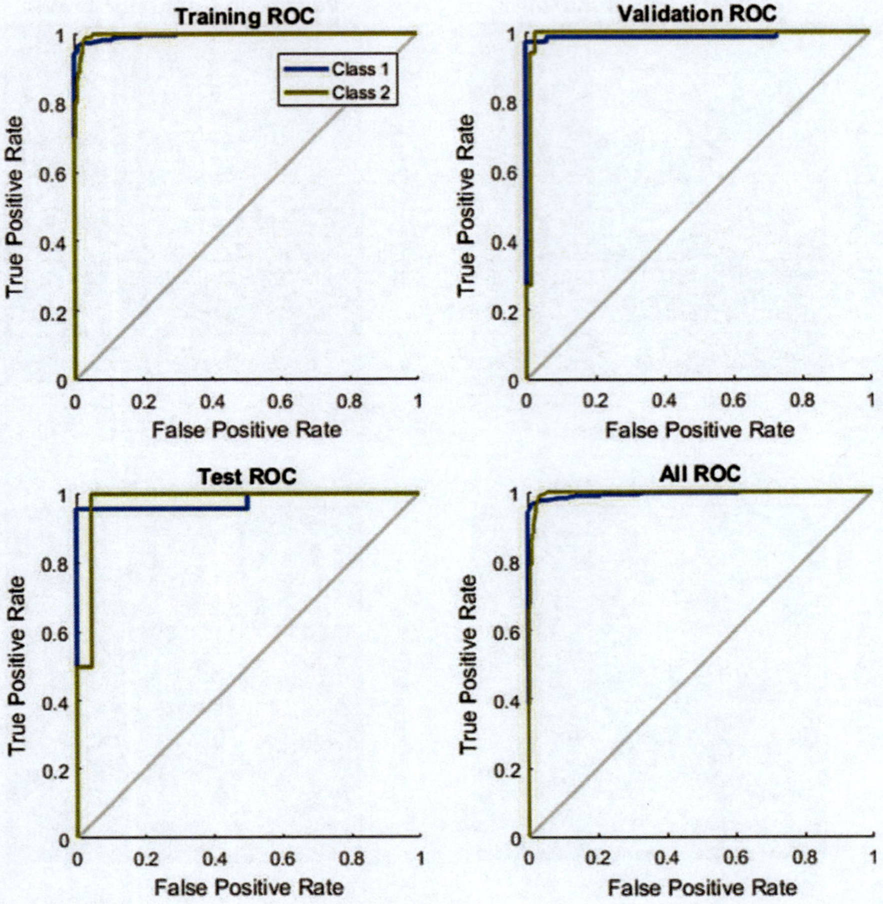

Fig. 4 ROC plot of training, testing, and validation data

Acknowledgements The authors are grateful to the Science and Engineering Research Board (SERB), Department of Science and Technology, Govt. of India, New Delhi for providing financial support (ECR/2017/003059).

References

Agrawal RK, Patel RK, Nainiwal L, Trivedi B (2014) Hydroxyurea in sickle cell disease: drug review. Indian J Hematol Blood Transfus 30(2):91–96

Colah RB, Mukherjee MB, Martin S, Ghosh K (2015) Sickle cell disease in tribal populations in India. Indian J Med Res 141(5):509

Khalaf M, Hussain AJ, Al-Jumeily D, Keight R, Keenan R, Fergus P, AlAskar H, Shaw A, Idowu IO (2016) Training neural networks as experimental models: classifying biomedical datasets for sickle cell disease. In: International conference on intelligent computing. Springer, Cham, pp 784–795

Patra PK, Chauhan VS, Khodiar PK, Dalla AR, Serjeant GR (2011) Screening for the sickle cell gene in Chhattisgarh state, India: an approach to a major public health problem. J Commun Genet 2(3):147–151

Report on Sickle Cell Disease (2016) American Society of Hematology, Published on: September 06, 2016. http://www.hematology.org/Newsroom/PressReleases/2016/6672.aspx. Accessed on 29 July 2017

SCIC (2017) Introduction, Sickle Cell Institute Chhattisgarh (SCIC), Raipur, Chhattisgarh (India). http://scic.cg.nic.in/aboutus.html. Accessed on 29 July 2017

Sickle Cell Anemia-Treatment (2017). http://www.mayoclinic.org/diseases-conditions/sickle-cellanemia/diagnosis-treatment/treatment/txc-20303509. Accessed on 29 July 2017

Steinberg MH, Lu ZH, Barton FB, Terrin ML, Charache S, Dover GJ (1997) Fetal hemoglobin in sickle cell anemia: determinants of response to hydroxyurea. Blood 89(3):1078–1088

Valafar H, Valafar F, Darvill A, Albersheim P, Kutlar A, Woods KF, Hardin J (2000) Predicting the effectiveness of hydroxyurea in individual sickle cell anemia patients. Artif Intell Med 18(2):133–148

A Survey on IoT-Based Healthcare System: Potential Applications, Issues, and Challenges

Kavita Jaiswal and Veena Anand

Abstract Advances in information and communication technologies have led to the emergence of the Internet of Things (IoT). In the modern healthcare environment, the usage of IoT technologies brings the convenience of physicians and patients, since they are applied to various medical areas (such as real-time monitoring, patient information management, and healthcare management). In the incoming world of the Internet of Things (IoT) for healthcare, diverse, and distributed devices will collect, examine, and transfer real-time medical statistics to open, private or hybrid clouds, making it possible to accumulate, pile, and analyze significant data streams in several new procedures, and activate context-dependent alarms. The current development of the traditional medical model toward the involved medicine can be improved by the Internet of Things (IoT) paradigm involving sensors (environmental, wearable, and implanted) spread inside local environments with the objective to monitor the user's health and initiate remote support. This state-of-the-art data gathering paradigm allows uninterrupted and pervasive medical data access from any connected device over the Internet and a novel health application ecosystem emerges. This paper represents a systematic survey of IoT-enabled healthcare system, their potential applications, issues, and challenges.

Keywords Internet-of-Things (IoT) · Health monitoring · Sensing network

K. Jaiswal (✉) · V. Anand
Department of Computer Science and Engineering, National Institute of Technology, Raipur, Chhattisgarh, India
e-mail: kjaiswal.phd2018.cse@nitrr.ac.in

V. Anand
e-mail: vanand.cs@nitrr.ac.in

A. A. Rizvanov et al. (eds.), *Advances in Biomedical Engineering and Technology*, Lecture Notes in Bioengineering,
https://doi.org/10.1007/978-981-15-6329-4_38

1 Introduction

Internet of Things (IoT) is the fundamental innovation to shape smart cities as they encourage daily items or individuals to deliver services and information to different users by cooperating and communicating with each other. The Internet of Things can be advantageous for medicinal services applications. Different kinds of sensors are used in IoT-based healthcare system to measure and monitor several well-being parameters in the human body. These devices can focus on observing a patients' well being when they are far away from everyone else or when the medicinal facility is out-of-reach. In this manner, they can offer a real-time response to the physician, relatives, or the patient. As many wireless sensing devices flood in the market bear medical sensor. These sensor helps to examine health parameter like heart rate, blood pressure, body temperature, respiration rate, pulse, and blood glucose levels. Medicinal services applications make free-living conceivable and more comfortable for the elderly and patients with severe medical conditions. Currently, IoT sensors are being utilized to consistently record and observe the well-being conditions and transmit alert in case any strange signs are found. If there should be an occurrence of a minor issue, the IoT application likewise has a facility to recommend a remedy to the patients.

Due to the proliferation of diseases, there remains a significant challenge before healthcare services to eradicate. The patients can avail reliable, effective, and smart healthcare support for chronic illness by combining IoT with available medical resources. Various advancements have been made in healthcare monitoring (Alemdar and Ersoy 2010). These accomplishments have exhibited the viability and likely eventual outcome of IoT in healthcare frameworks. In spite of the existent achievement, there is uncertainty, and the technical issues still exist about the inquiry of how to quickly and deliberately set up insightful IoT-based healthcare system (Yuehong et al. 2016). Targeting at exploiting the capabilities of IoT in healthcare systems, an ever-increasing number of researchers furthermore, associations have been committed to the improvement of IoT-based innovation (Sundmaeker et al. 2010; Bui and Zorzi 2011). Also, The IoT has given an opportunity to develop a smart house, hospitals, and many healthcare systems. The technology is still in its infancy where there exists a lot of challenge to obtain safe and secure useful healthcare applications as self-learning and self-improvement, standardization, privacy, and security. It is indeed a critical issue for the integration of IOT with the existing promising technology, services, and communication. The overall architecture and view of IoT-Cloud-based healthcare system are depicted in Fig. 1.

The functional component of the system is given below:

(1) *Sensing Network*: The sensing network creates the backbone for IOT framework. Gathering of information related to patient's health and distributing it using the wireless channel to IoT is the sole responsibility of IoT framework. For implementing this IOT uses specific protocols like WiFi, Bluetooth, ZigBee, etc.

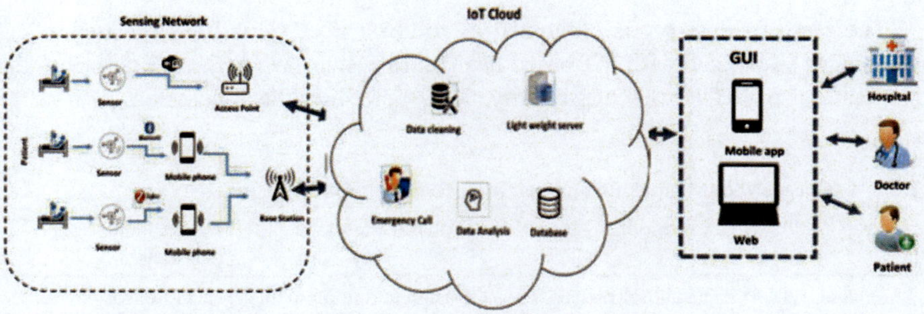

Fig. 1 Architecture of IoT healthcare system (Jaiswal et al. 2018)

(2) *IoT-Cloud*: Due to advancement in storage technology, the information collected through the IOT technique is stored, analyzed, and processed proficiently. It also proved the benefit of data evolution and analysis task using the influential server. The primary responsibility of this server is as follows:

- Data cleaning: The sensing data collected for diagnosis is profoundly affected by noise during the process of communication. The raw with noise may influence the diagnosis process. The cleaning process removes the noise from the data.
- Data storage: Patient's health information is stored in the database for analysis to find out the recovery process.
- Data analysis: Using machine learning or data mining methods the servers provide a platform for data analysis.
- Emergency warning: The edge server also has a provision to recognize the real-time health condition. Based on the results of data analysis the local caretaker receives warning message for providing first aid to the patient.

(3) *GUI*: Data visualization and management are done through the Graphical User Interface (GUI), which offers simple access to the information in the IoT-Cloud. Users can log in the IoT-Cloud and obtain patient's health-related information by just visiting the website or through mobile apps.

The existing literature for the application of IoT in healthcare system has been provided in Table 1. This paper examines the trends in IoT-based healthcare research and uncovers various issues that must be addressed to transform healthcare technologies through IoT innovation.

In this regard, the contribution of this paper is as follows:

- Classification of different IoT technologies available for the healthcare system.
- Bringing out related works in healthcare system using IOT technology.
- Providing open challenges for research work resource management.
- To facilitate novice researchers to work on the research problems.

The rest of the paper is organized as follows: Sect. 2 outlines the issues and challenges associated with IoT-based healthcare system, while Sect. 3 describes the applications of IoT Healthcare. Finally, Sect. 4 provides the conclusion.

Table 1 Overview of existing literature on IoT-based healthcare systems

Authors	Descriptions	Problem addressed	Devices and parameters
Yang et al. (2016)	The author proposes a novel method for ECG monitoring based on Internet of Things (IoT) techniques	Design and implement a wearable ECG monitoring system	ECG Sensor power voltage, ECG sensor output voltage, RR interval, PR interval, QT interval, QRS duration
Gope et al. (2016)	The author proposes a secure IoT-based healthcare system using Body sensor network	Data privacy, anonymity, data integrity, authentication	Body sensor, CPU cycle, execution time
Chatterjee et al. (2017)	The author focused on the aspects of IoT and decision support systems to proffer preventive and intelligent healthcare	Lowered cost of high quality healthcare, Best use of resources	Age, gender, systolic and diastolic blood pressure, g habits, Framingham score, risk score
Hossain et al. (2016)	The author presents a health IoT-enabled monitoring framework, where ECG and other healthcare data are collected by mobile devices and sensors and securely sent to the cloud for seamless access by healthcare professionals	Solution for IoT healthcare system	SNR PSNR, CPU memory and bandwidth utilization
Aminian et al. (2013)	This paper presents a monitoring system that has the capability to monitor physiological parameters from multiple patient bodies	Reduce the energy consumption to prolong the network lifetime, speed up, and extend the communication coverage to increase the freedom for enhance patient quality of life	Body sensor, BP level, end-to-end delay, coverage range, energy consumption

(continued)

Table 1 (continued)

Authors	Descriptions	Problem addressed	Devices and parameters
Yang et al. (2014)	The author presents an IoT-based intelligent home-centric healthcare platform (iHome system), which seamlessly connects smart sensors attached to human body for physiological monitoring and intelligent pharmaceutical packaging (iMedPack) for daily medication management	The proposed system takes the advantages of System onCchip (SoC) technology, material technology, and advanced printing technology, to build a patient-centric, self-assisted, fully automatic intelligent in-home healthcare solution	RFID tags, ECG sensor, controlled delamination materials (CDM) current

2 Issues and Challenges

With the exponential growth in IoT, the health sector is going to adopt it through the IoT application and devices. These applications and equipment might handle individual's private health information. The same can be accessed globally by the use of Internet resources. In particular, the security and privacy issues must be taken into considerations. This may include security requirements, threat models, vulnerabilities, etc. The following are a few challenges and issues need to be addressed concerning IoT-Cloud-based healthcare system:

- *Confidentiality*: It hides the patients' health record and medical information from unapproved users and eavesdroppers.
- *Integrity*: It helps in retaining the patients' health information unaltered while moving in an IoT-Cloud-based network. It never compromises the information with the eavesdroppers.
- *Authentication*: It helps in keeping the information secret by allowing only the authorized users to access the information.
- *Availability*: It allows the availability of patients' health information as and when the users require it. It is going to face a challenge of denial of service attack. This challenge is to be faced in IoT-Cloud-based healthcare systems.
- *Data freshness*: Data freshness mainly infers that individual data set is new and ensures that an adversary does not replay old messages. Since IoT provides estimates which is time varying in nature, so it is imperative to maintain freshness of a message.
- *Non-repudiation*: Non-repudiation ensures that the original sender of the message can't deny the message sent by it.

- *Authorization*: It ensures only the authorized person will access the services a resources provided by IoT-Cloud-based healthcare systems.
- *Security*: We need to deploy standard encryption and decryption method to provide security to the data and transaction takes place in IoT-Cloud-based healthcare system. This is the challenge to the research community to develop an efficient key management protocol. Mišic (2008), proposed an energy-efficient elliptic curve cryptography system. Another problem is to obtain authorization while the patient is in ICU or in an unconscious state, In this case, biometric may be used as a method to get authorization (Zhang et al. 2009). Bao et al. (2005) contributed to this challenge by developing a physiological signal-based authentication scheme. The technique captures ECG measures Photoplethysmogram (PPG).
- *Privacy*: Privacy of patient's health record should not eavesdrop. Ikonen et al. (2008) has proposed autonomy and control over the patient's health record. Besides image processing which is popularly adopted, other techniques should also be developed for the comfort of the patient's privacy. The encrypted image can be sent over the network to keep the privacy of information. Srinivasan et al. (2008) proposed Fingerprint and Timing-based Snooping (FATS) in WSN.
- *Resiliency*: It helps in protecting the patient's information even though the devices and interconnection have been compromised from any further attack.
- *Fault tolerance*: There should be some technology to be developed so that the system retains its functionality even in case of failure.
- *Robustness*: It ensures that when there is network failure or devices run out of energy. The network should have some technology to maintain the security level.
- *Computational limitations*: The IoT-Cloud healthcare uses lightweight protocols and devices which have limited power and computational speed. To run the security algorithm may require high power device which becomes a challenging task for the researchers for maximizing system performance with limited resource.
- *Memory limitations*: The IoT-Cloud-based devices have low memory. Their memory should be sufficient to run a security algorithm, or we should design such a lightweight security protocol to match the memory requirement of the available IoT-Cloud device.
- *Energy limitation*: The low-power devices save power by switching off the devices when not in use. The other way they conserve energy is by lowering the processing of CPU. It throws the challenge to the IoT-Cloud research community to develop energy efficient devices and protocols. The battery has limited power people may go for rechargeable batteries for the purpose, but this may be tiresome for elderly people. Battery rechargeable techniques should also be taken into consideration. This becomes another challenge for the researchers. Several works proposed an idea of using low-power sensing devices, still we need energy scavenging technique. The solar power can be thought as the alternative, but it is also tiresome because it always needs sunlight and cannot be implanted into the human body. So there is also a challenge for the researcher to develop a battery charger that takes the input from human motion, body heat, and other sources (Renaud et al. 2008).

- *Unobtrusiveness*: It is a challenge for the research community that while developing wearable sensors they need to maintain unobtrusiveness. According to Baker et al. (2007) when a patient carries sensors on their body then unobtrusiveness is a significant challenge. This is a hard task to put together different sensors into one platform vizLiveNet and PATHS. These body-worn sensor devices are substantial and highly obtrusive devices whereas ECG and watch-shaped activity recorder are much more comfortable to bear devices (Barth et al. 2009).
- *Sensitivity and calibration*: The sensitivity of the sensors devices is essential in a situation like fire situation or exercising. The sweat coming out of the body may affect negatively to the recalibration of the sensors. Puustjärvi and Puustjärvi (2011) proposed an algorithm for recalibration or self-calibration for the triaxial accelerometer. Still, improved techniques and algorithm should also be introduced for the pervasive healthcare system. Researchers should focus on low maintenance and highly sensitive devices.
- *Mobility*: In general, we deal with mobile devices in IoT-Cloud-based healthcare system. Through Wifi and Internet sensors devices can be connected. As an example, we can attach any wearable sensor and access the detail through the Internet. Body temperature, BP sensors, and so forth can be linked to the Internet to send the sensors readings to concern caregivers. These sensor devices are connected to the home network and these home networks further connected to the caregiver's network to provide service to the patient. The systems are heterogeneous concerning hardware, software, different configurations, and settings. Therefore a platform-independent device is a challenge for the researchers. The primary objective of IoT-Cloud-based healthcare system to people should have an independent life with high-quality healthcare services. The evolution of sensors and sensor-based network enabled the development of the application that ensures and encourages mobility of users. This way the WSN enabled the ubiquity of IoT-Cloud-based healthcare systems.
- *Ease of deployment and scalability*: The researchers always face a challenge to develop user-friendly IoT-Cloud-based healthcare system. With the increase in patient and caregivers in society, the existing system should support scalability and user-friendly application. Sensors, communication devices, and software should be used in parallel in the IoT-cloud-based healthcare system. With the heterogeneity in the component, the deployment of these devices becomes a challenge. We can use Software as a Service (SaaS) for the case of service and scalability with small, configurable sensor devices. The system should also support scalability in runtime to adapt to the ongoing technology. Atallah et al. (2008), Gao et al. (2005) proposed that the system software and distributed system can support the integration of the hardware and application level and interoperability among this different component. The sensor devices are increasing enormously resulting in the increase in devices in a global network. To develop such a scalable security scheme is a challenge for the researcher and academia.
- *User-friendliness*: Casas et al. (2008) propose a model to develop a natural system that provides the interface between IoT-Cloud-based healthcare system and diverse group of people. Jasemian (2008) proposed a survey which is conducted to

check the user-friendliness among a middle-aged group and conducted that they should revisit their work. They also found out that the different age groups should be differentiated clearly, which is a difficult task to be done. When we consider the interface between heterogeneous disease people like cognitive disability has different interaction characteristics than patient with high blood pressure. They must have gesture, visual animation, and voice interface for handicapped and the elderly avoid any particular kind of skills. In IoT-Cloud-based healthcare system should have the user-friendly and natural interface between healthcare professionals and caregivers.

- *Data acquisition efficiency*: This is a challenge for the research community to develop an efficient data processing technique as the rate of collection of data in IoT-Cloud-based healthcare system is increasing enormously. Sometimes the data collected from a few devices like ECG or 3-axis accelerometer may not be sufficient to debit all activities of the patient. This may be overcome by deploying more sensors to collect data. This, in turn, increases the data size. It is essential to implement some techniques to analyze and acquisition of data in realtime. Stankovic et al. (2005) proposed an open research challenge, viz, synchronization of different sensors, time stamping, and ordering of events. The research community should take care of designing a modular architecture which should support multisensory integration like wireless sensors, RFID tags, and printable body sensors.
- *Error resilience and reliability*: Low transmission power and small antenna sizes of wireless sensor devices cause reduced Signal-to-Noise Ratios (SNR) thus causing higher bit error rates and reducing the reliable coverage area. However, the reliable transfer of data in medical monitoring systems is vital. Therefore, error-resilient network coding schemes for medical data transmission should be developed for increasing network reliability. Marinkovic and Popovici (2009) propose a network coding technique for a TDMA-based protocol. It lets every sensor to transfer data through two relays and the relay nodes XOR the packets before sending. Although they have shown the improvements in the packet loss rate through simulations, the real deployments for measuring physiological signals, such as ECG and EEG and improving the proposed system accordingly are left as future works. The reliable data transmission should be studied thoroughly for low-power body area sensor networks.
- *Communications media*: There is a wireless device available such as ZigBee, Zwave, Bluetooth, Bluetooth low energy, WiFi, GSM, WiMax, and 3G/4G to connect health devices to a global network. Due to the evolution of these wireless devices, it becomes in appropriate to the use of wired security scheme to the IoT-Cloud-based healthcare system. It is the challenge before the academia and researchers that can treat wired and wireless channel characteristic equally.
- *The multiplicity of devices*: The devices used in IoT-Cloud-based healthcare network are diverse, this may include PC to low-end RFID Tags. The diversity in embedded software, memory, power, and computation among devices throws a challenge to the researcher to develop a security scheme which accommodates even the simplest devices.

- *A Dynamic network topology*: The IoT-Cloud-based Healthcare device may be combined together to form a network for accessing anywhere and anytime. At the same time, the network connection may exit abruptly or normally. The dynamic network topology is observed in healthcare devices with temporal and spatial admission characteristics. To develop a security model for this kind of environment is a challenge for the researcher and academia.
- *A Multiprotocol network*: The protocols used in healthcare devices are a proprietary in nature. The IoT devices use IP network to communicate with each other. It is a challenge for the research community to devise such a security solution for multiprotocol communication.
- *Dynamic security updates*: It is highly essential for the users to keep update security protocols to overcome the vulnerabilities. Therefore the security patches are required to update always. There is always a challenge to the research community to design dynamic installation of security patches.
- *Tamper-resistant packages*: Physical security is also a part of IoT-Cloud-based healthcare systems. Outsiders may tamper the devices to modify the software, program, extract cryptographic, security, and replace with malicious node. Tamper resistant is defended for this kind of attack which is a challenging task to implement in the real-time scenario.

3 IoT Healthcare Applications

In addition to IoT services, IoT applications are also getting tremendous attention. To develop applications services are used, whereas users and patients directly use applications. Consequently, services are developer-centric, while applications are user-centric. These days in addition to applications, there are many wearable sensing devices available in the market that can be viewed as IoT modernization to prompt different healthcare solutions and are discussed in this section.

- *Glucose level sensing*

Glucose level sensing is a way of testing the concentration of glucose in the blood. Particularly important in diabetes management, diabetes is one of the metabolic diseases. It is caused by the high percentage of sugar in the blood over a long period. Blood glucose monitoring aids in the planning of meals, activities to be carried out, and medication times and also give away patterns of blood glucose changes. For non-invasive glucose sensing an IoT healthcare configuration method is proposed in Istepanian et al. (2011) on a real-time basis. In this technique, sensors at patients' side associated with compatible healthcare providers via IPv6 Connectivity. The utility model proposed in Istepanian et al. (2011) based on IoT networks makes transmission device to broadcast the sensed data on blood glucose. This method includes a mobile phone, computer, processor, and a blood glucose collector. A related novelty is initiated in Wei et al. (2012). Also, a comprehensive IoT-Cloud-based medical acquisition detector is proposed in Lijun (2013) for monitoring the glucose level.

- *Electrocardiogram monitoring*

The electrical activity of heart is recorded by the electrocardiogram. The ECG data includes simple heart rate and determines the diagnosis of multifaceted arrhythmias, prolonged QT intervals, the basic rhythm of the heart, and myocardial ischemia (Drew and Funk 2006). The ECG data recording can be used to the fullest extent in IoT-Cloud-based healthcare system to record patients' data. There is numerous literature available which discuss the application of ECG monitoring for reading patients' heart status (Yang et al. 2014; Agu et al. 2013). Liu et al. (2012) proposed a model where they introduce ECG monitoring to patients' health monitoring that consists of a portable wireless receiving processor and wireless acquisition transmitter. They included alarm-based technique when there are abnormalities in patients' heart data it gives a warning on the real-time basis. In the application layer of the IoT network there exists an exclusive algorithm of ECG for data monitoring.

- *Blood pressure monitoring*

To read the data from the sensors Dohr et al. (2010) has proposed KIT, which is the combination of Blood Pressure (BP) meter and Near-Field Communication (NFC). And to remotely control the blood pressure of the patient an inspiring scheme is given in Puustjärvi and Puustjärvi (2011) by exhibiting the communications structure between a patients' hub and the medical service provider. Tarouco et al. (2012) addresses how the connection to a mobile computing device is operated within the blood pressure device. A mechanism is proposed in Guan (2013) to collect BP values and send it over IoT-Cloud-based healthcare network. This device consists of a BP machinery body with a communication module. Also based on IoT-Cloud a location intelligent end for blood pressure monitoring is proposed in Xin et al. (2013).

- *Body temperature monitoring*

In the healthcare system, body temperature plays a vital role as it is the sign of homeostasis maintains (Ruiz et al.2009). In Baker et al. (2007) IoT-Cloud-based healthcare scenario is verified using the TeloSB mote embedded with body temperature, and also it presented sample obtained from body temperature variations showing the successful operation of the developed IoT-based healthcare system. In Jian et al. (1028) a system for measuring the temperature is proposed based on a home gateway over the IoT. This home gateway with the help of infrared detection transmits the sensed body temperature to the concerned users.

- *Oxygen saturation monitoring*

For oxygen saturation monitoring, pulse oximetry instrument is used which continuously monitor the blood oxygen saturation in a human body without inserting the device into the patient's body, i.e., it noninvasively measures the oxygen saturation. The integration of the pulse oximetry with IoT technologies is profitable for medical healthcare application and services. In Khattak et al. (2014), the potential of IoT-Cloud drove pulse oximetry for CoAP-based healthcare services is reviewed. However, Jara et al. (2013) depicts the functionality of the wearable pulse oximeter wrist OX2. This device comes with Bluetooth connectivity to connect it directly to the money platform based on medical device profile. Furthermore in Larson et al. (2012) for monitoring the patients' health remotely an IoT-Cloud-based optimized low cost and a low-power pulse oximeter is suggested, which is used for constant monitoring of patients' health over an IoT-Cloud-based healthcare system. For telemedicine applications, an integrated pulse oximeter system is described in Larson et al. (2013), whereas in Larson et al. (2011) a wearable pulse oximeter is adopted in IoT-Cloud-based healthcare system for monitoring patients' health using WSN.

4 Conclusion

The healthcare industry is primarily moving toward inexpensive, accessible, and an excellent healthcare. IoT and adopt IoT-cloud driven systems and processes have the potential to mode this kind of healthcare, which heavily relies on patient participation. The survey presented here provides diverse aspects of IoT-based healthcare technologies and reviews on IoT-cloud usage in healthcare that provides access to the IoT backbone and enables medical data transmission and reception. Considerable R&D efforts have been made in IoT-driven healthcare services and applications. The recent technology for investigating, collecting, and switching data in IoT-cloud continues to smart, and in the future, the systems will move to improved IoT-cloud-driven healthcare applications.

References

Agu E, Pedersen P, Strong D, Tulu B, He Q, Wang L, Li Y (2013) The smartphone as a medical device: assessing enablers, benefits and challenges. In: 2013 IEEE international workshop of internet-of-things networking and control (IoT-NC). IEEE, pp 48–52

Alemdar H, Ersoy C (2010) Wireless sensor networks for healthcare: a survey. Comput Netw 54(15):2688–2710

Aminian M, Naji HR (2013) A hospital healthcare monitoring system using wireless sensor networks. J Health Med Inform 4(02):121

Atallah L, Lo B, Siegemund F, Yang G-Z (2008) Wirelessly accessible sensor populations (WASP) for elderly monitoring

Baker CR, Armijo K, Belka S, Benhabib M, Bhargava V, Burkhart N, Minassians AD, Dervisoglu G, Gutnik L, Haick MB, Ho C, Koplow M, Mangold J, Robinson S, Rosa M, Schwartz M, Sims C, Stoffregen H, Waterbury A, Leland ES, Pering T, Wright PK (2007) Wireless sensor networks

for home health care. In: 21st international conference on advanced information networking and applications workshops (AINAW'07), vol 2, May 2007, pp 832–837
Bao S-D, Zhang Y-T, Shen L-F (2005) Physiological signal based entity authentication for body area sensor networks and mobile healthcare systems. In: 2005 IEEE engineering in medicine and biology 27th annual conference. IEEE, pp 2455–2458
Barth AT, Hanson MA, Powell HC Jr, Lach J (2009) Tempo 3.1: a body area sensor network platform for continuous movement assessment. In: 2009 sixth international workshop on wearable and implantable body sensor networks, BSN 2009. IEEE, pp 71–76
Bui N, Zorzi M (2011) Health care applications: a solution based on the internet of things, In: Proceedings of the 4th international symposium on applied sciences in biomedical and communication technologies. ACM, p 131
Casas R, Marín RB, Robinet A, Delgado AR, Yarza AR, Mcginn J, Picking R, Grout V (2008) User modelling in ambient intelligence for elderly and disabled people. In: International conference on computer for handicapped persons. Springer, pp 114–122
Chatterjee P, Cymberknop LJ, Armentano RL (2017) IoI-based decision support system for intelligent healthcare applied to cardiovascular diseases. In: 2017 7th international conference on communication systems and network technologies (CSNT). IEEE, pp 362–366
Dohr A, Modre-Opsrian R, Drobics M, Hayn D, Schreier G (2010) The internet of things for ambient assisted living. In: 2010 seventh international conference on information technology: new generations (ITNG). IEEE, pp 804–809
Drew BJ, Funk M (2006) Practice standards for ECG monitoring in hospital settings: executive summary and guide for implementation
Gao T, Greenspan D, Welsh M, Juang RR, Alm A (2005) Vital signs monitoring and patient tracking over a wireless network. In: 27th annual international conference of the engineering in medicine and biology society, 2005. IEEE-EMBS 2005. IEEE, pp 102–105
Gope P, Hwang T (2016) BSN-care: a secure IoT-based modern healthcare system using body sensor network. IEEE Sens J 16(5):1368–1376
Guan ZJ (2013) Internet-of-things human body data blood pressure collecting and transmitting device. Chinese Patent 202(821):362
Hossain MS, Muham-mad G (2016) Cloud-assisted industrial internet of things (IIoT)–enabled framework for health monitoring. Comput Netw 101:192–202
Ikonen V, Kaasinen E (2008) Ethical assessment in the design of ambient assisted living. In: Dagstuhl seminar proceedings. Schloss Dagstuhl-Leibniz-Zentrum fur Informatik
Istepanian RS, Hu S, Philip NY, Sungoor A (2011) The potential of internet of m-health things m-IoT for non-invasive glucose level sensing. In: Annual international conference of the IEEE engineering in medicine and biology society, EMBC, 2011. IEEE, pp 5264–5266
Jaiswal K, Sobhanayak S, Turuk AK, Bibhudatta SL, Mohanta BK, Jena D (2018) An IoT-cloud based smart healthcare monitoring system using container based virtual environment in edge device. In: 2018 international conference on emerging trends and innovations in engineering and technological research (ICETIETR). IEEE, pp 1–7
Jara AJ, Zamora-Izquierdo MA, Skarmeta AF (2013) Interconnection framework for mhealth and remote monitoring based on the internet of things. IEEE J Sel Areas Commun 31(9):47–65
Jasemian Y (2008) Elderly comfort and compliance to modern telemedicine system at home. In: 2008 second international conference on pervasive computing technologies for healthcare. IEEE, pp 60–63
Jian Z, Zhanli W, Zhuang M (2012) Temperature measurement system and method based on home gateway. Chinese Patent, vol 102811185
Khattak HA, Ruta M, Di Sciascio E (2014) Coap-based healthcare sensor networks: a survey. In: 2014 11th international Bhurban conference on applied sciences and technology (IBCAST). IEEE, pp 499–503
Larson EC, Lee T, Liu S, Rosenfeld M, Patel SN (2011) Accurate and privacy preserving cough sensing using a low-cost microphone. In: Proceedings of the 13th international conference on ubiquitous computing. ACM, pp 375–384

Larson EC, Goel M, Boriello G, Heltshe S, Rosenfeld M, Patel SN (2012) Spirosmart: using a microphone to measure lung function on a mobile phone. In: Proceedings of the 2012 ACM conference on ubiquitous computing. ACM, pp 280–289

Larson EC, Goel M, Redfield M, Boriello G, Rosenfeld M, Patel SN (2013) Tracking lung function on any phone. In: Proceedings of the 3rd ACM symposium on computing for development. ACM, p 29

Lijun Z (2013) Multi-parameter medical acquisition detector based on internet of things. Chinese Patent 202(960):774

Liu M-L, Tao L, Yan Z (2012) Internet of things-based electrocardiogram monitoring system. Chinese Patent 102(764):118

Marinkovic S, Popovici E (2009) Network coding for efficient error recovery in wireless sensor networks for medical applications. In: 2009 first international conference on emerging network intelligence. IEEE, pp 15–20

Mišic J (2008) Enforcing patient privacy in healthcare WSNS using ECC implemented on 802.15. 4 beacon enabled clusters. In: Sixth annual IEEE international conference on pervasive computing and communications, 2008. PerCom 2008. IEEE, pp 686–691

Puustjärvi J, Puustjärvi L (2011) Automating remote monitoring and information therapy: an opportunity to practice telemedicine in developing countries. In: IST-Africa conference proceedings, 2011. IEEE, pp 1–9

Renaud M, Karakaya K, Sterken T, Fiorini P, Van Hoof C, Puers R (2008) Fabrication, modelling and characterization of mems piezoelectric vibration harvesters. Sens Actuators A 145:380–386

Ruiz M, García J, Fernández B (2009) Body temperature and its importance as a vital constant. Rev Enferm (Barcelona, Spain) 32(9):44–52

Srinivasan V, Stankovic J, Whitehouse K (2008) Protecting your daily in-home activity information from a wireless snooping attack. In: Proceedings of the 10th international conference on ubiquitous computing. ACM, pp 202–211

Stankovic J, Cao Q, Doan T, Fang L, He Z, Kiran R, Lin S, Son S, Stoleru R, Wood A (2005) Wireless sensor networks for in-home healthcare: potential and challenges. In: High confidence medical device software and systems (HCMDSS) workshop, vol 2005

Sundmaeker H, Guillemin P, Friess P, Woelfflé S (2010) Vision and challenges for realising the internet of things. Cluster of European Research Projects on the Internet of Things, European Commission, vol 3, no 3, pp 34–36

Tarouco LMR, Bertholdo LM, Granville LZ, Arbiza LMR, Carbone F, Marotta M, de Santanna JJC (2012) Internet of things in healthcare: interoperatibility and security issues. In: 2012 IEEE international conference on communications (ICC). IEEE, pp 6121–6125

Wei L, Heng Y, Lin WY (2012) Things based wireless data transmission of blood glucose measuring instruments. Chinese Patent 202(154):684

Xin T, Min B, Jie J (2013) carry-on blood pressure/pulse rate/blood oxygen monitoring location intelligent terminal based on internet of things. Chinese Patent 202(875):315

Yang G, Xie L, Mantysalo M, Zhou X, Pang Z, Da Xu L, Kao-Walter S, Chen Q, Zheng L-R (2014) Ahealth-IoT platform based on the integration of intelligent packaging, unobtrusive bio-sensor, and intelligent medicine box. IEEE Trans Ind Inform 10(4):2180–2191

Yang Z, Zhou Q, Lei L, Zheng K, Xiang W (2016) An IoT-cloud based wearable ecg monitoring system for smart healthcare. J Med Syst 40(12):286

Yuehong Y, Zeng Y, Chen X, Fan Y (2016) The internet of things in healthcare: an overview. J Ind Inform Integr 1:3–13

Zhang G, Poon CC, Li Y, Zhang Y (2009) A biometric method to secure telemedicine systems. In: Annual international conference of the IEEE engineering in medicine and biology society, 2009, EMBC 2009. IEEE, pp 701–704

Real-Time Data Augmentation Based Transfer Learning Model for Breast Cancer Diagnosis Using Histopathological Images

Rishi Rai and Dilip Singh Sisodia

Abstract The Real-time automated medical image diagnosis system could assist pathologist for speed-up diagnosis process for confirming the findings. In this paper, various issues involving Brest cancer histopathological image classification are considered. Presently, many models using traditional machine learning algorithms are used on the extracted features from the diagnosis images for the classification task. These models were able to archive high accuracies on a small dataset. However, the performance of classical classification models heavily relies on data representation and feature engineering, which requires prior expert domain knowledge and is extremely time-consuming. Many convolution neural network (CNN) models have been proposed to address this issue, but still, they face the problem of limited availability of authentic medical image diagnosis data. Therefore, it does not allow for highly accurate predictions and even may lead to overfitting on training data, especially when a deep CNN architecture is used. In this paper, a real-time data augmentation based transfer of learning model is proposed to overcome the existing limitations. Two well established and popular image classification models, i.e., InceptionV3 and Xception models were trained on an openly available Breast cancer histopathological image dataset called BreakHis. We first trained the models on learned weights transfer from the supervised training of the model on Imagenet and compared it with the results of the same models when they were trained from scratch. In our findings, it became very obvious that the model that was trained by the transfer of learning is more robust, as they significantly outperformed the same model trained from scratch. Also, the Xception model trained using the transfer of learning gave the best results with 90.86% accuracy on the classification task, exceeding the results obtained previously by the state-of-the-art system on the BreakHis dataset.

Keywords Histopathological images · Breast cancer · Classification · Transfer learning · Convolutional neural network

R. Rai · D. S. Sisodia (✉)
Department of Computer Science and Engineering, National Institute of Technology Raipur, Raipur, India
e-mail: dssisodia.cs@nitrr.ac.in

A. A. Rizvanov et al. (eds.), *Advances in Biomedical Engineering and Technology*, Lecture Notes in Bioengineering,
https://doi.org/10.1007/978-981-15-6329-4_39

1 Introduction

Nowadays, Cancer has become one of the most concerning health issues in the world (International Agency for Research on Cancer 2008), with breast cancer being the most common among women. Various efforts have been made to access the risk of breast cancer in women using machine learning techniques (Rodrigues et al. 2012; Hsu et al. 2015), but highly accurate and reliable diagnosis from histopathological images by machine learning models are still a long way to go. The diagnosis of almost all types of cancer from a histopathology image is considered a gold standard, including breast cancer (Rubin and Strayer 2008). However, the manual process of histopathological image analysis is highly time-consuming and requires a domain expert. Automation of this process can relive the workload from the pathologist and can speed-up the diagnosis, at least for easy cases.

In the past, the automated diagnosis of BC histopathological images was being carried out using a conventional machine learning algorithm on datasets of very limiting size, which is usually not available publically. These conventional machine learning algorithms required feature engineering which demands extensive domain knowledge and expertise to generate useful features (Sisodia and Verma 2011). However, some researchers recommend that the main weakness of this kind of models are the feature engineering step (LeCun et al. 2015; Bengio et al. 2013). The increasing availability of GPUs and advancements in representational learning caused increased attention toward Convolutional Neural Network (CNN). CNN's are a type of feed-forward neural network which can achieve a state-of-the-art result for different pattern recognition problems like face recognition (Schroff et al. 2015), multiple category image classification (Szegedy et al. 2015a; Krizhevsky et al. 2012), and medical image classification (Kowal et al. 2015; Lu et al. 2016; Ahmad et al. 2017). Also, CNN doesn't require the extra effort of feature engineering as one of the most important characteristics of CNN is its ability to learn relevant high-level features from low-level features in a hierarchal manner for a particular object.

But, training a deep ConvNet (CNN) usually requires a large amount of training data to attain high pattern recognition accuracy. The amount of labeled medical data is typically very limited. So, we used a learning technique known as transfer learning, in which learned knowledge is transferred across domain/tasks (Pan and Yang 2010). This type of learning technique has been proved to be beneficial especially when the training data is less (Ravishankar et al. 2016; Shie et al. 2015).

The rest of this paper is organized into the following sections. Section 2 discusses the related work reported on BCH images classification. In Sect. 3, the concept of transfer learning models is discussed in brief. Section 4 describes the methodology used in the classification of breast cancer images. Section 5 describes the evaluation parameters used in evaluating the models. In Sect. 6, all the results obtained in our experiment are states and explained. Section 7 summarizes the results of our experiment.

2 Related Work

In the past, traditional machine learning models were extensively used in which, nuclei segmentation followed by feature extraction and final classification were the steps generally followed. The paper by Kowal et al. (2015), proposed a two-stage segmentation which included testing four algorithms including fuzzy C-means, K-means, competitive learning neural networks, and Gaussian mixture models for fast nuclei segmentation. From these segmented nuclei, 42 features were extracted, which were used to classify them into benign and malignant using three classification algorithms. This method was implemented on 500 images, and the accuracies of these methods ranged between 95% and 100%.

Filipczuk et al. (2013), used circular Hough transform to approximate the position of nuclei in BCH images. Later these approximations were further screened using SVM for finding the most accurate approximation. Then the images are classified by 25 features extracted from the nuclei. They reported 97% accuracy on 737 images. Similar to this, George et al. (2014), classified cancer images into benign and malignant using four different classifiers namely PNN, Multilayer perceptron learning vector quantization, and support vector machine (SVM) based on 12 features extracted from the nuclei. As these were no large publically available data sets, Spanhol et al. (2016a) introduced BreakHis, a publicly available BHC image dataset containing 7909 images of 4 varying magnification factor. They also implemented the classification of these images based on textual descriptors of six different types. Their precision ranged between 80 and 85%. In another paper (Spanhol et al. 2016b), they argued that traditional classification models need robust feature creation and selection, which require prior knowledge of the domain and hence could be time-consuming. So, in this paper, they use a convolutional neural network, Alexnet, for the classification of breast cancer histopathological image into benign and malignant. They first split the images into a different number of patches, to increase the dataset size and then use these patches to train the convolutional neural network. Then for the classification task, the results of different patches were combined using four different fusion rules. Through this approach, they were able to archive 89% accuracy on 40× magnified images. But this approach lacks the learning of the overall feature from the entire image as it is being trained on small patches. In the paper (Ravishankar et al. 2016), Ravishankar et al. conducted a study on the automatic localization of kidneys in ultrasound images by using a varying degree of transferring knowledge learned from supervised training on Imagenet (Deng et al. 2009). They used CaffeNet (Jia et al. 2014) and 4096 features that were extracted from its "fc7" layer. The features were extracted from the model in three different ways, which includes Full Network adaptation (the entire model was fine-tuned), Partial Network adaptation (all the layers except the first two convolution layer was fine-tuned), and Zero Network adaptation (no fine-tuning was performed). In their experiment, the features extracted from Full Network adaptation performed the best of the three, even the features extracted from Zero Network adaptation outperformed the traditional Harr based features by 2%. This result was consistent with the one reported in (Shie et al. 2015). These results indicate that transfer learning is a strong learning technique with the best results being obtained when the entire model is fine-tuned. The paper by Shin et al. (2016), thoroughly explored the benefits of using transfer learning for medical images. They used three popular image classification model viz. CifarNet (Roth et al. 2016), AlexNet (Krizhevsky et al. 2012), and GoogLeNet (Szegedy et al. 2015a) for Thoracoabdominal Lymph Node and Interstitial Lung Disease detection. All the models were trained in three ways, viz. training from scratch, fine-tuning on the pertained model, and using "off-the-shelf" technique for feature extraction. For both the diseases, the best results were obtained by the models which were fine-tuned after pretraining on Imagenet. This further substantiated the use of transfer learning on medical image classification for obtaining more efficient results.

3 Transfer Learning

Because of the limited nature of the medical image data set, it becomes very challenging to archive very high accuracy on deep models. To tackle this, there have been many studies (Shie et al. 2015; Ravishankar et al. 2016; Song et al. 2017), which used transfer learning as a method to archive state-of-the-art results even with the limited dataset. As described in the paper (Pan and Yang 2010), transfer learning is the transfer of learned knowledge from one task to another task as shown in Fig. 1. A domain represented by D consist of X, a feature space and a marginal probability distribution $P(X)$, where $X = \{x_1, x_2, \ldots, x_n\} \in X$. A task denoted by $T = \{Y, f(\cdot)\}$, consists of a label space Y and a predictive function $f(\cdot)$, which is learned from the training data consisting of pairs $\{x_i, y_i\}$, where $y_i \in Y$ and $x_i \in X$. The predictive function $f(x)$ is used to predict the label for a particular value of x, it could also be written in terms of conditional probability $P(y/x)$. Transfer learning can now be

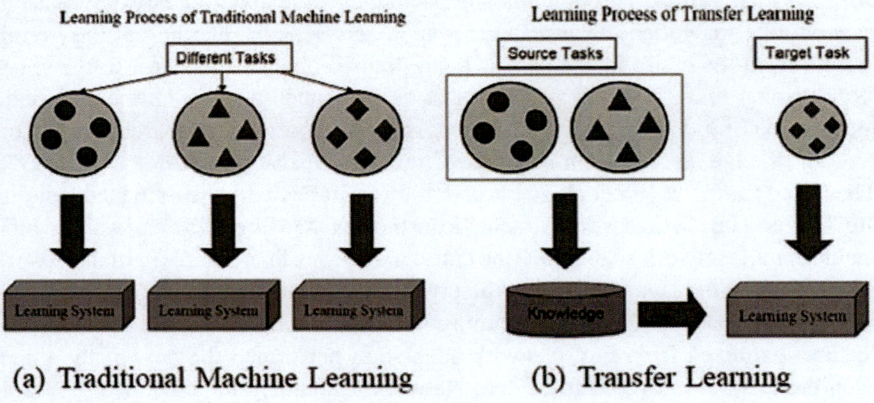

Fig. 1 Illustration of transfer learning (Pan and Yang 2010)

defined as, given a source domain DS and learning task TS, a target domain DT and learning task TT, transfer learning aims to help improve the learning of the target predictive function fT(·) in DT using the knowledge in DS and TS, where DS $\neq$ DT, or TS $\neq$ TT. In this paper, we performed a particular type of transfer learning known as inductive transfer learning, i.e., the source domain, and the task is different from the target domain and task. The application transfer learning has shown substantial improvement in performance on various tasks (Girshick et al. 2016; Shie et al. 2015).

In CNN, transfer learning can be done by first training the model on a bigger and general data set and later fine-tuning on the images for the specific task of the smaller dataset as shown in Fig. 1. We used Imagenet as our general image dataset for pretraining the classification models. Imagenet has more than 1.4 million images belonging to nearly 1000 categories. There have been many tasks, which has shown to perform well even without fine-tuning the models by using the feature vectors extracted from the learned model, also known as "off-the-shelf" CNN features(Sharif Razavian et al. 2014; Bar et al. 2015). Transfer learning on medical image diagnosis has proven to be very powerful as compared to other strategies (Song et al. 2017).

4 Methodology

The methodology adopted to carry out the proposed work is shown in Fig. 2.

In this work, we considered images of 40× magnification from BreakHis dataset as it gives the best result with CNN as reported in (Song et al. 2017; Spanhol et al. 2016a, b). Table 1 shows the number of malignant and benign images in 40× magnified images in BreakHis dataset.

4.1 Preprocessing

The dimension of images in the data set is 700 × 460 pixels. To reduce the amount of computation and for efficient use of the resources at hand, the images are cropped and resized to squares of 150 × 150 pixels.

4.2 Data Augmentation

Throughout the training, real-time data augmentation was used to improve the ability of the network to generalize well over the data. After every epoch every image is randomly augmented with the combination of the following:

- random rotation 0–270°
- Random horizontal flip
- Random vertical flip.

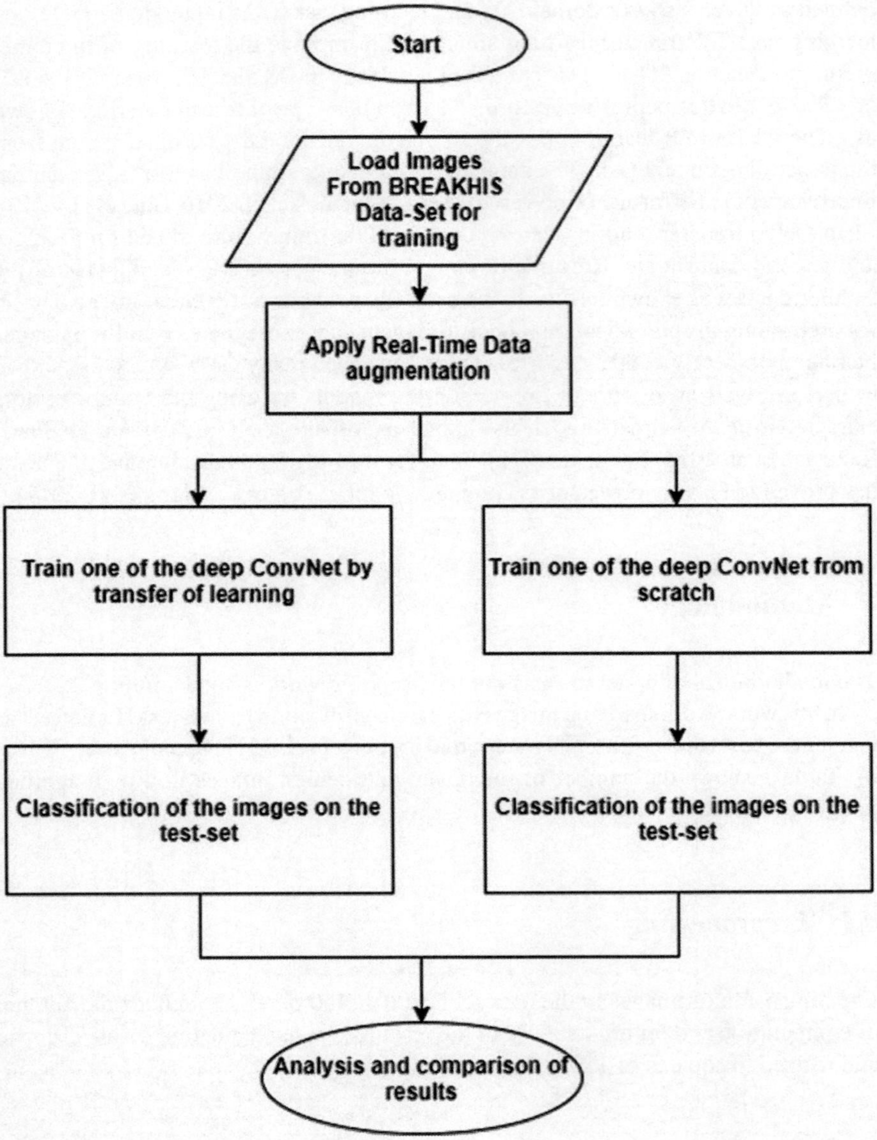

Fig. 2 Workflow diagram of a methodology

Table 1 Distribution of 40× magnified images in BreakHis

Magnification	Benign	Malignant	Total
40×	625	1370	1995

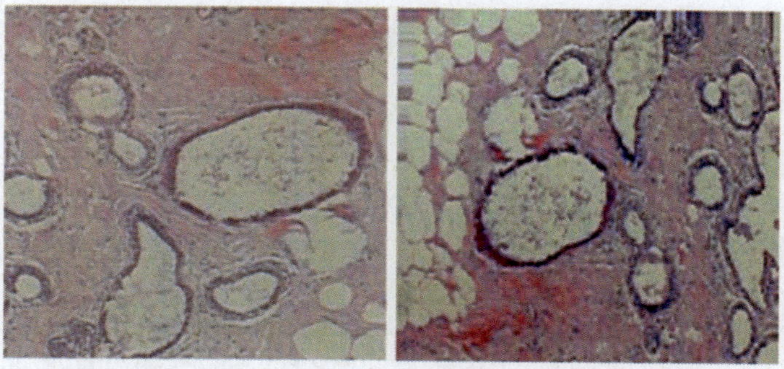

Fig. 3 Image before and after augmentation

This ensures that the model doesn't see the same image after each epoch hence, it doesn't allow overfitting (Pratt et al. 2016).

Figure 3 shows an example of data augmentation on one of the benign images. The image on the left is flipped and rotated to a certain degree when it is augmented. Due to such a vast range of augmentation performed on each image after every epoch, it becomes nearly impossible that the same image was used for training in any two different epochs.

Pixel Value Normalization: Before an image is given as input to the model, every pixel values present are first normalized. During the normalization, the RBG values ranging between 0 and 255 are converted to range between -1 and 1. This normalization helps in decreasing the huge variation that can be present in the image, which can lead to more efficient results.

5 Models Used for Training

For classification, we used two state-of-the-art well established deep Convolutional Neural Networks (CNN), i.e., Xception and Inception model.

5.1 InceptionV3

The inception V3 model is a deep convolution neural network that contains a particular type of inception module which is described in (Szegedy et al. 2015b). In simple terms, the Inception module is a network of smaller convolution layer that can learn better than a conventional convolutional layer.

Table 2 Top-1 and top-5 percent accuracy of Xception and ImagentV3 model on imagenet

Model	Top-1 accuracy	Top-5 accuracy
Xception	0.790	0.945
InceptionV3	0.827	0.941

5.2 *Xception Model*

The paper (Chollet 2017) considered Inception modules in CNN as being an intermediate step in-between the operations of depthwise separable convolution and regular convolution. They suggested a new deep convolutional neural network architecture inspired by Inception following the examination, where depth wise separable convolutions are used instead of Inception modules. This architecture, Xception, on ImageNet dataset slightly outperforms Inception V3 as shown in Table 2 (Even though Inception V3 was designed for Imagenet classification), and on a larger image classification dataset consisting of 17,000 classes and 350 million images, it substantially exceeds Inception V3 performance. The reason behind Xception outperforming InceprionV3 is because in Xception model parameters were more efficiently used.

5.3 *3-Layer CNN Model*

We also trained a three-layer shallow convolutional neural network whose architecture is shown in Fig. 4.

6 Evaluation Parameters

6.1 *Image Recognition Rate/Accuracy*

The image recognition rate is the percentage of images that are correctly classified of all the images in the test set. Equation 1 describes how the image recognition rate can be calculated.

$$\text{Image Recognition Rate} = \frac{C}{T} \tag{1}$$

Here, C is a number of correctly classified images in the test set and T the total number of images in the test set.

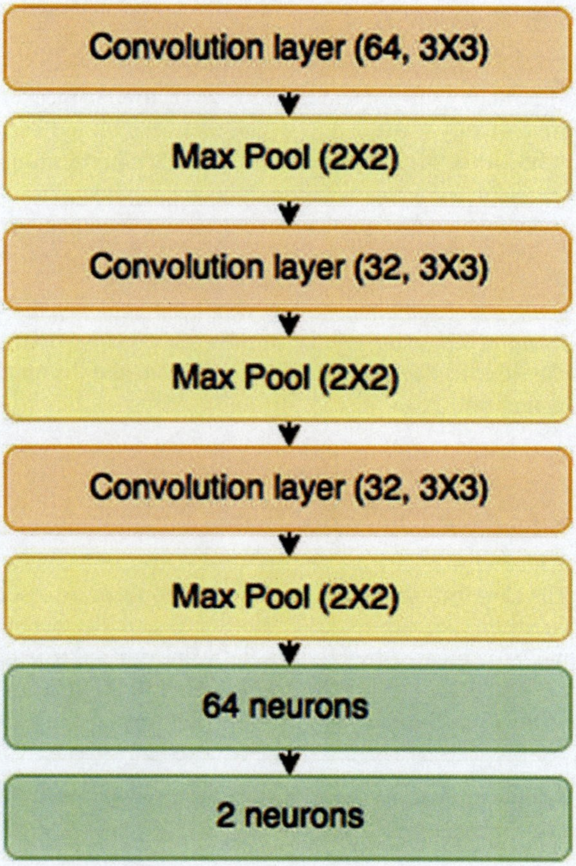

Fig. 4 The architecture of 3 layers CNN model

6.2 Sensitivity

Sensitivity (Altman and Bland 1994) is used to measure the potion of samples which are correctly identified as positives out of all the positives. This is considered to be one of the important measures when performing the medical analysis. It could be calculated using equation two given below.

$$\text{Sensitivity} = \frac{\text{TP}}{P} \tag{2}$$

Here, TP indicated the number of true positives (number of samples correctly classified as being positive) and P indicated the number of positive samples in the testing set.

6.3 Specificity

Specificity (Altman and Bland 1994) is used to measure the potion of samples which are correctly identified as negatives out of all the negatives. This is also considered to be one of the important measures when performing the medical analysis. It could be calculated using equation three given below.

$$\text{Specificity} = \frac{\text{TN}}{N} \tag{3}$$

Here, TN indicated the number of true negatives (number of samples correctly classified as being negatives) and N indicated the number of negative samples in the testing set.

6.4 Area Under the Curve

The classification capacity of a model could be assessed by drawing its ROC (receiver operating characteristic) curve(Subtil and Bermudez 2014). The ROC curve of a model is created by plotting the values of the true positive rate (TPR) against the false positive rate (FPR) (1 − specificity) at different threshold points. The plot for an efficient classification model will always remain above the line representing TPR = FPR. The area under the curve, also known as AUC, of a ROC curve, will be larger for a more robust model as compared to a less robust model.

7 Experimental Results

7.1 Training Strategies

We first split the dataset into two parts, 70% for training and 30% for testing, it is divided in such a manner that images of one patient are not in both the parts simultaneously to guarantee that our models generalize well over unseen patients. We trained the two models (inceptionV3 and Xception model) using two different Training strategies on the training set.

7.2 Training by Transfer of Learning

The training performed is similar to that performed in (Girshick et al. 2016). In both the models, we removed the last fully connected layers and attached our own

Table 3 Results obtained in our experiment

	3-layer CNN model	InceptionV3 trained from scratch	Xception trained from Scratch	InceptionV3 with transfer learning	Xception with transfer learning
Accuracy	77.08%	79.73%	79.90%	89.08%	90.86%
sensitivity	84.29	89.37	90.09	95.89	92.51
specificity	58.64	59.04	54.25	72.34	80.85
AUC	0.7520	0.8450	0.7990	0.9039	0.9120

fully connected (FC) layers to fit the model for binary classification. We first load all the layers trained weights of Imagenet except for the last FC layers. The trained weights are already available in Keras (Chollet 2015), which could be used directly. For training, we first freeze all the layer's weights except for the last fully connected layers, which we designed, for our initial training. This is because the randomly initialized weights would trigger huge gradient updates, which could destroy the convolutional base's learned weights. The initial training is done for 20 epochs. After the initial training, we make all the layers trainable so that the weights could be modified to fine-tune all the layers in such a way that the model could accurately classify the histopathological breast cancer Images. The model is trained for further 100 epochs after the initial training. Stochastic gradient descent optimization function was used with learning rate set as 0.001 and momentum at 0.9 with a mini-batch size of 16. The weights were updated using the backpropagation algorithm (Dreyfus 1990).

7.3 *Training from Scratch*

We train the model from scratch after randomly initializing the weights. The model is trained for 100 epochs. Stochastic gradient descent optimization function was used with learning rate set as 0.001 and momentum at 0.9 with a mini-batch size of 16. The weights were updated using backpropagation algorithm. Table 3, summarizes the results obtained in our experiments. In our work, we found that the Xception model, which was trained by transfer learning, performed the best out of all the three other models. Models trained by transfer learning substantially outperformed the same models which were trained from scratch.

7.4 *Image Recognition Rate/Accuracy*

To decrease overfitting in the models, we used real-time data augmentation, i.e., we apply data augmentation to our training data after every epoch, so that the model never sees an exact image more than once. From Table 3, it is clear that, when the transfer of learning trained both the models, they significantly outperformed the same models trained from scratch. In fact, the Xception model trained by the transfer of learning outperformed the state-of-the-art system.

Both the deep CNN models (InceptionV3 and Xception) outperformed our shallow three-layer CNN model, as expected. Since, both the models had more layers, they could work out even with higher level of abstract patterns among the images, helping them make better predictions.

7.5 ROC Curve

Figures 5 and 6 shows the ROC for InceptionV3 and Xception model, respectively, trained with and without using transfer learning. The graphs clearly show that the model trained using transfer learning is more robust and accurate that could more efficiently separate positive from non-positive images.

Figure 7, shows the ROC curve for our three-layer convolutional layer model. The value for the area under the cure is the least as compared to every model we trained; this is due to its shallow architecture, which doesn't allow for extracting higher level features.

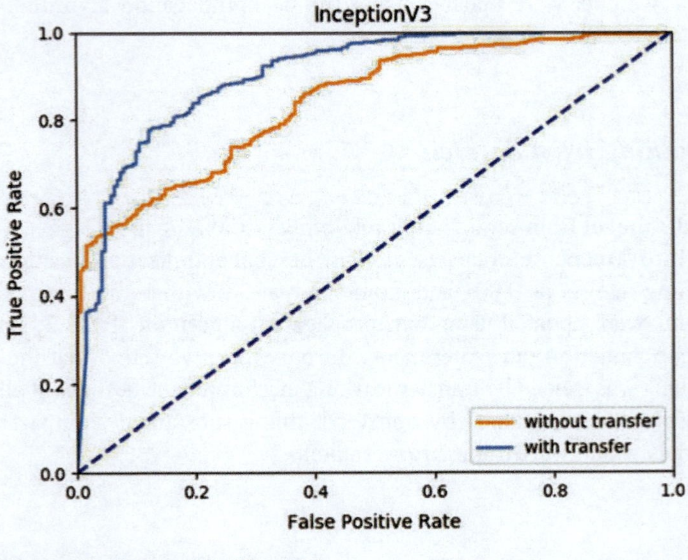

Fig. 5 ROC curve for the InceprionV3 model

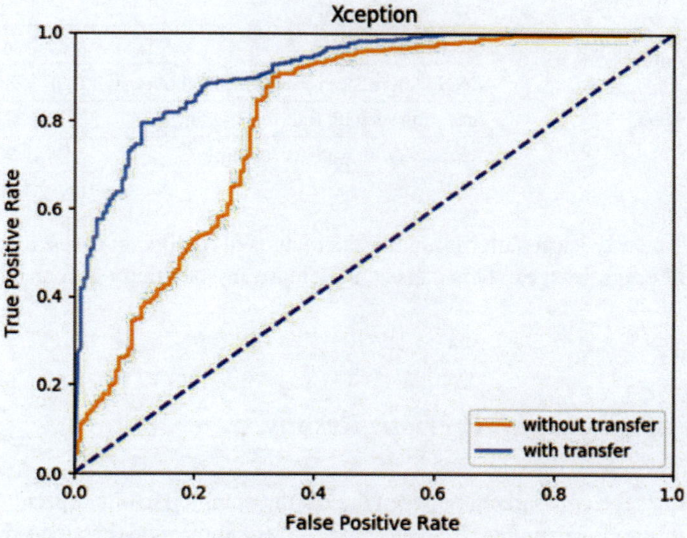

Fig. 6 ROC curve for the Xception model

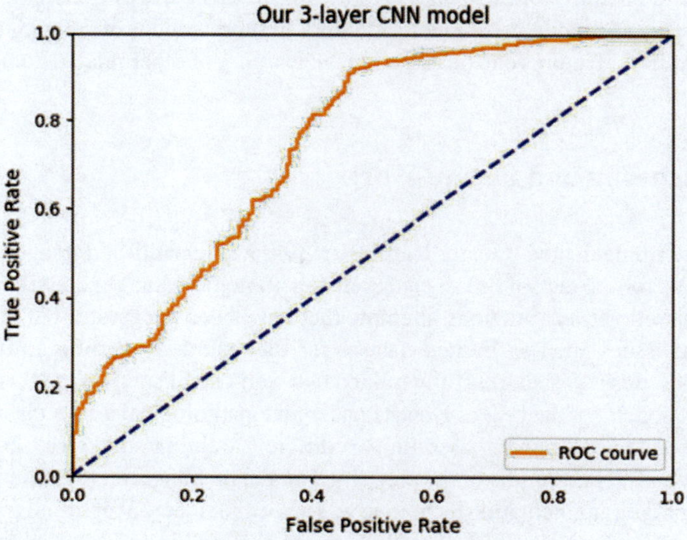

Fig. 7 ROC curve for 3-layer CNN model

7.6 Sensitivity and Specificity

As it could be seen in Table 3, the sensitivity is very high as compared to the specificity; this is because the BreakHis dataset is skewed. Nearly 70% of the total images are of malignant cases, and only 30% are benign. This difference causes the model

Table 4 Comparison of image recognition accuracy on BreakHis dataset by different models

Model	Accuracy
CNN with random patches (Spanhol et al. 2016)	89.6%
InceptionV3 with transfer learning	89.08%
Xception with transfer learning	90.86%

to more efficiently learn patterns among the images of malignant cases, as compared to that for benign images. This causes a decrease in specificity and an increase in sensitivity.

7.7 Comparison with Previous Results

Table 4 shows the comparison between the results obtained in our experiment to that obtained in previous studies. It shows that the Xception model trained by transfer of learning and real-time data augmentation shows a slight improvement over the previous state-of-the-art on a similar data set.

The main objective of this paper is to demonstrate the use of transfer of learning from a pretrained model to a new task with a limited amount of training data. The learning of basic features can be reused in the learning of other related domain tasks.

8 Conclusion and Future Work

This paper implemented transfer learning on two well-established image classification models for the task of breast cancer histopathological images classification. We also discussed various previous attempts that have been successful but have failed to address issues such as limited dataset for the task and overfitting. To address these issues, this paper successfully trained two well-established image classification ConvNet models for the task of Breast cancer histopathological image classification. These models were trained based on two different training strategies, i.e., training from scratch and fine-tuning of the pretrained model on ImageNet (transfer learning). A real-time data augmentation technique was also implemented for training to combat the overfitting of training data. The ROC curve obtained in our experiments clearly indicates that models fine-tuned on pretrained weights have higher evaluation ability as compared to the same model trained from scratch. This is also verified by the fact that, when the transfer of learning trained both the models, they significantly outperformed the same models trained from scratch. In fact, the Xception model trained by the transfer of learning outperformed the state-of-the-art system.

In the future, a wider range of image classification models can be tested for the BCH image classification task. Also, the effects of skewness in the data set could be reduced by implementing a more intense data augmentation on the benign images and using a larger batch size while training.

References

Ahmad J, Muhammad K, Lee MY, Baik SW (2017) Endoscopic image classification and retrieval using clustered convolutional features. J Med Syst 41(12). http://doi.org/10.1007/s10916-017-0836-y

Altman DG, Bland JM (1994) Diagnostic tests. 1: Sensitivity and specificity. BMJ: Br Med J 308(6943):1552

Bar Y, Diamant I, Wolf L, Lieberman S, Konen E, Greenspan H (2015) Chest pathology detection using deep learning with non-medical training. In: 2015 IEEE 12th international symposium on biomedical Imaging (ISBI). IEEE, pp 294–297. http://doi.org/10.1109/ISBI.2015.7163871

Bengio Y, Courville A, Vincent P (2013) Representation learning: a review and new perspectives. IEEE Trans Pattern Anal Mach Intell:1798–1828

Chollet F (2015) Keras: theano-based deep learning library. Code: https://github.com/fchollet. Documentation: http://keras.io

Chollet F (2017) Xception: Deep Learning with depthwise separable convolutions. In: Proceedings—30th IEEE conference on computer vision and pattern recognition, CVPR. pp 1800–1807, http://doi.org/10.1109/CVPR.2017.195

Deng J, Dong W, Socher R, Li L-J, Li K, Li F-F (2009) ImageNet: A large-scale hierarchical image database. In: 2009 IEEE conference on computer vision and pattern recognition. IEEE, pp 248–255. http://doi.org/10.1109/CVPR.2009.5206848

Dreyfus SE (1990) Artificial neural networks, back propagation, and the Kelley-Bryson gradient procedure. J Guid, Contr, Dyn 13(5):926–928. http://doi.org/10.2514/3.25422

Filipczuk P, Fevens T, Krzyzak A, Monczak R (2013) Computer-aided breast cancer diagnosis based on the analysis of cytological images of fine needle biopsies. IEEE Trans Med Imaging 32(12):2169–2178. https://doi.org/10.1109/TMI.2013.2275151

George YM, Zayed HH, Roushdy MI, Elbagoury BM (2014) Remote computer-aided breast cancer detection and diagnosis system based on cytological images. IEEE Syst J 8(3):949–964. https://doi.org/10.1109/JSYST.2013.2279415

Girshick R, Donahue J, Darrell T, Malik J (2016) Region-based convolutional networks for accurate object detection and segmentation. IEEE Trans Pattern Anal Mach Intell 38(1):142–158. https://doi.org/10.1109/TPAMI.2015.2437384

Hsu JL, Hung PC, Lin HY, Hsieh CH (2015) Applying under-sampling techniques and cost-sensitive learning methods on risk assessment of breast cancer. J Med Syst 39(4). http://doi.org/10.1007/s10916-015-0210-x

International Agency for Research on Cancer (2008) World cancer report 2008. Can Control 199:512. https://doi.org/10.1016/j.cma.2010.02.010

Jia Y, Shelhamer E, Donahue J, Karayev S, Long J, Girshick R, et al (2014) Caffe: Convolutional architecture for fast feature embedding. In: Proceedings of the ACM international conference on multimedia—MM '14. pp 675–678, http://doi.org/10.1145/2647868.2654889

Kowal M, Filipczuk P, Obuchowicz A, Korbicz J, Monczak R (2015) Computer-aided diagnosis of breast cancer based on fine needle biopsy microscopic images. Comput Biol Med 43(10):1563–1572. https://doi.org/10.1016/j.compbiomed.2013.08.003

Krizhevsky A, Sutskever I, Hinton GE (2012) ImageNet classification with deep convolutional neural networks. Adv Neur Inform Proc Syst:1–9. http://doi.org/10.1016/j.protcy.2014.09.007

LeCun YA, Bengio Y, Hinton GE (2015) Deep learning. Nature 521(7553):436–444. https://doi.org/10.1038/nature14539

Lu L, Shin H, Roth HR, Gao M, Lu L, Member S, et al (2016) Deep convolutional neural networks for computer-aided detection : CNN architectures, dataset characteristics and transfer learning deep convolutional neural networks for computer-aided detection : CNN architectures, dataset characteristics and transfer. IEEE Trans Med Imaging 35(5):1285–1298. http://doi.org/10.1109/TMI.2016.2528162

Pan SJ, Yang Q (2010) A survey on transfer learning. IEEE Trans Knowl Data Eng. https://doi.org/10.1109/TKDE.2009.191

Pratt H, Coenen F, Broadbent DM, Harding SP, Zheng Y (2016) Convolutional neural networks for diabetic retinopathy. Procedia Comput Sci 90(July):200–205. https://doi.org/10.1016/j.procs.2016.07.014

Ravishankar H, Sudhakar P, Venkataramani R, Thiruvenkadam S, Annangi P, Babu N, Vaidya V (2016) Understanding the mechanisms of deep transfer learning for medical images. In: Deep learning and data labeling for medical applications. LABELS 2016, DLMIA 2016. pp 1–8, http://doi.org/10.1007/978-3-319-46976-8

Rodrigues JJPC, Reis N, Moutinho JAF, De La Torre I (2012) Breast alert: An on-line tool for predicting the lifetime risk of women breast cancer. J Med Syst 36(3):1417–1424. https://doi.org/10.1007/s10916-010-9603-z

Roth H, Lu L, Liu J, Yao J, Seff A, Cherry K, et al (2016) Improving computer-aided detection using convolutional neural networks and random view aggregation. IEEE Trans Med Imaging PP(99):1. http://doi.org/10.1109/TMI.2015.2482920

Rubin R, Strayer DS (2008) Rubin's pathology : clinicopathologic foundations of medicine. Wolters Kluwer/Lippincott Williams & Wilkins

Schroff F, Kalenichenko D, Philbin J (2015) FaceNet: A unified embedding for face recognition and clustering. In: Proceedings of the IEEE computer society conference on computer vision and pattern recognition, vol 07–12–June, pp 815–823. http://doi.org/10.1109/CVPR.2015.7298682

Sharif Razavian A, Azizpour H, Sullivan J, Carlsson S (2014) CNN Features off-the-shelf: an astounding baseline for recognition. In: The IEEE conference on computer vision and pattern recognition (CVPR). pp 806–813

Shie C-K, Chuang C-H, Chou C-N, Wu M-H, Chang EY (2015) Transfer representation learning for medical image analysis. In: 2015 37th annual international conference of the IEEE engineering in medicine and biology society (EMBC). IEEE, pp 711–714. http://doi.org/10.1109/EMBC.2015.7318461

Sisodia DS, Verma S (2011) Image pixel intensity and artificial neural network based method for pattern recognition. World Acad Sci Eng Technol 57:742–745

Song Y, Zou JJ, Chang H, Cai W (2017) Adapting fisher vectors for histopathology image classification. In: 2017 IEEE 14th international symposium on biomedical imaging (ISBI 2017). IEEE, pp 600–603. http://doi.org/10.1109/ISBI.2017.7950592

Spanhol FA, Oliveira LS, Petitjean C, Heutte L (2016a) A dataset for breast cancer histopathological image classification. IEEE Trans Biomed Eng, 63(7), 1455–1462. https://doi.org/10.1109/TBME.2015.2496264

Spanhol FA, Oliveira LS, Petitjean C, Heutte L (2016b) Breast cancer histopathological image classification using Convolutional Neural Networks. In: 2016 international joint conference on neural networks (IJCNN), 2560–2567. https://doi.org/10.1109/IJCNN.2016.7727519

Subtil A, Bermudez PDZ (2014) ROC curve estimation: an overview. REVSTAT—Stat J 12(1):1–20

Szegedy C, Liu W, Jia Y, Sermanet P, Reed S, Anguelov D, et al (2015a) Going deeper with convolutions. In: The IEEE conference on computer vision and pattern recognition (CVPR). pp 1–9, http://doi.org/10.1109/CVPR.2015.7298594

Szegedy C, Vanhoucke V, Loffe S, Shlens J, Wojna Z (2015b) Rethinking the inception architecture for computer vision. In: The IEEE conference on computer vision and pattern recognition (CVPR), 2016. pp 2818–2826, http://doi.org/10.1109/CVPR.2016.308